Technical and Military Imperatives
Imperatives
A Radar History of World War II

Technical and Military Imperatives

A Radar History of World War II

Louis Brown

Carnegie Institution of Washington
Washington DC, USA

Taylor & Francis

Taylor & Francis Group
New York   London

Published in 1999 by
Taylor & Francis Group
270 Madison Avenue
New York, NY 10016

Published in Great Britain by
Taylor & Francis Group
2 Park Square
Milton Park, Abingdon
Oxon OX14 4RN

© 1999 by Taylor & Francis Group, LLC

No claim to original U.S. Government works
Printed in the United States of America on acid-free paper
10 9 8 7 6 5 4

International Standard Book Number-10: 0-7503-0659-9 (Hardcover)
International Standard Book Number-13: 978-0-7503-0659-1 (Hardcover)

Library of Congress Cataloging-in-Publication Data

Catalog record is available from the Library of Congress

Taylor & Francis Group
is the Academic Division of Informa plc.

**Visit the Taylor & Francis Web site at
http://www.taylorandfrancis.com**

CONTENTS

PREFACE WITH ADVICE TO THE READER

This book is about science and war. These are unquestionably the two most dissimilar manifestations of the behaviors that distinguish man from beast. Science—allow it here to mean all elements of our search to understand the universe and put that understanding to use—is the most remarkable of human occupations. It is also the most important. For us not to strive to understand the universe would be as foolish as for castaways not to explore the island on which they find themselves. Science, like art and literature, draws from the mind those inner, intangible feelings which we flatter ourselves give us our station on Earth. It discloses to us things of unparalleled beauty and wonder—the mathematical beauty of the laws of physics, the wonder that life is understandable. War is the negation of all this. It is a disease, like smallpox, that must be eradicated.

Yet war is almost as unique to man as is science. Other than ourselves, only ants organize their violence so that it can be called war. Violence is ubiquitous to the biological world, but it is man's doing to make it an innermost part of civilization. War is the great embarrassment of civilization, yet all great civilizations have maintained themselves by their military skills and have vanished when those skills no longer sufficed to protect them or sustain internal order. Moreover, from the dawn of civilization science and war have been inseparable companions, locked in a partnership that neither desires and that neither is capable of dissolving.

No one setting out to read this book needs the recitation of a catalog of invention and discovery that bears on military history to be convinced of science's significance, but if this relationship is so well known, what is the purpose of this book? It has a void to fill, for an important element in this story is not well told. Years ago I was struck by the absence of a comprehensive and international history of radar, of the kind that has appeared many times about the atomic bomb. Not that there were no books on the subject; on the contrary, there is almost an over-abundance, as the numerous references cited herein attest. There are among them some extremely good accounts of various parts of the story, and it is my hope that this study will impel many to read them for the things I have had to leave out. However, these histories are all parochial, for they invariably found their origins in the personal involvement of the authors in radar or on material gleaned from national or laboratory sources. The secret nature of radar did not give the engineers, scientists and military much chance to learn what took place on the other side of the hill, so national versions tell one little about other countries and all too frequently leave the reader with a distorted impression of the whole subject. This does not imply dispute about the facts. The German and British accounts each emphasize the development of their own equipment, of course, but describe the great radar struggle between the Luftwaffe and RAF Bomber Command without significant disagreement. They leave the curious impression, however, that World War II was a private fight between Britain and Germany.

There were in America four large radar development laboratories of comparable importance. The histories written about them are quite unequal and show poor understanding of the contributions of others. Three years into the project I was still encountering major surprises that changed my understanding in a substantive way. Most radar histories trace its origins to a few relatively well known and oft reported events. It has not taken any great depth of scholarship to recognize that radar was just in the air during the 1920s and 1930s. This is obvious when one learns that eight nations had radar projects in 1939. There is evidence that leads one to suspect that hundreds of communication engineers observed reflection and interference phenomena causing them to think about radio location without recording it. The strong similarities of the German and American programs alone speak volumes for its invention being a technical imperative.

'The bomb may have ended the war but radar won it.' This was the comment of many radar workers in August 1945, when the news of the atomic bomb upstaged the release of public knowledge of the MIT Radiation Laboratory, planned as a cover story for *Time*. This statement has been repeated many times since then, often with categorical statements concerning some engagements, but has never been examined for the war as a whole.

It is the object of this study to attempt to improve on these perceived deficiencies: (1) to describe the wide simultaneous development, (2) to examine the idea of it resulting from a technical situation that made it inevitable, (3) to determine how the severe restraints imposed by secrecy affected design and use and (4) to approach with the objectivity that should come with the passage of time the question of its effect on the war, both in deciding battles and altering the military leaders' approach to war. I shall have succeeded if my effort presents in balanced form the main elements of the whole story.

An appendix explains some of the rudiments of radar. Many will choose to skip it because their technical knowledge equals or, more likely, greatly exceeds what is presented. It is possible to gain a fair understanding of radar history by judiciously skipping technical matters because most of the story is about individuals and organizations, so readers for whom the terms megahertz, transmitter, antenna, bandwidth and ionosphere have only vague meanings need not set this book aside. They may prefer to skip certain sections whose absence will not detract greatly from understanding the general development of radar. These sections are 1.2, 2.1, 4.1, 4.5, 6.4 and 10.2. Other parts have technical matters that may perplex a lay reader, but I suggest adopting an attitude much as one must assume for science fiction—which cannot be understood by anyone. The lay reader will obviously appreciate that many readers will find the technical details meager.

In discussing these things with young people I have come to realize that events burned into the memory of my generation are no longer current, lying further back for them than the wars in South Africa or Cuba for us. Comments from readers of various draft sections have disclosed that many well read persons are not familiar with parts of the conflict. The British and Germans often have an uncertain idea of what happened in the Pacific, and many Americans know little about the air war against Germany other than it involved strategic bombing. This has required sketching in outline much of what happened from 1939 to 1945 in order to make the radar contribution coherent.

The subtitle, *A Radar History of World War II*, indicates that only equipment that was widely used and provided important results in the conflict is described in any detail. These were also the sets that established the basic principles. The mix of technical and military events is essential to understanding, for no weapon was ever designed with such intimate collaboration between inventor and warrior. Engineers frequently tried out ideas in combat, and some died there. No adequate history of radar can result from a knowledge of what transpired in the laboratories alone. Many radar sets were designed toward the end of the war—superb examples of the art—that had trivial effects on the outcome and so will either not be mentioned at all or be described only because the extent of the development project was so great or the device so unusual. None fits this description better than Cadillac, which could rightly have a book devoted to it alone, yet is a minor notation in the chronicle of the war even though it consumed a large part of the effort of the MIT Radiation Laboratory.

The same unevenness applies to the military evaluations. The presentation of World War II is highly skewed with campaigns treated according to the radar involvement. Thus the Pacific War is treated in detail whereas the titanic clash on the Eastern Front receives a single, brief section, and the Blitzkriegs, east and west, are hardly mentioned at all. Substantial space is given to battles in which radar was a unique and decisive weapon, especially early engagements in which commanders struggled to master this new technique and devise tactics to go with it. Little space is given to battles in which huge amounts of radar were employed routinely but without novel or decisive effect. Because of the lessons revealed, certain actions are treated in so much detail as to magnify them out of all proportion to the war as a whole or even to radar itself. Two examples of this are the radar aspects of the American carrier raids in early 1942 and the Dieppe Raid of the same year. One provided an invaluable, last-minute school for carrier operations that altered the way in which the US Navy conducted the Pacific war; the other showed the ineptitude of both sides in their use of radar during a major landing, lessons that had great impact on radar usage two years later in Normandy.

Sources properly belong to a section of acknowledgment, and they are listed there with many, very grateful words of thanks, but the nature of the sources has much to tell us about radar history. I began this study because the book I would have liked to read did not exist. As my studies progressed the amount of published material startled me and, more amazing yet, there seems to be no end to it. A great deal of very valuable information was to be found in publications of extremely limited circulation, much of it published privately. With the help of our librarian, skilled in the art of making inquiries by computer, many such books were located. He and his assistant found others by methods that must have been based on instinct supplemented by an intimate knowledge of the book trade.

There was, however, much more that was not accessible by such an approach. Early attempts to ferret these things from archives simply showed that, although they provided much help in following specific clues, they would have quickly become swamps from which I should have never extracted myself. Fortunately, I found another path on hard ground. There are radar-history enthusiasts to be found all over the world. They are generally electrical engineers or physicists and, more often than not, radio amateurs; they have squirreled away documents, books, photographs, journal articles and artifacts as well as written histories, and communicated with one another about their common interests. This rich source was opened through the old method of writing letters—many letters. These correspondents, found in eight countries, have the judgement and understanding to know what is important and have supplied me with a great part of the unpublished items used. Through their letters they have shown themselves to be astute critics possessed of a deep understanding of the background of events. A correspondence was also developed among military personnel with radar experience, but radar is not the common thread that links them, so this circle has not had the continually expanding radius that marks the other.

Louis Brown
July 1999

ACKNOWLEDGMENTS

From the beginning I decided against using interviews as a research tool, partly from memories of my past faulty attempts to gain information that way, but primarily from the certainty that such a course would skew the source material toward those groups for whom published memoirs and interviews were already available in significant quantity. In retrospect this has shown itself to have been a wise decision but for a different reason: I have come to find the interview a flawed method. Recollections of events half a century ago, called up in conversation with an interviewer uncertain of what needs to be asked and with neither party given sufficient time for reflection, generally extract little that is new and much that is misleading. There are numerous transcripts of such interviews and, though they certainly have value, they also disclose many errors, the result of the passage of years, the secrecy of the early times and the subsequent development of mythology. One must use these sources with caution and with emphasis on the subject's personal experiences and not his impressions of history.

My postal substitution for interviews generally began as a probe to follow a clue, asking the recipient to clarify some bit of puzzlement. The answers and subsequent exchanges frequently brought more than was requested, such as personal memoirs, documents, articles published in obscure magazines, out-of-print books and the like. The oft decried slowness of this method of communication has proved to be of value, as it assured both parties time to think over questions and answers. Information obtained in this way, including some acquired by less formal means, came from: H-J ALBRECHT, Niels Chr BAHNSON, Lori BERKOWITZ, David B DOBSON, John DUGGAN, Gerhard EBELING, Anna Maria Runge ELSTNER, George A EMERY, Pekka ESKELINEN, Helmut FAUDE, the late John FINDLAY, Ivan A GETTING, John M GODFREY, Charles R HABERLEIN, Frank HEWITT, Marvin HOBBS, Joseph HOPPE, Yasuho IMAI, F Y ITO, the late R V JONES, Harumasa ITO, Robert KAUFMANN, Gordon M KLINE, John B LUNDSTROM, Colin MACKINNON, John F MANNING, Allan A NEEDELL, Karl NEUMANN, Mark PEATTIE, Carsten PETERSEN, William R PIGGOTT, Charles T PREWITT, Henry PROBERT, the late Robert RALSTON, H-J RICHTER, J R ROBINSON,

Hans SORGENFREI, Rudolf STARITZ, Jane E TUCKER, Oliver VOLLAND and H W WICHERT.

In addition to providing information, some correspondents were kind enough to read draft portions of the manuscript, comment on the content and all too frequently point out errors. For such efforts I am indebted to Ralph B BALDWIN, David K BARTON, Ray BOWERS, the late Adam BUYNOSKI, Russell S CRENSHAW, Paul FORMAN, Richard B FRANK, Hugh G HENRY III, Axel NIESTLÉ, Hans PLENDL, Jr, the late Ernest C POLLARD, Peter J PRICE, Max SCHOENFELD, Merrill SKOL-NIK, P G SMITH, Harry STREET, John STREET, Kathleen WILLIAMS, the late Tom WINCHCOMBE and Thomas Y'BLOOD. Before any copy left the department it was subjected to careful reading by Merri WOLF, who removed many clumsy sentences and grammatical blunders and raised questions about clarity.

Radar history is by no means devoid of serious scholars, identified to me as the circle of my correspondence expanded. This resulted in exchanges of information as well as discussions in person, by telephone and mail. The late John BRYANT and I spent many happy hours arguing our varied viewpoints and exchanged large amounts of material. Michael S DEAN served through his maintenance of the Historical Radar Archive and in the organization of Royal Air Force radar reunions; his reading of the manuscript identified numerous errors, and he located many valuable photographs. F A KINGSLEY helped greatly in clarifying the contributions of the Royal Navy, some of which I had attributed to the Air Ministry. Russell BURNS and Sir Edward FENNESSY have not been content to accept some of my theses, which has provided lively correspondence, a few changed opinions and wonder that an otherwise rational man persists in error. Colin LATHAM and Hans JUCKER have provided important technical criticism, JUCKER adding an unexpected continental point of view. Frederick SEITZ proved a valuable companion in our combined efforts to trace early aspects of the German work. In addition to comments on the manuscript Robert BUDERI taught me most of what I know about postwar radar. Arthur O BAUER and Bernd RÖDE sent me their technical studies of the Würzburg of which I made frequent use. Australia and South Africa had their share of radar activity, and their veterans went beyond reunions to ensure a historical record, Ed SIMMONDS and Geoffrey MANGIN supervising. Andrew BUTRICA sent me his history of radar astronomy in draft, which helped remove the narrowness of my World War II point of view, and J Michel BLOM always seemed to find some new reference. Failing health and death prevented Fritz TRENKLE from providing the ever-ready answers to questions that the community had become accustomed to receiving, but his published volumes continued to guide us. Charles SÜSSKIND began a comprehensive study of radar history at a time when much less material was available, resulting in an unpublished manuscript that terminated at the time of

America's entry into the war. This limited study proved to be extremely useful.

At the beginning of the work the very important German side of the story was unsatisfactory. There were important published books and articles, but all suffered from the destruction of postwar records and the isolation imposed on the authors by the severe secrecy that characterized the Nazi state, and three years into the project had shown me no way of correcting this. Letters to museum curators and historians proved fruitless until Joseph HOPPE suggested that Conrad H von SENGBUSCH, an electrical engineer with a small electronics museum, might know some who had radar as a hobby, a suggestion that brought unparalleled riches as he introduced me to Hans Ulrich WIDDEL, physicist, radio propagation scientist and wartime radar operator. Not only did he have a wealth of documents, books and miscellaneous papers in his own files, he was connected through amateur radio to a group of like-minded individuals who began to discuss my needs during their routine contacts. Their individual efforts are included in the second paragraph and have completely transformed the German picture. The size of my correspondence with WIDDEL is larger than any of the others and has a corresponding breadth in technical and historical content. Equally important was the ten-year study by WIDDEL'S lifelong friend, Harry von KROGE, who was determined to learn details of GEMA, whose story had become distorted and much of it lost over the years.

Knowledge of Japanese radar had come to us primarily from the wartime and occupation technical intelligence reports and from articles by Shigeru NAKAJIMA, leader of their magnetron development. These sources, valuable as they are, miss many important historical details. Fortunately, Yasuzo NAKAGAWA wrote two books in 1987 that contained large portions about radar history. Don Cyril GORHAM translated relevant sections and thereby provided a more secure basis of study. Not surprisingly, this opened many questions, and I was greatly helped by Naohiko KOIZUMI, a retired electron tube designer with a wide knowledge of radar and a colleague of both NAKAGAWA and NAKAJIMA. He entered enthusiastically into primary historical research, uncovered many new items and explained the inevitable mysteries. He joined Bryant and me in editing the translation of Nakagawa's writings into a published book, which contains most of what is known about Japanese radar.

The modern library system with its computer-linked catalogs has been of such great help that one wonders how historical studies were done before. The librarian at the Department of Terrestrial Magnetism, Shaun HARDY, made the Inter-Library Loan a key research device, drawing on 28 libraries. Local libraries have been an important resource and their staffs courteous and helpful: the District of Columbia Public Libraries, the Naval Historical Center Library, the Bender Library of American University, the libraries of the Smithsonian Institution, the library (now called a Research

Information Center) of the National Institute of Standards and Technology (known by most as the National Bureau of Standards) and the Library of Congress.

Richard RHODES and Alfred PRICE resisted my urging to have them write this book; this is a pity, said even now that I have gained the pleasure of having done it, but they provided encouragement, something for which I shall ever be grateful, and PRICE supplied documents that only his long activity in the field of electronic warfare could have uncovered. William ASPRAY and Frederik NEBEKER at the Center for the History of Electrical Engineering provided early guidance and found competent critical readers at the half-way point.

At the beginning of the project there were tentative plans for an international museum exhibit on radar that involved me in meetings with E J S BECKLAKE, Oskar BLUMTRITT and Bernard FINN. Although the planned exhibit came to naught the discussions were productive. Thomas BALLARD, H Warren COOPER and Michael CROSS guided me through the archive and collection of the Historical Electronics Museum, a substantial resource for anyone studying electronic history.

By granting me emeritus status Dr Maxine SINGER, President of the Carnegie Institution, and Dr Sean SOLOMON, Director of the Department of Terrestrial Magnetism, significantly eased my labors by providing access to the important features of the modern office. Dr SINGER went beyond this in helping provide translations from Japanese.

To all of these persons and organizations my sincerest and most heartfelt thanks, extended also to the many persons from whom help has been received through casual discussion and whose names have slipped from an imperfect memory.

CHAPTER 1

PRELIMINARIES

1.1. RADIO VISION FOR WAR

In winter the Denmark Strait, the stretch of water between Greenland and Iceland, is hardly a favorite passage for mariners. Heavy fog is common, sustaining little or no vision during the very short days, and ice presses down from Greenland with bergs a danger in or out of season. Storms hardly count as a rarity. These unpleasantries seldom trouble a shipmaster as little commerce need pass these waters, but they were well lodged in the mind of Vice Admiral Günther Lütjens, who was passing the strait in February 1941 with the two German battle cruisers *Gneisenau* and *Scharnhorst*. His object was the destruction of British merchant shipping, but to do this he had first to bring his ships from German ports into the Atlantic. The trickiest parts of his task were caused by the geographical position of Great Britain and the pugnacity of the Royal Navy that compounded his navigational problems with cruisers, pickets set out for just this event.

Lütjens now felt confident because of his experiences of the previous few days with the electronic devices mounted on the forward and aft fire-controls of the two vessels—radar designed for the Kriegsmarine and bearing the name Seetakt. This was the best radar in the world for searching the ocean out to the horizon, and it had already paid for itself before they reached the Denmark Strait. They had sailed up the coast of Norway and headed southwest to pass south of Iceland, but the radar had picked up too many ships whose speed identified them as warships, and after several course changes Lütjens had turned back and headed north to rendezvous with a tanker to replace expended fuel. The motion of the British ships gave no indication that they had seen him, allowing him to make the correct assumption that they had no radar, at least none usable for surface search.

Meeting the tanker was no mean trick. With scant opportunity for a stellar fix in the fog, both he and his supply ship had been depending on dead-reckoning for position, using course and speed to estimate position and hoping ocean currents and wind had not made the inevitable errors too large. Without radar, finding a tanker on the fog-covered northern ocean might well have failed; instead it became a trivial exercise.

Now he was safely passing north of Iceland and had avoided ice and cruisers. His position was easily determined by observing Iceland's mountains with the radar. The way was free into the Atlantic where he ranged north and south, meeting supply ships and sinking convoys not fortunate enough to have capital ships for escort.

The early mist had cleared and the sunny prospects had alerted the Ventnor operators, in the forward-most station of England's system of radar early warning on the south coast of the Isle of Wight, that an air attack was probably in the offing. At 1100 hours small wiggles in oscilloscope traces, insignificant to any but the young women observing them, confirmed their expectation. Across the Channel formations of bombers were assembling and would soon be heading toward England. The operators were puzzled that the Luftwaffe adhered to the tactic of forming over Calais and Boulogne in full view of radar, as it gave an extra margin of warning. Indeed, indications kept coming that they simply did not know how the Royal Air Force was countering them. They seemed to know the significance of the huge towers that lined Britain's south and east coasts, because they had opened the heavy raids against England with attacks on several radar stations on 12 August 1940, a few weeks before. Fifteen Ju-88s of Kampfgruppe-51 had attacked Ventnor, destroying the majority of the buildings and creating circumstances that led to one operator earning the Military Medal—the first awarded to a woman. A delayed action bomb exploded later at the thick concrete wall of the transmitter block and severed the feeders to the antenna. Four days later another attack by six Ju-88s finished off the destruction and put the station out of operation for two months.

Ventnor's crew switched to a Mobile Reserve Unit at nearby Bembridge that replaced the more powerful main station and preserved most functions with shorter range. They saw the day's build-up, although later than adjacent stations that had full power. A tension reiterated itself. Had the Germans finally got the point? Were they finally going to take out the stations? But the objective was neither the radar stations nor the bases of Fighter Command, Churchill's 'Few', which had been severely mauled during the preceding weeks, many of which could have been put out of service completely by one or two more raids. No, their destination was London and decisive defeat, for this was 15 September 1940, remembered as Battle of Britain Day.

The three men in the open cockpits of the biplane seemed in that summer of 1941 to be in the wrong war. Their aircraft was slow even by the standards of the previous war and was armed even more poorly, having only a single Vickers gun firing forward and a single Lewis for someone in the rear cockpit, which was half the guns a comparable machine would have had in 1918. It had, however, design elements that compensated for such remarkable deficiencies. There was the 690 hp Pegasus engine that allowed it to carry 1000 kg loads, generally either of bombs or a torpedo, for in addition to reconnaissance the Swordfish was the Royal Navy Fleet Air Arm's dive and torpedo bomber. To perform such duties required an extremely strong airframe. To attain this strength yet allow the wings to be folded for storage on a carrier required struts and wires in an arrangement that defied perfunctory understanding.

On this particular night one of the men in the rear cockpit was studying the traces of an oscilloscope, for their plane had been outfitted with radar, and they were hunting for convoys of Axis ships bound from Italy to Tripoli with supplies for Rommel, who was menacing Egypt. They were skilled at night operations and appreciated this new electronic eye. A convoy duly appeared on the scope, and the pilot followed the operator's instructions so as to drop a flare that illuminated it for the other aircraft of the flight to launch their torpedoes. The ships not hit during the night could look forward to a reception by submarines after daybreak, when the Swordfish's infirmities made a return to Malta prudent before the Luftwaffe began to seek them out. Rommel was never to secure his supply line.

The United States Navy was having a difficult time in the Solomon Islands during summer and fall of 1942. They had suffered a humiliating, and what could have been a decisive defeat during the night of 8/9 August, had the Japanese followed their victory with the destruction of the transports and supply ships unloading marines onto Guadalcanal. As night closed on 11 October circumstances were similar: a Japanese squadron from the northwest was intent on destroying vessels reinforcing the island that a screen of American cruisers was equally intent on protecting. Night action was the specialty of the Imperial Fleet, and their mastery of it had led to the earlier loss of three American and one Australian cruisers. Radar was supposed to have given the Allies compensating night fighting skills, but clumsy use of it in August had, if anything, contributed to the disaster. There had been subsequent improvements, but command understanding of the new technique was still marginal. The skipper of USS *Boise*, Captain Edward J Moran, appreciated the power of radar and had studied it. He also had a competent radar officer in Lieutenant Philip C Kelsey to look after the fire-direction radar for his main battery of fifteen 6 inch guns and a recently installed microwave surface-search radar that allowed the scene of battle to be presented as a maplike presentation.

Naval battles are often confusing, and night actions are almost never capable of accurate reconstruction after the event. The naval actions in the Solomons were night-time gun battles where it was difficult to establish who was friend and who foe, but in the fight of 11/12 October the beginning at least stands out in remarkable clarity: the *Boise* initiated the battle by opening fire in the darkness with radar control alone, and her first salvo scored direct hits on an astounded enemy. Kelsey had first sighted them, and Moran had maneuvered to obtain the best firing position. The loss of six warships and the death of the commanding Admiral sent a strong message to Tokyo about the importance of their lagging radar development.

The nights of late fall 1943 were particularly hard for the men who fought one another in the sky over Germany. The Royal Air Force pressed attacks on German cities to break the will of the civilian population and injure the means of producing and transporting the goods on which a modern war depended. Air Chief Marshal Arthur Harris, chief of Bomber Command, had initiated a series of enormous bombing attacks on Berlin in November with the objective of destroying the war potential of this sprawling metropolis. Key to this effort was a radar system called H2S that gave the navigators of the Lancaster four-engine bombers something purported to be a radar map of the ground. Its purpose was to allow the formations to find their way and aim their bombs. Both attacker and defender were continually faced with the consequences of not being able to see in the dark. Both attempted to 'see' with radar. Neither was conspicuously successful.

After a year of war the RAF and the Luftwaffe had come to realize that large daylight bombing attacks could not be sustained, and both had changed to night bombing. When Hitler made war on the Soviet Union, strategic bombing, which is the euphemism given to bombing civilian targets, became primarily a British function. German antiaircraft (AA) guns, first with searchlights and optical sighting, then supplemented with radar control, had forced the bombers to their maximum altitude. By the end of 1943 the bombers carried one, sometimes two kinds of radar plus devices to warn them of the enemy's radar and to identify themselves electronically to their countrymen and allies. The defending night fighters carried radar sets for locating the bombers as well as homing devices that allowed them to locate either the enemy aircraft by their radiations or their own landing fields by radio beacons. They were aided in their search through the darkness by an extensive ground radar control system that tried to place them close enough to the bombers for their airborne radar to pick up a target. Success then turned on the skill of the air-crew radar operator in guiding the fighter to a position close enough for the pilot, depending on whatever light might be present from the stars, Moon or burning cities to identify it and open fire.

These technical tasks, each requiring analytical thought best attained with a relaxed mind, had to be performed in extreme cold and noise while breathing oxygen through masks and in the presence of electronic interference designed to thwart them. The bombers threw out bundles of aluminum foil cut to resonate at the frequency of the AA and older airborne radars. Both sides transmitted electronic noise intended to overwhelm their opponent's radar. There were elements of humor in the attempts to confuse the night fighters by interjecting false information and distractions into the instructions of the ground controllers, but the extreme casualty rates and the apparently unending number of sorties required of the participants made humor a rarity and dampened any thrill of combat in all but the most devoted warrior.

When war broke out in September 1939, South Africans had not yet been informed by Britain about radar. They quickly received a briefing covering the technical essentials but learned that the Dominions could not expect to receive any equipment of British manufacture for the indefinite future. A group at the Bernard Price Institute in Johannesburg, under the direction of Dr Basil Schonland, built a set of their own design from available components and by December were tracking planes with it. Events had not yet established South Africa's part in the growing conflict, other than the responsibility for protecting the valuable sea routes around southern Africa, but modern weapons would be essential. This home-made radar, called JB for Johannesburg, allowed training, and its designers formed a cadre for future radar needs, needs that became fixed when Italy sent forces into recently annexed colonies to attack adjacent British and French territory. South Africa countered these moves and JB radars went north to help protect Nairobi, the capital of Kenya and nerve center of the vast region. These sets, each somewhat different from its predecessor and accompanied by one or more of its builders, moved on to help in the defense of the Suez area after the Italians were disposed of.

In 1942 when submarine activities began to take a toll on Cape shipping, the JBs returned to join other locally manufactured sets along the southern coast. Their presence forced the U-boats out of range, which in turn kept merchant shipping dangerously close to an unmarked shore, and radar frequently prevented steamers from running aground. Although they seldom saw the conning towers of U-boats on their screens, they did see much of the debris that covered the war-racked ocean and sent aircraft to investigate, occasionally forcing down or sinking a submarine. Sorties that did not lead to an attack on a U-boat were compensated many times over when the target proved to be neither a conning tower nor floating wreckage but a lifeboat full of exhausted survivors.

On 30 July 1943 American warships began the preliminary bombardment of Kiska, one of the western islands of the Aleutians that had been occupied by Japan more than a year earlier. Retaking the other island, Attu, had required the better part of the preceding May and had led the attackers of Kiska to expect the worst, hence the extravagant expenditure of explosives that became the standard prelude to a Pacific invasion. But when the 40 000 American and Canadian soldiers went ashore no one was to be found. The Navy was confident the garrison could not have been evacuated, so they must still be on the island. It was obviously some grand trick, and patrols cautiously sought them out. But no one was there. All had been evacuated.

Japan had come to realize there was nothing to be gained by holding this Aleutian real estate, which had become difficult to supply. Experience on these islands had also convinced them that they were unsuitable bases for bombing the homeland, one of the causes for the original occupation, so why hold them longer? Rear Admiral Masatomi Kimura had entered the harbor with two cruisers and six destroyers, loaded all personnel in less than an hour and departed. What the US Navy had ruled impossible had been made possible through use of the microwave radar that had allowed Akiyama to find the island and its harbor in the eternal Aleutian fog. A bit of luck in the dispositions of the blockading American fleet helped, but being able to see in the fog and thereby move rapidly was the secret. The Allies did not suspect that Japan had such shortwave radar. Such beliefs naturally influenced the actions of the blockaders.

The search for variety of experience has sent more men to war than patriotic zeal, ideological fervor, high pay or hopes of plunder. The preceding incidents, to which many more could be added, have sufficient variety in them to gratify this longing in all but the most voracious appetites, but they have military experience that confounds anything ever encountered before. In the endless struggles of the biological world, vision is by far the faculty of greatest value. Certainly hearing and smell contribute advantages, but their power is almost trivial in comparison with sight. Other than a few crude uses of infrared, the First World War saw no improvements in this all-important sense beyond the telescopic equipment that had evolved since Galileo. The introduction of radar, a completely new way to see, in the Second World War altered the basis of warfare more profoundly than any of the inventions that had marked the industrialization of combat. To be able to see an adversary in the dark, in fog, at distances difficult or impossible even with the best optical equipment under the best conditions was unprecedented. This was radar, and it is its story that I propose to tell.

1.2. ELECTROMAGNETIC WAVES

Ask an electrical engineer or a physicist to name the most momentous occurrence of the 19th century, and the answer could well be 'Maxwell's formulation of electromagnetic theory'. More likely, the answer would be the clipped 'Maxwell's equations'—four equations that one can write without crowding in the message portion of a picture post card. In them reside all our knowledge of electricity and magnetism except that lying in the domain of quantum theory, and even there they have furnished reliable guidance. They are to the engineer Truth. They are to the physicist matters of veneration that form an ever present model in the search for the description of subatomic forces. Maxwell's formulation was the second great fundamental law of physics, following Newton's laws of mechanics by two centuries.

Newton's laws united the mechanics observed on Earth as described by Galileo with the motion of the planets as set forth by Kepler, the work of just two men. By way of contrast Maxwell's theory united the experimental and theoretical work of many investigators, great and small, and concluded a striving to understand mysteries that had intrigued our ancestors back into prehistory. Amber, the fossilized resin whose Greek name, electron, has so enriched our vocabulary, has been found in the remains of European lake dwellings where the certain presence of fur must have led to the discovery of its strange, attractive powers. Amber was an object of trade in the ancient Mediterranean civilization and is well remembered in its literature. The discovery of magnetism probably does not lie so far back, as it must have taken place early in the iron age, very likely by a smelter who encountered a lodestone in his ore. Legend has it that a shepherd, Magnes, discovered the lodestone. If he made this discovery as reported by finding the nails of his boots magnetically extracted, it reflects badly on his cobbler. But the legend improved, and soon had nails pulled out of ships to the consternation of the crews.

The ancients pondered these two forces and found it puzzling that amber and magnets were so similar and yet so different. Centuries passed with knowledge of the two forces stagnant, generally in the custody of wizards and charlatans. From the 12th century, two millennia after the earliest records of magnetism, one finds confused reports of the north-pointing characteristics of a magnet, and a century later Peter Peregrinus invented a compass by which a mariner could steer, with immediate consequences for navigators and the basis for the opening of the Age of Exploration. Queen Elizabeth's physician, William Gilbert, published an exhaustive treatise on magnetism, based in good part on his own experiments, that gave the subject a sound foundation. Henceforth magnetism could be considered a science.

Electricity, more correctly electrostatics, advanced during those years no further than separation of charge by friction, and knowledge of it was left to industrious mechanics, who devised ever cleverer ways of building up charge. The discovery of electrostatic influence and the capacitor, going by the descriptive name of the Leyden jar, gave rise to much improved electrostatic generators that allowed some rather spectacular demonstrations in courts of the 18th century. The discharge of a capacitor charged to tens of kilovolts gained the respect that it retains to this day. This was the state of the subject to which Benjamin Franklin, a Philadelphia publisher, applied his remarkable skills. Out of his experiments and study came our complete qualitative knowledge of electrostatics, but his most notable and well remembered discovery was the identification of lightning as an electrical discharge.

The flow of electric current was discovered from its physiological effects, proceeding from the twitching of frogs' legs through chemistry to the Voltaic cell, which allowed the chemically generated potentials to be identified as electric. In 1819 Hans Christian Oersted, a professor of natural philosophy in Copenhagen, made the great discovery that the current from a chemical cell passing through a wire deflected a compass needle. The suspicion, so long held, although without basis, that electricity and magnetism were in some way related had been proved, and the search for an electrical effect generated through magnetism led to Michael Faraday's discovery of the magnetic induction of electric currents a dozen years later.

The end of the 18th century saw the initiation of quantitative measurements on matters electric with the inverse square force law being the first established. From it grew a system of units for electrostatic quantities, followed quickly by a similar system for electrodynamic phenomena. It was, of course, realized that quantities measured in units defined by electrostatics could have their values expressed in units defined by electrodynamics. A conversion factor, c, quantified the conversion, and although its value was unknown it was known to have the dimensions of velocity. The system of units now universally used by engineers, which includes the familiar ampere and volt, eliminates the need of the conversion factor and its dimensionality by adding current to the three mechanical quantities taken to be fundamental: length, mass and time.

In 1856 this ratio, c, was measured experimentally by W Weber and F Kohlrausch [1]. A Leyden jar of known capacity was charged to a potential determined with an electrometer, thereby establishing the charge in electrostatic units. The jar was then discharged through a galvanometer, a current-measuring device calibrated in magnetic units. The ratio of the charge calculated for the Leyden jar in electric units to the charge measured by the galvanometer in magnetic units proved to be the same as the velocity of light, which had been known accurately for only a few years. The historical records do not tell of the psychological effect of this on the theorists of the day, but it must have been strong. Faraday had shown a little earlier that polarized light was affected by magnetism in some experiments furnishing a hint that light and magnetism were related, but this new result certainly gripped the mind. Its significance was hardly lost on James Clerk Maxwell, a professor at King's College, London, who wrote 'we can scarcely avoid the inference that light consists in the transverse undulations of the same medium which is the cause of electric and magnetic phenomena' [2].

Maxwell's achievement was not only to unify the laws governing electromagnetic fields but to change entirely the way in which physicists look upon the universe. Maxwell wrote sets of partial differential equations that stated the three laws determined earlier by Gauss, Faraday and Ampère plus the mathematical statement that there are no isolated magnetic poles analogous to electric charges. From this he obtained no mathematical indication of the wave motion so necessary if light was of electromagnetic nature, as strongly suggested by the recent experimental value of the unit conversion factor, c. He noted a curious lack of symmetry in his formulation for Ampère's law, which had no time derivative to match the magnetically similar statement of Faraday's law. He remedied this by boldly postulating the existence of a quantity, named the displacement current, that gave his equations the desired form. When applied to space having neither static nor moving charges his equations then reduced to the wave equation with a velocity equal to c.

The effect of this formulation, published in 1864, was in no way commensurate with its importance. Physicists found the introduction of the displacement current artificial and unnecessary and the mathematics impenetrable, primarily because the equations were not given in the crisp notation familiar to modern readers. Several eminences of physics went to their graves without understanding. Nine years later and five years before his untimely death, Maxwell published his treatise on electricity and magnetism, which attempted to clarify his ideas; it was generally respected but generated little enthusiasm, and waves of electric and magnetic fields were not immediately produced in the laboratory. Two champions appeared on the scene who altered the situation dramatically.

When the treatise was published a copy came into the hands of a telegrapher, Oliver Heaviside, with no education beyond what had been available to a child of the London lower class. Heaviside decided that to understand electricity he would have to read the book and set about to teach himself calculus and differential equations, both ordinary and partial. In order to comprehend the material he invented a new branch of mathematics (vector analysis) and in so doing reduced Maxwell's page of equations to the four recognizable by the modern reader. In the course of this remarkable achievement, during which he lived off his kin, he invented another branch of mathematics (operational calculus, transformed by rigorists to a subject honoring Laplace), derived the telegrapher's equation, which finally solved the problems of transmission lines to the great relief of the owners of the new telephone systems, and engaged in delightful and acrimonious controversy. He lived and died in poverty.

Heinrich Hertz also found inspiration in Maxwell's treatise and fought his way through to an understanding using mathematical notation not greatly improved over the original. This understanding guided him to the production of electromagnetic waves in his Karlsruhe laboratory in 1887. He produced waves of length from a few decimeters to meters and demonstrated reflection, refraction and polarization. He measured their velocity and found it as predicted. The response of the scientific world was rapid and positive with numerous confirmatory experiments undertaken. German scientists and engineers began to use his name for the unit of frequency, and it was finally taken up by the International System of Units in 1960. The Nazi regime attempted to have it replaced because of Hertz's want of racial purity but surprisingly failed. He died before the Marconi experiments showed the communication possibilities of his waves.

The scientific basis of radar was not complete by the turn of the century, but the fundamental understanding of what have come to be called radio waves was complete. Small libraries were to be written about their interaction with antennas, wires, transmission lines, waveguides, dielectrics, the ionosphere, the sea, clouds, meteors, nuclear explosions, aircraft, ships and more, but the fundamental principles were well established and the dramatic predictions verified. By the turn of the century wireless telegraphy found ever wider application for marine communication, and in 1905 a major naval battle was fought in which wireless had much to do with the outcome. The fundamental advances leading to radar and all of modern electronics were now to be made in other parts of science.

Hertz's waves were not originally seen as suitable for communication, and the first experimenters concentrated on demonstrating properties that were recognizably similar to light. The virtues of much longer wavelengths than Hertz had used were noted by Guglielmo Marconi, an Italian with strong family ties to Britain through his mother. Marconi became interested in these waves as a student and began increasing the dimensions of the dipole, giving him ever greater range. Family connections and capital allowed him to form a company for providing communication between ships, an important application that he recognized and dominated commercially for the first decade of the century. The subject naturally attracted the mentally active, and invention followed invention all through this century bringing rapid changes in the design of transmitters and receivers. Its story becomes the history of electronics from which we shall be able to examine only a small part of the riches.

Hot filaments of carbon and refractory metals were found to emit electrons, which could be made to flow from a hot filament to a positive electrode, a vacuum diode. It was John Fleming of Marconi's company who applied the vacuum diode as a sensitive detector for wireless signals and to which Lee de Forest added a grid between cathode and anode to make the triode, an element capable of amplifying an electric signal. By 1920 huge arrays of triodes were generating easily controlled kilowatts of radio frequency power and rapidly replacing mechanical alternators as well as arc and spark transmitters with vacuum tube oscillators. The triode oscillator had been invented at about the same time by de Forest and Edwin H Armstrong, who fought for a score of years for a court judgement of a priority long since granted both by their colleagues. They also fought about the regenerative receiver that came about when the inductive coupling of the oscillator was reduced below some critical value.

During the 1920s the vacuum tube transformed 'wireless' into 'radio', and broadcasting burst on the public as a new and almost universal form of entertainment. The new circuit elements quickly replaced the amateur's spark equipment, and the demands of commercial wireless companies for low frequencies drove them to wavelengths shorter than 200 m, with which they soon began attaining astounding ranges with tiny fractions of the power required by the big stations. In November 1923 a French amateur in Nice established a two-way connection with two amateurs in America [3]. Within a year amateur contacts were worldwide. These spectacular feats were recognized to result from multiple reflections between the ionosphere and the earth's surface, but success in making a connection depended on the frequency used and time of day with other apparently random effects thrown in. An empirical understanding of the ionosphere and the way it affected long-distance transmissions was not long in coming, with commercial and government use not far behind. The very long wavelengths, which had been quickly overcrowded, were left for transmissions for which the location of the receiver relative to the transmitter was not well known, generally maritime. This pioneering by amateurs forced them to relinquish some desirable frequencies when the commercial value of 'short wave' became known. It was not the last time they would be 'rewarded' in this manner for their contributions to radio science.

1.3. PERCEPTIONS OF AIR POWER, 1919–1939

Jan Christian Smuts rode onto the pages of history during the third or guerilla phase of the South African War as the leader of a particularly successful Boer commando, one which rampaged through Cape Colony, even menacing Capetown. With defeat Smuts, along with several other prominent Boers, made peace with Britain and worked out a satisfactory position for the Transvaal within the Empire. By the outbreak of war in 1914 he had become a trusted imperial advisor, and it was in this capacity that Prime Minister Lloyd George called on him for recommendations about a highly political military matter. On 13 June 1917 a group of German bombers killed 162 (including 16 children at school) and wounded 426 in London and departed without loss despite the efforts of a much larger group of defending aircraft [1]. A large number of air squadrons and AA guns had already been deployed for defense against the attacks of airships, which no longer dared daylight attack and had begun to suffer losses high enough even at night that their end as an effective weapon against Britain was at hand. The bombers had struck in daylight. That the air defense of Britain already made use of a force much larger than was reasonable given the possible damage that the raiders could do, and that the civilian deaths, however regrettably distributed among women and children, were insignificant compared to the daily attrition of the Western Front, had no calming effect on public reaction, which showed signs of panic. The Cabinet met to consider the problem, and Smuts was asked to study it.

A conventional answer was obvious: more defending aircraft, more AA guns, more ground observer stations and above all unified command, all of which were incorporated in Smuts's July report. Smuts had a broad education, a penetrating mind and wide experience in war and peace and followed the first recommendation with another a month later that was certainly not conventional. He recognized that the air attacks, though not militarily serious at the moment, were a completely new form of warfare, unrelated to either the Army or the Navy, and proposed that the Royal Flying Corps and the Royal Naval Air Service be combined and reorganized as the Royal Air Force with status equal to the other two services, including having a Secretary of State for Air in the Cabinet [2]. This would remove the subordination of aviation to the Army and Navy and allow it to organize air offense as well as defense. He further recommended forming an independent bomber command to carry attacks to Germany. It was his opinion that direct bombing of cities would become the principal means of waging war in the future. The plan had enough political appeal to be enacted quickly. After all, if sturdy Englishmen had been thrown into a panic by a few bombs, the Germans would be so terrorized as possibly to make peace. In 1918 cities in the Rhineland and beyond were bombed and the first four-engine bombers capable of reaching Berlin began operation, although the Armistice postponed that city's experiencing air attacks for a quarter of a century. The German defense against these attacks had not been successful and had left a residual belief that the bomber could always get through [3]. The attacks on London were stopped by mid-1918 but at a cost of deploying 376 airplanes, 469 AA guns, 622 searchlights, 258 height finders, ten sound locators and a balloon apron, requiring 13 400 men and women [4].

When British fliers set about in the post-war world to protect their infant and rapidly shrinking RAF from being devoured by its parent services, Americans of the US Army Air Service returned from France filled with enthusiasm for the future of military and civilian aviation. Their leader was Brigadier General William Mitchell, who had commanded them in France and who had ideas about air power that went far beyond reasonable extrapolations of his wartime experiences. As the son of a US Senator he inherited natural political instincts to further his cause, which he enveloped with evangelism. His cause was the formation of an American Air Force, independent of Army and Navy. His extravagant claims, demands and charges soon led him to face a court martial and to enter the hagiology of aviators.

British and American airmen came to accept similar views about aerial warfare. Simply stated these views reduced to the idea that strategic bombing would decide future wars. Exactly what strategic bombing meant was never entirely clear but what it did not mean was clear. It did not mean attacking the enemy's armed forces as had been done in 1914–1918, something to be resisted at all costs. It did, of course, include attacking enemy air power as a vital part of the 'knock out blow' delivered at the opening of hostilities, but strategic bombing generally meant attacks on the enemy's industry, transport, communication and cities. It was assumed that attacks on cities would lead to panic and immediate demands for peace. It would have an ugly aspect that the airmen regretted but would insure a rapid end to the war with far fewer casualties than even a few days of trench warfare would accrue. Sir Hugh Trenchard, Mitchell's British counterpart, valued the morale effects of attacks on cities as 20 times the material.

Civilians began to expand on the ideas of strategic bombing during the inter-war years. There was a curious alliance of thought in Britain between airmen and pacifists on the terror and death that would come from the sky, with the airmen emphasizing the terror and the pacifists the death. Both assumed the usage of liberal dosages of poison gas. From a widely read book [5] of the time: 'But first we must make up our minds on one very important point, namely, that gas *will* be used. Let there be no mistake about that!'. A spectacular movie issued in 1936 by Alexander Korda entitled 'Things to Come' foretold war beginning on Christmas 1940 with the dreaded bombs falling on London. The screen writers had missed the teaching of Mitchell and Trenchard about a quick decision and prophesied instead a quarter-century of war that almost brought an end to civilization. For whatever value it had as prophesy, the scenes of attacking waves of bombers and of destroyed cities showed remarkable prescience. The film did little to calm the fears of society. The resulting expansion of Fighter Command was the result of political pressure and viewed by the RAF as a sop for the public.

If British and American airmen thought in terms of strategic bombing, their air comrades on the continent thought in terms of army support. This is curious because they showed great interest in the writings of Giulio Douhet, commander of the Italian air service during World War I, who stressed strategic bombing in a book that had been published in translation in French, German and Swedish by 1936 but not into English until 1942 [6]. Thus the writing that best expressed the dogma of the Anglo-American fliers was absent from their bookshelves during the 1930s, although translations had been studied at the US Air Corps Tactical School. The founders of the new, independent Luftwaffe often mentioned Douhet's effects on their thinking, but in 1919 the German air service began an analysis of military aviation, concluding that their ground support had been valuable but their strategic bombing not. By their evaluation the RAF's strategic bombing of Germany in 1918 had cost the British more in aircraft lost than in damage inflicted [7], and the Luftwaffe became, in fact and spirit, subservient to the army. Unlike the RAF, which organized according to mission—Bomber Command, Fighter Command, Coastal Command—the Germans organized Luftflotten, air fleets, that had a balance of all types of machine and even included anti-aircraft and airborne infantry units. Germany did not reject strategic bombing as such and made extremely good use of British, French and Czech fears of it during the months before the outbreak of war, but neglected it during the pre-war planning. There was some interest in long-range bombers, and two prototypes flew in the mid-1930s, but the costs of such aircraft let them slip out of the production plans for the new air force [8]. When development resumed, a terrible design resulted, the Heinkel 177, the worst production airplane of the war.

The new Luftwaffe had few aviators in its top ranks. Albert Kesselring, who commanded Luftflotte II in the Battle of Britain and in the invasion of the Soviet Union, was transferred to air service against his will in

1933 and learned to fly at the age of 48 [9]. It is hardly remarkable that he and his comrades carried with them an army point of view. Kesselring even commanded ground forces again as the war proceeded, holding the Allies at bay in Italy with very little air power. The superb army support techniques that characterized the Blitzkrieg came from the experiences of the members of Legion Condor in Spain, not from higher levels of command [10]. Germany's professional military caste also looked on war as a clash between military forces, not as a clash between nations. Their ideals of service grew out of studies of Frederick the Great and the War of Liberation against Napoleon. 'The professional soldier was very particular about the proper observance of the conventions—the usages and disciplines of war. These conventions were rather like a set of trade union rules, designed to make the profession of soldiering tolerable' [11]. Unnecessarily involving civilians repelled them. That their careers became tied to a government entirely without moral scruple is one of the ironies of the Second World War.

Thus we find a most curious situation as war crept closer. Britain, both her military and civilian components, put every available resource into protecting herself against strategic bombing, which Germany was not seriously planning. That the Luftwaffe lost the forced and unforeseen Battle of Britain came about in no small part because they fought it ill considered, whereas RAF Fighter Command had planned it for four years.

While Britain and America were specifying long-range heavy bombers in the mid-1930s the Luftwaffe was testing prototypes of excellent army support planes. Particularly valuable in that role was the Ju-87, the Stuka dive bomber—valuable, that is, so long as the Luftwaffe maintained command of the air. Dive bombing had evolved in ground support aviation on the Western Front in 1918 where it had found limited special application. The American Navy and Marine Corps developed it in the 1920s but the RAF and the US Army Air Corps (upgraded from Air Service in 1925 but not to be independent of the Army until after the war) were not interested in techniques or equipment that deflected them from their ability to win the next war alone. In 1939 dive bombing was standard for ground support only in the Luftwaffe [12]. This had come about in part from the accuracies demonstrated by it in comparison with the results of the inferior German sight for horizontal bombing [13].

Japanese attitudes toward the use of air power were better expressed by their actions during 1938–1941 than by theory. The war with China that began on 7 July 1937 soon bogged down along the coast and railway lines. The army could take land but was unable to force the Chinese army into a decisive battle. This exasperation caused them to attempt to reduce the Chinese will to fight by bombarding cities, beginning mid-1938. In May 1939 Chungking, the Nationalist capital, suffered 8000 killed in two days and by 1941 experienced 'fatigue bombing' in which aircraft were overhead all day, there being no air opposition left [14]. The Japanese army and navy maintained separate air arms that were not known for inter-service cooperation.

The legal and moral aspects of strategic bombing were widely discussed in Britain during the inter-war years but essentially ignored in Germany because of the attitudes of the professional military and later by the suppression of critical thought by a dictatorial regime. In the United States any strategic bombing was expected to be accomplished with the miraculous Norden bombsight, hence precisely dropped on key targets of the enemy economy, thereby minimizing civilian deaths—the combustible cities of Japan perhaps excepted [15]. But strategic bombing was not the official mission of the Army Air Corps. It had gained the responsibility of coast defense in 1931, allowing it to order the much wanted four-engine bombers, which American fliers were sure could sink surface ships under operational conditions and be ready to fulfill the unexpressed strategic mission if war came [16]. But British thought was thoroughly exercised by the legal and moral problems of bombing, and many articles and books appeared.

The legal case for strategic bombing was perhaps best made by J M Spaight in 1930 [17], who based his arguments on the examination of the history for the previous two centuries and on international law. Spaight considered an air force to be similar in function to a navy, and one of the time-honored functions of navies had been to raid enemy ports. The Royal Navy made several raids of this sort during the War of 1812 with towns being bombarded and destruction carried out by shore parties. Other port cities had been bombarded during the 19th century with severe damage to the civilian population as a result of inaccurate gun fire and the desire to destroy civilian property of value to the war effort. These depredations had been found to fit international law, such as it was, and to be accepted practice. Spaight noted that during the American Civil War destruction of cities and the civilian economy became routine and quoted General Philip Sheridan: 'Reduction to poverty brings prayers for peace more surely and more quickly than does destruction of human life, as the selfishness of man has demonstrated in more than one great conflict' [18].

Efforts were made around 1932 at Geneva to place some kind of control on air war with proposals generally running toward either banning the bombardment of cities or prohibiting air forces outright. None of the proposals was adopted, and September 1939 saw no international law concerning the employment of air power at all. There were Hague rules for warfare on land and sea but nothing for the air. The use of gas was banned, and it was not employed significantly during World War II. It is an open question whether such a ban on air power would have been effective.

The airplane was not the only novelty to have emerged in 1918. The tank had done more to gain victory for the Allies than the airplane. It was not a particularly radical device. A mobile soldier protected with armor had been the reason behind the knight of the Middle Ages, brought to an end of his dominance by firearms. It did not require a lot of imagination to replace the horse with a tractor that could carry sufficient armor to stop bullets.

The first employment of tanks in 1916 and 1917 had not been extraordinarily successful, but by 1918 their tactics had been mastered and the effect at the opening of the Amiens offensive was dramatic—'the black day for the German army' that convinced Ludendorff that the war was lost. The Germans had recognized by then the danger of the new weapon and had fielded anti-tank guns that quickly helped reduce the rate of advance to one typical of the Western Front. The Air Force attempted to counter this, as a report from the beginning of the campaign describes: 'The tanks suffered heavily from gun fire, and one of the lessons of the battle was that it was imperative to allocate a special fighting squadron to the Tank Corps in order to develop liaison and offensive plans for dealing with the anti-tank guns in the next attack' [19].

The lesson was carefully studied but the results were not greatly improved. The aircraft attacking battlefield targets were unarmored and took heavy losses from ground fire. An analysis by Wing Commander John Slessor questioned whether air actions had had any effect on the Amiens offensive at all other than preventing with an overwhelming air superiority German reconnaissance observing the build-up and thereby gaining the crucial surprise. Specifically, the Air Force was unable to cut the rail and road supply lines, which allowed the penetration to be contained within a few days [20]. That was the lesson, a consequence of primitive aircraft, that was fixed in the minds of the RAF after the war. It meant they required a mission other than army support to justify their existence. That mission was to be strategic bombing.

Most tank use in 1918 had had them moving forward at the speed of advancing infantry, eliminating the dreaded machine gunners and often terrorizing the enemy, although there had been more imaginative and successful tactical variations that led to local breakthroughs. Post-war military doctrine saw no reason to change the direct infantry support mission, and it was still the doctrine of the British and French commands in 1940. It was certainly not the doctrine of younger officers, especially those in the Royal Tank Corps. They, speaking frequently through the voice of B H Liddell Hart, saw matters from a radically different point of view. That they were overruled led to the great disaster for France and Britain in 1940.

Hart had been gassed on the Somme while serving as an infantry officer. He wrote the Army's official Infantry Training manual in 1920 and began a long period of military scholarship after being invalided out of the service in 1924. He was military correspondent for the *Daily Telegraph* and later for the *Times*, functions that kept him in close association with his old comrades. He observed and criticized military exercises and gained acceptance into the ruling stratum of British society.

Liddell Hart recognized the dominant effect of mechanized forces on the outcome of the past war, which he formulated into a widely discussed book, *Paris and the Future of War* [21], playing on Paris as the Greek hero, the city as a military objective and as a capital. Out of these generalized concepts grew his ideas to maximize mobility and surprise with a fast moving armored breakthrough supported by low flying ground attack planes and accompanied by motorized infantry. These tactics were tested by J F C Fuller with the Royal Tank Corps on the Salisbury Plain and convinced the tank men of their validity. The aging, horse-loving leadership of the Army rejected them, and the RAF wanted nothing to do with army support. The writings of Hart and Fuller were translated into German and found a strong disciple in Heinz Guderian, who put them into practice in Poland and in the Ardennes breakthrough that defeated the British and French armies with such startling suddenness. He generously gave credit for the foundation of his victories to Hart and Fuller but noted that he had had first-hand instruction at Cambrai [22]. The RAF had to learn—or was it relearn—those lessons over the desert battlefields of North Africa.

The aircraft carrier emerged out of the First World War as well as the tank, but its contribution to the struggle was close to zero. Planes had landed on and taken off from platforms added to warships even before the war, and during the war Britain converted a merchant ship to the first aircraft carrier, HMS Argus, which was used to make a raid on a German coastal air station. The part the carrier was to play in future warfare was much less well defined than that of the tank. Indeed how air power itself would affect sea power in the future was a wide-open question, one that greatly agitated naval thinking during the immediate post-war years. The mental ferment about tank warfare found its great debate in England, but the equivalent about carrier warfare took place in America. The contrast in the style of the discussions was characteristically different. The British disputes took the form of articles in service journals and newspapers, books, heated talk at table over brandy and cigars, and exercises carried out on the Salisbury plain, the interpretations of which left the opposing parties as far apart as before. The American disputes were much less tidy, with factions contending within the Army and Navy, superimposed on fights between the two services. It saw Admiral William Moffett, ardent builder of naval aviation, the fierce enemy of General Mitchell, to the extent of testifying against him at his court martial. It saw General John J Pershing, Commander of the American Expeditionary Force, announcing that the battleship was still the backbone of the fleet and Admiral William Sims, Commander of the US Atlantic Fleet in the recent war, asserting that no more battleships should be laid down. It involved Congressional committees and sensational newspaper headlines. At the middle of it stood Mitchell, whose rhetorical craft was unfortunately limited to hyperbole [23]. Few of the participants can be considered, given the vantage point of time, to have demonstrated great sagacity. There was no American Liddell Hart for naval aviation, but Hart's message was ignored by his own people whereas sound naval doctrine emerged from the American cacophony of the 1920s.

Could aircraft sink surface warships? This was the question about which the controversy turned. Scarcely returned from France, Mitchell called for a chance to sink a battleship with Army bombs. Equipment for the tests was at hand: a number of warships ready for scrapping and a few impressively large bombing planes. All parties called for tests, but each had its own view of how the tests were to be carried out. When all the explanations and interpretations for the results of the tests, which began in November 1919 and ended five years later, had been voiced, one fact remained uncontested: a battleship of the most modern design could be sent to the bottom by a well placed bomb carried by an airplane.

What did not and could not come from these tests was the reality of the situation. The ships were at anchor; their watertight integrities were not maintained; they were undefended by other planes or AA fire (which Mitchell dismissed then and forever as hopeless [24]) and there had been no opportunity for damage control. But he and his Army comrades concluded that land-based heavy bombers could dispose of any surface vessel and that the nation's coasts could be so protected. Surface vessels were obsolete, expensive and no longer needed for national defense. Regardless of the protests voiced by the admirals about the meaning of the tests, they made the Navy conscious of air defense.

While these shouting matches raged Moffett called for carriers. The first, USS *Langley*, the conversion of a collier, was commissioned in 1922. Her function with fleet maneuvers was so impressive that two uncompleted battle cruisers were finished as the carriers *Saratoga* and *Lexington*, commissioned in 1927. Tactical doctrine began to take shape, and the three kinds of carrier plane that were to be decisive were recognized: dive bombers, torpedo bombers and fighters. Mitchell had only scorn for dive bombing and aircraft carriers [25]. Others saw with insight that their thin decks made carriers very fragile, and the British soon began to build them with armored decks.

Britain and Japan launched carriers in the 1920s and set about similar armament programs, although without the level of contention that marked America's entrance. Britain commissioned HMS *Eagle* in 1923, *Furious* in 1925 and *Glorious* in 1930; Japan commissioned *Akagi* in 1927 and *Kaga* in 1928. At the outbreak of war the Japanese carrier force was the best, whether measured in number of ships and aircraft or in the quality of equipment and training. The American force was smaller, had poorer planes and crews that had not been pushed in their capabilities to the level of the Japanese—but it had radar, the ability to read Japanese signals and a comprehensive training program. Britain was not free of carrier acrimony. The Royal Navy had lost airplanes and pilots to the Air Force in 1917 and found this increasingly exasperating. When the Air Force suffered the inevitable cuts, the carriers lost much of their bite [26]. The Navy continued to press for full control over the Fleet Air Arm and succeeded in August 1937. Force of argument had eventually combined with a diminished fear in the RAF of being eaten alive by its parent services to decide the outcome [27]. Their carriers had obsolete aircraft but fortunately were to have only a brief, if bruising, encounter with their Japanese counterparts. The German Navy never gained control of an air arm.

If any of the navies gave thought about how aircraft might affect the conduct of submarine actions, it is not apparent. In so far as Britain and America were concerned the U-boat problem had been solved by convoys in 1918, and any residual difficulties would be taken care of by asdic or sonar, the underwater sound detection methods that the two navies trusted. All navies had to re-learn the submarine lessons of 1918. In 1938 only Dönitz understood, and he had not yet convinced his superiors.

The Second World War was to put to the test all of the theories concerning air power. All had been conceived without so much as an inkling of radar or its possibilities. In strategic bombing radar would dominate all aspects of the conflict, and almost everything was to proceed differently than expected, the changes ever confounding the planners as the conflict evolved. Radar would favor the defender one month, the attacker the next. Radar would affect naval warfare to just as high a degree. It would provide fragile carriers with electronic armor. Admiral Dönitz was to write a completely new chapter on the tactics of submarine warfare, but aircraft and electronics would be employed to negate it. Aviation's use for ground support would, under the force of circumstances, evolve from the tactics established in 1918, and radar would not affect it until late in the war.

1.4. NAVIGATION IN 1939

Navigation is now a commodity, something that can be purchased for a remarkably small sum. The mariner who wishes to know his place upon the sea need only glance at the liquid crystal display of latitude and longitude to be found on the bridge, and it is irrelevant whether he runs in fog or even a storm. A few hundred dollars can provide him his position at any time to an accuracy of tens of meters. Signals from orbiting satellites are received by radio and interpreted with an attached computer, replacing at one the skill and art that had always set navigators apart from the rest of mankind. For years now aircraft have followed courses set by devices sensing their accelerations and accurately calculating whence they have come and how they must steer. The completely automatic pilot is a technical reality.

The contrast between the present state of things and that of 1939 is greater than that between 1939 and the age before the invention of the ship's chronometer. Just the changes brought about during the Second World War completely remade the world of air and sea travel, changes resulting from radar and its companion electronic techniques. To appreciate the accomplishments of radar and how they affected the nature of the war requires a knowledge of the state of navigation at the opening of the struggle.

In 1939 the word 'navigation' usually carried the modifier 'celestial', for it was by observing the stars that sailors and more recently airmen had found their way. Latitude was the easiest to obtain; with an ephemeris one could have it quickly from a measurement of the altitude of the sun or any other listed star as it crossed the observer's meridian, the north–south great circle in the sky that lies directly overhead. Longitude was equally simple in principle but far more difficult in practice. One must know the time at which a given star passes through the meridian and compare it with the time when it had passed over the location of zero reference, taken internationally as the Royal Observatory at Greenwich. In addition to the ephemeris that gives the time a given star passes over Greenwich one needs a clock that runs on Greenwich time. It was the absence of a clock that would run accurately aboard a rolling, generally damp and uncomfortable ship that led to countless vessels breaking up on unexpected rocks.

A voyage from England to Jamaica and return in 1762 with a chronometer manufactured by John Harrison showed an accumulated error in longitude of less than 2 minutes, which met the standards set forth by the Board of Longitude for the award of a prize of £20 000. The skills of horologists soon provided seamen with excellent instruments at affordable prices, and the ticking of three chronometers became one of the many sounds to be heard in the cuddies of deep-water sail. The substantial prize for a practical means of determining longitude had not been intended for a practical man, however; the Board had had an astronomer in mind, and Harrison waited ten years to receive his reward [1].

The period before 1939 saw a number of practical aids for navigators in the forms of tables easing calculation and the issuance in 1914 of the *Nautical Almanac abridged for the use of Seamen* and in 1933 the *Air Almanac*. Agreement to begin the day of Greenwich Mean Time at midnight rather than noon ended a source of unnecessary confusion [2].

Celestial navigation presented the navigator with a sufficient variety of difficulties as to transform what appears to be scientific into the realm of cunning. Foremost was the need to see celestial objects and the horizon at the same time. When the horizon cannot be seen, as is often the case in an airplane, a spirit level had to be attached to the sextant. When the sky was overcast one had to rely on the unfortunately named method of dead reckoning, which predicts successive positions from the course and speed of the vessel or aircraft. This involved also knowing the effect of the motion of water and wind on the craft. For a steamer this is not so difficult, although days without a fix could produce tension on the bridge. For an aircraft moving at 20 or 30 times a steamer's speed and much affected by unknown, possibly high-speed winds the results could be very serious. The need to make observations and calculations correctly and rapidly in a noisy and bitterly cold airplane did not contribute to their reliability.

Fliers navigated during the day by comparing their view of the ground with their map while keeping an eye on the compass, and they especially liked to follow the railroads. Unexpected clouds often meant serious trouble. In 1939 Royal Air Force bombers were unable to find their targets in daylight 40% of the time [3]. Night flying became sufficiently common in the 1930s that beacon lights began to add their rotating beams to the evening stars that rural people watched as they relaxed at the end of day. Flying over the open ocean presented an obvious extension of difficulty.

Radio brought some degree of help after the First World War in the radio direction finder and the radio compass. Direction finders for low-frequency signals proved to be rather good, and various shore stations could determine the direction to whatever ship or aircraft made request. The information from two or more such stations yielded an accurate fix. Ships and planes with loop antennas could determine the direction to known stations and establish their positions without the need of time-consuming interrogations. This was much more difficult for aircraft because only small loop antennas could be used, and adjusting the loop to minimum or maximum signal was difficult in the noisy conditions that usually prevailed in the aircraft of the time. The high frequencies that were convenient for communication had poor directional qualities, owing to the altered polarization of the wave reflected from the ionosphere and its irregular reflecting surface.

During the 1920s a system of radio navigation based on a 1907 proposal by Otto Scheller came into wide use in equipment designed by the German firm of Lorenz that was manufactured abroad under license and known as the 'four-course range'. (Here 'range' is aviation parlance and does not infer range as a measured distance, as the four-course range provided only direction or course.) In one configuration widely used before the war, four vertical antennas located at the corners of a square, about 100 m on a side, received the same audio-modulated radio frequency on the 100 m band. Alternately feeding opposite antenna pairs with the Morse-code dit–dah or dah–dit produced two overlapping figure-eight radiation patterns that identified one pattern with 'A', the other with 'N'. Four directions emanated from the station where the intensities of the of the two patterns were equal, and the signal received by the pilot was a pure tone. If he strayed to the left, he got an 'A', to the right an 'N', or the reverse, depending on the course [4]. By 1939 radio ranges had led to the phrase 'on the beam' entering the language of the young and technically minded.

The known imperfections of aerial navigation do not seem to have troubled those planning strategic bombing and the wildly inaccurate bombing of Britain by the Germans in the First World War motivated no intense navigation studies [5]. The traditional methods were taught, but few exercises attempted to evaluate this crucial aspect of American and British war plans. Before war broke out both thought in terms of daylight bombing. If clouds obscured vision, navigation suddenly became quite problematic and bombing impossible. Navigation over a completely blacked-out enemy land left the attacker entirely dependent on the stars or whatever radio signal he might devise. The Germans, who were not planning strategic bombing, were nevertheless concerned about being able to guide their planes over the enemy at night. The firm Lorenz AG had marketed a beam navigation system and were given the task of constructing one that would be good enough to drop bombs accurately at night or through clouds. An excellent beam system was ready at the outbreak of war, but its similarity to known systems was to make countermeasures relatively easy.

Aircraft introduced a third dimension into navigation, altitude. It was not so difficult to determine as latitude and longitude, but the accuracy of its value was extraordinarily critical. Uncertainties by amounts trivial for the other two coordinates could be fatal to a flier. The method used initially was the decrease in atmospheric pressure as one ascends, discovered by Galileo's secretary, Evangelista Torricelli, the first person to create a sustained vacuum. Aneroid barometers were installed in the earliest aeroplanes and were unrivaled during the first four decades of flight.

Such instruments have two serious flaws. In an ideal atmosphere they give the height of the aircraft above mean sea level, but fliers must deal with real atmospheres in which weather dominates and not with ideal ones. Discrepancies can run to several hundreds of meters as a result of local barometric pressure variations. This can be corrected with an adjustment at the time of take-off, if the altitude of the field is known, but such a correction does not necessarily apply to a location far removed from the origin of the flight or even at the origin hours later. More important than these difficulties is the uncertainty about where the ground is. Height above mean sea level is useful, but it is height above the surface that is critical. When flying over poorly mapped terrain or when lost, height above the ground can be a matter of life or death. A bombardier must know his altitude above the ground accurately, if he is to hit the target. In either case the inaccuracies inherent in the aneroid altimeter can be serious.

As a consequence of these problems inventors sought other methods, and radio formed a favorite approach. During the 1920s and 1930s patent clerks were busy sorting out the tenuous differences in the many suggestions for radio altimeters that were proposed. Heinrich Löwy was apparently first, suggesting in 1923 the use of pulsed high-frequency waves by timing the arrival of the reflections from the Earth's surface. The electronic techniques of the time did not allow realization of his suggestion, but that did not prevent it being followed by many others employing reflection phenomena, either with pulsed or continuous waves [6].

The Aircraft Radio Laboratory of the US Army Air Corps and the Naval Research Laboratory worked on the problem with a Navy design, which came out of their radar research. A contract was given to the Radio Corporation of America, and a satisfactory pulse altimeter prototype was demonstrated in 1937. Its price of $20 000 caused much Air Corps unhappiness, but by haggling and scouring ledger books for money two sets were bought. By April 1940 a 40 kg set of good accuracy was designated SCR-518 and put into production, ultimately reduced in weight to 12 kg and functioning to an altitude of 12 km [7].

1.5. ANTI-AIRCRAFT ARTILLERY, 1914–1939

Anti-aircraft artillery dates from 1871 when the Prussian forces besieging Paris countered the French balloons, used to communicate with forces outside the city, with a hurriedly designed Ballon-abwehr-kanone, a Krupp 36 mm breech-loading gun allowing high elevation and complete traverse [1]. By 1914 ballistics had made great strides since that beginning but aiming had not changed and was quite incapable of hitting the fast, agile aircraft that presented themselves as targets. The problem was similar to that confronting naval and coast defense gunners, a target that would change its position significantly by the time a projectile fired at it arrived. But it was vastly more difficult, for the speeds were much greater and the targets much smaller; furthermore, ships are confined to the surface of the ocean, making the calculation of gun orders a two- rather than a three-dimensional problem. The small size of the target meant that a direct hit by an artillery projectile was unlikely to an impossible degree, so the shell had to be made to burst with a time fuze set to explode when the gunner calculated it should be close enough to be destructive.

If the target could be seen, either in daylight or illuminated by searchlights, a series of three independent measurements had to be made of its present position to predict its future position. The target had to be tracked in the cross- hairs of telescopes to obtain its present azimuth (horizontal direction) and elevation angle, relatively simple to do unless angular speeds were very high or visibility poor. The third and key measurement was the slant range to the target, which was harder to obtain. It required either a stereoscopic optical range finder operated by scarce observers with exceptional vision or triangulation from a base line of at most a few meters. Neither could determine ranges accurately enough for calculating firing data. These three measurements had to be converted into gun orders of elevation, azimuth and fuze time, and the calculations had to be made simultaneously with the observations. The computer, called a director or predictor, was a mechanical analog computer that used cams of extraordinarily high precision. When all of this was working well and the target conditions were ideal, the results were good enough to make the fliers worry, but the conditions were seldom good, and extravagant amounts of ammunition were consumed with little effect.

Raids at night required that the gunners find and hold the attacker in the beam of a searchlight, a difficult problem in its own right. Sweeping the sky was of little use, but if the weather was favorable, an array of listening horns could give clues to put skilled searchlight operators onto the target. In fog or above the clouds nothing worked, but then the same was true for the fliers.

During a high-altitude bombing run there was a period when the pilot of the attacking airplane had to set a level, straight course to allow the bombardier to aim. This course was also a perfect opportunity for AA gunners, as close to ideal as they would find, and provided a strong inducement to the fliers for reducing this time to a minimum, with de-creased bombing accuracy generally the result. Good fire could drive the bombers off the selected path. One of the reasons the Luftwaffe favored dive bombers was the efficiency demonstrated by their Flak on these kinds of run.

An alternative to directed fire was used by the British well into the war: barrage fire. A region of bursting shells was placed through which the attacking aircraft had to pass if they were to reach their target. What it lacked in accuracy it sometimes made up in morale effect on the pilots encountering the curtain of shell bursts. Scientists in operational research had little good to say about this technique. Professor A V Hill dismissed it 'as based on sloppy thinking and bad arithmetic' [2].

Anti-aircraft gun fire became ineffective at close range because of the guns' inability to track rapidly, and this region fell naturally to machine guns. In the 1914–1918 conflict these were adapted from existing types and included the 37 mm Maxim pom-pom. During the inter-war years more powerful automatic weapons appeared. John Browning contributed the 0.50 inch and a high velocity 37 mm. The 20 mm Oerlikon came from Switzerland and the 40 mm Bofors from Sweden. The US Navy developed a 1.1 inch. All contributed to the streams of tracers that filled the air during World War II, but few were to shoot with radar direction.

1.5.1. *Germany*

The German Army entered the First World War with six truck-drawn and twelve horse-drawn 75 mm Ballon-abwehr-kanonen, organized as sepa-rate guns, not as batteries. A year after the outbreak of war the number of AA guns had increased by a factor of ten in addition to field guns set up in various improvised mountings. By fall 1916 air units and air defense units were organized into a single branch and anti-aircraft received the abbre-viation that was to become internationally recognized, Flak, standing for Flugabwehrkanonen. By 1918 there was a total of 2900 AA guns in service. Fifty batteries [3] had computing directors of some kind, none entirely sat-isfactory, but the Schönjahn-Gerät was good enough to become the basis for the director used in World War II. Evidence for greatly improved tech-nical capability is found in the numbers of Allied planes brought down by AA gun fire: of the 1590 total for the war, half fell during the last ten months [4]. A new 88 mm gun saw service before the end of the war.

German anti-aircraft left the war with a tactical experience denied to the Allies that was to prove of value in the later conflict. When tanks encountered AA guns in 1918 they came out the worse for it, the result of the high velocity and ease of traverse of the AA guns. Revisions of the design in 1936 and 1937 turned the 88 into the 'triple threat', useful for anti-aircraft, field or antitank artillery. The last capability would prove extraordinarily useful in the steppes of Russia and the deserts of North Africa, places where long, unrestricted fields of fire made such a weapon very effective.

The Treaty of Versailles forbade the German Republic both an air and an anti-aircraft arm, although clandestine development of both took place: in Russia for air, in Sweden for anti-aircraft. Whether this had a reverse effect during the rearmament of the 1930s is a psychological question, but for whatever reason Flak was given high priority and held in high esteem as part of Hermann Göring's new Luftwaffe. By the fall of 1935 Flak had organized 15 heavy and three light battalions; 12 months later these numbers had doubled. Naturally there was a special Flak school [5].

German anti-aircraft profited by experience in the Spanish Civil War as did other branches. General der Flieger Sperrle led an air contingent for Legion Condor that contained Flak units. By the end it had grown to nine batteries of 88 mm, which were credited with destroying 61 Loyalist planes. The versatile nature of the 88 proved itself, especially as field artillery in Franco's artillery-poor forces. On return to Germany the Legion Condor veterans were distributed throughout their branch [6].

Assigning Flak to the Luftwaffe brought problems with the Heer (Army) because this arrangement suggested poor protection for front-line troops. This led to the organization of heavy machine gun battalions in the fall of 1938 that were armed with the 20 mm automatic gun and under Army control.

When war broke out in the summer of 1939 Germany had far and away the most advanced anti-aircraft force of any nation, with 107 000 men on the rolls [7].

1.5.2. United States

The US Army not only had no anti-aircraft units in 1914 but had none when America entered the war in April 1917, although the matter had been studied as early as 1913, airships being considered the possible raiders. In 1915 the Ordnance Department began the design of a 3 inch AA gun, which reached prototype stage by 1917 but never reached troops in France. Experience with Curtiss JN-2s in the expedition to Mexico in 1916 'indicated that airplanes had a sufficient tendency to fall unaided out of the sky' so that an air defense arm hardly seemed to be urgent, but the nature of the war in Europe quickly dispelled such ideas [8].

That anti-aircraft artillery was assigned to the Coast Artillery Corps came about in part from the success of the Royal Navy in holding the German fleet in its harbors, making the possibility small that the heavy guns defending the east coast cities would be needed [9]. The Coast Artillery also had had experience shooting at moving targets even if an order of magnitude slower and confined to two dimensions. Thus it was a detachment of Coast Artillerymen who embarked for France to study European AA techniques, establish a training school and organize American units. They found that French gunners had evolved a more successful technique than the British, who mistrusted 'technical shooting' and preferred to shoot by eye [10]. With French instruction and French guns the Americans deployed eight batteries in combat, credited with having downed 19 enemy aircraft. Eight machine-gun batteries were credited with having downed 41 [11].

The National Defense Act of 1920 assigned air defense to the Coast Artillery, which organized four battalions in 1921, expanded three years later into four regiments of two battalions each. These were followed by the organization of AA regiments in the National Guard and the Organized Reserves. The Coast Artillery pressed for modern equipment and secured from the Ordnance Department in 1925 a satisfactory plan, which came to nought because of the budget restrictions of the 1930s. The Coast Artillery School at Fort Monroe emphasized technical quality.

In 1937 tension between the United States and Japan resulted in the War Department giving top priority to air defense. The number of 3 inch guns was to be increased from 135 to 472, which would equip 34 regiments. As the characteristics of the four-engine heavy bombers became known in the mid-1930s the inadequacy of the 3 inch gun to reach them became apparent, and by 1938 a successful prototype 90 mm gun was ready as a replacement for the 3 inch. As part of a general strengthening of the arm a 37 mm automatic gun, which had been under development for more than a decade, was to replace machine guns [12].

Despite the rejection of the Air Corps thesis that land-based bombers sufficed to dispose of ships, the Navy took anti-aircraft defense seriously and as funds became available in Fiscal Year 1938–1939 began improving these capabilities in existing vessels. A memorandum to the Chief of Naval Operations summarized AA firing as generally ineffective, resulting in the designation of Admiral Ernest King as head of the Navy Department Anti-Aircraft Board. The dual purpose, semi-automatic 5 inch/38-caliber gun had been introduced in 1934 on the *Farragut* class destroyers. It was judged to have adequate range and power and was to be mounted wherever space allowed on all heavier ships as well. Fire control of matching excellence was absent. The earlier decision to rely on the combination of 1.1 inch and 0.50 inch machine guns for close defense was rejected for the 40 mm Bofors and the 20 mm Oerlikon [13].

1.5.3. Great Britain

The preparation of the anti-aircraft defense of Britain fared substantially worse than its American counterpart during most of the pre-war years. At the time of the Armistice the homeland had 286 guns and 387 searchlights in position; by 1920 there were none. Soon the gunners were reduced to a small Regular Army cadre and an AA school. Such troops as existed were in the Territorial Army, which had occasional evening drills and summer exercises using inadequate equipment. Anti-aircraft suffered, then and later, from a social sensibility the Americans would not have understood. It was not a branch of service to which an Englishman, defined by his club and his regiment, wished to be assigned, whereas graduates of West Point often selected the Coast Artillery for assignment because of the technical nature of anti-aircraft. That the men and women who fired on the Zeppelins, Gothas and Giants received no War Medal tells us much of the attitude toward this arm. Among some there was even the romantic idea that by not facing their foes the gunners were distastefully comparable to guerillas attacking from ambush [14].

Many in Britain shared Billy Mitchell's opinion that AA guns were useless and would remain so, there being no immediate cure for their deficiencies. Vocal in this was Frederick Lindemann, later Lord Cherwell, Churchill's friend and scientific adviser. During 1914–1918 attacks on cities had generally been made at night, and the AA guns of both sides had not shown themselves particularly lethal—but then neither had the raids. In America the Coast Artillery on assuming the duty of anti-aircraft sought the help of the Ordnance Department and the Signal Corps for methods of locating aircraft at night, yielding active, if modest research in both services; no such pressure was felt in Britain, where radar for AA gunners came along as an afterthought in the radar project, as the poor quality of its gun-laying equipment was to show.

There were committee studies, however. In April 1923, when there was uneasiness about the size of the French Armée de l'Air, the Steel–Bartholomew Committee reported that Britain's air defenses were non-existent. The effect of the report was negligible. By 1934 the absence of air defense began to cause serious concern, this time with the worry directed towards Germany, and Sir Robert Brooke-Popham, the Air Force officer in charge of the Air Defence of Great Britain, began serious study of the matter. The committee that he headed recommended a first stage of preparation be attained by 31 March 1940 of 17 AA batteries and 42 searchlight companies, and this part of Britain's arms build-up slowly began [15].

In the Army exercises during the two-decade armistice General Sir Frederick Pile was one of the few Army officers who gave evidence of understanding the significance of the changes that technology had brought to the profession of arms. A hereditary title interfered not at all with his enthusiasm for things mechanical, which brought him to the Royal Tank Corps where he was remembered by Liddell Hart: 'One of the few bright spots in the exercise was the activity and ubiquity of the reconnaissance group of armored cars and tankettes under Tim Pile' [16]. As mobilization began to accelerate in 1937 there was talk of creating an armored division, and Pile wanted very much to command it. One tends to speculate on the course of the future had it been formed then and Pile made its commander, but it was not to be. Instead he was asked to accept command of the 1st AA Division, which was being formed. He had had no experience with anti-aircraft and came to the assignment because of the unpopularity of his tank warfare ideas with the upper levels of the Army and because of his obvious technical ability. There is reason to believe that Pile accepted the command with small pleasure, but he was a fine soldier and devoted himself with all the drive and ability that he possessed to what was in the British Army a military backwater. The job was expanded in April 1939 to leadership of AA Command, which shared the Battle of Britain with RAF Fighter Command; but where Fighter Command received the praise and honor it so richly deserved, AA Command more often received the abuse of a public that did not understand. Pile's nature was not one to show resentment, but he was bitter about criticism that touched the skill and devotion of his people, whom he commanded until the end of the war.

Pile took over a pitiful group of Territorials and appeared before town meetings trying to bring his division to authorized strength through appeals for volunteers. Only after Munich, which delighted Pile for the time it gave him, did recruits begin to join in substantial numbers [17]. The 3 inch guns from 1918 were unable to reach the altitudes of modern bombers, and the new 3.7 inch guns did not begin to come off production lines in quantity until after the Munich crisis. The old predictors required six skilled operators and were not satisfactory at best. On one target course flown under conditions ideal for the gunners but tactically unrealistic, only two hits could be scored out of 2935 rounds fired. Most of Pile's comrades looked on AA fire as a way of making the enemy fliers nervous and making noise to reassure the civilians. Tactically it was seen to have value in breaking up the bomber formations, thereby giving the fighters a better chance. For the first years of the war Pile had to soldier with this kind of technical support and for most of the war with these kinds of attitude.

The Royal Navy also lagged their Yank cousins in anti-aircraft. A report of January 1939 described the alarming vulnerability to air attack of carriers and cruisers and found destroyers almost defenseless. The situation was so serious that foreign guns were ordered and licenses for manufacture purchased in March, the 40 mm Bofors from Sweden, the 20 mm Oerlikon from Switzerland [18].

Of all the weapons in general use in 1939 none were improved in such a startling way by radar as was anti-aircraft artillery. When war broke out in 1939 the effectiveness of anti-aircraft fire ranged from thousands of 'rounds per bird' for daylight shooting to tens of thousands for night. Automatic weapons firing at low-flying planes did much better. The accuracy of heavy guns began to improve rapidly as gun-laying radar and better directors came into use. By 1945 an airplane caught in the range gate of an SCR-584 radar feeding the data to an M-9 director controlling an automatic tracking 90 mm gun using proximity fuzes simply meant that the plane was finished. It might save itself by violent evasive maneuvers, but if the first rounds were not placed accurately enough to produce bursts, there was no warning.

PHOTOGRAPHS: PRELIMINARIES

A 1931 transmitter–receiver antenna pair from an experimental 17 cm communications link across the English Channel. It failed because the circuitry to allow the simultaneous transmission of many telephone channels had not been developed, leaving it inferior to submarine cable. Such wavelengths were the basis of early American and German radar experiments but had to be given up because no transmitters of suitable power were at hand. Photograph courtesy of Bernd Röde.

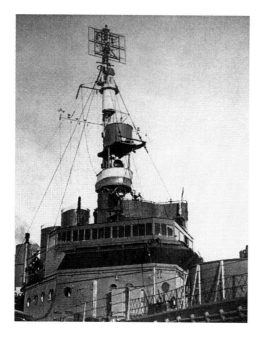

Prototypes of the SC and SG radars mounted on USS Semmes in May 1941. The dipole array at the top of the foremast is the SC, a 1.5 m air-warning set based in design on the XAF or CXAM but smaller for use on destroyers. The dark cylinder just below the crow's nest contains the paraboloid antenna of the Radiation Laboratory's 10 cm surface-search radar. Final design replaced the full with the cut paraboloid soon to become familiar to seamen. National Archives photograph 80-G-700311.

The antenna of an experimental 1.5 m radar being tested aboard USS Leary by the Naval Research Laboratory in April 1937. Mounting it on the destroyer's deck gun allowed pointing the beam at targets. The next stage of development was the XAF, which used a dipole array rather than the Yagi antenna shown here. The success of XAF in the fleet exercises in early 1939 put the Navy on a strong radar development program. National Archives photograph 80-G-700710.

Failed intelligence. In August 1939 Colonel Wolfgang Martini, Chief of Luftwaffe Signals, chartered the airship LZ-130 for a flight along British coastal areas in the hope of locating possible radar defense. There are various published versions of this flight, each with abundant error. Determining the true course of events and the reason for the failure to observe the radar that followed the ship during the entire flight have engaged the detective instincts of four investigators in Britain and Germany. The airship broke through the clouds in the vicinity of Aberdeen, causing people to call the Dyce Station of the Auxiliary Air Force from which a plane with a photographer took off and made this photograph. Royal Air Force Museum, Hendon, London, photograph P26. Crown Copyright.

31

Failed intelligence. A German publisher issued annually a pocketbook describing the ships of the world's navies. When the 1939 volume appeared it caused consternation in German radar circles because of a photograph dated 1938 of the black Torpedo School Ship G 10 displaying a prominent Seetakt antenna just forward of the foremast. The photograph was passed for publication by naval authorities, all kept in the dark about the new technique and, of course, unable to recognize the apparent mattress as the mark of a secret weapon. There is good reason to assume that the British naval attaché in Berlin purchased the book and that naval intelligence in London studied it, but there is no record of them having grasped the significance of the antenna either, very likely for the same reason that the picture had escaped in the first place—radar was too secret. Photograph through the courtesy of Hans Sorgenfrei.

CHAPTER 2

ORIGINS

2.1. ELECTRONIC COMPONENT DEVELOPMENT

So much of modern technology owes its origin to military requirements that it is somewhat startling to learn that the pace of radar development until 1939 was set instead by civilian electronic use. Except in Great Britain the pre-war radar engineers worked on restricted, sometimes very tight budgets. They could not have considered such extravagances as having special vacuum tubes or other devices designed for their needs and had to make do with the components available from civilian electronics suppliers. But electronics was a strong industry even during the depression, for radio broadcasting had boomed in the 1920s, and the public spent lavishly buying radios, paying to have them repaired, replacing rapidly aging sets with new models and in the United States buying the products that broadcasting advertised. When hard times hit, the family radio replaced other pleasures that had to be foregone and kept farmer and rancher advised of the latest market fluctuations. Radio was good business throughout the interwar years, and broadcast companies dreamed of the coming market for television, and television—like radar—required high frequencies.

Military services did improve their communication sets during the inter-war years, but these projects, with rare exception, came from the normal evolution of communication electronics of the civilian world, adapted by the small military service laboratories to their special needs. Development contracts to private corporations, now the principal method of designing new equipment and the source of many post-war electronic marvels, were completely unknown.

2.1.1. Cathode-ray tubes

Ferdinand Braun, Professor at the University of Strassburg, was one of the eminences of early wireless. He saw the need for waveform diagnostics and applied the newly discovered cathode rays to this purpose [1]. In the anatomy of a radar set the cathode-ray tube is the heart. Cathode rays—streams of electrons—had been observed when high voltages were first applied to electrodes in vacuum tubes with effects dependent on the degree of vacuum (by present standards scarcely a vacuum at all), the voltage applied and the configuration of the electrodes. Julius Plücker discovered them in 1859 when he succeeded in reducing the pressure in his discharge tube to new limits and demonstrated that they could be deflected with a magnetic field. The next two generations of physicists spent many happy hours in the laboratory experimenting with them. In 1897 J J Thomson produced cathode rays in well defined beams and demonstrated that they were made up of grains of electricity, electrons, by deflecting them with electric and magnetic fields. At about the same time Braun constructed a tube with a cathode at one end followed by a positive electrode with a small hole in it. When high voltage was applied and the gas pressure adjusted to the right value, electrons passed through the hole and struck the glass at the far end of the tube, causing the glass to glow. By extending the path from the positive electrode to the target glass, Braun was able to deflect the beam with magnet coils applied at the outside. Waveform analysis made use of rotating mirrors to view the tube or of an external magnet rotating in synchronization with the waveform being studied [2]. The electronic time base or sweep circuit lay in the future.

Use of the technique was slow in establishing itself because the operator needed to be a master experimentalist, and just operating the tube and its peripherals was a major undertaking, certainly not the kind of thing for which the wireless men had time. In 1922 J B Johnson and H J van der Bijl of the Bell Telephone Laboratories [3] produced a low-voltage cathode-ray tube that Western Electric manufactured as the WE-224. It used two sets of internal electrostatic deflection plates and had a modern appearance. Its success resulted from something called gas focusing, which also set its limits.

The need for the extreme high voltages in the early tubes was primarily to produce a tight beam that made a small, bright spot on the glass target, but the high voltages made the tube less sensitive and troublesome to operate. If one used lower, more convenient voltages the beam diverged and produced less light from the glass, resulting in a large, dim spot. Johnson produced a bright spot by having a copious electron beam taken from a thermal filament impinge on a phosphorescent coating. By introducing some gas into the tube he could produce a counter-effect to the space charge of the electrons that caused the beam to diverge by forming a core of positive ions at its center that canceled the effect of the negative electrons. The problem was that the electron beam was no longer light and nimble but heavy and sluggish, as it had to carry the positive ions with it to remain focused, making the response time of gas-focused tubes about 80 000 times slower than high-vacuum tubes, argon being a favorite gas. Such performance was out of the question for the future television receivers, so major efforts went into correcting this while preserving the simplicity of the low-voltage tubes. Such tubes were also what the radar men needed, because the gas-focused tubes were much too slow to display a pulse of only a few microseconds in width, a minimum requirement. The ease of operation of the gas-focused tubes was strong motivation for correcting the gas problem. Focusing the beam of a high-vacuum tube required a new branch of engineering called 'electron optics'.

Two names are deservedly associated with these developments: Vladimir Kosma Zworykin in the United States and Manfred von Ardenne in Germany. By 1930 the crucial element for radar and television was at hand and by 1935 was incorporated into handy laboratory oscillographs with frequency responses limited only by the amplifiers driving the cathode ray tubes.

Zworykin was one of the many refugees who have enriched American life. He fled the new Soviet state at the age of 30 and became Director of Research at RCA in 1929, the first of a series of leading positions that he held with the company. At the same time Zworykin demonstrated his proto-television [4] there were almost identical activities in Germany by von Ardenne [5]. Indeed, Zworykin and von Ardenne duplicated one another's inventions with near simultaneity for more than a decade: the electrostatically focused cathode-ray tube, the flying-spot television camera and the electron microscope. But whereas Zworykin worked for the growing giant RCA, von Ardenne had his own private laboratory at Berlin-Lichterfelde, which he had established at the age of 21 with capital loaned to him on the basis of his youthful successes in radio. He was 23 when he publicly demonstrated his all-electronic television system. To exploit the market for cathode-ray tubes and oscillographs that was about to overwhelm his laboratory he entered into partnership to form the firm of Leybold und von Ardenne; to exploit the greater potential market for television he entered into an agreement with the firm of C Lorenz [6].

2.1.2. *Multielectrode tubes*

Electron tube designers of the 1920s soon encountered the parameter of the new triode that checked their plans: the capacity between the control grid and the positive electrode, called the anode or plate. If one incorporated this capacitance into the tuned circuit of an oscillator or tuned amplifier, one could work frequencies of tens of MHz, but if one wished to build a high-frequency broad-band amplifier, this inter-electrode capacitance drastically limited the frequencies attainable. The envisioned television sets required radio-frequency amplifiers with broad pass bands, and triodes alone could not do the job. The solution was obvious: drastically reduce the grid–anode capacitance.

In 1919 Hiroshi Ando applied for and was eventually granted a Japanese patent for a tube with a fourth electrode, another grid, introduced between the control grid and the anode specifically intended to prevent 'the undesirable influence of the inter-electrode static capacity between electrodes almost entirely or substantially' [7]. No use was made of this specific property, however. In 1926 Albert Hull and N H Williams of the General Electric Company described a similar tube [8]. (Double grid tubes had been around since 1916, the invention of Walter Schottky, but for them the second grid was connected to the anode in order to neutralize space charge with the object of being able to use lower supply voltages.) In another paper Hull presented examples of circuit design with tetrodes, as the tubes came to be called [9]. At about the same time Bernard Tellegen of Philips Research Laboratories patented a similar design [10].

If used for large-amplitude signals, the tetrode showed an unpleasant characteristic caused by secondary electrons released at the anode being attracted to the new screen grid whenever the anode potential went below the screen–grid voltage. Tellegen quickly solved this problem by inserting yet another grid between screen grid and anode that was connected to the cathode and that reflected the low-energy secondaries back to the anode. In his patent he acknowledges Ando's work. The pentode reigned supreme as amplifier for three decades, although marketing arrangements kept it out of the United States for three years [11]. The multi-element tubes made possible the necessary broad-band amplifiers, and completely functioning television broadcast systems of current-use definition were tested at the beginning of the 1930s [12].

Electronics engineers found a multitude of uses for pentodes in almost every kind of electronic circuit. Radar would have been impossible without the pentode.

2.1.3. Electron velocity effects

The speed of an electron accelerated by a potential difference of a few volts is pretty fast. For voltages typical of electron tube usage it can be calculated as a fraction of the speed of light to be 0.001 98 times the square root of the potential difference in volts. If the signals being generated or amplified have wavelengths below one meter, these speeds, slow relative to the speed at which electric potentials are propagated, cause complications and limits in electron tubes that rely on control grids. Sluggish electrons are not the only problem. Reactance of, radiation from and resonances of the conductors making up the circuit cause problems of their own, and designers of the 1930s found themselves blocked at almost every turn in their quest for the highest frequencies.

The first tube capable of working centimetric wavelengths was the Barkhausen–Kurz tube [13]: a triode with a grid placed at a positive and an anode at a slightly negative potential relative to the cathode. The electrons are attracted toward the grid but generally miss it, are reflected by the anode and head back toward the grid, resulting in oscillations about it. Changes in direction by the electrons require radiation of energy, which will occur primarily at wavelengths favorable to a resonant structure formed by conductors connected to the grid. The radiating electrons spiral down onto the positive grid. Such tubes generate microwaves but at very low power because the electrons spend their excess energy in heating the grid, which can radiate only small amounts of heat, owing to its tiny surface area. The Barkhausen–Kurz tube was at once both a step forward and an end.

Another approach to the velocity effect began in 1916 at the General Electric Company as a way of circumventing de Forest's patent on the triode. This invention by Albert W Hull, called a 'magnetron' at General Electric, controlled the electrons leaving the filamentary cathode with an axial magnetic field. Its function was to replace de Forest's grid and had nothing to do with high frequencies; it was capable of high power, and the complicated trajectories followed by the electrons on their way to the anode attracted theorists. Alterations in the anode, specifically dividing it into more than one piece and using a static magnetic field, soon brought the production of very-high- frequency oscillations in experiments in Europe and Japan, the latter the result of work by Hidetsugu Yagi, a student of Heinrich Barkhausen, and his own student, Kinjiro Okabe [14], a discovery shared with August Zacek of Prague [15]. Magnetrons were capable of generating very short waves at power levels typical of small transmitter tubes and much greater than the Barkhausen tube because the anode had sufficient area to radiate the heat produced. Large tubes required large magnets, putting a practical end to growth. A formidable literature grew concerning magnetrons.

A direct approach to the problem of electron velocity effects was the reduction in dimensions between electrodes, which was undertaken in 1934 by RCA in three miniature tubes called 'Acorns' intended for experimental use at extremely high frequencies. A triode, type 955, was followed by pentodes, types 954 and 956. To reduce inductance and capacitance the electrode connections were placed at the top, bottom and sides of the tiny glass envelopes. These tubes were quickly prized by radar men and copied by German and British manufacturers [16]. They were also difficult to manufacture, requiring highly skilled personnel [17].

2.1.4. Transmitter tubes

Initially radar engineers worked with commercially available transmitter tubes. American designers, both Army and Navy, saw great merit in Eimac tubes, a product intended exclusively for radio amateurs, who wanted tubes that could take punishment [18]. It was often said they wanted to be able to read from the glowing anode. The reputation of Eimac tubes spread during the war, remembered even in an Australian poem [19]; they were produced for decades after the war. Early British radar work relied on silica tubes manufactured by His Majesty's Signal School. Fused silica has remarkable physical properties, specifically its extremely low thermal

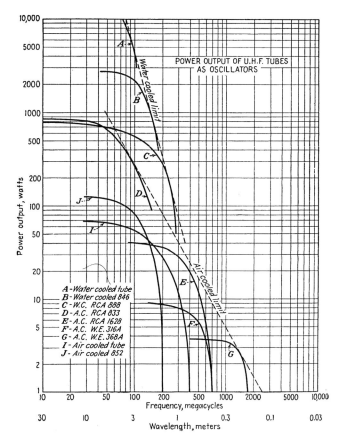

High-frequency operation of transmitter triodes. Inter-electrode reactance and electron velocity limit the maximum frequency of conventional vacuum tubes. This figure plots the maximum continuous output power of selected transmitter triodes against the frequency in MHz with wavelength shown on the lower scale. The WE316A was typical of a number of triodes used for decimeter-wave equipment, such as the British NT99 or the German LS180. When operated in a pulsed mode, the peak power was several kilowatts. From Karl Spangenburg, Vacuum Tubes.

expansivity, high melting point and strength that allowed much smaller high-power transmitter tubes to be made. The Navy incorporated them into their equipment but, there being no demand for them outside the Navy, had to make them at the Signal School [20]. By 1938 special tubes for radar were being made in the United States, Britain and Germany.

The very-high-voltage operation in radar transmitters caused serious positive ion bombardment of the cathode, which rapidly destroyed oxide cathodes. Tungsten cathodes were sufficiently robust for this service but required heating to 2500 K, consuming a lot more power and adding significantly to the heat load. Thoriated tungsten, the discovery of Irving Langmuir in 1922, came to the rescue. It was equally robust and functioned at 2000 K [21]. Driven to the limits of what one could do with triodes these tubes had very short lives, usually less than 100 hours.

2.1.5. Polyethylene

A troublesome problem for electronic engineers throughout the 1930s was the lack of an insulating material for high-frequency high voltages that was flexible and could be shaped. Ceramics insulated well enough but were certainly not flexible, and machining ceramic was sufficiently difficult to remove it from consideration for production. There were, of course, many insulators that were flexible or machinable or both—rubber, resin, amber, paraffin, wax but especially the then new plastics.

The industrial manufacture of plastics can be said to have begun in 1869 with the invention by the American John Hyatt of Celluloid, made from nitrocellulose but unfortunately not losing its inflammatory relationship to gun cotton in the processing. By the turn of the century advances began to come ever more rapidly with polyvinyl chloride in 1912, Bakelite in 1918, acrylates in 1927 and nylon in 1928 [22]. But all had relatively high dielectric loss, i.e. at high frequencies they absorbed energy, which attenuated any signal passing them and, given enough power, seriously overheated.

Expedients had to make use of glass and ceramics. Transmission lines for early high-power sets had balanced stiff copper conductors, which always radiated a bit, were a hazard and were clumsy. Coaxial cables, now ubiquitous in the electronics laboratory, had to have the central conductor held in place by beads and were generally unsatisfactory both mechanically and electrically. A German co-ax used cup-shaped ceramics linked together in a daisy chain that was remarkably flexible and did not allow the center conductor to become misaligned, something that causes unfavorable transmission characteristics [23].

Important help came by accident, at least from radar's point of view. In March 1933 E W Fawcett and R O Gibson at Imperial Chemical Industries in Northwich, Cheshire found a thin layer of white waxy solid in a vessel in which ethylene had been subjected to 1400 atmospheres at 170 °C, one of a series of high-pressure experiments being carried out [24]. They were not particularly pleased, as they had hoped to produce a lubricant [25]. The repeated experiment took a violent end, and nothing more was done until May 1935 when better high-pressure apparatus yielded 8 g of white solid. An improved compressor, not constructed for the project but available to it, allowed larger amounts of the substance to be produced by December 1935, but the question of its use had not been considered. They named it polythene, called polyethylene in America.

Someone noticed its similarity to gutta percha, long used for insulating submarine telegraph cables but which had troublesome losses even for the audio frequencies of telephony, so polyethylene's electrical properties were soon determined and found very satisfactory. It proved to have excellent high-voltage and high-frequency properties. Coaxial cables holding in excess of 50 000 volts became available to the radar men. On the day Germany invaded Poland a plant began producing hundreds of tons per year, and use of it in field installations is reported within months [26]. In 1941 DuPont and Union Carbide equipped themselves to produce huge quantities of the wonder insulator [27], and pre-polyethylene electronics went the way of spark wireless.

The German electronic use of polyethylene came about through radar equipment taken from a British bomber downed near Rotterdam in February 1943, an incident much more famous for the discovery of a cavity magnetron, the microwave generator that so greatly altered radar. It was quickly ascertained that the material used in the British high-frequency cables was a plastic manufactured since 1938 by I G Farben under the name Lupolen H [28]; they had missed its remarkable electric properties.

After the war the United States sent investigators to evaluate the German plastics industry and found a plant in construction hidden away at Gendorf, Austria to escape the air attacks that made production in Ludwigshafen impossible [29]. The Germans were manufacturing the substance by a more advanced process than the ICI–DuPont method. Further investigation showed that this was the same process developed by the Liquid Nitrogen Division of what is today Union Carbide and transmitted to their affiliate, I G Farben [30]. Given the secretiveness of the chemical industry, one must assume Union Carbide accepted the task of making polyethylene from ICI without comment.

2.2. BEGINNINGS, 1902–1934

For convenience the mind likes to associate inventions with inventors, even though most readers realize that things are often too complicated for such a simple correspondence. Most new devices have long pedigrees with persons other than the titular inventor furnishing important elements. Radar may provide the extreme example of multiple origins. Certainly one cannot attribute radar to any single person. Indeed one cannot even honestly say who first thought of building such a device, for it was an idea that was incipient in Hertz's experiments, which demonstrated reflections of short waves quite prominently, as well as the disturbing effects of nearby objects. A description of relevant technical work in the years preceding the first prototype radar sets should remove the idea of a single inventor without much further evidence.

Most remarkable of the early suggestions was in fact an experimental demonstration. Christian Hülsmeyer patented and built a device that he demonstrated to numerous witnesses. It used all of the elements of Hertz's experiments: a spark-dipole transmitter with a cylindrical parabolic reflecting antenna mounted next to a similarly constituted receiver. The assembly was to be mounted aboard ships and give indication of objects with which the vessel might be in danger of colliding. Hülsmeyer tried to convince ship owners to buy it, but the primitive state of the technology of the day, especially the receivers of 1902–1904 and above all the absence of the range data, which required accurate timing, easily explains why such a thing would not have appealed to mariners [1].

The Hertz parabolic reflector used by Hülsmeyer was not a bad radar transmitter, although a modern engineer would find design elements to criticize. Each spark created a short wave train for which the peak power was very high, and the time between sparks was of the same order as the time between typical radar pulses. The receiver, of course, was woefully inadequate, having orders of magnitude less sensitivity than modern equipment and no time resolution.

Those familiar with early wireless history will find it no surprise that Nikola Tesla made an early suggestion for using reflected radio waves to detect distant objects, but given the complete lack of guiding detail it is hardly surprising that the idea did not inspire the engineers of the time. That Tesla was already noticeably on the path from brilliance to self-delusion did little to advance interest, and the idea was lost among the myriad proposals coming from his fertile imagination.

The World War of 1914–1918 brought wireless techniques to many projects, but the closest things to radar were attempts to locate aircraft from the radiation generated by the ignition systems of their engines. Edwin H Armstrong, serving as a major in the American Signal Corps, worked unsuccessfully on the project. The interest generated in him for sensitive receivers led him to the fundamentally important superheterodyne principle [2].

The World War brought various experiments for locating aircraft by observing the infrared radiation produced by the motors, and this approach was continued during the postwar years but with insufficient improvement for use in aircraft detection. In a world in which the doors of public buildings are opened, television sets are switched, the deaf are allowed to hear the actors at the theater, intruders are detected and missiles are guided by infrared, one may well question the lack of application in the 1939–45 conflict. Infrared detection depended then on its heating of a substance, thereby changing its electrical properties. Postwar semiconductors extended the sensitivities of these effects and added the photoelectric effect, which provided startling improvements in sensitivity. Infrared is unsuitable in any case for many of radar's functions because of its inability to penetrate fog and rain.

Probably the most often cited suggestion of radar is from a review paper on radio telegraphy delivered by Guglielmo Marconi at the meeting of the Institute of Radio Engineers in New York in 1922 [3]. The idea proposed was Hülsmeyer's without spark and with vacuum tube. The earlier work was not mentioned, most probably being unknown to Marconi. For completeness one must note that King George V made a similar suggestion to the Admiralty Director of Scientific Research in 1925, probably without knowledge of any previous proposals. He asked whether a radio method for locating aircraft analogous to the acoustical methods used against submarines could not be devised and was told 'no' [4]. The Royal suggestion evidently caused afterthoughts because the Navy's Signal School entered a comprehensive patent a few years later for radio location [5]. While these experiments of the mind were going on, electrical engineers were beginning to bump into radar as they mastered the techniques of tens of megahertz. All through the 1920s and 1930s many papers were written and patents issued describing methods of determining distance by radio [6].

Engineers examining the propagation of high-frequency waves during the late 1920s, especially those planning television, found much to tickle their imaginations. Their experiences were much the same as those of the reader who has noted the effect of human movement in a room in which a television or fm radio is receiving a weak station. Strong reflection and interference are easily observed, and with care one can note the passage of an airplane. Such disturbances are reduced in modern equipment by automatic gain control and were far more prominent with the early detectors. In 1931 Marconi experimented with a 50 cm transmission link between the Vatican City and Castel Gandolfo and noted a disturbance in the signal that he traced to the motion of a steam roller, initiating an interest that led to Italian prewar development [7]. It was the kind of knowledge that must have spread informally and widely among high-frequency experimenters.

2.2.1. The United States

In September 1922 Hoyt Taylor and Leo Young, engineers from what soon became the US Naval Research Laboratory (NRL), were studying various characteristics of 5 m equipment with which they hoped to design sets that might be less easily overheard than the long-wave sets that dominated maritime communication at the time [8]. They noticed interference and standing waves resulting from reflections from buildings, followed by an observation guaranteed to arouse the interest of naval men: a steamer passing on the river produced strong variation in the signal recorded by the receiver from the transmitter on the opposite shore. Memoranda were written, ideas for harbor defense discussed.

Three years later NRL was to furnish a transmitter for an experiment with pulsed radiation. Fifteen kilometers north by west of NRL is the Department of Terrestrial Magnetism, Carnegie Institution. At that time one of its staff, Gregory Breit, and a Johns Hopkins student and future staff member, Merle Tuve, devised a successful demonstration of the existence of the ionosphere by measuring its height. They had an NRL transmitter modulated with a sine function and measured the phase lag between the modulation of the wave received directly at their laboratory and of the wave reflected off the ionosphere. The transmitter was soon changed to pulsed modulation, which greatly improved accuracy. The aircraft traffic of a nearby field often disturbed their measurements [9].

In June 1930 Young and Lawrence Hyland carried out experiments at NRL observing reflections from airplanes with 9.1 m equipment, and succeeded in attaining Navy authorization for two low-priority radio detection programs. One was to develop superfrequencies, as most thought that very short wavelengths were highly desirable, possibly necessary, because the radio searchlight that was envisioned meant the antenna would have to be large compared with the wavelength. If anything, all the work around 10 m emphasized how difficult direction determination was going to be with long wavelengths. The other program was to concentrate on aircraft detection with designs that relied on circuits known to work. There were more reflections, big ones from the airship *Akron*.

Young and Hyland devised with Taylor a system of widely spaced transmitters and receivers based on these observations that would signal the passage of aircraft into the area. It was at best a qualitative effect with little or no localization of the plane possible, something of a 'radio screen'. It was also obviously of little value at sea, so Taylor tried to interest the Army Signal Corps in the idea as a means of protecting cities. The Director of the Signal Corps Laboratories at Fort Monmouth was Major William Blair, who was quite interested in a radio or any other technique for the detection of aircraft and was invited to a demonstration of the method at NRL. Blair's comment that there was really nothing new being demonstrated—which tells us something about the general perception of the principles involved at the time—followed by his inability to see any use for it in air defense so long as position was not determined led to Taylor breaking off the meeting abruptly. Blair was never invited to NRL again and remained bitter over the incident [10]. In late 1935 the Bureau of Standards suggested the same idea to the Signal Corps [11].

Blair's and Taylor's discordant points of view about there not having been anything new in the NRL work are not revealed in the source of this dispute, but can be reconstructed with some assurance. Blair worked at Fort Monmouth in metropolitan New York where Bell Telephone and RCA had their laboratories. In the 1920s engineers from Bell had observed strong reflection and interference from New York's tall buildings and noted that the patterns altered greatly with slight changes of frequency [12]. In July 1931 engineers measuring signal strength in the second floor of the RCA Building produced by all-electronic television transmissions from the top of the Empire State Building found they could monitor the motion of the elevator in their building and the automobile traffic in the street below, stop and go conditions being clearly discernable [13].

No engineer working with meter-wave amplitude-modulated television could have possibly failed to observe the effects of airplanes, and Bell had seen them too. All of this was obviously the subject of informal discussion among communications engineers in the New York area, discussions in which the NRL men seldom, if ever participated. That communication existed between the Signal Corps Laboratory and the television engineers is evidenced by a Fort Monmouth engineer having published a paper on television matters in 1933 [14]. One is inclined to suspect that the RCA people did not add aircraft to their reports of elevators and automobiles out of deference to their colleagues in the Signal Corps and the obvious defense aspects of radiolocation.

In March 1933 men at Bell Labs published a paper describing how their high-frequency work had been affected by airplanes [15]. In July 1934 the Chief of the US Army Signal Corps recommended a radio echo technique for the Army, and a modest program got under way at Fort Monmouth. Blair thought microwaves, wavelengths significantly shorter than the dimensions of antenna or target, were the reasonable approach to the problem. At RCA Irving Wolff had built 10 cm equipment with paraboloid antennas that he demonstrated to a meeting of the Institute of Radio Engineers in 1934, and Blair invited him to demonstrate it to the Signal Corps on Sandy Hook, where numerous ships entered New York harbor regularly. He demonstrated the ability to determine the angular location of objects from which reflections were received, but the reflected energy was much too weak for any practical purpose. Their continuous-wave reflections showed the Doppler effect produced by moving motor vehicles. They were ready for traffic police duties if not for war[1].

Wolff continued his work with microwaves and by 1937 had constructed a remarkable radar set, which he called 'radio vision'. It used 10 cm waves generated with a split-anode magnetron[2] and modulated to 1 μs pulse width. A transmitter and receiver were set up on the roof of the RCA Laboratory in Camden, New Jersey that received reflected signals from ships in the Delaware River and from the Philadelphia skyline, but the short range still left it inherently uninteresting to military designers [16]. RCA made no further significant contributions to microwave radar during the war.

2.2.2. France

Pierre David, an engineer at the Laboratoire National de Radioelectricite, had pressed before 1930 for a program of radio detection of aircraft, but nothing was undertaken until impelled by the published reports from Bell Labs of the reflection of 4 m waves from aircraft. In 1934 he observed airplanes at Le Bourget Field when they passed between widely spaced receiver and transmitter. This system eventually became the 'barrages' that

---

[1] See 'Reflected signals' in the Appendix (pp 468–9).
[2] See 'Electron velocity effects' in Chapter 2.1 (pp 36–7).

were soon installed to protect the ports of Cherbourg, Brest, Toulon and Bizerte from ships as well as aircraft. The equipment was simple, inexpensive and easily mastered with a little instruction, but the observations were difficult to interpret [17].

Camille Gutton had experimented with 16 cm waves in 1927 using Barkhausen tubes that had the resonant dipole as a part of the internal electrode structure. The tubes, whether used as oscillator or as regenerative detector, were placed at the focus of parabolic reflectors. These experiments encouraged Gutton to try the system on aircraft, done in the summer of 1934 by Société Française Radio-Électrique (SFR), a company run by his son Henri Gutton. In addition to the 16 cm set, which had less than 1 W power, they tried 80 cm equipment that used a split-anode magnetron that gave a few watts. The results were failure to detect aircraft but they caused the son to think of detecting obstacles to ocean navigation.

The owners of the new trans-Atlantic liner *Normandie*, probably thinking about the *Titanic*, requested that the ship be equipped with the system for her maiden voyage. As a preliminary Gutton mounted one on the cargo steamer *l'Oregon* and in July 1935 on the liner, although too late for the initial crossing. The set, which had twin-paraboloid transmitter and receiver antennas, made only the one crossing; it did not like the sea and generated no enthusiasm among the deck officers [18]. It did generate news reports that attracted the attention of engineers working on secret projects.

2.2.3. Great Britain

During World War I a young physicist, Robert Watson Watt invented methods for the location of thunderstorms by triangulating the directions of the lightning discharges detected at two or more stations. By 1917 he could locate storms as distant as 2500 km. He continued this work, which had obvious peacetime application, and in 1925 improved it with a technique that would have two critical applications during the next war. The original direction finding technique followed that which had been used successfully to keep track of the German fleet. Maritime wireless of the time only used long wavelengths, which allow accurate directions to be determined by means of manually adjusted loop antennas, and lightning generates large amounts of long-wavelength noise, one of the disadvantages of such wavelengths for communication. In locating thunderstorms the operator had to contend with the various lightning strikes of a storm coming from an ill defined area, so obvious inaccuracies resulted, with the adjustments leading only to an average direction. Watt's improvement eliminated the need for the operator adjustment. At a given station he arranged two antenna coils at right angles. The relative strength of the signals from the two coils depended on the direction of the incoming signal. The trick was to measure their ratio for an individual strike.

The element that allowed this was the new cathode-ray tube, WE-

224, recently invented at the Bell Telephone Labs by J B Johnson [19]. This rather modern looking tube could be wired into a laboratory setup without the horrible complications attendant to other cathode-ray tubes. Signals from the receivers connected to the two antenna coils were applied to the horizontal and vertical deflection plates of the oscilloscope. Each lightning strike produced a streak across the screen that gave the direction to the strike—with a 180° ambiguity, of course.

The WE-224 was not suited to radar because of its slow response, but a later design was to become the heart of the radar system Watt would later design that would change the course of history—Chain Home. It was also to become the heart of the radio direction finders that functioned well on an extremely short transmission. During World War II Adm Dönitz's signals specialists, who were evidently not well read in the literature of lightning observation, assured him that direction finding was too slow for locating transmitting submarines, which had reduced the length of their messages to extreme limits.

Watt was a Scot, a tribal attribute that gave him satisfaction almost as great as his conviction that he had invented radar, two pieces of information that seldom escaped the attention of others. He was descended from the inventor of the external-condenser steam engine yet wanted that illustrious name improved by adopting a double-barreled version on receiving honors.

2.2.4. Germany

Research in radio engineering attracted German engineers and scientists from the beginning. Competition between Germany, Great Britain and the United States was keen and generally even, as the nearly equal development of television during the early 1930s demonstrates. Hülsmeyer is usually given credit, if late in recognition, for having first built anything approximating a radar set. He was followed by Hans Dominik, who built a similar set during 1915–1916 that apparently did not work as well as Hülsmeyer's and excited no interest among military authorities. Heinrich Löwy, an Austrian, had a much more important effect on the thinking of the men who made German radar. He was well known among physicists and devoted much effort in using electromagnetic waves to probe the earth's interior. While studying their transmission in rock (attaining distances of a few kilometers) he proposed in 1912 to time radio wave reflections from conducting strata of the earth using an antenna with a motor-driven mercury switch that alternately connected it to transmitter or receiver. The switch's function was that of a Fizeau toothed wheel in velocity of light measurements [20]. He continued these experiments over the years but with little success. In 1923 he made the first of the many proposals for a radio altimeter, and the citations of his patent lead one to believe that it was frequently discussed. Löwy seems to have been the first to try timing the propagation of pulsed high-frequency waves [21].

The German Navy established the Nachrichtenmittel-versuchsanstalt (NVA, Communications Research Laboratory) at Kiel in 1923 to investigate what one now calls the electronic methods of warfare. In 1928 Dr Rudolph Kühnhold, having recently received his degree in physics from Göttigen, became the group leader for acoustics that had the goal of determining all possible uses for underwater sound: communication, listening for vessels on and below the surface and determining direction and range to surface and underwater targets. Experiments soon showed that underwater sound had little chance of locating surface ships accurately, except at very close range, and certainly did not form the basis for a method of directing gunfire when optical means failed. This failure impelled Kühnhold beyond the bounds of his assignment to consider doing the job with radio waves above the water. A change of research direction proved no administrative problem, because he had been named scientific leader of NVA in 1933.

He envisioned doing this by forming a tight beam of a few centimeters wavelength with a parabolic mirror and bought an oscillator and receiver for 13.5 cm from the Berlin firm of Julius Pintsch. Attempts in fall 1933 failed to observe reflections from a large building less than 2 km distant, not surprising when one considers he was using continuous waves for which the reflected signals were not Doppler shifted and his transmitter power was only 0.1 W [22]. He then began to cast about among electrical engineers for a means of realization, which fatefully led him to two dynamic young engineers.

Paul-Günther Erbslöh and Hans-Karl Freiherr von Willisen were two childhood friends, at the time less than 30 years of age, who had followed their hobby of amateur radio into the ownership of their own company manufacturing phonograph records and recording equipment. They were the best in the new techniques of transforming electrical signals into mechanical vibrations and found interest in any new electronic application. Kühnhold had already used their little company, Tonographie, for his naval underwater sound work. They were immediately intrigued by his ideas and had already began to experiment with high-frequency radio for telephonic communication [23].

2.2.5. *The Soviet Union*

The Red Army of the between-war decades cannot be faulted for ignoring anti-aircraft defense and its attendant problem of air warning. The Main Artillery Administration (GAU) devoted much effort to trying to eliminate the deficiencies inherent in devices based on sound, infrared and searchlights, coming to the same conclusions of their brothers-in-arms in other lands. The failure of these methods left them with some sort of radiolocation method as the only hope. Working carefully within the prescribed administrative framework they saw their proposals rejected as unrealistic by two organizations: the Scientific Research Institute of Communication Engineers of the Red Army (NIIIS-KA) and the All-union Electrical Institute (V3I), which was the greatest repository of radio science in the country.

Demonstrating foresight and tenacity, the GAU refused to accept these rejections and went to the Central Radio Laboratory (QRL), a research institute for the electronics industry where they found a positive response in late 1933. Yu K Korovin had been successful during the preceding months in designing decimeter-wave communication links and was eager to try the ideas for radio location that came to all engineers working this wavelength range. On 3 January 1934 using 50 cm waves of 0.2 W (presumably from a Barkhausen tube) and a 2 m paraboloid, he succeeded in receiving Doppler signals from an aircraft at a range of 700 m [24].

At the same time interest in radiolocation grew independently of the GAU in the mind of a remarkable member of the Air Defense Forces (PVO), the organization of the Red Army that employed air-defense weapons. Pavel K Oshchepkov was a waif of revolution and civil war who received his first schooling in 1920 and graduated from Moscow University 11 years later as an electrical engineer. Soon after graduation he joined a research section of the PVO where he worked on sights for anti-aircraft guns. In August 1933 he was sent to confirm reports from the Soviet Academy of Sciences about some kind of interest in a 'radio-technical' approach to aircraft detection. His discussions with academician Abram F Joffe so impressed this eminence that Oshchepkov found himself in charge of a special project. Joffe rejected methods using very short waves, probably because of a view held by many theorists world-wide that such waves would be reflected but generally not back toward the interrogating device. It was a view that was to last in some quarters beyond the 1930s and has enough truth in it to form the basis of modern radar-evading aircraft design.

The outcome was a warning system of the kind Taylor had proposed and the French would construct. A transmitter of 200 W on 4.7 m was placed 11 km distant from the receiver. Tests performed in August 1934 detected the presence of an aircraft as far away as 75 km but without revealing where. The system, called Rapid, and referred to as an 'electromagnetic curtain', produced elation among its creators and derision among others. Useful or not, it was a start on radar research [25], and Oshchepkov was already working on the next design step. He saw the importance of pulsed signals both for determining range and for greatly enhancing output power, techniques he saw leading to panoramic scanning and maplike display [26].

Thus by the beginning of 1934 the Soviet Union was well along the path to producing first-rate radar. One competent group was moving in directions that should have led to a good searchlight and gun-laying set; another group should have quickly produced an air-warning set. Without exaggeration one can say they were unsurpassed at the time. But it was not to be.

Radar was in the air; indeed it had been almost since the Hertz experiments. There is even an excellent description of radar in a 1911 piece of science fiction [27], and numerous patents were issued, although none reached prototype stage. Contributing to the atmosphere were reports of 'mystery rays' and 'death rays' that appeared in the popular press with predictable regularity. A report in the *New York Times* of some infrared experiments caused the Signal Corps minor difficulties in 1935 [28], and a 'death ray' question was to have an important effect on British radar.

The parallel approaches taken in America, France and the Soviet Union emphasize how ideas spring up together when conditions are ripe. The multi-station 'radio screens' of Taylor, the 'barrages' of David and the Soviet 'Rapid' were clearly independent, all having been secret with no hint of espionage. Japan put a similar system into operation, as we shall see, at the start of the war. All were the same as Taylor's suggestion, which had led to animosity between him and Blair because of the latter having rejected the idea out of hand. There is, however, nothing in the French, Soviet or Japanese experience that contradicts Blair's snap judgement.

The low-power microwave equipment examined during the early 1930s in the United States, France and Germany was not secret, the techniques being available in the open engineering literature. A cooperative venture between International Telephone and Telegraph and Le Matériel Téléphonique introduced a microwave communication link between Dover and Calais on 31 March 1931 using 18 cm radiation from Barkhausen–Kurz oscillators of 0.5 W with parabolic reflectors. The modulation circuitry to allow the simultaneous transmission of many telephone channels had not been developed, leaving the link inferior to submarine cable, but the extraordinary ability of a directed beam to function at low power with very low noise or interference left a strong impression on communications engineers [29]. The microwave attempts at radar were all failures, forcing everyone to longer wavelengths, and microwave research was dropped by those actively pursuing radar in order to put scarce resources behind something having a good prognosis. Microwave radar was to come later when an adequate generator was invented. Similarly, most of these groups had tried infrared as a means of detecting aircraft or ships and continued research throughout the 1930s. All failed. Detection was to be with radar, and for the immediate future it was to be half-meter waves and longer.

2.3. BRITAIN BUILDS AN AIR DEFENSE SYSTEM

With Hitler's coming to power the doubts that had filled the minds of thinking Englishmen about the peace of 1919 turned into serious concerns. The treaty had not long been signed before the tensions that it provided became all too obvious. Any desires by statesmen to emulate the end of the

Napoleonic Wars had been quickly swept aside by the demand for revenge on the defeated by the populations that had suffered much and that had been fed steady diets of propaganda asserting that the enemy bore complete responsibility for all wrongs. The model was to be Brest-Litovsk, 1918, not Paris, 1815. All this had engendered a sympathetic counter-reaction for Germany's immense postwar problems among many in Britain, creating difficulties with France and generating the most pervasive pacifist movement that Europe had ever seen. The general reaction of the people to the Nazis was initially one of distaste and indifference rather than fear; the depression had given them enough problems close to home to worry about, but there were a few who perceived the danger, regardless of the chain of events that had brought the dictator to power, and realized the pitiful state of Britain's arms [1]. The difficult economic situation prevented the Government from initiating a rearmament program had they wished to do so—and they did not.

Britain's vulnerability to air attack was present in the minds of many, and out of these concerns in the Air Ministry came the Committee for the Scientific Survey of Air Defence, almost immediately called the Tizard Committee after its chairman, Henry Tizard. Seldom has a committee of any kind done so well.

The first order of business was to clear out files of plans and ideas accumulated over the years in order to be able to answer without wasting time the questions that would be posed when the public became concerned [2]. One of these was the 'death ray'. The turn of the century marked the beginning of a series of discoveries that bore on the public's mind in a hopelessly confused jumble: radio waves, x-rays, radioactivity, and in 1932 Hitler had had to share the headlines with the news that scientists had split the atom. What all this meant to laymen varied, but to some it meant that scientists could manufacture a ray, of what kind they knew not, that would kill the King's enemies, dared they invade the sanctity of his realm. A casual reading of the newspapers told Harry Wimperis, Director of Scientific Research at the Air Ministry, that he must be able to deal with death rays, so he requested Robert Watson Watt, the superintendent of the Radio Research Station at Slough to examine the matter of a radio death ray.

Watson Watt and Arnold F Wilkins calculated how much they could raise the temperature of an attacking pilot's body, if they concentrated the most powerful wireless signals available onto him. The answer was, as they and Wimperis knew in advance, trifling. Watt did not leave the matter there, however; he changed the question. If one were to irradiate an airplane with a high-frequency wave, currents would be induced into its metal structure and these oscillating currents would then re-radiate, and the question was how much. Was it enough to offer a means of locating an airplane? In sum, substitute radio detection for radio destruction. To make the problem simpler an ideal approximation was assumed, which is the physicist's first step in approaching any problem. Watt and Wilkins assumed the irradiation had a wavelength twice the wing span of the target aircraft, which was replaced in the calculation by a wire and which was found to re-radiate a most satisfactory amount. On 28 January 1935 Watson Watt sent to Wimperis his 'anti-death-ray' memorandum in which he outlined a radio location scheme capable of giving Britain warning of air attack.

Why Wimperis approached Watt rather than Edward V Appleton, Wheatstone Professor at University of London, chairman of the Radio Research Board's Propagation of Waves Committee and world authority on radio matters, is not known. It may have turned on the matter of secrecy, as Watt was a government employee, Appleton not. Whatever the reason, it generated a fierce animosity between the two. Matters were certainly not helped by Appleton not being told of the radar work until summer 1936. Watt's theoretical approach to reflections from aircraft, a phenomenon long observed by Appleton and his students and certainly by Watt, who had worked extensively with Appleton on the ionosphere, adds an additional puzzling element [3].

The next step led to action by Air Vice-Marshal Hugh Dowding, Wimperis's chief. Since 1930 Dowding had sat on the Air Council as Air Member for Supply and Research, a position that had recently been split leaving him responsible only for Research and Development. Dowding had seen the importance of aircraft in a 1912 Staff College exercise and had learned to fly the following year at his own expense. By 1918 he had had just about every kind of flying experience that it was possible to survive and had risen from subaltern to brigadier general on the basis of technical competence and ability. As a training officer he had resisted to the bounds of direct disobedience sending inadequately trained pilots to the front, which made his relationship with Marshal of the RAF Hugh Trenchard, who ruled the Air Force during and for the decade after the war, disputatious but eventually respectful [4]. Trenchard was an all-out believer in the bomber dominating aerial warfare, and Dowding believed it could be countered with properly handled fighters. His first important work in Supply and Research was to put all his influence behind procuring the engines and airframes that became the 'Hurricane' and the 'Spitfire'. His second important work was radar [5].

Trenchard and most Air Force officers were convinced that there was no effective defense against bombers. This had been amply demonstrated in exercises. It required a minimum of 20 minutes after an alarm for a fighter squadron to reach the expected location of the attackers, provided one knew where to send them. So much time was not available, if the warning came from visual ground observers, and for it to come from standing off-shore patrols would require by any calculation an enormous number of planes, so when Dowding learned of the new idea he listened with more than casual interest, which was fortunate because it was from Dowding that the first funds had to be obtained.

Dowding was interested but wanted something more than 'speculative arithmetic'. Watt agreed to give a demonstration, on the condition that a negative result would not be held against him, because the apparatus would be hurriedly assembled. (Because of Watt's prior knowledge of aircraft reflections, Appleton considered this 'pure theater' [6].) At Daventry, about 20 km southeast of Coventry, was the large Empire transmitting station. Watt and Wilkins set up on 26 February 1935 a receiver a few miles from the Daventry station with a double antenna so balanced that the direct signal from the 50 m transmitter was greatly reduced, the object being to make it sensitive to the same wavelength coming from a different direction. The approach of a bomber produced an unequivocal signal, with A P Rowe, secretary of the Tizard Committee, serving as impartial witness.

What Watt and Wilkins had produced was identical to the earlier experiments in America, France and Russia, but what came next was completely different. Watt quickly saw how to make the system give the position of the target aircraft, something that had escaped the others. His solution was to incorporate two electronic techniques long used at his station: the pulsing of radio waves for determining the height of the ionosphere with the echoes and the instantaneous determination of the direction of lightning strikes using crossed antennas. Earlier investigators had only thought in terms of continuous wave operation for their 'radio screens'.

Another very important difference distinguished the British approach to radar from that of other countries. Where foreign radar men had had to contend with moderate interest and weak support, British radar men received strong support from the highest level. Dowding immediately sanctioned £12 300. The possible location of the enemy 75 to 100 km off the coast would give just the amount of extra time he needed to scramble fighters, and his fervor for radar remained steadfast. Fighter Command, a military organization destined to join such immortals as the 'Old Contemptibles' and the 'Army of Northern Virginia', emerged out of a reorganization of the Air Force the next year and Dowding became its first commander.

He and Tizard realized that early warning was of no value unless it yielded a suitable response. What was needed was a system for analyzing the information gained by the radar and ground observer stations and for its communication to the fighters—a process that must be accomplished in a very few minutes. While Watt assembled his design crew Dowding began to build his control system and train his pilots to use control from the ground. The latter task was not easy, for these young men had seen 'Hell's Angels' and 'The Dawn Patrol' several times and had been reared on the romances of Scott and Tennyson. Like their German cousins they thought in terms of knights of the sky, not policemen responding to a call, but Dowding succeeded, and the Germans, who were to enter the war with technically better equipment, had no organization to make use of radar for air defense and little comprehension of its tactical significance.

Although the basic design was obvious almost immediately after the

Daventry experiment, the creation of equipment required much detailed work. To do this in extreme secrecy demanded a secret place, and one presented itself. Orfordness [7], a long spit of land 15 km up the coast from Felixstowe, had been used for secret aviation research during the war and had been reactivated in 1929. It had the rudiments of shelter and electric power, and Watt soon began his experiments. By the end of July 1935 they had tracked aircraft to 60 km. In September the Air Defence Research Committee had approved the construction of Chain Home, and in December (the year is still 1935!) the Treasury allocated £60 000 for five stations for the Thames Estuary [8]. Rowe gave the device the code name RDF, intended to imply 'radio direction finding', which was a good name to hide behind because of Britain's success during the First World War with radio direction finding and Watt's previous work.

Equipment terminology can be confusing as new designs and modifications of old ones are encountered. British radar was initiated by the Air Ministry, which found it necessary around 1940 to designate equipment as AMES (Air Ministry Experimental Station) Types. The original design was AMES type 1, generally referred to as CH. Other AMES sets received similar initial names, which will be used here so long as clarity is preserved. The Army and Navy soon began to develop radar and adopted designations of their own. As sets other than AMES type 1 began to take their place along Britain's coasts the term 'Home Chain' came to mean that entire radar network.

The first months at Orfordness were spent in experiments with lashed-up equipment to verify the basic ideas of the design of Watt and Wilkins and modify it before production orders were placed. The design that eventually evolved used two arrays of horizontal dipoles slung between two 100 m steel towers, constituting a minimum transmitter station[3]. These radiated over a wide angular front more like a floodlight than like the searchlight that other radar designers strove for. Reflectors were placed behind the arrays in order to prevent wasting power and generating confusing signals from the rear. The pair of arrays allowed the vertical lobes of the radiation pattern to be altered. Initial plans had foreseen jamming, hence stations had multiple towers that allowed dipoles of different lengths to be slung, as a dipole antenna allows only slight variation of the frequency determined by its size.

They immediately discovered what all other radar designers learned when working with pulse transmission: that the maximum ratings for vacuum tubes applied to the average power, determined by the time it took to heat the anode to some critical temperature. In pulsed operation anode voltages could be sent way beyond what was normal, from kilovolts to tens of kilovolts. The Royal Navy manufactured a silica triode, type NT46, that could be driven very hard, which was used in early stations, but its

---

[3] See 'Antennas' and 'Vertical lobes' in the Appendix (pp 469–72).

specialized production could not meet all the demands for it, so other output tubes had to be found. Special tetrodes, Metropolitan–Vickers type 43, were ordered that had water cooling and that were vacuum pumped rather than sealed so that they could have hard-driven filaments replaced or otherwise repaired [9].

Receivers used pairs of crossed dipole antennas mounted on 75 m wooden towers with one for the E–W signal component, the other for the N–S signal. The operator could adjust the relative amplitudes of the two signals to determine the direction. It was a copy of the equipment used for the determination of the directions to lightning strikes with the difference that dipoles had been substituted for the loop antennas because of higher frequencies. Antennas were located at three levels on the receiving towers. Estimates of the height of the target resulted from the comparison of signal strengths from dipoles located at different levels.

Originally wavelengths of 50 m had been planned to allow resonance with the wing span of a typical bomber, but airplanes have complicated structures and this special wavelength was not found to be important. In-terferences with communication signals soon forced them to shorter wave-lengths with a band from 7.5 to 15 m becoming typical. These wavelengths produced quite a lot of unwanted reflections, sometimes from the iono-sphere, sometimes from distant continents, ships and cities, the result of ionospheric bounces, because CH was from the beginning an 'over the horizon' radar, although incapable of making sense of the jumble of those distant returns. For that one had to wait for more advanced signal pro-cessing. (Long-wave sets having focused beams did observe and identify targets at extreme range during the war, but these incidents came from exceptional atmospheric circumstances.) All this meant that a much lower pulse repetition rate had to be used in order to eliminate these extraneous signals instead of the hundreds of Hz that were reasonable for ranges of 100 km or so. The value selected was 25 Hz, half the British power fre-quency. This also allowed adjacent stations to be synchronized with each other using the electric power grid [10].

For a station to be able to determine direction reliably and height at all, aircraft were required to fly preset paths while calibration teams labo-riously took huge amounts of data, a mind-numbing and costly procedure that had to be repeated at regular intervals for each station. The results of an observation were readings of dials and goniometers, which had to be transformed into three-dimensional coordinates for the information to be of any use to fighter directors. Not only did this require trigonometry but had to incorporate the calibration corrections for the station: simple but it had to be done quickly. G A Roberts of the Bawdsey staff devised an electric analog computer for this task, soon called the Fruit Machine, that allowed the operators to enter the data plus their estimates of the number by pressing keys at their desks [11].

The year 1938 marked the transformation of CH from an interesting radio experiment into a functioning air defense system. The previous May the first CH station had been turned over to Air Force personnel [12] and other stations followed as they went on the air. To Squadron Leader Raymond Hart fell the responsibility for organizing the training school, and he quickly became the leading figure in applying the new techniques to war by methods soon to receive the name of 'operational research' [13]. For practice, Fighter Command began on 1 January 1938 to intercept 'discretely' but in the beginning not very successfully KLM and Lufthansa airliners arriving from the continent [14]. By July five stations were ready for the August air defense exercises.

The equipment that made up the system that began to grow in 1937 was, with three exceptions [15], unlike any other radar systems. The characteristics of CH were so different from those of the German design that it baffled completely the first attempt at electronic espionage sent against it. It was in 1940 an obsolescent system, but it was crucial to winning the Battle of Britain and earned the undying affection of its operators and mechanics, who often referred to it in later years as 'steam wireless' [16].

While work began at Orfordness the Tizard Committee proceeded to examine all aspects of scientific air defense and came up against a vexing problem, one which would remain one way or another until the end of the war. Churchill's party was out of power and had not a great deal more interest for rearmament in 1935 than Labour. In calling alarm Churchill was very much alone but also very vocal and spoke for most of those engaged in radar work. Owing to his insistence that something be done about air defense he was added to the Imperial Defence Committee, which certainly did not upset the growing radar community, but Churchill insisted that his personal friend and scientific advisor, Professor Frederick Lindemann be added to the Tizard Committee. On the surface this might have seemed useful because he and Tizard had once been fast friends, but they had become estranged for reasons that remain obscure and the estrangement was complete. Lindemann, the later Lord Cherwell, had by most measures an abrasive personality and quickly transformed a highly productive group into one that wasted its time fighting the newcomer.

His scientific ability had initially been sufficiently great for him to have been included among the 24 who made up the Solvay Conference of 1911, joining such lights as Planck, Sommerfeld, Jeans, Rutherford, Einstein, Lorentz, Curie and Poincaré, but he lacked the soul of an engineer and proposed various ideas for air defense, all of which he wanted seriously investigated and all of which the committee rejected. His favorite, aerial mines, was regarded as nutty by almost every technically competent person who evaluated it. He was not enthusiastic, many say hostile towards radar. Matters reached such an impasse that he had to be removed from the committee by an administrative trick with Appleton replacing him. In 1940 Lindemann became the Prime Minister's advisor, and for five years the British technical community had its hands full. The controversy

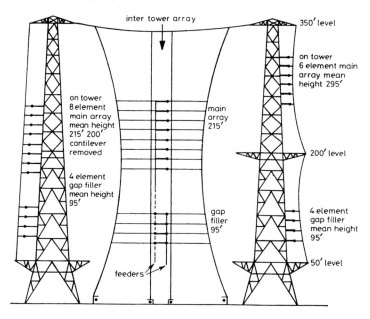

Transmitter towers from Britain's CH. This radar, remembered for its service in the Battle of Britain, retained the same basic design but had numerous antenna configurations, some of which are illustrated in this composite of two towers. In some cases dipoles for the 10 to 15 m radiation were mounted at the sides of the steel structures, as shown on the left and right, but the radiation patterns were often unsatisfactory, leading to them being suspended as a curtain array between two. In the expectation that there would be jamming, stations usually had more than two towers so that arrays with different wavelengths would be immediately available to the operators. The array of eight dipoles formed a relatively narrow radiation pattern in vertical extent but very wide in its horizontal. Reflection off the ground and sea produced a vertical lobe structure with blind regions. These were filled with the lower array, called the gap filler. Adapted from the Radar Supervisor's Handbook and taken from Sean Swords' Technical History of the Beginnings of Radar. IEE Copyright.

has endured. The voice of the prosecution comes from C P Snow [17], that of the defense from the Earl of Birkenhead [18].

However all this may be, Cherwell deserves the credit for instilling in Churchill a high opinion of the value of science in warfare, one of many qualities that set the Prime Minister apart from his German counterpart. As Churchill's advisor he was able to have his aerial mines tried in 1941, confirming experimentally the judgement of his critics.

Orfordness suited research needs in only the most rudimentary manner and soon began to suffer growing pains, so the first of four changes of station took place in March 1936, this first time to the Bawdsey Research

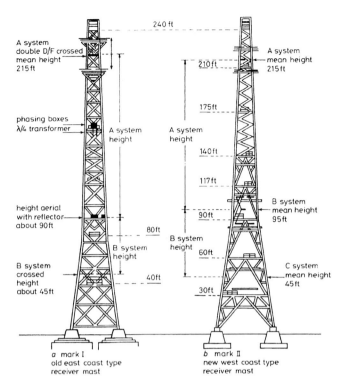

A system
double D/F crossed
mean height
215 ft

A system
mean height
215 ft

240 ft

210 ft

175 ft

phasing boxes
λ/4 transformer

A system
height

A system
height

140 ft

117 ft

B system
mean height
95 ft

height aerial
with reflector
about 90 ft

90 ft

80 ft

B system
height

B system
height

60 ft

B system
crossed
height
about 45 ft

40 ft

30 ft

C system
mean height
45 ft

a mark I
old east coast type
receiver mast

b mark II
new west coast type
receiver mast

Receiver towers from Britain's CH. Antennas for transmitter and receiver were separated by a few hundred meters. The receiver antennas were pairs of crossed dipoles tuned to the frequency of the transmitter with pairs mounted at two different heights. The comparison of signal amplitudes from the two dipoles of the pair allowed the direction of the target to be ascertained. Comparison of the amplitudes of the signals from pairs at different height together with a third non-directional antenna allowed estimates of the target height to be made. Wood construction was required in order to eliminate the interferences caused by a metal structure. Receiver towers went through minor variations as did the transmitters. Adapted from the Radar Supervisor's Handbook and taken from Sean Swords' Technical History of the Beginnings of Radar. IEE Copyright.

Station. This was a manor, suitably isolated from the overly curious, with sufficient space for laboratories and quarters and with extensive grounds for experimental antennas. 'It was a grand place for the work that had to be done and the only feature inappropriate to the new conceptions of defence which were born there was the motto above the door of the Manor, "Plutôt mourir que changer"' [19]. With its timbered hall and roaring fire it resembled an Oxford or Cambridge college rather than an engineering laboratory. The university atmosphere increased after a visit by Lord Rutherford, whose natural and decisive support soon had students and faculty from nuclear physics laboratories joining. The Bawdsey days were long remembered for the relaxed mode and the long hours. Many an electronics problem was thrashed out in the manor house, at swimming, on walks through the grounds or at some other recreation.

Watt proved an exemplary leader of this collection of young physicists. It was probably the peak of his life, and Hanbury Brown remembered his first year at Bawdsey as 'one of the happiest I ever spent' [20]. Watt did not initially want communication men, as he expected more ingenious ideas from scientists, but he paid a price in some early failures caused by the lack of rudimentary radio knowledge, especially of what test equipment was available and needed. In designing airborne radar Brown later worked with EMI television engineers and was shocked to learn how much more advanced their circuitry and equipment were [21]. One is led to speculate on whether the design of CH would have survived the criticism of such professionals had they been present from the beginning.

It is difficult to escape wondering what the course of British radar might have been had EMI been approached in 1935 rather than the Radio Research Station, because in 1934 they had assembled for television '32 graduates (nine with PhDs), 32 laboratory assistants (of first year BSc standard), 33 instrument and toolmakers, glass blowers and mechanics, eight draughtsmen designers, and nine female assistants' [22]. This extraordinary source of electronic talent was unaware as late as 1939 of the radar project. EMI had declined contract work for military communications equipment, giving their deep commitment to television as the reason [23]. Whether they would have responded differently to a radio-location project, magnificent perhaps but ill defined and with uncertain long-term government support, will never be known. One suspects the start would have been time- consuming negotiations.

Watson Watt was promoted to Director of Communications Development at the Air Ministry in May 1938 and turned Bawdsey over to A P Rowe, an excellent administrator with intimate knowledge of the Air Ministry but essentially no technical knowledge and little patience for the antics of his charges. Bawdsey's extraordinary environment was soon swept away by the inevitable growth of 'the official way'. Impressed by the manorial surroundings, he even entertained the idea of requiring the staff to dress for dinner [24]. Rowe had one supreme virtue that more than compensated for his stiffness: he strongly believed in the free exchange of ideas. This eventually led to his 'Sunday Soviets', meetings of scientists, engineers, technicians, serving officers and administrators where one could say, indeed was expected to say anything that came to mind. They proved most rewarding and did much to forge the strong bond between laboratory men and fighting men. If Rowe was known as a stickler for the rules, he was also known for his consideration for the staff.

The possibilities of radar having functions beyond those of CH had presented themselves from the beginning, which led to the establishment at Bawdsey of an 'Army cell' on 16 October 1936 under the leadership of Dr E T Paris with half a dozen design engineers, including W A S Butement and P E Pollard, who had written a memorandum on the construction of a radar set five years earlier for the Inventions Book maintained by the Royal Engineers [25]. It was a good design that envisioned pulsed 50 cm waves and parabolic antennas and lacked only the elusive microwave transmitter and administrative support.

Their first task at Bawdsey was to build a smaller, portable version of CH, the Mobile Radio Unit (MRU) that used smaller antennas suspended from 20 m field masts with consequent shorter range and poorer height-determining capability. It received belatedly the designation AMES type 9 and came in various arrangements indicated by at least three sets of initials. Further reference will be to MRU without semantic subtlety. Responsibility for this equipment was transferred to the Air Force in 1938, its function being early warning, and the Army cell then occupied itself with shorter wavelengths. They set about constructing 1.5 m equipment, of a kind very much like what was being made in America and Germany, with coast defense fire control the object. They had been stimulated by the success of E G Bowen, who had been using this wavelength with airborne radar in mind [26]. They gave up the use previously made of the Yagi antenna because of the poorly understood criticality of dimension and went instead to a dipole array, which allowed lobe switching. By May 1939 they could determine the range of a 2000 ton vessel at 15 km from a set placed 18 m above sea level and determine direction to an accuracy of 15 minutes of arc at 10 km with lobe switching[4]. Practice shooting disclosed echoes from the explosion of shells, allowing adjustment of fire [27].

Out of Butement and Pollard's Coast Defence set grew two projects of importance for the air war that was coming: Chain Home Low (CHL) and Ground Controlled Interception (GCI). A characteristic of CH was the large vertical lobe patterns that resulted from surface reflections and that were used for height determination. Unfortunately, these patterns were such that very-low-flying aircraft could escape detection until alarmingly close. The angle at which maximum detection occurs can be approximated by dividing the wavelength by four times the height of the antenna above the Earth's surface. A small angle, the result of short wavelength and necessary to detect ships, gave the Coast Defence set a significant advantage over CH in detecting low-flying aircraft, for which it became CHL. Placing it on a high tower allowed it to perform both functions as CD/CHL. By fall of 1939 there was great pressure to deploy it, and the first station was on the air by 1 December 1939 [28].

In parallel with this Pollard designed a rudimentary radar to allow anti-aircraft batteries to determine range accurately and to help put their searchlights on attackers. It had a small antenna and a long wavelength,

---

[4] See 'Lobes' in the appendix (p 471).

from 3.5 to 5.5 m, and was ironically named Gun Laying I (GL mark I); Gun Assisting would have been better terminology. It gave accurate range but poor horizontal direction (azimuth), no elevation angle at all, and multiple targets could completely fog even experienced operators [29]. It was quickly designed, production was cheap and more than 400 were made; of the slightly improved GL mark II, which provided elevation data at large angles, over 1600 were made [30]. Elevation angle is difficult to determine for meter-wave radar because the vertical lobe pattern is unpredictable unless the ground is a perfectly flat conducting surface, approximated by wire netting. Locations by the sea, common in the Pacific but where GL mark II was not employed, allow good use of meter-wave equipment. The meager effort that went into the design of GL probably had its origin in the extremely unsatisfactory state of the directors available to use radar data [31], making the expenditure of time for developing a set capable of blind fire not worthwhile. Using it for blind fire was a waste of ammunition: if set up under ideal conditions, it returned one hit for 4000 rounds expended [32].

Butement took on another electronics problem dealing with antiaircraft (AA) artillery, the proximity fuze, but it failed because of low priority. Butement did design the elegant circuit for it that would later be used in America to make the device a success [33]. The primitive character of GL marks I and II, and the virtual abandonment of proximity fuze work contrasted markedly with attitudes in the US and Germany at that time and demonstrates the low esteem that AA artillery had in Britain. It was, however, perfectly consistent with the wooden-headed attitudes of the Army in technical matters during the first years of the war. The War Office use of scientists was the perfect inversion of that of the Air Ministry. They did not wish scientists to initiate projects but to do what they were told; they did not wish scientists to work closely with serving officers to gain understanding of the weaknesses of their inventions; they did not wish to hear the opinions of scientists on how best to employ the new weapons. In the case of GL, excellent engineers were told to provide a better range finder. The AA predictor was incapable of using the continuous stream of data provided by a radar set and remained so until the arrival of the American M-9 director in 1944 [34].

The 1.5 m equipment on which the Army cell built their Coast Defence set came from the first attempts to construct airborne radar. Looking forward to the success of early warning radar Tizard had predicted that the enemy would necessarily turn to night bombing for which there was no effective defense. If searchlights could be brought to bear on an attacker, then fighters could intercept it or AA guns fire on it, but it was no small trick to get a light beam on a plane and hold it. Large listening horns gave some indication of the direction of the plane but were strongly affected by weather and confused by large flights. The regions serviced by searchlights would always be small, greatly reducing the amount of time for fighters or guns to engage. What was clearly needed was airborne radar. Given the gargantuan size of Britain's initial radar equipment this seemed a tall order.

Bowen had taken to the idea of this project and had pressed hard for it, convincing Watt to begin it after the move was made to Bawdsey. The problem was poorly defined, and the constraints imposed by installation in aircraft only vaguely understood, so Bowen discussed the matter with two communications engineers at Martlesham Heath, the nearby Armament and Experimental Establishment airfield from which Bawdsey drew aircraft cooperation. These discussions led to self-imposed limits of 100 kg weight, 500 W power and dipole antennas not more than a meter in length. At just the right moment appeared the EMI Company's 6.7 m television straight-vision receiver [35], a small but very sensitive unit that amplified the radio-frequency signal without heterodyne action and that had been well engineered, so the first experiments were carried out on this wavelength.

After satisfactory ground tests a receiver and indicator were transferred to a Heyford bomber, which was large enough for the 6.7 m antenna and had an electrically shielded ignition system. Aircraft echoes produced by the transmitter, which was much too heavy to be carried aloft, were satisfactorily received along with the transmitter pulses, of course, which furnished a not particularly helpful time reference. Bowen was a strong advocate of further developing this ground–air system, called RDF 1.5. (RDF 1 refered at the time to CH and RDF 2 was to be the 1.5 m airborne set.) Watson Watt could not be convinced, and the approach was dropped, a decision Bowen always regretted.

With RDF 1.5 eliminated Bowen pushed to shorter wavelengths, settling on 1.25 m. The EMI receivers became the intermediate-frequency amplifiers that followed frequency conversion with an RCA Acorn tube. The Handley Page Heyford bomber was replaced with two Avro Ansons, excellent machines for flying laboratories, and with them came the personnel of D Flight, 220 Squadron. In August 1937 the team had clean echoes from ships and the following month created somewhat of a sensation in fleet exercises by locating the carrier *Courageous* through thick clouds.

At the end of 1937 the airborne project had two goals, Air to Surface Vessel (ASV) and Air Interception (AI), and an important new member, Robert Hanbury Brown. The details had to be worked out for sideways and forward lobe patterns in the ASV set and for acceptable maximum and minimum ranges for AI. The two ASV modes were for maximum search and head-on attack. The need for a maximum range in AI was obvious enough, but the minimum range was even more important and much harder to bring off because it meant very short pulse duration. Radar had to bring the pilot close enough for him to see his target, and a 1 μs pulse produced a wave train 300 m long, which was greater than the minimum distance for visual contact, and sub-microsecond pulses were not so easy to produce. Both ASV and AI were well advanced when war came but far from ready [36].

The Tizard Committee had discussed air defense problems with Charles Wright, the Admiralty's Director of Scientific Research and G Shearing, Chief Scientist of His Majesty's Signal School's Experimental Department before disclosing to them the approach taken by Watt. Shearing volunteered that radio should be carefully considered, pointing out that the Post Office had had reception interferences caused by aircraft when using 5 m waves. Wright was soon invited to Orfordness and was able to bring Shearing with him to a second demonstration in July 1935. In as much as the Signal School was the only source of the silica transmitter tubes that Watt wanted, discussions tended to be harmonious.

The Navy did not want their radar research conducted at Bawdsey, and the Admiralty initiated work in its own existing laboratory at HM Signal School in Portsmouth, to the irritation of Watson Watt, who wanted all radar research concentrated in one place. The Navy's reasons were that naval use required that designers at every level know about conditions at sea, the effect of gunfire, the interference from masts and other structures, the mutual interaction of various shipboard electronics and the problems of maintaining equipment at sea. HM Signal School was no newcomer to high-frequency work, having led the world in equipping vessels with high-frequency direction finders. In contrast to Bawdsey, they conducted research in microwaves [37]. The Navy won the dispute and formal Treasury approval at the end of 1935 [38].

The results of the next 18 months were not impressive. R A Yeo conducted experiments on 4.0, 1.5 and 0.6 m and in July 1937 produced some not particularly impressive range-only experiments with a set designated type 79X installed on an old minesweeper. Secrecy had much to do with the poor progress, as Yeo could speak about the work with only two outside of his immediate group. He was frequently unable to get time on the minesweeper, which was used for school functions, because he could not tell its keeper why he needed her. The contrast with Bawdsey was extreme. A new commander of the Signal School, Captain A J L Murray, was appointed on 25 August 1937; he took personal interest in the radar project, including attending their staff meetings. The change was abrupt and very helpful.

Alfred W Ross took over the design of type 79 and immediately changed the wavelength from 4 to 7.5 m. This had three advantages: (1) the Navy's silica transmitter tubes had too much inductance in the leads for satisfactory operation at 4 m, (2) 7.5 m was in the band of British Broadcasting Company's television, allowing commercially developed receivers to be used, a trick probably learned from Bowen, and (3) this wavelength fitted well with techniques used in the Navy's communications transmitters. The result was that Ross tested a successful prototype 79 in early 1938 with copies installed on the battleship *Rodney* and the cruiser *Sheffield* near the end of the year.

The type 79, soon improved to type 279, used synchronously rotating transmitting and receiving arrays on separate masts. The antennas were small, hence the beam broad; this was not considered serious in an air warning set but was a problem for observing other ships. This led quickly to the design of type 286 for 1.5 m for the same reason Butement had selected this frequency for his Coast Defence set. The 286 had non-rotatable antennas with the result that one had to swing ship to gain directional information [39]. The equipment did not compare favorably with the sets with which the American and German navies were being equipped, but they provided essential information and had been designed by engineers who understood what was required to make electronics work at sea.

J F Coales had proposed in February 1937 developing 50 cm equipment for gunnery. After various considerations he decided to use a specially designed triode made by H C Calpine. By the end of 1938 transmitter, receiver and antenna (a Yagi backed by a cylindrical paraboloid) were ready for test. Sea trials aboard HMS *Sardonyx* in June 1939 observed ships at 8 km and low-flying aircraft at 4.5 km. The 50 cm equipment was to evolve into a series of gunnery sets [40], all characterized by pairs of Yagi antennas, which came to be called fishbones, that used lobe switching. By 1940 designs were complete for the type 284 for main armament and type 285 for anti-aircraft; both went into production the following year. The Navy's enlightened attitude concerning gun-laying radar contrasted with the Army's near indifference.

By 1939 Dowding's air defense system was reaching a remarkable stage. He began a series of summer exercises immediately after taking over Fighter Command and quickly recognized that the first problem was confusion. Air Controllers were overwhelmed with too much information, some of it contradictory or just wrong. The first step was to route the various radar and ground observer sightings to a filter center at Bentley Priory, Fighter Command Headquarters, to evaluate them for veracity. Filtering these observations, which required instinctive as well as scientific skill, produced the information sent on to the three Fighter Command Groups. There the air situation was also presented by markers designating enemy and friendly units, moved on great map-tables by women and men members of the Air Force who were connected to the filter center by telephone. Above, watching the movement of the pieces on the table, sat the controllers, themselves fliers, who sent orders to the pilots by radio-telephone. Command functioned with Dowding making decisions day by day, Group Leaders hour by hour, and squadron controllers minute by minute. Communication between all radar stations, observers, filter centers and squadron headquarters was through an independent wire network. Communication with the pilots was through the new very-high-frequency (VHF) radio-telephone link, the only one in the world. It gave low-noise transmissions, provided plenty of channels and had range limited by the horizon, making eavesdropping difficult.

Women did not serve just as croupiers at the great tables but in many functions throughout the Air Force. In 1937 it had been realized that Britain would face a severe manpower shortage, and the successful employment of women in air defense in the previous war pointed to an obvious solution, the Women's Auxiliary Air Force (WAAF) resulting. Three women from the Bawdsey clerical staff were given training in operating the equipment of a CH station, and they quickly demonstrated a keen ability, especially for the wily nature of the receiver oscilloscope displays [41]. The value of women as radar operators was soon widely appreciated. A succinct explanation came from Australia: 'Women *did* make the best radar operators, because they watched the screen' [42]. For those with insight into character it came as no surprise that they remained calmly at their posts when stations were attacked.

The months between the Munich conference and Chamberlain's Declaration of War on 3 September 1939 saw a remarkable alteration of attitudes in Britain. After the euphoria of 'peace in our time' came an understanding of the price that had been paid for it. The people and their government decided that the next aggression meant war, and Britain gave a guarantee to Poland over an issue that even had elements of justification for Germany. The people accepted war with a determination unmatched in France or Germany. France would soon collapse and Germany would not confront the grim reality until 1942.

Britain was the least prepared for war of the three powers, except in radar. In the summer months of 1939 the physics departments and technical schools began emptying to fill the ever-growing need for radar men and women. The number of people working in radar one way or the other was large by then, probably greater than the combined totals of all other nations, and the number was growing rapidly. The engineering quality of British sets was poorer than those of Germany or the United States, reflecting Watson Watt's dictum, 'Give them the third best to go on with; the second best comes too late and the best never comes' [43], but in the quality of the radar operators and mechanics it was another matter, one that more than offset any deficiencies of equipment. They were expected to cope with all manner of problems, and their training approached that of electrical engineers. This core of technicians was the narrow margin of defense.

2.4. AMERICANS AND GERMANS BUILD PROTOTYPES

2.4.1. *US Naval Research Laboratory*

Albert Hoyt Taylor was born nearly a decade before the demonstration of electromagnetic waves that were to shape his life. He was born in an age that viewed the atomic theory of matter as just an interesting hypothesis. He died in an age that feared the power latent in the atom might destroy us all. During his school years there arose in him an interest in all things electrical that led straight to a PhD in applied electricity from

Göttingen. The First World War deflected him from an academic career into the Navy where he remained after the war as a civilian employee and directed research in radio at the newly created Naval Research Laboratory. The contrast of Taylor's German degree and Leo Young's high school diploma disappeared in their common love for radio; both were active amateurs—Taylor, with call letters 9YN, and Young, W3WV. From the time they observed the effects of the passage of a river steamer between their 5 m transmitter and receiver they never had radio location far from their minds.

These thoughts came to nothing during the 1920s because the necessary high-frequency electronic components simply did not exist. The combination of the Young–Hyland observations in 1930 and the availability soon after of high-vacuum low-voltage cathode-ray oscilloscopes and multi-element electron tubes changed things, but their preliminary results did not provide them with a high priority by the Bureau of Engineering, in spite of the strong support for radio location by the Laboratory Director, Captain Edgar Oberlin. The publication in 1933 by three investigators at Bell Telephone Laboratory of picking up high-frequency reflections from airplanes [1] did not make the low priority sit any better with Taylor.

At the time their primary concern was an attempt to remove the Laboratory's research function and make it in Oberlin's words 'a glorified test shop' for the Bureau of Engineering, which thought that NRL could not or anyway should not compete with commercial laboratories, a change that would have put an end to radar work. Oberlin was not lacking in the skills needed to protect his command and succeeded in retaining NRL's research freedom. He was soon backed by the appointment in May 1935 of Rear Admiral Harold G Bowen, a strong believer in NRL research, as head of the Bureau of Engineering [2]. That the Laboratory's allotment for 1934 decreased by a third from the preceding year did not make matters any easier for the birth of Navy radar.

Despite all, the radar program received a better priority in early 1934. The first important effect of this was the assignment of Robert M Page full time to the project, which had been the part-time job for Young and two others. Hyland had left naval service, somewhat acrimoniously [3]. Like Taylor and Young, Page was a mid-westerner from the kind of family that of necessity prepared children to make their own way in the world. He came to NRL with a bachelor's degree in physics from a small church college in St Paul. Navy radar was to grow around Page.

His first task was to try pulsed waves. Young had convinced Taylor that this had to be done, for continuous waves had produced nothing but the unpleasantness with Major Blair of the Signal Corps. The idea had two obvious origins in NRL experience, the sounding of the ionosphere with radio and of the ocean depths with sound, but both were in fact quite different. Both used pulse durations measured in fractions of milliseconds and had signals reflected from huge surfaces. The size of an airplane or ship could hardly allow the use of a train of waves 300 km long that a millisecond pulse would generate. These plans differed from Watson Watt's Chain Home by envisioning from the first a searchlight-like beam with target direction ascertained by rotating the antenna for maximum signal.

The first attempt with radio waves pulsed for 10 μs took place in December 1934 and had little to recommend it other than a clear indication of where the problem lay—the receiver.

Page had successfully used a multi-vibrator, an electronic switch capable of turning the transmitter on or off very quickly, to modulate the transmitter, but the receiver was a slightly modified communications set. It had high gain and the sharp tuning for which such circuits are noted. The transmitter and receiver had separate antennas arranged to have a minimum transmitter signal picked up by the receiver, but enough was picked up to cover the screen because of the ringing caused by the narrow pass band[5]. Nevertheless, there was a definite effect from signals reflected from the test aircraft [4].

Page thus faced the same problem that faced all other radar engineers and set about to solve it his own way from basic principles using a French paper as his guide [5]. This problem was then being faced not only by radar but by television engineers, who had to contend with changes in the strength of their video signal that were just as rapid as the radar pulses. It is curious that no one at NRL seems to have noted the significance of the descriptions of RCA's prototype television system that appeared with descriptions of the receivers [6], more evidence for NRL's isolation from the New York electronic community of RCA, Bell Labs and Signal Corps.

Page solved the design and equally formidable construction problems by November 1935 using funds illegally diverted from another project. Taylor realized that radio location had to have more help—indeed Page had had to leave the work for a while during the year to help out on another job—and in early 1935 visited Mr James Scrugham, the most influential member of the House Naval Appropriations Committee, to explain his concerns. The result was a telephone call next day telling Taylor that the Committee had agreed to an additional $100 000. Among other things this allowed Robert Guthrie to work with Page, who put him to work on the transmitter. They formed a congenial working pair as they brought the device on line.

The next test in April 1936 achieved what was desired, with planes observed at ranges of 8 km, quickly extended to 27 km, and Adm. Bowen gave the project the highest priority. Taylor immediately made two demands of Page: that a common antenna be used for both transmission and reception and that a shipboard test be made at the earliest possible date.

For meter waves the antenna had to be an array of dipoles, and the larger the array the narrower the beam and the greater the intensity of the

---

[5] See 'Receivers' in the appendix (pp 475–6).

projected power. Using a single antenna that had the area of two separate ones yielded improvements in angle and intensity of a factor of four. The difficulty lay in preventing the transmitter with its enormous power from burning out the receiver. Page arranged fractional-wavelength transmission lines in such a manner that the receiver input tube grid current, an effective short circuit, reflected a high impedance to the transmitter. It was a neat idea that formed the basis of several variations. Page called the device a 'duplexer'.

In April 1937 a 1.5 m set was installed on the destroyer *Leary* with a Yagi antenna mounted to the barrel of a deck gun to allow pointing. The results were successful in so far as shipboard service went, but the transmitter power had to be increased. The solution to that problem was a ring of Eimac 100TH transmitter tubes that allowed power to be built up in phase. The Signal Corps adopted this tube and transmitter design. A target date of 1 September 1938 was set for a prototype set, XAF, to be mounted on a battleship and tested in fleet exercises.

In mid 1937 two electronics corporations were brought into the planning: Radio Corporation of America and American Telephone and Telegraph. RCA had had an interest in radar through Wolff's 10 cm work, but for the men from Bell Telephone Laboratory, representing AT&T, it was all new and quite astounding [7]. RCA was interested in producing the XAF but wanted to build a design of their own for the fleet exercises, designated CXZ. The Bell Labs people returned home to think about these new ideas.

By January 1939 XAF was mounted aboard USS *New York* and CXZ aboard USS *Texas*, and exercises in the Caribbean began. The results of XAF were spectacular and made instant converts to the new weapon of all line officers select enough to have encountered it. Ships were observed at 16 km and aircraft at 77 km. Shells could be followed in flight and their splashes noted. Night destroyer torpedo attacks were thwarted. An unexpected but very important use of the new equipment soon became apparent. While operating in the vicinity of the Virgin Islands at night the ship's position became poorly known, and Page located it by ranging mountain tops on distant islands [8].

XAF liked the sea and did not mind gunfire at all, something that could not be said for CXZ. Taylor found this useful when insisting that RCA build exactly to NRL specifications, especially keeping Eimac tubes in the production model. The Chief of Naval Operations immediately asked for 20 sets in their current form. This was to become CXAM, which would later evolve into the SK, the Navy early warning set for the entire war, referred to by all as the 'flying bedspring'.

2.4.2. *Bell Telephone Laboratories*

On returning from their radar demonstration on 13 July 1937 the Bell Labs engineers initiated serious discussion about the part AT&T was to have in this new technique. There was another visit in November after which the directors decided to convert the field station at Whippany, New Jersey and place it under the control of W C Tinus. It had been established in 1926 for trans-Atlantic telephony and broadcast development, and its isolation, 50 km west of Bell's New York City labs, made it ideal for secret work. The company decided to enter the field at its own expense rather than solicit a government contract because the rules of the time made it difficult to undertake something of a highly speculative nature, and the Bell engineers wanted to try shorter wavelengths [9].

One reason for selecting short wavelengths was the recent invention at Bell Labs by A L Samuel of triodes for very high frequencies [10], which placed them in a marginally better position to do something with very short waves than their numerous predecessors. Their designs paralleled the RCA Acorns[6] but also included a transmitter tube, 316A, whose shape caused it to be called a 'doorknob'. Work began on 40 and 60 cm equipment intended for gun laying or fire control as the Navy preferred to call it. A single horizontal cylindrical paraboloid antenna, the kind used by Hertz and Hülsmeyer, and Page's duplexer formed the basis of the set.

Bell engineers demonstrated their first prototype, the CXAS, more correctly a pre-prototype, in July 1939 to representatives of both Navy and Army. They selected a position on a bluff 25 m above the sea at Atlantic Highlands, New Jersey, which presented a rich field of targets in the ships moving into and out of New York [11]. The 40 cm set could pick up some ships 15 km distant. The accuracy in range was satisfactory enough—and very important for naval fire control—but angular accuracy was inadequate for radar-directed fire. Work was already proceeding on eliminating this fault using the lobe-switching technique that they had learned from the Signal Corps[7]. In the meantime the Navy began to plan radar's integration into current fire control systems and ordered ten to be delivered designated CXAS, then FA [12]. The first set was installed aboard USS *Wichita*, but not until June 1941. The Army was interested in the device for coast defense, but delayed ordering what they designated as SCR-296 until late 1940 [13]. For the Army it proved to be a dead-end design, for the Navy the beginning of all fire control, both surface and air. For Bell Labs and AT&T it was the beginning of a gigantic undertaking.

Bell Labs introduced horizontal lobe switching to alter the design for main-battery fire control in the FC and horizontal and vertical switching for anti-aircraft in the FD. Production began in 1941 [14]. Thus the US and Royal Navies independently introduced similar equipment at about the same time: FC corresponding to British type 284, FD to type 285. In production Bell greatly improved output power over the original design

---

[6] See 'Multi-element tubes' in Chapter 2.1 (pp 35–6).
[7] See 'US Army Signal Corps' in the next section and 'Lobes' in the appendix (pp 71, 471).

that used triodes, which the British continued to use, but that story must be delayed until Chapter 4.1.

2.4.3. *US Army Signal Corps*

William R Blair was about the same age as Hoyt Taylor. He too earned a PhD degree but in physics and from the new University of Chicago rather than from an old and famous German school. His interest in electromagnetic waves originated in his thesis work for Professor A A Michelson during which he generated microwaves in the range from 10 to 40 cm with much the same techniques as Hertz. He noted, possibly for the first time, that these waves were reflected by materials other than conductors. His work with Michelson also included measuring the speed of light. He thus filled his mental storehouse in graduate school with two key elements of radar.

Radar first occurred to him in the spring of 1926 when attending the Command and General Staff College where he heard lectures about locating aircraft by sound detectors. There were so many problems apparent to Blair in such a method that he immediately thought of the problem in terms of microwaves and the velocity of light. After graduating from this preparatory school for generals he was assigned as Chief of Research and Engineering directly under the Chief Signal Officer and soon became Director of the Signal Corps Laboratories [15].

The Signal Corps had become responsible for Army radio research in February 1919 and located the work at Camp Vail, soon renamed Fort Monmouth, in central New Jersey. Their mission was differently defined than the Naval Research Laboratory. Whereas the Navy laboratory had a wide range of freedom in seeking out subjects for study, the Signal Corps Laboratories were restricted to the expressed needs of the various arms. In so far as radar was concerned this did not present a problem because the Coast Artillery Corps, to which anti-aircraft defense had been assigned, was very much interested in any way of locating aircraft, although the Air Corps was not.

A great deal of effort was devoted to detection with infrared, work not completely given up until 1939. Blair never believed infrared would be the answer, primarily because of its inability to penetrate fog and rain, but his approach was positive. When he learned that the country's leading expert for infrared, Dr S H Anderson, had lost his job as a result of depression cut-backs, he hired him to push the technique. To be the best in the technique also fit his wish to build a first-rate laboratory. They learned as did investigators world wide that it had no value for aircraft detection [16].

Blair was set to avoid a trap with infrared but was ill prepared to give up microwaves, despite the poor results that Wolff's equipment and the Signal Corps work carried out by William D Hershberger had shown. At the moment when important decisions needed to be made, a new player entered the game, Lt Col Roger B Colton, as Chief of Research and Engineering. So secret was radar that he was told he could not see what NRL was up to, and it even took a while for him to learn that the Corps was already on the project. His first actual encounter with radar-like apparatus was in the summer of 1935 when he was invited to observe a demonstration by General Electric of a method for the radio detection of airplanes by Dr C W Rice, who was experimenting with microwaves using split-anode magnetrons. It was a failure but impressed Colton as having failed because the antenna was too small; his interest increased on visiting Rice's laboratory where he was shown the detection of moving vehicles at half a mile [17], essentially the same experiment that Wolff had conducted at Sandy Hook the year before.

Colton was quite a different personality from Blair. Blair had in a few years made the Signal Corps Laboratories first rate, but Colton was a radar zealot who wanted to build equipment. He was also an inspiring leader and soon gained the respect of everyone at the Corps Lab, military or civilian [18]. He was also known for preferring to settle an argument by drinking his opponent under the table [19].

An unexpected incident in December 1935 marked the turning point in the Corps' path to radar. A proposal for the radio detection of aircraft was sent to the Signal Corps by the Bureau of Standards and came naturally enough into Colton's hands. Colton requested of Blair that he send Hershberger to evaluate it and visit NRL, where Hershberger had worked on sound a few years before. (The secrecy curtains that separated the two services were not particularly hard to draw when there was a need.) It was a propitious decision, not that there was anything in the Bureau's proposal— it was the Taylor–Young–Hyland 'radio screen' again—but it required a report by Hershberger, who was heartily fed up with microwaves by then and had suggested pulsed meter waves two years earlier. Hershberger's visit with Page allowed him to see the results of the first pulsed-wave experiment and the preparations being made for improving the receiver. His report gave the reasons for the Army's lack of progress as preoccupation with microwaves and lack of support for radio detection. Blair had lost hope for microwaves by then too and forwarded the report with a strong recommendation to Maj Gen James B Allison, Chief Signal Officer. Allison asked the War Department for $40 000 for radio location and was refused [20].

Allison gave the project the highest priority, asked for the funds again and was again refused. He then agreed that Blair should use $75 741 appropriated for a top priority communications project for fiscal year 1937. It was illegal, but Allison said he would stand behind Blair. Blair promised to do his best to have something to show by 1 June 1937. Days followed with many unpaid Sundays and overtime hours [21]. In August 1936 Colton was assigned as Blair's Executive Officer. In October Hershberger left to take a doctorate in physics at the University of Chicago, and Paul Watson was named chief engineer [22].

Events then moved swiftly. In December 1936 some lashed-up equipment was transported to Princeton Junction to track the steady flow of air traffic in and out of New York. The transmitter was placed a mile from the receiver to keep it from being overwhelmed, and the reflections were quite pleasing. They then built large antennas consisting of arrays of dipoles to allow direction to be determined, and five months later they were ready to demonstrate a 3 m set with direction and ranging capability to the Secretary of War and the Chief of Staff. To the astonishment of the visitors the strange arrangement of wires on wooden framework succeeded in detecting the arrival of an unlighted night-flying bomber and putting a searchlight on the attacker in plenty of time for AA guns to open fire. It was almost too successful because the Chief of Staff wanted the present design put into immediate production, something the Signal Corps was loath to do.

Allison was now offered $50 000 by the War Department but demanded $250 000, which required a Congressional appropriation that was thought impossible. Colton insisted. Congress was asked and gave. The money famine was over [23].

The next step was to prepare for a demonstration of the prototype of what was to be SCR-268 (Signal Corps Radio-268). It was in fact only an electrical prototype by then using 1.5 m; the mechanical design went more slowly and could not be completed until the electrical specifications were fixed. It was operated by a detachment from the 62nd Coast Artillery Regiment commanded by Lt A F Cassevant, to learn how soldiers would adapt to such a device; they also furnished the security for the equipment, then on isolated Sandy Hook. With time Cassevant became an accomplished radar engineer and transferred to the Signal Corps. Colton insisted that SCR-268 should have lobe switching for both the horizontal and vertical antennas and put James Moore on the design. Engineers from Westinghouse and Western Electric took up residence in order to facilitate production later [24]. Blair became hospitalized in the summer of 1938, Colton was named Acting Director, and when Blair had to retire because of his health, Colton became Director. The tribulations in getting SCR-268 to the tests for the Coast Artillery Board at Fort Monroe included severe damage of the equipment by hurricane, the near capsizing of a ferry over the Delaware and a complete disruption of highway traffic to a degree unknown to the innocent motorists of that day.

The tests were made at the end of November 1938 with the usual notables plus Brigadier General Henry Arnold, Assistant Chief of Staff of the Air Corps. The feats of the previous test were to be repeated but at greater ranges and with greater accuracy. Almost as if following stage direction it ended with an incident of high dramatic value. A B-10 bomber was to make night approaches to Fort Monroe at 6000 m altitude from the west. The 268 was to pick it up and point the searchlights, as they had done at lower altitudes before. The Air Corps liaison announced the plane was overhead. No radar contact. After the third failure the radar men began to look at places where the plane was not supposed to be and found it—well out over the Atlantic, for a very strong west wind had completely fooled the experienced night navigators. The return flight took an hour, and when the plane came within searchlight range a beam of light ended on clouds. But the stage directions seemed well worked out, because when the errant aircraft passed out of the clouds into clear sky, it was illuminated by the searchlight. Arnold wanted an early warning radar [25].

The troublesome details of the mechanical design that made the 268 portable enough to accompany, to be sure rather clumsily, troops in the field were overcome by midsummer 1940. The tests of 1937 and 1938 had used three separate antennas: a transmitter, an azimuth receiver and an elevation receiver. These had to be incorporated into a single unit suitable for field service. A contract was let to Western Electric from which deliveries began in February 1941 [26].

Following the May 1937 tests the Air Corps had expressed interest in an early warning set, and work started in parallel with that of the 268 on what was to become SCR-270, a mobile set, and SCR-271, for fixed installations. Accuracy of direction was sacrificed for range so the wavelength was increased to 3 m and lobe switching dispensed with but duplexing incorporated using a design by Dr Harold Zahl. Whereas the 268 used Page's ring oscillator, initially of eight, later 16 Eimac 100THs, Westinghouse agreed to make a special water-cooled output tube, WL-530, that could deliver the desired 100 kW power with a simple push–pull circuit, by then a standard electronic design. The difference in wavelength, 1.5 to 3.0 m, made the difference between a 16-tube ring oscillator and a two-tube push–pull. In June 1939 an engineering model of the 270 tracked an aircraft consistently to 125 km and occasionally as far as 240 km and a flight of bombers consistently to that range. A contract was let to Westinghouse in August 1940, who delivered 112 sets before the US entered the war. The fabrication of a 271 was completed at Fort Sherman, Panama in June 1940 and a number of 270s were shipped to Hawaii in the latter half of 1941. Six were spotted around the perimeter of Oahu on the morning of 7 December 1941 [27].

The 268 remained the only radar for Army AA artillery until early 1944, when it began to be replaced by the incomparable SCR-584. The Signal Corps was aware of the Bell Labs equipment and its great potential for searchlights or gun laying but rejected it in favor of the 268 because of the latter's much greater range. This gave an AA battery a single piece of equipment useful both for acquiring a target and directing guns or searchlights onto it. In the Pacific it would serve in a function unsuspected in 1939—ground-controlled interception, about which more later. The 270 remained the standard early warning radar until the end of the war and even saw service in the Korean War [28].

Thus by the time war had broken out in Europe the American Army and Navy both had excellent prototype radars that would be in production well before the time of the attack on Pearl Harbor. As is so well known, an SCR-270 secured for the sleeping garrison nearly an hour's warning, but the organization necessary to make use of this intelligence was only just being formed.

2.4.4. GEMA

For those persons who delight in irony the development of microwave radar has an example. The British gave such short wavelengths the least thought during the 1930s, yet invented the revolutionary cavity magnetron that delivered prodigious power at 10 cm. The Germans pushed research in the region shorter than 50 cm wavelength all through the decade and stopped only when a wartime shortage of technical manpower and the absence of a microwave oscillator of sufficient power forced termination, yet they missed the cavity magnetron and ended far behind the Anglo-Americans in this critical technique.

Serious radar work in Germany came from the zeal of Dr Rudolf Kühnhold, the Technical Head of NVA. His work in underwater sound led him in 1933 to the idea of using radio waves for locating and ranging targets, and his failure to observe reflections with 13.5 cm waves impelled him to consult with a leading electronics company, Telefunken, where toward the end of the year he discussed the problem with Dr Wilhelm Runge, Telefunken's acting laboratory director. Runge was not interested because he saw in it the need for vacuum tubes that he thought lay years in the future. To others Runge voiced the opinion that the whole project was 'utopian' or worse [29]. Kühnhold found Runge rude and did not forget. This rebuff helped divide German radar development and production between three companies: Telefunken, Lorenz and one not initially in existence.

By this time Kühnhold's dealings with Tonographie concerning underwater sound had become so agreeable that he dropped attempting to gain the interest of the communication industry. He was particularly pleased with the resolute spirit of Tonographie's owners, Paul-Günther Erbslöh and Hans-Karl Freiherr von Willisen, and they were equally pleased with Kühnhold's openness. Kühnhold loathed bureaucracy and relished the absence of it in company operations, an appealing contrast with the procedures for doing anything at his own NVA. Their business in phonograph recording was prospering, which provided capital for taking risks, and, in seeking new fields for their talents, the two entrepreneurs were experimenting with 95 cm directed-beam communication, which they thought might interest the Kriegsmarine for secure communication between ships. It was an application for which the low transmitter powers of the very short wavelengths were no problem and was also compatible with Kühnhold's interest in radio location. To aid them they sought two high-frequency experts as consultants: Dr Hans E Hollmann and Dr Theodor Schultes, both of the Heinrich-Hertz Institute [30].

In November 1933 von Willisen noticed an advertisement by Philips for a magnetron that delivered 70 W at 50 cm and ordered one immediately, news that raised Kühnhold's spirits measurably. In February 1934 while waiting for the magnetron equipment to be finished von Willisen produced a failure of his own at 75 cm. Using the techniques being tried for directed beam communication, he connected his transmitter to a Yagi and attempted to observe the reflections with a dipole whose receiver was overwhelmed with the direct wave signal, but the completion of the magnetron equipment swept aside any regrets about that experiment. He designed a 48 cm transmitter and connected it to a Yagi; Hollmann used a Barkhausen tube as a regenerative receiver connected to a dipole antenna. All attempts to keep the direct signal out of the transmitter failed, even substitution of a paraboloid for the Yagi.

This failure was tempered by a successful use of the magnetron for communication. The magnetron had proved to be extremely unstable in frequency, especially in the push–pull form first used, so von Willisen replaced it with a single-output-tube design and made a careful study of the settings for voltages and magnetic field to ensure best performance.

A successful experiment by Kühnhold in June with the Pintsch equipment that had now attained 0.3 W caused von Willisen to try his improved circuitry with a paraboloid for the transmitter and a dipole array designed by Schultes for the receiver. This proved to be a twin success; a large steamer that passed by was tracked by the 13.5 cm equipment to 4000 m and by the 48 cm equipment to 2000 m [31]. An important difference was, of course, using a moving target that generated a Doppler interference signal.

Before these critical experiments were carried out Erbslöh and von Willisen incorporated a new company in order to separate these military applications from Tonographie. They called it Gesellschaft für elektroakustische und mechanische Apparate, a name intended to deflect the curious by its reference to previous activities of the owners and, for those who might have known, those of Kühnhold. The acronym GEMA came into immediate use, and the seven-word name seems never to have crossed human lips again. They hired Schultes to run the high-frequency laboratory and another Heinrich-Hertz man, Dr Walter Brandt, to run the low-frequency lab, thereby completing the top design levels of the company for several years [32]. Hollmann withdrew from the company when, owing to his association with them, he was forced to delete portions of a book he had just completed, which soured him on military work. Despite the omitted parts, it became the ultimate authority for microwaves until the end of the decade [33].

The GEMA people carried out continuous-wave experiments through October 1934 and saw that nothing useful could be done with them, although a strong reflection from an airplane that happened to pass through their beam opened their thoughts to targets other than ships and helped open the pockets of the Kriegsmarine for a grant of Rm70 000 ($16 500). The problem with continuous waves was the near impossibility of keeping the transmitter signal out of the receiver, and on the advice of Hollmann and Schultes they undertook to try pulse techniques. They had by then set up their 'GEMA tower' on the NVA experimental grounds at Pelzerhaken near Neustadt on the Lübecker Bucht and from it had a notable success. The transmitting antenna had ten dipoles backed with a reflecting mesh, the receiving antenna three dipole pairs. Transmissions were on 52 cm with 2 μs pulses at 2000 per sec. The receiver was a broad-band heterodyne that used the new RCA Acorns. Success of a form came when the set first operated in May 1935; they were surprised by seeing the wood on the opposite shore showing up at 15 km, but ranges on the 400 ton research boat *Welle* were not satisfactory. The receiver was redesigned to use two intermediate frequencies, and by 26 September they could follow the little boat to 8 km. It was time to show the results to Admiral Erich Raeder, commander of the Kriegsmarine, and lesser notables [34].

This demonstration secured adequate funding for the project but disclosed the first of many difficulties that were to accrue from working for the Kriegsmarine. The demonstration had incorporated lobe switching that had given a directional accuracy of 0.1°, and to their astonishment the engineers were told this was too complicated and to leave it off present designs [35]! Next they were told that a cathode-ray tube was too delicate for use aboard ship and that a substitute must be found. The second objection was soon removed and the decision reinforced later in a rather macabre affair: the little test boat *Welle* was lost with all hands, but the cathode-ray tube that was in a prototype set on board was recovered and still functioned, which convinced the naval authorities that it was an acceptable component [36]. The first objection was not overcome until it was too late to provide blind-fire capabilities in any important naval action. Radar was to provide range and search capability only.

These results led to disagreements about what to do next. Kühnhold wanted the 50 cm work to continue, but Erbslöh and von Willisen wanted to increase the wavelength to gain radiated power and thereby range. Kühnhold planned to continue his 13.5 cm work and strongly favored pushing toward shorter wavelengths, marshaling all of the valid arguments for microwaves. (He and Blair would have understood one another.) After discussions colored with asperity they decided to do all these things. GEMA's 50 cm work became a surface-search radar, DeTe-I, the first of the Seetakt series, and the long-wave work became an air-warning radar, DeTe-II, which soon settled on 2.4 m. The shorter wavelength was needed for surface targets because the vertical lobe structure severely cut the range to surface targets as wavelength increased; the longer wavelength was needed for the extended range to pick up distant air targets. Von Willisen replaced the unstable magnetron in the 50 cm set with a high-frequency triode, TS1, of GEMA's manufacture, and Kühnhold replaced the hopelessly weak Barkhausen tube in the 13.5 cm equipment with the magnetron. He continued microwave work at NVA until a British microwave radar came into German hands in February 1943.

The TS1 triode had to be driven very hard to meet power requirements, had a very short useful life and was replaced by an improved model, TS6, which did somewhat better. It was the same story in Britain and America. The short lifetime of triodes in a British 50 cm set would alter the course of battle with the Bismarck in May 1941[8].

Raeder's eagerness for the new device was tempered by a greater eagerness for squeezing as much as possible of the generous flow of money coming from the new regime into preparing more tangible armaments for a surface war with Britain, to take place according to Hitler's assurances in 1944 at the earliest. Furthermore, he emphasized to Kühnhold that NVA's highest priority must be under-water sound. Nevertheless, on 18 April 1936 the Navy decided to equip all cruisers and battleships with the Seetakt surface-search radar. The Navy specified that the sets were to be mounted on the rotatable housing of the optical fire control with its primary function to be the accurate determination of range, always the weak aspect of optical systems, with detection of ships and obstacles in darkness or fog a secondary purpose [37]. Blind-fire capabilities were certainly important but were to be postponed until made simpler for the operators.

During the remainder of 1936 and 1937 there was a great deal of experimentation on the naval sets, including sea trials. The inadequate range of the 50 cm equipment forced the wavelength to 60 cm, used in the set installed aboard the pocket battleship *Admiral Graf Spee* in January 1938. The next sets took the final Seetakt wavelength of 80 cm.

In adopting the first operational naval radar the German Navy appears to have been highly progressive, and surface commanders were to make excellent use of the new weapon, but this was more the result of the individual initiative that came to all commanders who encountered the powerful new eye, rather than from high-level interest or understanding. Radar had come from below, not been ordered from above. The sets were accepted and installed but GEMA had to learn by experience about the harsh environment of a warship with many early failures resulting. Worse yet, the Navy's operating and maintenance personnel were poorly trained and instruction manuals, even circuit diagrams, were, initially at least, not found aboard ship—the result of the obsession with secrecy and the attitude that equipment installed aboard ship had to be foolproof, would not fail other than through enemy action and would not require extensive technical knowledge by the operators [38]. No officers were assigned to supervise the operation and use of radar on German ships until near the end of the war [39].

Kühnhold continued to work closely with GEMA, contributing fre-

---

[8] See '*Bismarck* and *Prinz Eugen*, 18–27 May 1941' in Chapter 3.3 (p 124).

quently to design but never as a member of the company. He remained at the naval laboratory, which expanded with many new functions, moved to Kiel-Dietrichsdorf and changed its name to Nachrichtenmittel-Versuchs-kommando (NVK) at the beginning of 1939 [40].

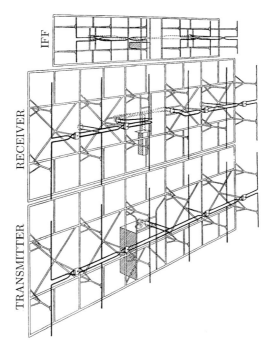

The feeder arrangement for a German Freya. This shows the antenna connections of three dipole arrays such as were typically found with modification on dozens of designs by all nations. The dipoles are spaced horizontally one-half wavelength apart and placed one-quarter wavelength in front of a conducting screen. German usage differed from other nations in using full-wave dipoles, which allowed them to be mounted without an insulator to the grounded frame at the voltage null found one quarter wavelength from the end. The receiver array is divided into left and right arrays to allow the increased accuracy of lobe switching. The upper-most array is for IFF; in this case the interrogation was by means of the radar signal with the reply on a shorter wavelength to which the IFF antenna responded. Air Ministry Air Scientific Intelligence Report No 34, April 1945. Crown Copyright.

The Kriegsmarine planned to use DeTe-I for coast-defense batteries as well as for ships, and this proved to be where the greater number of sets went. DeTe-II was the air-warning radar for their land installations and was not installed in warships. This was a curious attitude when compared with that of the US Navy, and may have had its roots in German naval thought not having been through the bomber-ship controversy of the 1920s (which at times gave the impression that the next war would be fought between the US Army Air Corps and the US Navy). Both models for land use were transportable. The existence of an excellent air-warning radar was, of course, secret, even from the other branches of the Wehrmacht.

The first knowledge of it obtained by the Luftwaffe came from a visit by Hitler, Göring and Raeder to the Torpedoversuchsanstalt (a general naval weapons laboratory) in July 1938 where DeTe-II was demonstrated. Göring was furious that his chief of signals, Col Wolfgang Martini, had neither been informed of the existence of this device nor invited to the demonstration. The response was that this was a Navy weapon; invent your own. The Kriegsmarine did all it could to keep the Luftwaffe from buying equipment from GEMA, a circumstance that no doubt contributed to the widespread belief that the firm was a front for the Navy. Until the end of the war they used all possible ways to undermine GEMA's growing and very pleasant relationship with the Luftwaffe [41]. Martini was delighted by radar and immediately ordered some DeTe-IIs, which acquired the name Freya [42].

In September 1938 Martini deployed two Freyas with GEMA personnel for possible use during the invasion of Czechoslovakia: one on Geisinger Berg in the Erzgebirge and another on Grosser Schneeberg in Silesia. There the restrictions of siting were first encountered. The mountain-top locations chosen resulted in a clutter of reflections from nearby peaks, the lesson being that the best location for such long-wave equipment was at the bottom of a shallow bowl-shaped depression [43].

Martini had organized the Luftnachrichtentruppe (Air Signal Corps) for the new Luftwaffe and would remain a competent and enthusiastic user of radar throughout the war. After graduating from school in 1910 he became a cadet in Telegraph Battalion 1 and acquired experience in military communications during the 1914–1918 conflict. His grasp of the use of radar was instinctive, but he repeatedly failed to make those higher up comprehend the full significance of such technical matters. Watson Watt possibly explained the matter:

> I have a very dear postwar friend in General Wolfgang Martini, a shy, modest, charming, very perfect gentleman. His many claims on my affectionate respect include his failure to endear himself to Göring from whom the qualities I have just tried to summarize may have concealed General Martini's very high technical competence, wisdom, and resource [44].

2.4.5. *Telefunken*

When Runge dismissed Kühnhold's suggestions that Telefunken enter the radar business it was not because he was not interested in very short waves. On the contrary they had become his main interest. Shortly before Kühnhold's visit the management of Telefunken had been reorganized; among other changes the Jewish director, Emil Mayer, had had to step

down. Jewish engineers, having marketable skills and international connections, began leaving the company during the next couple of years despite the protests of their less perceptive colleagues. One of their best, Otto
Böhm was designing antennas for Marconi within a year and later for the
British Admiralty [45]. The new leaders favored adding some research to
Telefunken's consumer-goods orientation, so Runge was ordered to begin
work on decimeter waves, incorporating a group currently active in the
field. Lacking other instructions they set about trying to develop navigational aids as well as communication radio. During these experiments
they noticed the disturbing effects of ships and aircraft passing into their
beams but took no great notice of the significance.

They settled instead on what was to become for the German forces in
the war a valuable communications technique, decimeter-wave relays between line-of-sight stations (Richtfunkverbindung). Such links were much
easier to set up in the field and used far less material than wire lines and
were almost as secure. These transmissions used frequency modulation
and proved especially useful in the steppes of Russia and deserts of North
Africa [46].

Wilhelm Tolmé Runge, who made Telefunken a great radar company,
was descended from a Bremen commercial family that provided him with
English uncles. He was the son of the eminent Göttigen mathematician
Carl Runge and the brother-in-law of the equally eminent Richard Courant
of New York University. The natural expectation that Wilhelm followed
the scholarly path such a family had established is mistaken. After his
father had placed him in three schools in the hope of finding one in which
he could succeed, he graduated only as the reward granted the excited
young men who volunteered in 1914 and received a war diploma. He escaped the fate of so many in the raw student battalions who demonstrated
the futility of rushing entrenched infantry with fixed bayonets, but his
younger brother, Bernhard, did not, and fell in October 1914 [47]. Wilhelm
transferred to signals and ended the war as an instructor at the Nachricht-
enschule (Communication School) at Spandau [48]. His wartime signals
experience steadied his interests, and his postwar studies at the Darmstadt
Technische Hochschule resulted in a dissertation that attracted attention
in the radio industry [49].

By the summer of 1935 Runge began to wonder what might be done
with reflected signals and set one of his 50 cm transmitter antennas directed
straight up with a receiver nearby. He then had one of Telefunken's Ju-52s
fly directly overhead while he watched the signal from the receiver. The
result was a large Doppler-effect meter deflection. His immediate superior
showed no great interest, which acted as a great stimulant on Runge. The
company did make a press release, however, that was picked up in America
by *Electronics* as 'Microwaves to Detect Aircraft' and later in *Popular Science*
[50]. Pulsing and duplexing (common antenna) with a paraboloid reflector
were quickly introduced. Tests conducted on the Baltic coast in the summer

of 1936 with the help of Wilhelm Stepp, who soon assumed the primary design responsibility, picked up an airplane but at only 5 km. The set was named Darmstadt for Stepp's home town and was the first in a series of geographical code names used by Telefunken.

This work was being carried out with the rather naive idea, possibly rooted in Runge's droll sense of humor, that it was navigational, and he found himself in hot water for giving an open lecture about the pulse technique at the Lilienthal-Gesellschaft in early 1939 that also caused Kühnhold to complain to Telefunken directors about this 'unauthorized work', which was futile anyway because of only 15 W of transmitter power. The result of this scolding on Telefunken's administration was to have their tube designers provide Runge the output triode that he needed, the LS180. It was the design that was popping up everywhere that positioned leads to minimize reactance at the price of inconvenience when installing the tube. It was 140 mm long, 50 mm in diameter and gave 8 kW, extending the range to 40 km for aircraft. Someone stuck a pin in the map, selecting in this way the name Würzburg. It was demonstrated to Wehrmacht authorities in July 1939. They soon used a rotating dipole feed, which Runge and Stepp called the Quirl, at the focus of the 3 m diameter paraboloid. This gave them a conical form of lobe switching that produced very accurate directional pointing and that made the various Würzburg models the best AA gun-laying radars until displaced by 10 cm equipment three or four years later [51].

The Würzburg had all of the key elements of World War II radar: a well defined beam with high directional accuracy, excellent range accuracy and a common antenna.

2.4.6. *Lorenz*

Telefunken and GEMA were not the only German companies to enter the radar business on their own. The Lorenz Company were not strangers to high-frequency work and had developed in 1935 a series of very high-frequency triodes. In 1938 Dr Hermann Berger left the laboratory Hollmann had founded on leaving GEMA and became head of Lorenz's tube development laboratory, leading to their RD12Tf becoming a better decimeter transmitter tube than either GEMA's TS6 or Telefunken's LS180, the result of having mastered the use of oxide cathodes in this demanding service [52]. They had marketed a beam navigation method of overlapping lobes designed by their engineer, Otto Scheller, whose remarkable 1907 patent encompassed every element of this widely used system [53]. It had been licensed widely abroad, even used by the Royal Air Force. These extensive international connections caused a certain amount of hesitation to entrust them with secret work. They had started collaboration with GEMA very early, but the Kriegsmarine had required GEMA in April 1936 to terminate it because of security concerns [54]; the Luftwaffe eventually overcame these concerns, given that the secrets were Lorenz's to start with.

The interest once aroused was not to be stilled, and Gottfried Müller insisted they press on. By the beginning of 1936 they had constructed a 70 cm, 400 W, pulsed set with separate, rotatable 'mattresses' for transmitter and receiver. From the top of their laboratory at Tempelhof they obtained reflections from the Berlin Cathedral, 7.4 km distant. During the course of the year the transmitter tube, DS320 (soon replaced by RD12Tf), that generated 1 kW allowed them with 1 μs pulses to extend the range to 14 km. In observing a windmill they noted the 'propeller modulation' on the returned signal, something to be of use in the not too distant future. They replaced the mattresses with paraboloids of 2.4 m diameter, changed the wavelength to 62.4 cm, increased the range for aircraft to 30 km, demonstrated it to Col Martini and gave it the name Kurfürst. Martini had it tested at the Anti-aircraft Training School and was sufficiently impressed to order a few sets of more advanced design for field evaluation: 20 each of Kurpfalz and Kurmark [55].

As work progressed, the design by Runge and Stepp began to show clear superiority over the Lorenz device for gun laying as well as the designated function in air warning, although the RD12Tf gave Kurfürst a greater range. Even without the Quirl the Würzburg A, demonstrated to military authorities in July 1939, proved sufficiently good to be used for firing on targets obscured by clouds or fog. Telefunken's design with rotating dipole, common antenna and simpler circuitry [56] was clearly superior, and Lorenz ceased production after a total of 40 sets [57]. Telefunken engineers then began working on equipment intended from the start for gun laying, which eventually became the Mannheim. In the meantime the basic Würzburg was provided with a Quirl to become Würzburg C.

German radar was concentrated for the moment in two companies, GEMA and Telefunken, each with a basic design. GEMA had two bands, 80 cm for Seetakt and 2.4 m for Freya. Telefunken used 50 cm for the Würzburg and the decimeter communication relays. Both companies brought out modifications and new models but remained true to the original, open-end designs for which wide variety became possible, from gargantuan early-warning to diminutive airborne sets. By this means they made economical use of their limited engineering personnel. Lorenz was not finished with radar, however, and would return with an excellent airborne sea-search set.

When the Luftwaffe became an independent arm of the Wehrmacht it obtained from the Heer (Army) the AA artillery. Initially, it did not have an organization to evaluate new weapons other than aircraft, so it relied on an army agency, the Heereswaffenamt (Ordnance Department) for judgement about its AA guns and, when the matter arose, for radar. This office arbitrarily classified the early sets into three types. For various reasons they referred to GEMA's Freya as A-1, to Lorenz's Kurfürst as A-2 and Telefunken's Würzburg as A-3. Initially this coincided with

A-1 for early warning, A-2 for searchlight direction and gun laying and A-3 for local observation and tracking respectively. Names were retained through numerous design changes and adopted for subsequent new models. A uniform type designation was introduced indicated by FuMG (for Funkmessgerät) followed by model number; Kurfürst became FuMG 38L (L denoting Lorenz), for example. As the importance of radar became more obvious, Göring wanted it added to his bureaucratic empire and had it moved from the Heereswaffenamt to the Reichsluftfahrtsministerium (National Air Ministry), at which time new designations were applied. The Reichsluftfahrtsministerium was not responsible for equipment of the wildly jealous Kriegsmarine, but the Marine nevertheless adopted the uniform nomenclature for radar.

2.4.7. Comments

It is scarcely necessary to point out to the reader the parallels in American and German radar work. The earliest work started in service radio laboratories with heavy emphasis on microwaves. Both dropped these wavelengths in their prototypes for want of transmitter power, although retaining some research. This resulted in excellent meter-wave equipment: XAF/CXAM for the US Navy, SCR-270 for the US Army, Freya for the Luftwaffe and Seetakt for the Kriegsmarine. The approach to decimeter waves by Bell Telephone Labs is remarkably similar to the path followed by Telefunken and probably came about because both had tube laboratories. The Bell FD/mark 4 was the equal of the Würzburg, indeed in design its cousin; it was with modification the US Navy's AA gun-laying radar throughout the war. One might speculate that this similarity among these industrial engineers resulted from close association of the circuit-design engineers with tube designers, thereby learning much earlier of the latest in tubes, whereas the engineers at the American service laboratories worked with generally available components. The Würzburg was a better gun-laying set than the SCR-268, and the American equivalent, FD/mark 4, was used only by the Navy. On the other hand the SCR-268 functioned also for distant target acquisition, which the Würzburg did not. The Germans were generally about a year ahead of the Americans.

In 1939 the German and American prototypes were superior to the British except for CD/CHL, which was a typical dipole array on 1.5 m. The heart of the radar defense of Britain was Chain Home, a design not admired abroad, but it worked, and here lies the principal difference between the German–American and British approach. Britain knew danger in 1935 and had the wisdom to realize that intelligence gained by radar was worthless unless promptly interpreted and acted upon. Speed was vital—as history has so amply shown—so a design was selected for which construction could begin almost immediately and with it the integration with Fighter Command. British production began in 1936, and the Royal Air Force took over an operational station in May 1937. Neither Germany nor the United States had a significant number of operational sets in September 1939 and neither had any organization to use them for air defense in any way comparable to Great Britain. Britain had by the start of the war a functional air defense system for the homeland.

Another distinction between the German–American approach and the British was the relationship of the radar men to their governments. Radar grew out of the experiments and plans of electrical engineers in both America and Germany and came to the attention of high officials through demonstrations. In Britain the pattern was inverted. The Tizard Committee actively sought new scientific weapons and, on finding radar and seeing the first examples of its power, procured for them a blank check. Possibly even more important was the interest at cabinet level from the beginning, which had important direct and indirect effects on both the technical development and the application. The German engineers had to contend with Hitler and Göring at the top, both anti-intellectual and scientifically illiterate. Churchill took an active, if not always helpful interest in the scientific conduct of the war. The American Secretaries of War, Henry L Stimson, and of the Navy, James Knox, were alert to the importance of radar and gave it high priority.

As the new technique moved toward deployment it carried various designations. A P Rowe named it RDF, thought to throw off the overly curious as 'radio direction finding'. The Signal Corps referred to it as 'RPF' for 'radio position finding', and the Air Corps, overcome with the need for secrecy, called it 'derax'. The name 'radar' was invented by S M Tucker, an American naval officer, and adopted officially by that service in November 1940 [58], and it seems that the world was just waiting for it. It was a composite with *ra* for *radio*, *d* for *detection* or *direction finding*, *r* for *range*, and the second *a* for *and*. The British ordered its use after 1 July 1943 [59]. The Australians have their own way with the language and called the new device a 'doover'.

2.5. FIVE OTHER NATIONS

2.5.1. Japan

Japan left the Washington Naval Conference of 1921 with limitations that required them to build fleets substantially smaller than those of Great Britain and the United States and was determined to offset this by improving the quality of the Imperial Navy. One response was the creation of the Naval Technical Research Department (NTRD, Kaigun Gijutsu Kenkyusho), construction of which was begun in 1922 and completed in September 1930, part of the delay resulting from the great earthquake of 1923. Its first director was Yuzuru Hiraga, a senior naval officer known as 'the god who creates warships' [1].

Yoji Ito was a career naval officer who was recommended for advanced study in electrical engineering in Germany in 1926 where he came under the tutelage of Heinrich Barkhausen at the Sächsische-technische Hochschule at Dresden, an appointment that also introduced him to Hidetsugu Yagi, Barkhausen's first Japanese student and by that time an internationally recognized authority on high-frequency radio. Barkhausen was loved for his genial nature, his admirable pedagogical ways and his exceptional skills in radio matters, things that attracted excellent students. By a curious turn of fate Ito and his brother Shigeru Nakajima visited him in June 1937 when the son of A Hoyt Taylor, Ito's American counterpart, was one of the Professor's students [2]. From the time of Yagi's student years Barkhausen maintained very friendly relations with his Japanese colleagues, culminating in he and his wife arriving in September 1938 for a long-remembered visit. The affection of the Japanese for him is found in Yagi proposing the 'bark' as the unit of amiability [3].

On his return trip to Japan in 1929 Ito chanced to meet Commander Ryunosuke Kusaka, who was returning by sea from a flight from Japan to the United States aboard the *Graf Zeppelin* (LZ-127). As a result of this meeting Ito was assigned to the newly opened NTRD where he took up work in radio-wave propagation [4] and soon became the Navy's leading electronics expert, a capability he successfully carried over into the postwar world. In conducting ionosphere soundings he noted the effects of the passage of aircraft, which started him thinking about radio location [5].

During the early 1930s various reports of electronic weapons came to the laboratory. In 1935 the proposal by an American to sell his invention for the radio detection of aircraft was discussed at length but rejected. When Ito visited him in 1937 Barkhausen told him that the German Navy had apparently succeeded in developing a device for measuring range at night. At about the same time he learned that Marconi was working on similar equipment in Italy. Ito passed this information immediately to the military attaché in Berlin for transmittal back to Japan, where it raised no interest that could be noticed. But some imaginations were primed; the commander of a Japanese cruiser sent to participate in the naval ceremonies for the crowning of King George VI observed some British exercises in which searchlights were brought onto aircraft uncannily fast, attributed by him to some radio location method that, of course, did not exist [6]. None of these alarms resulted in the beginning of a radio-location program at the NTRD.

Japan was to be the host of the Olympic Games in 1940 and wanted to make a better international impression than had Germany in 1936, which meant that even better television coverage would be required. In order to bring this about Masatsugu Kobayashi, an engineer heading the vacuum tube section of Nippon Electric (NEC, Nippon Denki), left in May 1938 on a tour of American and European television laboratories. He judged the British work to be the best but was not allowed to visit their laboratories. While in London he observed the strong effects on television reception caused by the passage of an airplane nearby. On returning to Japan he learned that the Olympics were off as a result of the war with China.

Kobayashi had hardly returned home when he repeated the London experiment unintentionally when examining signal strengths of a very high-frequency transmitter with a portable receiver; he noted a beat-frequency when airplanes passed between him and the transmitter. He quickly perceived what was happening and constructed the last of the 'radio screens' [7] that had already come forward in the US, France and Russia, but it was the only one that proved to be useful. It was developed as the Army's Type A Bi-static Doppler Interference Detector and was able to detect the passage of an aircraft at ranges exceeding 500 km [8]. Such equipment, which delivered neither range nor direction data, was useless in Europe and America where the density of flights would have saturated it but was of use in China, where few Chinese aircraft were encountered.

The Japan Radio Company (Nihon Musen) began research in microwaves in 1932 using the Barkhausen tube. The work was criticized for want of applications as being a waste of the company's resources and would probably have ceased had not Ito, who sought outside help whenever he could, offered a Navy Lab collaboration. This led to abandonment of the Barkhausen tube for the split-anode magnetron that Kinjiro Okabe had invented in 1927 [9]. This work culminated in a unique magnetron design capable of 500 W at 10 cm in April 1939. We shall return to this astounding discovery in Chapter 4.1 dealing with microwaves. For the moment suffice it to say that Ito finally got a radar program started, although one without high priority.

2.5.2. *The Soviet Union*

Whoever wishes to learn how governments fail in the duties of protecting their peoples from disaster should study the history of the Soviet Union; whoever wishes to learn how competent engineers can best be thwarted in their efforts to provide weapons vital for defense should study the history of Soviet radar. It is always the case, that in large projects those close to the details are vexed by the confusion and mismanagement they perceive in their leaders and, above all, in the administrative machinery that attempts to carry out their leaders' instructions. Such was the case in all radar development in the Second World War, but those who toiled in the laboratories of Britain, America, Germany and Japan and who suffered in this way little knew that their work places were ruled by reason and benevolence when compared with their counterparts in Muscovy.

There was every reason to believe that the Soviets might have surpassed the west in this new craft. They began first with high-level support, had the influential interest of academician A F Joffe, had a brilliant young electrical engineer and veritable model of the new Soviet man, Pavel Oshchepkov, and a radio engineer who had proved himself with decimeter waves, Yu K Korovin, as enthusiastic leaders, and had obtained financial support in 1934 of 300 000 roubles, which dwarfed that provided by any other power [10]. The initial work pointed toward the development of a Freya and a Würzburg, but by 1940 the resulting radar designs were poor, inferior to the Japanese, and left Russia dependent on Britain and America for much of her needs during the war.

To understand the history of this place and period one must learn the identities of the contending bureaucratic agencies, and in order to keep this cast of characters straight it is best to know them by their identifying abbreviations.

GAU The Main Artillery Administration, an engineering service of the Red Army concerned with the design of weapons.

PVO The Air Defense Forces, the service to which Oshchepkov was assigned and that had the responsibility for the employment of AA troops; they had interests in weapons design.

SKB (also KB-UPVO). A special construction office within the PVO to produce radar, opened in 1933 with Oshchepkov in charge.

VNOS The Aerial Observation, Warning and Communication units of the PVO, which were to be the immediate users of radar.

VTU The Military Technical Administration, a part of Red Army headquarters.

LFTI (also LIPT). The Leningrad Physical–Technical Institute, Joffe's organization, which included D A Rozhanski until his death in 1936.

LEFI Leningrad Electro-physical Institute, a GAU laboratory, led by A A Chernishev.

QRL (also TsRL) The Central Radio Laboratory, another GAU laboratory, led by D N Rumyanysev.

NII-9 Scientific Research Institute 9, another GAU laboratory that absorbed LEFI in fall 1935. The renowned radio engineer Professor M A Bonch-Bruevich became its director after the purges and attracted good men. Unfortunately, he died in March 1940.

UFTI The Ukrainian Physical–Technical Institute, a laboratory organized by Rozhanski and where research in magnetrons was conducted. Later directed by A A Slutskin.

NKTP The Research Sector of the Commissariat for Heavy Industry, the supervisory organization for both LIPT and UFTI.

NIIIS-KA Scientific Research Institute of Communication Engineers of the Red Army, a group with its own program for the development of signals equipment.

VEI All-Union Electrical Institute, a competent research organization with a laboratory for ultra-short waves led by Professor B A Vvedenskiy.

SRI (also NII-RP). Scientific Research Institute of the Radio Industry headed by A B Stepushkin.

The Academy of Sciences (mercifully seldom referred to by initials). An organization consulted at the highest levels that concerned itself with all manner of scientific and engineering problems relevant to the Soviet state.

NKVD The People's Commissariat of Internal Affairs, the secret police, the name of whose chief, Lavrenty P Beria, carried terror to millions.

At this point one might well let the reader form in his mind the kind of radar that was to come from the machinations of these agencies, all of which participated, and his construction would probably be close to the mark. Yet the poor Soviet product resulted as much from the purges that Stalin initiated in 1937 as from clumsy, bickering agencies—which knew how to use the NKVD for their bureaucratic ends and to take care of a few personal matters along the way [11]. Fear concentrates thought—but on survival, not on the subtle intricacies of electronic circuits. This no doubt lies behind the marked deterioration in design encountered in the second stage of Soviet radar development. Important parts of this story will remain unknown to us.

The PVO had responsibility for early warning and had sponsored the early work on the radio screen Rapid that was done at LEFI until that group was absorbed into the Television Institute, the combination becoming NII-9. Despite the objections from many that Rapid gave precious little data of value about intruding aircraft, PVO had LEFI build under the supervision of B K Shembel a model suitable for army deployment, which had its first tests in July 1934 [12].

The GUA also wanted radio AA gun-laying equipment and had been sufficiently impressed with the experiments QRL had conducted in January 1934 with a 50 cm set that they wished the idea exploited, and NII-9 undertook the task of providing a suitable prototype. The work was started under Shembel. By early fall 1936 NII-9 had produced an experimental continuous-wave twin-dish set, Storm, which operated on 18 cm using early magnetrons from UFTI that gave about 6 W of continuous wave power. The detection range was only 10 km, and the directional accuracy only 4°, neither adequate. The range problem was a compound of magnetrons with too little power and frequency stability and a noisy receiver that also picked up too much of the primary transmitter signal. Shembel devised a solution for the direction problem analogous to lobe switching and presaging mono-pulse radar. He used four dishes, one a transmitter and the other three paired off in horizontal and vertical coordinates [13]. The first trials failed, and he was unable to bring the concept to fruition before being separated from NII-9 in 1937 [14].

The skepticism that had met Oshchepkov's Rapid soon hardened into hostility to radio location generally in the form of a report in 1935 by the Red Army Chief of Signals, which asserted on the basis of studies by his own NIIIS-KA that radio location was unrealistic and a waste of time. M N Tukhachevskii, Chief of Ordnance, had been impressed with the possibilities of the new technique, even if it was not satisfactory at the moment, and decided in favor of retaining the infant radar program after a rousing fight. It was, as those familiar with the ways of bureaucracies will recognize, not the end but only the beginning. Life was becoming complicated—and dangerous [15].

In 1937 Army Commander A I Sedyakin conducted a large air defense exercise using conventional acoustical and optical methods that had a most unsuccessful outcome. He became acquainted with the new radio-location methods during discussions with General M M Lobanov of the GAU, who convinced him of the need to pursue this kind of work. This happy state of affairs came to a quick end in June 1937 when the purge swept military and technical ranks. Both Tukhachevskii and Sedyakin were quickly eliminated, and the NIIIS-KA instigated investigations of NII-9 and SKB with the resulting arrest of NII-9's chief and the dismissal of Shembel. Bonch-Bruevich, who had attracted Lenin's favor for his early radio work and who stood highest in electronic prestige, appealed directly to Central Committee Member Andrei A Zhdanov, who used Party influence to preserve the activities of NII-9. SKB was cleaned out and Oshchepkov, along with other radio engineers, went to the Gulag for ten years; he survived, thanks in no small part to academician Joffe, who sent him food packages and letters [16].

NIIIS-KA stepped into this to absorb NII-9 and SKB, and once the hated radar project was theirs their attitude changed; they completed the transformation of Rapid into an army prototype called RUS-1 (Rhubarb). It had a truck-mounted transmitter and two truck-mounted receivers that were normally placed about 40 km from the transmitter. These sets, of which only 44 were manufactured in 1940 and 1941, were technically not significantly advanced over the way Oshchepkov had left them and proved to be of little value [17].

Following completion of the design of RUS-1, LFTI set about building a pulsed air-warning radar, RUS-2 (Redoubt). It was bi-static equipment of 50 kW working on the 4 m band; transmitter and receiver were mounted on separate trucks having Yagi antennas that tracked one another in direction, although they had to be located about 1000 m apart, the obvious result of not having solved the common-antenna problem. The experimental set started in 1936 was not completed until late 1939, just in time for it to be tried in the Russo-Finnish War where it was successful enough for ten sets to be ordered on a crash basis [18]. RUS-1 failed the same test completely [19]. RUS-2 provides an exception to the general rule for design of meter-wave air-warning sets of the time in using Yagis rather than dipole arrays backed by conducting screens, the directional antenna immediately and instinctively adopted by others.

Victorious NIIIS-KA also continued the GUA gun-laying project, replacing NII-9 with UFTI. The principal deficiency of Storm had been the use of continuous rather than pulsed waves. Bonch-Bruevich had held the project to continuous waves, despite having used pulsed waves in early ionosphere studies; he even terminated pulse work at NII-9 when he became director in 1935 [20]. UFTI turned their efforts to a new, pulsed wave 64 cm design called Zenith. It combined every bad feature one could reasonably imagine in one set. It reported the coordinates of range, azimuth and elevation only every 17 s, making it useless for directing an AA gun, and had a dead zone extending out to 6 km, the result of the receiver being unable to recover from the transmitter pulse, although it could observe aircraft to 25 km. A pulse length of 10 to 20 μs gave correspondingly bad range accuracy [21]. Work continued and by the middle of 1940 the range had been extended to 30 km, but the equipment had such a catalogue of ills that it was given up [22]. The technical reasons for failure are not apparent. It would appear that the designers were unable to master the techniques of microwave electronics and thereby profit from the magnetron that N F Alekseev and D D Malairov had invented [23][9].

The purges had at least made one agency responsible for radar, NIIIS-KA, but in the process had removed good engineers from the laboratories and the most supportive top military commanders. Soviet radar entered World War II a low priority project with equipment inferior to all the major powers. Yet it need not have been so. The early start with high-level support, capable engineers and the cavity magnetron could easily have made the Soviet Union the leader in radar.

The reader must consider these simplified attempts at recounting relevant events in Stalin's state with suspicion. The material available is limited and was written before the collapse of communism opened secret files—and by men not indifferent to what history would record.

2.5.3. *France*

Pierre David's experiments with the 'radio screens' or 'barrages électromagnétiques' left him with ideas about how to make this kind of system work for air warning, and he devised a method of using multiple stations to determine direction and speed. For a pair of stations the observed Doppler shift, which can be measured accurately, depends on the aircraft course, speed, altitude and the angles from the airplane to transmitter and receiver—assuming a single target and straight, level flight. With multiple pairs of stations David proposed to use these data to determine direction and speed [24]. The system necessarily covered a large area, indeed this was considered its important virtue, but a large area meant it could be seriously confused if more than one plane or formation was present. It is difficult to see what was expected of a device that gave such rudimentary data acquired with such restrictive assumptions.

---

[9] See Chapter 4.1 (p 157).

Despite these restrictions Société Française Radio-Électrique (SFR) agreed that David's meter-wave 'barrage' had more possibilities for air defense than their own 16 cm equipment and submitted a proposal in March 1935 for using the longer wavelengths. It received a Navy appropriation of F70 000 in April. David tested this Doppler system, which was easy to manufacture, in the summer of 1936 on the Loire to the south of Paris. It detected aircraft and estimated their speeds and indicated the directions of some, results good enough that the Army decided in September to adopt the scheme. The result brought F60 000 from the Army for twelve stations placed along two lines around Rheims and in the Argonne [25]. The Navy constructed stations at Cape Martin and Cape Camarat and planned stations at Cherbourg, Brest, Toulon and Bizerta. In a manner unique to radio location projects world-wide the equipments were made by the same firms for both Navy and Army: transmitters by Etablissements Kraemer, receivers by Societe Anonyme des Francaise Radiotelegraphiques (SADIR). Twenty mobile sets were used in the 1938 maneuvers.

David was well aware of the shortcomings of the Doppler method and proposed building a pulse echo system in October 1938. There was not much enthusiasm for this project until the British disclosed their radar work to their prospective French allies through a visit by Watt and Wilkins in April 1939. Orders were placed for pulse radars with a 6 m, 12 kW transmitter produced by SADIR in October, which detected aircraft at 60 km [26].

Henri Gutton continued the work with 16 cm radiation. From his experience at sea on the *Normandie* he concluded that ranging information was essential and set about constructing pulsed equipment, which he did at a land station. He replaced the Barkhausen tube with a split-anode magnetron and by early 1938 had constructed a station on the coast using the paraboloids from the ship with 16 cm wavelength and 6 μs pulse duration. By March 1939 he could range a 3000 ton vessel at 6 km to an accuracy of about 250 m. The outbreak of war canceled plans to mount the improved equipment aboard the *Normandie*. It served as a component of the harbor defense of Brest but was too weak for detecting aircraft at any reasonable range [27].

2.5.4. The Netherlands

Dutch radar can be said to have originated from 'death ray' concerns much as had the British. Such concerns had raised questions in Parliament in 1926 and had helped secure the establishment of the Physics Laboratory of the Netherlands Armed Forces at The Hague. Such fantasies were quickly shunted aside to be replaced by research in communication and in various forms of aircraft detection. A radio for artillery spotting that worked on the 1 m band gave rise to what must now be considered a traditional accidental observation: a transmitter and receiver separated by a sand dune were unable to communicate until an airplane passed overhead. These 1937 observations led almost immediately to the construction of a 70 cm pulsed set that used Telefunken type RS297 output tubes driving a 16-dipole array with common transmission and reception. Its power of 1 kW sufficed for the ranges needed to aid searchlights.

It was a neat little set. The antenna moved in azimuth and elevation and was mounted on a small, easily transportable mount that allowed elevating it with a handle and turning it with a bicycle drive, so arranged that the operator could see the indicator tube. The first prototypes were ready for inspection when war broke over The Netherlands.

The Philips Physics Laboratory at Eindhoven began working in 1933 on applications for the split-anode magnetron that they had invented and that had got Kühnhold at the German Navy's NVA thinking seriously about radar. The result was experiments with paraboloid reflectors that began to provide reflection phenomena of diverse sorts to stir the engineers' thoughts. Trials of pulsed radar went off and on during subsequent years and were taken up seriously again in 1939, too late to provide anything for use in defending the lowlands.

As soon as the invasion was seen to be fatal the two Dutch radar men, C H J A Staal from Philips and J L W C von Weiler from the Armed Forces Lab, were sent with critical components to England where von Weiler joined the staff of HM Signal School [28].

2.5.5. Italy

Italy had established in 1916 the Regio Instituto Elettrotecnico e della Communicazioni della Marina (Royal Institute for Electro-technics and Communication, RIEC). Inspired by Marconi's work with very short waves Professor U Tiberio submitted a report in 1935 proposing that RIEC institute a program for radio location. An experimental 1.5 m continuous-wave set, EC1, was tried in 1936 with improvements the following year. These tests showed that one must use pulsed waves for anything practical, which led to such a set, EC2, that proved to be unsatisfactory. Work was discontinued until the shock of conducting war without radar drove the belated lesson home [29].

PHOTOGRAPHS: AIR WARNING—EUROPE

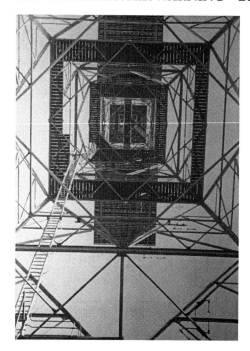

Looking up the axis of a CH transmitter tower. The platforms and gantries are evident. The ladder, beginning at the lower left, required a bit of mental conditioning for some of the men and women who worked these stations but was an exhilaration for others. Historical Radar Archives, RAF photograph. Crown Copyright.

Interior of a CH receiver-operations room circa 1943. The receiver is located at the left rear with one operator, those at the right evaluate the data before transmission to the Filter Room. Historical Radar Archives, RAF photograph. Crown Copyright.

Fighter control with radar: Britain. It was realized early that radar sightings of enemy aircraft were of no value, if the information was not evaluated within minutes and transformed into orders for fighter squadrons. It was in this that Britain was far ahead of any other country in September 1939, when a complete functioning air-defense system was in operation. Shown here is the Operations Room of Fighter Command. It also illustrates how Britain used women in air defense. Historical Radar Archives, RAF photograph. Crown Copyright.

German Fahrstuhl Freya (FuMG 43). This version of the Freya was devised in the field by Lieutenant Hermann Diehl in 1940 to study experimentally by lifting the dipole array and thereby altering the vertical lobe structure. 'Fahrstuhl' can be translated as 'elevator' or 'lift'. When used near the ocean it provided height data. National Archives photograph 111-SC 269066.

German Freya LZ (FuMG 401). This 2.4 m set, known originally as DeTe-II, appeared in numerous modifications that varied as application required, this one was referred to by British intelligence as 'Pole Freya'. Its electronics formed the basis for nearly all of Germany's air-warning radar. The basic Freya shown here had 12 vertical dipoles on the lowest (transmitter) antenna with an equal number on the middle (receiver antenna). At the top is an IFF antenna. Freyas quickly incorporated lobe switching for directional accuracy. This model is transportable by air, LZ standing for Lufttransport-zerlegbar. Reports are that the operator's cabin was designed for maximum discomfort. National Archives photograph 111-SC 269043.

The German meter-wave phased-array radar Wassermann M II (FuMG 402), one of various models. The series was referred to by British intelligence as 'Chimney'. This set used the same 2.4 m electronics as the Freya but gained range and height-finding capability through the use of a very large dipole array. This immense array formed a thin, horizontal fan-shaped beam that could be moved up and down by altering the phase shifts in the transmission lines feeding the dipoles. This allowed a direct and rapid way of determining elevation to an accuracy of 0.75° for aircraft lying between 3° and 8°; lobe switching gave an azimuthal accuracy of 0.25°. At 100 km this localized the target in a box-shaped region 1200 m high, 300 m long and 435 m wide for planes above 5200 m. Aircraft 8000 m high could be detected at 210 km, and very high fliers at 300 km. It was capable of panoramic operation. National Archives photograph 111-SC 269031.

The German meter-wave phased-array radar Mammut (FuMG 52), one of various models. The series was referred to by British intelligence as 'Hoarding'. Like the Wassermann, this set used the same 2.4 m electronics as the Freya but gained range and directional accuracy through the use of a very large dipole array. This immense array formed a thin, vertical fan-shaped beam that could be moved from side to side by altering the phase shifts in the transmission lines feeding the dipoles. This allowed a direct and rapid way of determining direction to an accuracy of 0.5° over a ±50° sector. There was a dipole array on both sides that allowed front and rear coverage, which gives the antenna a complicated appearance. National Archives photograph 111-SC 269022.

CHAPTER 3

FIRST CLASHES

3.1. WAR IN EUROPE

By August 1939 Chain Home was providing continuous operational coverage for portions of the coast that offered probability of air attack from Germany. At the beginning of that month, so heavy with foreboding, what must have seemed a ghost of the previous war approached Britain. At an extreme range a very large blip appeared on the CH screens. It was an airship, but no air raid alarms were sounded nor fighter squadrons scrambled. For whatever fears it might have raised a quarter century earlier, it was no longer a bombing threat, but its mission was military and the radar operators knew what it was.

It was LZ-130, named *Graf Zeppelin*, the sister ship of the *Hindenburg* (LZ-129), which had ended commercial flying for the International Zeppelin Transport Company in a catastrophic fire at Lakehurst, New Jersey in May 1937. As a consequence the older *Graf Zeppelin* (LZ-127) had been removed from the incredible Frankfurt–Rio de Janeiro service that had fascinated the world, and passenger service that had started in 1910 came to an end. LZ-130 was completed while arguments raged about what to do with it and the older *Graf*. In any case, test flights were made, but the one approaching England that August was on charter for Colonel Wolfgang Martini for electronic espionage. There had been other such trips, east and west, and now it was Britain's turn. Evidence of radar would definitely interest the colonel.

Martini selected this mode of transport because of its range and weight lifting capacity, for he loaded it with radio equipment and technicians under the supervision of Dr Ernst Breuning. He also liked its leisurely pace, which allowed time to examine any intriguing signals. A gondola with antennas could be hung far below the main airship body in order to reduce the effects of the huge aluminum frame on the radiation patterns and had been used on three earlier flights but does not seem to have been used on this one [1]. After following a course well clear of the Lowlands it turned north at a point about 45 km east of Lowestoft, a consequence of the weather in the Channel, and proceeded up the coast using clouds and overcast to stay out of visual sight [2].

Flight Lieutenant Walter Pretty was on duty at Fighter Command Operations Room and recognized the data being reported to him as coming from an airship. The notion that the flight was espionage occurred quickly, but the Air Force personnel already knew a fundamental law of electronic warfare: when you see a suspected electronic snoop on your radar, he has already seen you, so if you turn off your set you have given him proof of your function.

All stations within range followed the ship, which was above cloud cover, as it proceeded up the coast, and Pretty noted it straying over land at Hull. This was certainly unintentional, but given the difficulties of aerial navigation at the time and the extreme uncertainties placed upon the navigator by the effect of unknown winds, it was not an excessive error. When the ship made a transmission in the clear giving the position that CH had shown seriously at fault, Pretty amused himself with the impossible idea of sending a helpful message to LZ-130 informing them of their true position. Course was then corrected, presumably through a solar fix, and the trip continued off shore [3].

A newly established station near Dundee, equipped temporarily with a Mobile Radio Unit, picked up the large-amplitude signal in its turn at a range greatly in excess of 150 km and well to the south on 3 August. They first thought it to result from a large group of planes but ruled that out because such a target would generate a highly variable return and this one was quite steady. Their report of the strange phenomenon yielded nothing more revealing than the command to keep plotting. The airship then broke through the overcast near Aberdeen and became visible to civilians. Squadron Leader Findlay Crerar of Dyce Station of the Auxiliary Air Force took off with an observer and returned with an excellent photograph of the beautiful craft (the tail insignia excepted) [4].

The trip continued far enough north for examining any signals that might have come from the Scapa Flow naval base but none of the Thames estuary. The original route would have covered these stations and the entire southern coast, but the weather—England's old ally—intervened, causing the airship to pause near Bawdsey as it changed course. Rowe accepted this as proof that Bawdsey was marked for destruction when war began [5].

There have been a dozen versions of the zeppelin espionage flight, frequently conflicting and that have been eliminated by subsequent research, but all agree that Martini found no sign of British radar. The suspected reason is that no one in Germany conceived of radar on such long wavelengths, of such a low pulse repetition rate or of radiating over such a wide front as Chain Home operated rather than forming a 'searchlight' beam. There is reason to believe that Breuning's receivers were in fact overloaded with what he thought was the German ionosphere soundings that worked on 50 Hz, easily confused by non-linear receiver response with the 25 Hz repetition rate of CH [6]. Indeed, the original proposal for CH intended that its radiations be perceived as ionospheric soundings [7]. Had CH ceased operation on sighting the dirigible, the puzzling nature of the radiation received might have been revealed. When questioned about the incident after the war Martini was startled to learn that LZ-130 had been tracked, as Bruening had reported no radar [8]. He did observe some test transmissions of the new VHF (greater than 20 MHz) radio equipment being installed in British fighters [9].

When Britain declared war on 3 September 1939 CH was providing radar coverage from Aberdeen to Southampton, but the expected air attack did not come, although a confusion in the filter center, a consequence of excessive back-radiation from a CH station, led to a fight between Hurricanes and Spitfires three days later in what became known as the Battle of Barking Creek [10].

The greatest radar activity, however, was a precipitous evacuation of the Bawdsey Research Station. The move had been foreseen by Rowe and Watt as war approached because of their certainty that the activities had been sniffed out, were considered vital and would be blown to bits. A committee had been formed to look for a new location and suggestions discussed, although no site inspections undertaken, but their ideas became irrelevant because earlier Watson Watt had in a casual fashion extracted an agreement from the Vice Chancellor of the University of Dundee (Watt's alma mater) to give the laboratory emergency accommodations, a place that had never even been considered by the committee. Rowe's arrival at Dundee met an incredulous Vice Chancellor who had forgotten or misinterpreted Watt's request and had only two rooms for hundreds of people and many truck loads of equipment. There had been no advance party, no proper planning, just headlong flight. The effect of this move was to set back a number of projects seriously. Unable to work, many sat around completely dispirited. Space was eventually found in the Teachers' Training College, which had no alternating current power [11].

As it proved, the fear was unjustified. Bawdsey had an operational CH station and was not the recipient of an attack until mid-1940. An administratively correct name replaced Bawdsey Research Station: the Air Ministry Research Establishment (AMRE), making another break with a heroic past. Located with them was the former Bawdsey 'Army cell', now the Air Defence Research and Development Establishment (ADRDE), which avoided the Dundee problems by moving to Christchurch.

There was one important exception to the Dundee move. Activities at Bawdsey were split between research and responsibility for the engineering and construction of CH. Edward Fennessy, representing the chief of this latter group, H Dewhurst, protested with vigor that placing the base of their activities at Dundee would make continued construction as well as vital maintenance and support nearly impossible, owing to the constricted transport and communications that connected the north of the island with the south. Given two days to find an alternative, Fennessy selected Leighton Buzzard, a market town 80 km north of London and near the Air Force communications center having land lines to CH [12].

The airborne radar work being carried out at the Air Force base at Martlesham Heath had increased by summer 1939 to include ten experimental aircraft with more being outfitted and a substantial complement of civilian and Air Force personnel. To the enormous anger of E G Bowen they too had to decamp for a civilian airfield at Perth based on arrangements also made by Watt. The station manager was already hard pressed for space, for he was giving basic flight training to Air Force men and had no room to spare. He and Bowen soon worked something out that allowed both to get on with their duties until a promised new station could be found for airborne radar. The new station was the Air Force field at St Athan on the south coast of Wales about 20 km west of Cardiff, which was occupied at the end of October. It had space but no heat, being hangars, often open to the elements. Fabric partitions established working areas, and overcoats and gloves provided what warmth was to be had [13]. They escaped the effects of a 500 kg bomb through a defective detonator [14]. So much for the wisdom of evacuation.

It is doubtful whether Rowe's bureaucratic personality ever clicked with Bowen's, but after the move from Bawdsey it reached a low from which it continued to sink with Bowen transferring his radar work to America. After nearly 50 years Bowen's anger spilled over the pages of his memoirs. '... we were the principal Air Defence Establishment in the country. We were supposed to know when the bombers were coming and where they were heading and we had already devised a credible system of sending fighters against them. If we were the first to cut and run, what about the fate of the rest of the population [15]?' Bowen's remarks stand up well given the course of the war but would never have been written had German intelligence known Bawdsey's function and had Göring had the understanding to act upon it. The tarrying of LZ-130 should have given anyone pause.

The work conditions at St Athan were appalling but were marked by one singular piece of good fortune. The first polyethylene co-axial cables became available during those dark months, simplifying the construction of airborne radar tremendously. A second morale boost came from a visit by King George VI and Queen Elizabeth, made all the better by a royal understanding and appreciation of radar.

The AI and ASV engineers also learned about that time of the first 'micropup' design, a high-power, high-frequency triode formed by joining copper electrodes with metal–glass junctions (Housekeeper seals) that allowed external air cooling. The first, VT90, was tried in an experimental airborne set that generated 5 kW peak power at 1.5 m and engineering enthusiasm. By early 1941 the most widely used model, NT99, began production initially at a rate of a few per week. These tubes used oxide-coated cathodes, which allowed much higher emission compared with tungsten or thoriated tungsten. Experimenters learned a very happy and unexpected characteristic: oxide-coated cathodes gave substantially more current for microsecond operation. This resulted in a push–pull circuit that generated 150 kW and found wide application both in the air and on the ground [16].

During that time Bowen and his staff also trained squadrons in the use of AI (airborne interception), the equipment to play such an important part in the Blitz—although not effectively until May of 1941—but these first squadrons were not allowed to hone their skills with the difficult equipment or devise tactics for a completely new weapon. One particularly adept squadron was sent to help the Finns hold off the Russians and was wiped out. The other was thrown into the RAF sink that Belgium and France became in 1940 [17]. The introduction of AI marks I and II also suffered from having been put into production before they were adequate for the job. Training and maintenance were even worse with no technical literature or spare parts, a condition Bowen blamed on the inadequate support given the airborne work by Rowe. The situation caused Hanbury Brown to compose an addendum to Watson Watt's rule of 'give them the third best' that ran 'but don't give them the fourth best because it encourages them to throw the whole thing out' [18]. The design quickly proceeded to AI mark III, which eliminated some of the problems but still had an unacceptable minimum range of 330 m.

By 1939 the engineers at Electrical and Musical Industries, Ltd, a name long forgotten and replaced by the initials EMI, realized that something was going on in the matter of radio location, possibly through Bowen's use of their tuned radio- frequency, amplitude-modulated television receivers, and this brought Alan Dower Blumlein into radar design. Blumlein was Britain's foremost and certainly most versatile electronics engineer, the person most responsible for London having the first high-definition broadcast television in 1936. On its own initiative the company quickly offered the government a 5 m gun-laying set with characteristics similar to GL mark I. This disposed of any constraints on EMI people designing radar, and in December 1939 the company was asked to help with AI [19].

Blumlein saw the difficulty in attaining an acceptable minimum range, which was the problem of producing sufficiently narrow pulses. The method initiated at Bawdsey was to form the pulse of radio frequency signal and then amplify it, and it was this amplification that did not allow sufficiently rapid rise and fall of the pulse. Blumlein's method was to switch the high direct voltage that supplied the oscillator, and rapid switching of high voltage was a part of the television engineer's stock of circuits [20]. His transmitter–modulator reduced the minimum range to 130 m and was incorporated into AI mark IV with first installation in August 1940.

Air-to-surface-vessel radar (ASV) progressed more satisfactorily. By January 1940 this set, using the 1.5 m wavelength band as did AI, was sufficiently well advanced for three squadrons of Coastal Command to be outfitted with equipment manufactured by E K Cole Ltd (transmitters) and Pye Radio Ltd (receivers). It could spot a surfaced submarine at a tactically reasonable range, although not the simplest piece of equipment for new operators to use. Brown arrived at Leuchars, a station south of Dundee whose planes patrolled the North Sea, thinking he would be 'welcomed with open arms', but such was not the case. Military people in times of mobilization are sufficiently harassed that anyone who attempts to add to their responsibilities is seldom welcomed with open arms. Brown succeeded in equipping the squadron's Hudsons and instructing the crews in their operation, but it was the squadron commander who caused the equipment to be valued and used.

Aerial navigation in the habitually bad weather of the North Sea made even returning to the base problematic, and planes were frequently lost. Squadron Leader Sydney Lugg set up at the base a radar beacon, the modification of an IFF mark II, a device designed to return a strong signal to an interrogating radar, originally intended to identify an aircraft as friendly[1]. The result was immediate, and Coastal Command made wide use of it with crews giving studious attention to mastering their new equipment [21]. The beacon was immediately named 'mother'—home to mother! By the end of the war radar beacons were important elements of aerial navigation. ASV mark I also proved very helpful in locating convoys that were to be guarded and in finding coastlines, but its effectiveness against submarines was poor [22].

For whatever acceptance ASV mark I had acquired it was a poorly engineered set and St Athan was a terrible place for engineering design, so the work was transferred to the Royal Aircraft Establishment at Farnborough in February 1940, the beginning of the disintegration of Bowen's group [23], which finally ended when Bowen went to Swanage in June to work on microwave AI[2].

Gerald Touch at Farnsworth, helped initially by Brown, completed the design of ASV mark II, one of the outstanding radars of the war. The antennas were so arranged that two lobes interrogated for searching, one to the left, the other to the right. This allowed a swath 40 km wide to be observed. When a target was located the pilot turned 90° and was guided to it by forward-looking antennas operated with lobe switching to give accurate direction. Pye and E K Cole received orders for 4000 ASV mark IIs in the spring of 1940, but German bombs and the pressing need for AI sets delayed production [24].

The doctrine of the Royal Air Force required that a war be initiated by a 'knock-out blow', but all that was achieved was a gesture, one certainly not noticed by the Poles, who were fighting a desperate, radarless war.

---

[1] For details of IFF, see Chapter 3.4 (pp 129–35.
[2] For details of the move to Swanage see Chapter 3.1 (pp 99–100.

Minutes after the declaration of war a reconnaissance flight disclosed warships near Wilhelmshaven, which were then attacked by three squadrons, only two of which found the target. Bombs struck but did not explode, and half of the attacking planes failed to return [25].

Neither side wanted to initiate bombing civilian targets, so Bomber Command restricted itself to night flights over Germany dropping propaganda leaflets and continued daylight attacks on warships in the Helgoland Bight. This was where radar drew first blood. Leutnant Hermann Diehl of the Luftnachrichten-Versuchs-Regiment Köthen (Experimental Air Signal Regiment Köthen) had supervised the erection in October of a station on the island of Wangerooge and begun experimental work, specifically examining the vertical lobe structure of his radiation patterns, which he decided was the cause of the peculiar disappearances of a target as it moved toward or away from the station [26]. Everyone was learning about vertical lobe structure, but only Watt and Wilkins had put it into an original design.

On 18 December he observed bombers at 114 km, but because of earlier false alarms this vital information was not taken seriously until the 22 Wellingtons were also reported by a Navy Freya on Helgoland. The scrambled fighters then opened communication with Diehl, who successfully guided their pursuit [27]. Of the attacking aircraft only ten returned, of which three made forced landings [28]. This incident made a distinct impression on certain elements of the Luftwaffe and initiated the development of a system for vectoring fighters into attack positions.

Diehl became very active in this work. The 25-year-old lieutenant had an abundant supply of confidence in himself and the fighter control method he had discovered and found Martini at his side when obstructionists blocked the way [29]. He invented, after devoting a lot of thought to the consequences of reflections off the ocean surface, the Freya-Fahrstuhl (Freya Lift) in which the antenna could be raised 20 m on a vertical track so as to alter the lobe pattern and allow better estimates of the target height [30]. He also worked with GEMA to install lobe switching for accurate horizontal direction [31]. At upper levels of the German command radar was given little thought—it was a defensive weapon and Germany was on the offensive—but Diehl, Martini and their associates at GEMA were doing a lot of thinking about early warning radar. Freya was to evolve into Wassermann and Mammut, unquestionably the best early warning radars of either side until mid-1944 and probably for the entire war. The approach to estimating height did not, however, make use of vertical lobe structure. Wassermann was given an enormous vertical array of dipoles that formed a beam so thin that it could fix the altitude of the target more accurately and with much less confusion than by noting the change in echo amplitude as the target entered and left vertical lobes. In August 1944 Diehl was an Oberleutnant and a very successful fighter control officer in the west, a curious advance for one so technically gifted [32]. The American Army never put into practice any of these improvements on SCR-270 and 271, which were very similar to the first model Freya.

By the end of 1939 GEMA had produced eight Freya and four Seetakt sets [33]. By March 1940 the Luftwaffe had established a chain of 11 Freya stations for air warning along Germany's western frontier, starting at Stylt in the north and ending at Freiburg in the south, and had integrated them into the visual and acoustical observer system. These Freya stations practiced their skills during the Phoney War by vectoring fighters onto enemy aircraft over France, little realizing how valuable this knowledge would be during the coming years. The Freyas were entrusted to the ground observer units in the Luftnachrichtentrupp (Air Signal Troop), which had gained non-radar experience in the Spanish Civil War [34].

If the repulse of the December RAF attack opened a few Luftwaffe eyes to radar, it did nothing to alert Britain that her enemy possessed such equipment. Indeed, it was not until 24 February 1941, when aerial photographs of a Freya had been added to electronic evidence, that Lord Cherwell, the Prime Minister's scientific advisor, was convinced that the Germans had radar [35]. R V Jones, a young man assigned to scientific espionage, had become convinced much earlier, in no small part as the result of a mysterious document mailed after dramatic preliminaries to the British Embassy in Oslo. This Oslo Report from a 'German friend' described a number of German technical secrets including radar, although it incorrectly ascribed RAF losses in a North Sea attack in September to radar, a time when none was installed. Other British intelligence experts thought it a plant to deceive, but Jones continued to use it successfully as a guide [36]. Jones, naturally curious as to the identity of the 'German friend', finally met him years later through a chain of events stranger than those that led to the receipt of the report itself [37]. He was Hans Ferdinand Mayer, who had been a physicist for Siemens and Halske and who was imprisoned later in the war for incautious talk.

If Bomber Command was not disposed to see the December disaster as evidence of German radar, it was disposed to see it as evidence for the futility of daylight attacks, for this had been simply the worst of the handful of raids made over the North Sea, and the others had entailed losses too heavy to be sustained. At the same time Bomber Command had been covering the German countryside at night with leaflets and experiencing very light losses. The lesson seemed clear: strategic bombing will have to be carried out at night. Whether this extra ration of toilet paper had landed on the cities for which it was intended was, however, unknown [38].

A sortie of the German battle cruisers *Scharnhorst* and *Gneisenau* [39] for a few days in late November, both of which had just received Seetakt sets, was the way Admiral Raeder wanted to deal with British shipping. Konteradmiral Dönitz had a different vision, but it was a view not shared by many at the Kriegsmarine because the technical capabilities of the new U-boats were not significantly improved over those of 1918, and the U-boats had been roundly defeated in that year. There was much more confidence in surface raiders, which had the additional advantage of sinking or taking their prizes according to accepted international rules. Hitler did not want to bring the United States into the conflict as unrestricted submarine warfare had done 22 years earlier.

The sortie succeeded in sinking the armed merchant cruiser *Rawalpindi* with accurate initial fire, which should have been ranged by radar, but the new equipment had failed shortly before the engagement [40]. The two raiders evaded their pursuers, who were not radar equipped [41]. For the two battle cruisers to have sunk a much weaker adversary was a bit of historical inversion. In 1914 two cruisers of Vizeadmiral von Spee's squadron carried the names of the two modern vessels and had been sunk off the Falkland Islands by much heavier British naval artillery.

The pocket battleship *Graf Spee* with attendant supply ships had put to sea in late August and was well into the South Atlantic when war was declared. This class of warship had much greater cruising range than ships planned for bruising fleet actions in the North Sea, and she was to demonstrate this characteristic in sinking nine merchantmen before being brought to action off the River Plate by one heavy and two light cruisers on 13 December 1939. In that action her fire was quite accurate throughout the engagement and the attacking cruisers were badly damaged without inflicting serious harm on their stronger adversary. The consistent accuracy of the *Graf Spee*'s fire indicates that the crew had mastered the arts of maintaining their radar and had adapted it to the harsh environment of a warship, a reasonable assumption given that she was the first warship equipped with operational radar.

It had not been easy. On 3 October the set had failed and had resisted attempts by the operator to repair it—a task not made easier by his very brief training and the absence of technical manuals and circuit diagrams— they were too secret to be allowed aboard ship! The importance this new device had assumed is made clear by the captain's order to his chief radio mate to drop everything until the radar worked again, which succeeded after about a week [42].

The *Graf Spee* took the temporary refuge allowed in the neutral harbor of Montevideo where her captain decided from various sources that a much greater force was assembling outside the harbor. An erroneous sighting of a presumed heavy unit confirmed his fears and caused him to scuttle his ship [43].

The Admiralty sent L H Bainbridge-Bell, one of the early Bawdsey team, to examine the wreck; he noted the radar and made a full report to the Admiralty, where its significance seems to have been lost on everyone but Jones [44]. There is evidence that it was suppressed for fear distribution might compromise Britain's great secret [45].

The Phoney War, known to the other side as the Sitzkrieg, ended on 9 April 1940 with the invasion of Denmark and Norway. Britain had naturally been displeased that iron ore was passing into Germany from

Sweden through the Norwegan port of Narvik thence by a sea route that followed the coast within Norway's own territory. Churchill, at the time First Lord of the Admiralty, had proposed planting mines in these waters as early as September 1939, thereby forcing the German vessels out of neutral water where they could be attacked. This had been suggested in 1918 but rejected on the grounds of violating the sovereignty of a friendly, neutral state. By spring the Cabinet had been brought around to Churchill's view. It was assumed that this action would provoke German intervention, and to forestall this British troops prepared to invade first. Admiral Raeder had alerted Hitler in October of the advantages of having Norwegian naval bases, but the dictator had preferred to leave Norway neutral. Diplomatic and military activities during the ensuing months made it clear that Swedish iron was going to bring on trouble, so German war plans proceeded accordingly. Hitler did react—and rapidly—to the British moves [46].

The invasion of Denmark and Norway furnishes the historian with naval action in which German skill and audacity together with British slowness and bungling brought Norway under Nazi domination in the face of overwhelming naval superiority. The campaign has much of interest, including the first airborne assault, but little in the way of radar. It did include the air lift of three Freyas to southern Norway: at Stavanger, Mandel (near Kristiansand) and Bergen [47].

Radar's part was restricted to the early morning of 9 April. The *Gneisenau* and *Scharnhorst* were covering a group of destroyers transporting the troops that occupied Narvik. At 0450 hours the radar of the *Gneisenau* reported a target to the southwest at a range of 25 km, but the commander waited for optical sighting, as the Seetakt equipment could not provide direction accurate enough for blind fire; he was also a little skeptical, as there had been trouble with the set earlier. Visual contact clarified the matter but also gave the advantage of light to HMS *Renown*, the target, which fired the first, well ranged salvo despite the want of radar [48]. The German vessels replied accurately, but lost the use of their Seetakt sets to the first shock of their own artillery [49]. The landing was safely ashore, so the Germans broke off action. Neither side suffered serious damage. The two battle cruisers encountered the carrier HMS *Glorious* returning from Narvik and sank her and two escorting destroyers quickly with accurate initial fire at 25 km. Presumably Seetakt was functioning.

The Royal Navy had radar installed on three anti-aircraft (AA) cruisers that were used to protect the British troops put ashore at Narvik from air attack, the landing party having no AA units. They found the cliffs of the fjords made their radar useless. They also found this a poor way to use these specialized warships, one of which was sunk [50].

When the British Expeditionary Force went to France at the outbreak of war, air defense had two components: one to function in the forward areas in support of the Army, the other to defend ports and industrial centers in cooperation with the French. For the latter a method of control similar to that of the Home Defense was sought, and to this end a left-flank group of MRUs began their emplacement in November just behind the Belgian border. The stations at Boulogne, Calais, Lille, Arras, Cambrai, Aresne and Sedan together with the station at Dover reported to a filter center at Arras. As equipment became available a right-flank group was extended to Verdun, Mount Haut, Bas le Duc and Troyes reporting to Rheims.

To function effectively these stations had to be linked to the filter centers that in turn had to be connected to Headquarters Fighter Command and the War Room in Paris. This proved more difficult than setting up radar stations. The telephone lines were inadequate both in number and quality; they were noisy, overhead lines that were easily put of action or sabotaged. It was a sorry contrast to the excellent system used in England.

The French radio screens and the ground observers of both nations were tied into this system, but there was neither IFF nor the substitute used in England that located friendly fighters by a grid of direction-finding stations that triangulated them as they emitted the radio carrier for 14 s once every minute. The mobile units gave very poor height data; their design had evolved out of CH, which had had the vertical lobes determined essentially by the ocean, conditions not well approximated by the local terrain. The result of this was a radar contribution to stopping the Blitzkrieg that was trivial, very likely zero [51].

3.2. THE BATTLE OF BRITAIN AND THE BLITZ

The only virtue of a battle is that it sometimes renders a decision, and it is the nature and importance of this decision that elevates or degrades the bravery and sacrifices of the warriors. The Battle of Britain was not, like Waterloo, immediately recognized as decisive. Indeed, it took the passage of weeks for people to realize that there had been an important victory; it took months for them to realize that it ranked with Trafalgar in preventing invasion by a foreign despot. It was like Trafalgar in being the result of one man forging and exercising the required weapon during the preceding years and then using it with skill. But whereas the leader at Trafalgar was immediately raised to the pinnacle of national glory, Dowding, the leader of the Battle of Britain, was relieved of his command, sent out of the country to do he knew not what and, after being assigned a task of finding economies in the Air Ministry, transferred to the retired list. He did not attain the highest rank in the Air Force although he did receive a barony three years after the Battle, a noble rank three degrees below that conferred on the victor of Waterloo.

It was a unique battle. It is the only set-piece battle fought entirely in the air. It is the only air struggle that resulted in an unequivocal outcome. It was fought with an absence of hate in a war dominated by hate. The age of chivalry had no battle that surpassed it in such ideals, and it was fought by the knights in far greater proportion to the commons than any medieval clash. Measured by the total number of combatants killed, it was a small battle; measured in the fraction of the combatants who survived it, it was a slaughter. Far more people watched from the ground, chancing the stray bullet or shell splinter, than fought in the air, making it the most observed battle of all time. Yet the ardent public rarely knew whether to cheer or cry as they watched a plane burning its way to the ground. It was a bizarre battle in which a parachuted escape from death was sometimes followed minutes later by being guest of honor at a pub.

But out of this chivalric battle grew, as inevitably as the seasons, a form of warfare as savage as any fought by the Iroquois. The opposing fliers conducted until the end of the war an honorable if brutal fight between one another, but the war was not to be won in the air but on the ground, and the war that began—at least in Europe—with careful consideration of the civil population slowly became one of attempted mutual extermination.

If Dowding was the Horatio Nelson of his time, Churchill was William Pitt, the Younger, and having to face a much more desperate situation than Pitt, Churchill emerged the greater historical figure. Defeat in France left Britain in an impossible position, but Churchill placed the English language in the front rank and so matched his words to the race that the defeat of France was met with a 'wave of lunatic relief'. The people faced the future exhilarated! 'Now we are on our own. Now we know where we stand.' Churchill's words earned the right to be remembered with those given by Shakespeare to Henry V. That they imparted strategic nonsense troubled the Islanders not at all.

On 16 May, less than a week after the start of the German offensive against France, Dowding had comprehended the situation and Britain's position in it perfectly and had set forth his analysis in a letter to the Air Ministry that ranks with any of the battlefield writings of Caesar or Grant for incisiveness. He had seen that France was almost certainly defeated and that Britain would quickly have to face Germany alone. Through his summer air exercises he had learned what his fliers with their electronic eyes could do; he had determined how many squadrons he would have to have and had repeatedly given the number as 46. The French Armée de l'Air had collapsed as rapidly as had the Polish air arm eight months earlier, leaving the Royal Air Force to face the Luftwaffe in France essentially by itself. The Hurricane squadrons were unable to protect the outclassed Allied bombers in doing the army support the Air Force had continually rejected for them and were too few to prevent the Stukas from doing the army support for which they had been designed. Fighters were the key to everything—in blatant disregard of Air Force doctrine—and their losses were eating up Dowding's reserves. From Paris, where he learned the shocking state of French defense on the same day as Dowding's letter, Churchill called for six more squadrons of Hurricanes to bolster the French, causing an immediate confrontation and rescission of the order. The debacle in France cost the Air Force a quarter of its fighter strength and irreplaceable pilots.

Britain's defiance after France's defeat left Hitler at a loss. The Soviet Union had just expanded its borders to the west by annexing the Baltic states and part of Rumania, steps going beyond their assignment to the Soviet sphere of influence indicated in the secret protocols of the 1939 German–Soviet treaty, steps undertaken without consulting Berlin. In her path to independence Finland had received substantial German aid in 1918, and the two nations had maintained friendly relations subsequently. The Soviet attack on Finland took place just weeks after the pact had been signed and had struck Berlin decidedly the wrong way. Furthermore, the attack's objective was obviously to strengthen Soviet defense against Germany. When all this was added to more than a decade of fulmination against the bolsheviks it left Hitler disposed to see a much greater enemy in the east, one which he thought should be dealt with quickly. He was, therefore, willing to come to terms with Britain and spent the first weeks after the French defeat pursuing that goal. In the meantime the Wehrmacht gave home leaves generously. There was a general feeling that the war was over. There was certainly no feeling of urgency in the Luftwaffe.

Hitler's confidence in speedy victory or his lack of comprehension of the nature of technical warfare (probably both) is reflected in the oft encountered statements that he ordered in early 1940 military research not completed for 1940 or shortly thereafter to be discontinued [1]. It did not, of course, apply to research done by private corporations and was ignored by GEMA, Telefunken and Lorenz.

When it became obvious that Britain could not be subdued by a de-based diplomacy, two alternatives were open: invasion or strangulation. The latter seemed too slow, so the former was selected—without war plans or prior staff work—for the summer of 1940. The Kriegsmarine had suffered losses in Norway that made it incapable of escorting the invasion flotilla across the channel in the face of the Royal Navy, making the only solution the countering of the Navy by the Luftwaffe. To do this meant that the Royal Air Force would first have to be defeated, which Göring assured Hitler could be done. Thus the battle was set, almost with the formality of the Middle Ages after the last futile exchange of couriers.

Technically, CH was ready. The east and south coasts of England and Scotland had adequate warning with long waves, and the most vulnerable sections had warning against low fliers with meter waves. In the plotting rooms markers on large map boards indicated formations of aircraft, hostile or friendly, with the time of the last fix shown by a color to match that on the five-minute sectors of the clocks that were standard throughout the Air Force and Observer Corps. The women and men on duty were well trained and ready.

Radar gave valuable information about the horizontal position and speed of the bombers but was much less reliable about height and numbers. Once they were over land, information had to come from the ground observers with all the infirmities that weather and lighting could bestow, although some CHL (Chain home, low cover with 1.5 m) stations could follow planes inland depending on station location and siting.

Particularly difficult for the observers was distinguishing friend from foe, something of obvious importance for the controllers. Inland radar coverage with electronic identification was how it should have been done, but in this British radar was not ready, so a substitute was used that had been devised years before for directing fighters. It employed a network of direction-finding stations that tracked radio signals emitted by the fighters. This system of triangulation had allowed fighter control to be developed. The last modification, called Pip Squeak, had the aircraft radio automatically emit the transmitter carrier for 14 s out of every minute with the pilot able to override this when he needed the set for communication.

Tactically, Chain Home was also ready. Stations had blast proof buildings, and the large towers were not easily knocked down. Radar mechanics were well trained in repair and had necessary spares. Mobile Radio Units (MRUs) were available to be erected, were a station knocked out. The crews expected and were ready to face the worst.

Colonel Martini's men were examining (unimpressed) the wreckage of British MRU and GL sets left at Dunkerque before the smoke of battle had cleared. It soon became apparent to them from listening to radio transmissions that the RAF was sending fighters aloft too soon to have been the result of visual observation, so they set up directional antennas on Cape Gris Nez and ascertained what the strange towers across the channel at Dover emitted [2]. But a detailed Luftwaffe intelligence briefing in July makes no mention of radar [3]. It was but one of many failings of Luftwaffe intelligence that summer.

British intelligence did much better. By July 1940 virtually every radio transmission of the German forces within range was overheard, and the information gleaned from it put to immediate use. The extremely secret Ultra was beginning its cryptanalysis of German signals, but since the Luftwaffe used land lines for much of its communication not directed to and from fliers, Ultra's part was not comparable to radar in conducting the battle. In those cases where details of an attack were so obtained the information was of small value to the fighter controllers, who needed to know where the enemy was to be expected with positional and temporal accuracy not obtainable from a tactical order. Nevertheless, the information Ultra provided Dowding was helpful, and Göring's Order of the Day on 8 August initiating the battle with the command to 'wipe the British Air Force from the sky' was in Dowding's hand within an hour of its transmittal [4].

The attacks that began the main Battle on 13 August 1940 were preceded by fighter sweeps and raids on Channel shipping intended to wear down Fighter Command. These were not strongly countered in order to conserve strength for the serious fighting that lay ahead and were wrongly interpreted by the Luftwaffe as weakness. They helped finish for the RAF the painful lesson started in France that their fighter tactics based on tight formations were no match for the loose ones used by their opponents in Me-109s. They brought home to the Luftwaffe the weaknesses of the long-range fighter, Me-110, which had been intended to accompany and protect the bombers and of which so much had been expected but which proved to be a liability in this kind of war. The Stuka dive bomber, Ju-87, which had terrorized ground forces east and west and which was needed to sink the Royal Navy, fared even worse. Both were incapable of holding their own in a fight with Spitfires or Hurricanes.

When battle was joined on 13 August it came with a bruising crash, and five days later produced the greatest number of casualties for a single day [5]. The attackers concentrated primarily on air defense targets in southern England with civilian targets avoided. Dowding's system was well exercised by then and its radar component proved of inordinate value by giving the defenders frequent opportunities for ambush. Attacks on five CH stations on the 12th knocked out Ventnor on the Isle of Wight and damaged the others [6], but repairs succeeded quickly either by returning them to service, setting up an MRU or at least by generating transmitter emissions, leading the Luftwaffe signal troops to interpret the attacks as failures and Göring to suspend them. Poling, near Portsmouth, was heavily damaged on the 18th, requiring the placement of an MRU for a period. This was probably a target of opportunity because Göring had suspended further attacks directed at radar.

Radar was not without fault. Estimates of the number of aircraft in an attacking formation had to be discerned from the magnitude of the radar pulse returned and the degree with which it fluctuated, and unlike the skills that attempted to extract height from the ratio of the signals received from the dipole pairs at various heights, these skills had not been calibrated on realistically sized formations numbering hundreds. Furthermore, accurate knowledge of the numbers in an attacking formation were seldom known at the time, so operators were unable to gain much from their experience and the estimates were frequently off by a factor of two or more.

Göring's important subordinate commanders were both Field Marshals: Albert Kesselring for Luftflotte 2 and Hugo Sperrle for Luftflotte 3 with a boundary running north-north-west through the middle of England roughly allotting the east to Luftflotte 2, the west to 3. Kesselring was to prove a skillful opponent for the Allies until the last day, and it is probably fortunate for Britain that Göring and Hitler made strategic decisions during that summer of 1940—and later too, for that matter—rather than he, although it is really incorrect to speak of a German strategic plan at all that summer, for plans were made on a day-to-day basis.

Dowding's important subordinate commanders were two Air Vice Marshals: Keith Rodney Park for 11 Group and Trafford Leigh-Mallory for 12 Group, with 11 Group covering southeast England and the Thames Estuary and with 12 to protect the Midlands and serve as a tactical reserve for

11 Group. Dowding's strategic plan was the result of five years' thought. Park believed in it with heart and soul and executed it with consummate skill. Leigh-Mallory had other ideas about the conduct of the defense, and as the tension of August and September weighed upon these men, these ideas took on forms that skirted perilously close to insubordination. That Leigh-Mallory was Park's senior yet Park's command was the first line of defense did nothing to lessen a dispute that eventually went to cabinet level and possibly to Dowding's replacement.

The few thousand young men that fought one another were remarkably similar. Both sides showed courage, skill and daring in extreme measure; both took great losses stoically; both went to the limits of exhaustion. But whereas one leader cared for his men with an affection that earned them the name of 'Dowding's chicks', and the Prime Minister gave them perpetual honor as 'the Few', Göring heaped abuse on his men, blaming them for the Luftwaffe's defeats and on occasion implying it was the result of cowardice.

Radar's part in the battle appears rather less than heroic at this point because it functioned just as expected. It had its greatest moment on 15 August. On that day Luftflotten 2 and 3 carried out a maximum effort just before an attack from Norway over the North Sea against Scotland and the Midlands by Luftflotte 5, initiated after it was thought the first two had tied up Fighter Command. These attackers from the north-east had expected to meet little resistance, expectations that show clear evidence both of how the importance of radar had not yet been grasped and how faulty was the knowledge of the RAF order of battle. The distance was too great for the bombers to be protected by the Me-109s and great enough to allow the controllers an almost leisurely placement of their fighters. The result was serious losses for the attack and few for the defense. Luftflotte 5 found other occupation for its skills during the Battle of Britain.

As the days passed, the German attacks on RAF bases began to tell and, had they been continued, might have produced the desired result, but a change altered the tactics. On 25 August German bombers had, in violation of Hitler's orders, dropped bombs at night on central London through navigational error, hardly surprising given the difficulties of that art even without the distraction of defensive fire. Churchill seized on this as an intentional attack on civilian targets, which he almost certainly knew was not the case, and 'suggested' night attacks on Berlin by Bomber Command, which followed the next night [7]. They had little effect on Germany's war-making potential and were, because of necessarily poor navigation, nearly all on civilian targets. The inability of the Luftwaffe to do anything to stop the attacks added to the embarrassment of Hitler and Göring, who allowed such personal feelings to override whatever judgement they were bringing to the matter. Beginning 7 September London was to feel the full brunt of air attack as retribution. This decision gave the RAF bases the respite they sorely needed and simplified the tasks of the air controllers by giving them a single, obvious target to defend. It also allowed Leigh-Mallory's ideas to be fitted perfectly into the scheme of things. The result was an epic struggle subsequently commemorated as Battle of Britain Day.

On 15 September the Luftwaffe put into its attacks on London everything that its exhausted crews could fly, who had been bolstered by intelligence reports that Fighter Command was at the end of its strength. As they headed for their target Park's squadrons were onto them with their usual fury, apparently unaware of the reports of their demise, but the real shock came on reaching the city, for Leigh-Mallory had had time enough to form his squadrons into the big wings he favored (over attack by individual squadrons) and to send them straight onto the already harried bombers. The Germans could scarcely believe the number of fighters in the air. The Royal Air Force was clearly not defeated, the invasion was postponed indefinitely, and the British people began to experience a less exhilarating phase of aerial warfare, the Blitz.

A division of the air war that enveloped Britain from July 1940 through May 1941 into the Battle of Britain and the Blitz is, of course, arbitrary but is useful here because radar functioned differently for the two parts. In the former it executed a well performed drill; in the latter a desperate makeshift. There were attacks at night before 15 September and in daylight afterwards with the last great daylight fight on 30 September, but the nature of the conflict took on a completely new form with large attacks the rule at night and those in daylight usually as low-flying fighter–bomber sweeps for which CHL became almost the only warning device.

If the Germans lagged behind the British in realizing the importance of radio techniques for defense, they were well in advance in understanding the need for improved navigational aids for attacks at night or through clouds. In the summer of 1937 Oberstleutnant Friedrick Aschenbrenner, commander of the Luftnachrichten-Versuchs-Regiment Köthen, described to various notables radio navigational equipment being developed since 1934 by Dr Hans Plendl of Deutsche Versuchsanstalt für Luftfahrt (German Aeronautical Research Establishment) at Rechlin in cooperation with Kühnhold of NVA [8]. It was based on the Lorenz Company's blind approach techniques that were well known in aviation circles: invented by Dr E Kramar, introduced for Lufthansa in 1934, and sold worldwide. In it a pilot received a pure audio-frequency tone if he were flying where the left and right lobes defining the radio beam had equal amplitude. If he strayed to either side, the pure tone became dominated by either a Morse code dot, indicating he was on one side, or a dash indicating the other; added together and heard sequentially the two codes, which were transmitted on the same radio frequency carrier, formed the reassuring continuous pure tone.

This system used a single dipole transmitting an audio-modulated carrier on the 10 m band. At each side of this dipole was located a reflector dipole with relay-operated switches in the center. Closing the switch of one of them threw the radiation pattern to the other side. A motor-driven cam alternately switched between the two, longer for the dash reflector than for the dot [9]. Depending on the sensitivity of the detector, be it the navigator's ear or some electronic instrument, there was a range of position for which the equipment would indicate that the flier was on the prescribed course. This range was 8 km at a distance of 100 km but when used to guide the aircraft towards the transmitter for landing it became narrower as the distance decreased, so the relatively large lobes of this simple antenna could be tolerated.

Guiding a bomber over a distant target where positional accuracies of 200 to 400 m were demanded at distances of as many km required narrow and accurately directed lobes. Plendl followed a straightforward way of accomplishing this. Using dipole arrays he could direct beams of sufficient accuracy that were detectable at great distances, depending on the height of the aircraft. The pilot would fly out on a beam, constrained in motions left and right until he intersected a second beam directed from a station positioned to give good triangulation. The accurate topographic maps, of which the European states were justly proud, allowed a completely blind bomb release. In winter 1938 a bomber formation, KGr (Kampfgruppe) 100, was formed out of the Köthen Regiment to apply the technique with Aschenbrenner, by then a Colonel, in command.

The range allowed by this method was restricted by the curvature of the Earth's surface because meter waves travel very nearly—but not exactly—line of sight. Substantially greater range than line of sight had been reported in the open engineering literature in 1933 [10], and Plendl had either read this paper or in all probability made the discovery himself, as he was a propagation specialist. Ignorance of these deviations caused early confusion for the British in realizing what was happening.

The first equipment along these lines was called X-Leitstrahlbake (direction beacon) and was a straightforward enhancement of the blind-landing method. This equipment showed good results in 1937 blind-bombing tests but had a clumsy antenna. At about that time Kühnhold designed an ingenious antenna that was much smaller yet generated equally narrow beams—unfortunately, it generated 14 of them, all of equal intensity [11]. This system, called both X-Gerät and Wotan 1, was adopted, and by 1939 14 stations were in operation. The confusion the air navigators might encounter owing to the multiple beams, especially from the cross beams that determined bomb release, was to be countered by training.

Plendl expanded the X-Leitstrahlbake into a heavy, rigidly mounted dipole array 30 m high and 90 m wide, bent in the middle with a dihedral angle of 165° to form two principal lobes. The shape of the array caused it to be called Knickebein (bent leg). The whole arrangement rotated on an accurately positionable turntable and was easier for the fliers to use because each array had but one prominent lobe. It was transportable with difficulty [12] and used the Lorenz navigation wavelengths, double the 5 m wavelength used in X-Gerät. This allowed the bombers to use Lorenz navigation receivers that had been given increased sensitivity.

The difficulties of moving Knickebein and the confusions in using X-Gerät prompted the design of Y-Gerät, also called Wotan 2, which eliminated the need of cross beams, an obvious operational complication of the other two. In Y-Gerät the array of dipoles and reflectors was so fashioned that a single lobe, larger than those of X-Gerät but still using lobe switching, predominated over the side lobes. A more sensitive detector was used to locate the equilibrium point between the switched beams and thereby overcome the inaccuracy inherent in the larger beam. The radio-frequency carrier was modulated with an audio-frequency signal, which was used to determine the distance of the aircraft from the transmitter. This modulating signal was retransmitted by the aircraft on another radio frequency and received at the transmitter where the shift in phase of the audio signal was measured. This phase shift gave a measure of the distance of the plane from the transmitter, which allowed a signal for the bomb release to be sent. It was significantly easier to use and did not require the cooperation of two greatly distant stations [13]. The smaller dipole array of both X- and Y-Geräte meant, of course, less precision in directional accuracy than Knickebein.

X-Gerät was used in the war against Poland in bombing a munitions factory in Warsaw, but the results could not be evaluated: the target had been bombed by other units. The system was employed in Norway and in the invasion of France, often in the daytime. One is forced to assume that X-Gerät was not found to be particularly successful, because Knickebein stations were quickly built in Holland and France [14]. X-Gerät was first used against Britain in a successful night attack of 13/14 August on a machine-tool factory at Birmingham, although not recognized by the defenders as a radio navigation flight [15].

R V Jones, a young protégé of Lindemann at Balliol College, Oxford, was sufficiently concerned about Britain's defense that he chose in 1936 to become a Scientific Officer in the Air Ministry rather than continue studies at the Carnegie Institution's Mount Wilson Observatory in California. He became Deputy Director of Intelligence Research and can be said, without stretching things too far, to have invented scientific military intelligence. His first interest lay in evaluating what the enemy had by way of radar, and he began to use the Oslo Report with ever increasing confidence as a guide.

One of the first major problems he had to face was the bits and pieces that pointed to the German's having radio-beam navigation equipment. The first clues came from conversations overheard from prisoners in which X-Gerät had been mentioned. As information began to come from Ultra at about the time of the fall of France other indications came to be seen with Knickebein making its way into the reports.

Lindemann alerted Churchill to these concerns, and Jones found himself in a cabinet meeting with Churchill on 21 June 1940, a moment of great national peril, which tells us the degree of high level concern about the beams and certainly demonstrates the seriousness with which the Prime Minister approached technical matters [16]. (That Air Marshal Arthur Harris, soon to lead Bomber Command in its night attacks on Germany, was of the opinion that celestial navigation sufficed well enough at night and that the concern with the German beams was misplaced tells us about the state of mind in some Air Force circles [17].)

There now began a fascinating struggle, the first electronic warfare, at least of a form that current practitioners of that art would recognize. Jones began putting together every possible bit of evidence that he could gather: prisoner interrogation, decrypts from Ultra, searches for the beams in aircraft or perched with receivers on top of the huge CH towers, statistical analyses of bombing attacks, even studies of Norse mythology for clues in the names given the systems by the Germans and in the background ever the Oslo Report.

Two defenses offered themselves: destroying the stations and interfering with the beams to make navigation either impossible of deceptively wrong. Both were tried. Dr Robert Cockburn of TRE was assigned responsibility for inventing electronic countermeasures, a project he was to carry far beyond the 'Battle of the Beams', and 80 Wing of the Air Force was organized to put these measures into practice.

This initiated a long period of espionage having some of the thrilling elements that go into that kind of activity, which Jones has recounted in his extremely readable memoirs. It was a confusing story as the two adversaries thrust about for one another in the dark. To be effective the countermeasures had to be carried out on exactly the right frequencies— both radio and audio—that had to be determined in advance of the attack and for which Ultra was to be of great value. It was not made easier by the poor calibration of the British receivers used at the beginning. In evaluating the success of the jamming it was difficult to distinguish results caused by interference from those caused by the deficiencies of the German equipment, deficiencies unknown to the British at the time and with which KGr 100 struggled. Furthermore, the most accurately delivered attack, one in which the jammers knew they had failed, took place on a clear moonlit night [18]. How successful was the interference? It must remain an unanswered question.

With the onset of the Blitz, Fighter Command and Anti-aircraft Command, their headquarters next to one another at Bentley Priory, faced the problem that they had expected for years. While CH was in construction Tizard had pronounced that it would be a success and that this success would bring on night attacks requiring airborne radar for night fighters. The predicted moment had arrived but the AI radar for night fighters had not; it was in the form of unsatisfactory prototypes.

The maximum range of AI was blocked by the enormous ground return at a range equal to the height of the fighter. Hanbury Brown saw that the immediate solution to this, until microwave AI became operational, was a night version of the daytime system in which ground controllers guided the fighters into attack positions on the basis of ground radar signals [19], data CH could not supply. A completely new radar system had to be constructed based on CHL, which had demonstrated its ability to follow planes inland from the coast when station siting permitted. Out of this design grew Ground Controlled Interception (GCI) with prototypes having to be tested in combat during the winter of 1940–1941.

The 1.5 m band was the obvious one to use until power was available at shorter wavelength. The first sets used separate antennas for transmitter and receiver, each having its own man-powered pointing gear with the transmitter following the receiver. Height information was as important as ever, so the receiver had two dipole arrays positioned one over the other and from which the operator used the relative amplitudes in determining the angle of elevation. The determination of horizontal position used the new Plan Position Indicator (PPI) in which the oscilloscope trace began at the center of the tube face and extended outward in whatever direction the antennas were pointing with targets shown by a brightening of the range trace [20]. This was the practical beginning of the most widely used radar display but differed from what has become universal in that initially the antennas were being pointed by the operators, not rotated steadily so as to produce a maplike picture. A common-antenna version of GCI quickly replaced the separate antennas, and operations rooms began to be organized around the functions of this new craft [21]. The early GCI suffered from an inability to deal with more than one interception at a time, although skilled operators could conduct two simultaneous interceptions. Later stations had multiple displays with a controller for each.

An entirely new technique of radar observation and fighter control had to be developed and personnel trained for it—while fighting off the attacker. The lesson of proper siting had to be learned first, as the designers of Freya had learned a couple of years earlier. Placing the set in terrain having the shape of a shallow bowl produced strong nearby ground returns but had clear traces beyond its rim, which had to provide uncluttered ranges for tracking bomber and fighter. The collecting of multiple observations and filtering them for the controllers proved quite incapable of functioning as it had during the daylight; the night controller had to deal more rapidly and with much more accurately defined positions. CH and CHL alerted the GCI stations to the incoming raiders, but the fighter controller now sat at his PPI scope rather than looking down at a map table as he followed the two adversaries and guided them to their deadly encounter by moving his antenna back and forth from bomber to fighter while talking the fighter pilot along his way until the AI radar picked up the prey. The first GCI interception was in the night of 18 October 1940 by a Beaufighter equipped with AI mark IV [22]. This was the beginning of a successful defense, but one which required months of frantic effort to bring to fruition, as GCI and AI sets were built and deployed, and fliers, radar operators and controllers were trained in this new endeavor. A dear price was now paid for the time lost through the moves from Bawdsey to Dundee to Swanage with side trips to Perth and St Athan for the airborne work[3].

By the time US Army Air Corps officers visited Britain in mid-1941 GCI and AI worked well enough for them to insist that they be copied by the Signal Corps, GCI becoming SCR-527 and AI mark IV SCR-540. This went more slowly than hoped, in no small part because the British design was not yet fixed, and the first of the 527s were not produced until the spring of 1943 and the 540s were never deployed because microwave equipment was ready first [23].

If Fighter Command had troubles during the Blitz, AA Command faced the attackers with equipment quite incapable of even giving the bombers a proper fright, the consequence of attitudes about AA artillery dating from the First World War. Although the 3.7 inch gun was bal-listically up to the task, the director was of poor quality and dependent on optical sighting. This was difficult enough during daylight raids, but meant at night that the attacker had first to be picked up by searchlights, and putting a searchlight beam onto an aircraft at high altitude was a tall order just in itself. The gun-laying radar sets, GL marks I and II, were of more use than the acoustical horns in helping the searchlight operators, and they could provide accurate range, providing there was not a confus-ing number of planes within their huge beams, but General Sir Frederick Pile, the chief of AA Command, reported that not a single aircraft had been brought down with the help of radar by October 1940 [24].

Pile sought technical help desperately and found it in P M S Blackett, who took as his first task securing personnel capable of instructing troops in the operation and maintenance of the new equipment so that the sets worked at all. This he accomplished by recruiting school teachers and bi-ologists as radar officers—the pool of physicists and engineers had already been taken up. They were given a brush-up course in electricity, a crash course in radar and distributed to the batteries where they were quickly appreciated for their competence [25]. The next problem was to make the best use of the radar data acquired other than helping the searchlight op-erators; the directors were not capable of incorporating the output of the GL sets in their calculations, and he finally settled on plotting the positions by hand so as to predict the optimum location for a gunfire barrage [26]. It was a wretched method but superior to firing barrages on no data at all. Blackett left AA Command in April 1941 for Coastal Command where his newly invented 'operational research' was to play a much more decisive part.

---

[3] For details of the move to Swanage see Chapter 4.1 (p 153).

As the Blitz proceeded the gunners shot away the rifling of their gun tubes so that projectiles followed unpredictable trajectories and fell on the towns supposedly being protected because the time fuzes were not activated, owing to the lack of spin. When, in order to save usable guns, Pile restricted fire on courses where it could not be effective he received a personal call from Churchill demanding fire [27]. Few besides Pile looked on the guns as having any value other than sound effects for the public. What Pile needed were Würzburgs or SCR-268s. AA Command eventually became a deadly arm, but not in time to affect the Blitz. Even as evidence accumulated that excellent gun-laying radar could be made, Lord Cherwell, Churchill's scientific advisor, was opposed to investing much effort on such things [28].

At Pile's strong, early recommendation AA Command also had women from the Auxiliary Territorial Service (ATS). They operated searchlights, radar sets and optical sighting equipment but masculine delicacy did not allow them to load and fire the guns. They lived in field conditions distinguished by cold, mud, rain and snow, often endured in the absence of heat. Whereas women in the Air Force (WAAF) and Women's Royal Naval Service (WRNS) received steady recognition, the AA Command ATSs fared little better than the gunners, who were frequently reviled in the pubs after nights without gunfire when Pile and Dowding tried using the searchlights to illuminate the bombers for the fighters—not with outstanding success.

The Blitz set the moral course of the remainder of the air war. The attack on Coventry during the night of 14/15 November 1940 resulted not only in significant industrial damage but in severe damage or destruction to some 50 000 houses and to St Michael's Cathedral; more than 400 were killed and double that number wounded. British propaganda broadcast to the world the barbarity, and as if to emphasize the point, Dr Goebbels's own propaganda trumpeted it as a harbinger of things to come, if Britain persisted in defying German war aims. He promised that other cities would be 'coventrisiert' [29] and in so doing removed future cause for complaint when the situation was reversed.

May 1941 generally marks the end of the Blitz, for in that month the GCI-AI system began to work effectively with 102 bombers shot down that month, some during the day [30] and leaving the Luftwaffe to try their luck on the Eastern Front. The skies over England were not to be free of bombing planes until near the end of the war, but the Luftwaffe never returned again in force.

Lord Dowding observed final victory as a bitter observer. The mode of his relief from command agitates his countrymen to this day. It was by no means a simple matter, being composed of ample portions of anxious concern for the conduct of the war and old animosities within the Air Ministry. The Blitz was going hard against Fighter Command in November when Dowding departed, and his enthusiastic support for the development and deployment of GCI-AI counted for little, as their success came later. Dowding was one of the most scientifically minded military commanders of the Second World War. His analysis and conduct of the air defense of Britain has stood the test of hindsight. His failing was as a commanding officer. Two of his lieutenants ended the Battle of Britain almost incapable of cooperating with one another. Dowding either ignored the situation or was unaware of it, which is difficult to imagine, but then Dowding seldom visited his subordinates. A dispute of the kind between Park and Leigh-Mallory is not unusual among military commanders. War is a trade that nurtures strong ambition and firm opinion, characteristics that must inevitably lead to clashes, but it is the responsibility of a commander to deal with that sort of thing. Dowding did not.

And radar? Did this miracle weapon turn the tide? Of course radar did not win the Battle of Britain. Brave men and women won it. But could they have won it without radar? That question has been posed many times and has, more than for any other engagement of the War, received a negative answer. It is difficult to imagine a satisfactory outcome for Britain had the Air Force not had radar. Had the Germans proceeded with a carefully thought out strategy instead of day-to-day makeshifts, there is good reason to believe that they might have won in spite of radar. With such a narrow margin of victory one is compelled to hand a victory medal to radar, but it is a medal that the Prime Minister would not have awarded. He gave radar but a single condescending sentence in the chapter of his memoirs devoted to the Battle of Britain [31], in remarkable contrast to the eight pages about the German beams [32]. He was an active partisan in the Lindemann–Tizard fight, and Tizard left high council before Dowding.

Churchill's failure to realize or give credit to radar's part in this epic battle slips into insignificance when viewed against his own contribution. If it is difficult to imagine a positive outcome without radar, it is even more difficult to imagine it without Churchill's leadership, for it was he who inspired a whole people. It was expressed simply in the response to his visit to a poor London neighborhood that had been very hard hit during the Blitz: 'Good old Winnie! We knew he'd come'.

On 18 June 1941 Lord Beaverbrook, British Minister of Aircraft Production, released in a radio broadcast that a 'radiolocator' was an important part of the country's defense and called for volunteers with radio skills [33]. The United States revealed something about its own radio location work at the same time and called for electronic volunteers [34].

3.3. THE ATLANTIC, 1941

The complete defeat of the U-boat in 1918 had left Admiral Erich Raeder with little interest in using that arm for the commerce raiding he planned for the expected war. He and most of the top ranks of the Kriegsmarine were convinced that surface vessels could better take on this task. Only

Dönitz held back from this viewpoint and obtained limited support for his submarines.

Surface raiders were of two classes, armed merchant ships and warships, both the modern embodiment of naval tactics as old as ships. Whereas the warships went about their work in a direct and straightforward manner, the armed merchant ships brought to their task disguise and guile. For the disguised raiders there were sailings in exotic waters, captured ships sent back under prize crews, supply ships met on lonely stretches of ocean, mysterious communications with agents on shore, repairs at island bases of the quasi-ally Japan, selected prisoners with whom to enliven evening conversation and the excitement of being chased by and evading the Royal Navy. They conducted it—with one notorious exception—according to the rules of warfare, indeed in an almost gentlemanly manner. They emulated the exploits of the *Emden* and the *Seeadler* of the First World War and, like them, caused comparatively little loss of life. Seven such vessels sank a total of 39 ships and removed 590 000 tons of shipping from British usage. One, the *Atlantis,* known to the Admiralty as Raider C, put to sea on 31 March 1940 and cruised until sunk in November of the following year [1]. The story of these ships is an interesting one, all the better for being generally devoid of the viciousness which the war by U-boat became, but we must leave it—these ships used no radar.

The warship raiders were different. They were big ships, 8 inch gun cruisers being the smallest. More often than not they were unable to capture ships but sank them with gunfire. Their crews certainly felt the excitement of being chased by and evading the Royal Navy, but despite their size and power they had to be timid in the face of serious force, for even slight damage could seldom be repaired at sea, thereby forcing the cruise to be terminated, so speed was essential. The captains of these ships recognized the value of radar from the start, for it allowed them to locate their prey at night or in fog and to distinguish, usually by size and speed, the escort vessels or any other ships that meant trouble. Radar proved to be an exceptionally important aid to navigation, and to this the 80 cm Seetakt was well adapted.

After the loss of the pocket battleship *Admiral Graf Spee* six warship raiders followed that calling. The *Admiral Scheer* was a pocket battleship that mounted 11 inch guns, as did the two battle cruisers, *Scharnhorst* and *Gneisenau.* The *Admiral Hipper* was an 8 inch gun cruiser and sister ship to the *Prinz Eugen*, which would enter upon surface raiding briefly but eventfully as consort of the battleship *Bismarck* in May 1941. All were equipped with radar.

Both classes of raider were dependent on world-wide rendezvous with supply ships and tankers to remain at sea for long periods, and theirs is a good story too but one without radar except as used by the warship needing supply.

3.3.1. *Admiral Hipper, 1–27 December 1940*

Of the warship cruises this one by the *Hipper* was the least successful, primarily the result of mechanical problems that forced curtailment, but her radar worked perfectly. After a false start in September 1940, terminated by engine troubles, she departed German waters at the beginning of December. Her entry into the North Atlantic was through the Denmark Strait, which lies between Greenland and Iceland and was a favorite point of exit for warship sorties because of the dependably foul weather, and her escape went unnoticed. Her orders were to attack convoys only, not independently routed shipping.

On 19 December she easily found with Seetakt the tanker spotted for her refueling despite a very uncertain position after days without an astronomical fix. Following this she used radar to avoid the forbidden encounters with single ships, finally observing a convoy at 1700 hours on the 24th at a range of 15 km. Soon eight to ten ships were observed, one as distant as 22 km—a maximum for Seetakt. The commander's plan was to track the convoy until morning and then attack by gunfire, but during the night he thought it worthwhile to try a torpedo attack based on radar data. At midnight *Hipper* closed on the convoy and launched a spread of torpedoes at a range of 6 km. No hits. Investigation disclosed the torpedo officer had ignored the radar data and launched at a shadow he was convinced was the convoy.

When *Hipper* prepared to attack at 0630 hours on Christmas Day, radar reported intruders that proved to be a heavy cruiser with other escort vessels following. Fire was exchanged from which *Hipper* received little damage, but mechanical problems again set in and she had to make for Brest, the first large German ship to make use of the new French base [2].

3.3.2. *Scharnhorst and Gneisenau, 22 January–22 March 1941*

These two vessels left Kiel under the command of Vizeadmiral Günther Lütjens. The plan was for the *Hipper* to sail out of Brest whenever the two battle cruisers had been detected and had attracted the hounds, leaving much of the Atlantic unprotected, and it worked out pretty much according to plan.

The battle-cruiser pair followed the Norwegian coast, left it where it turns north-east, and attempted to pass to the south of Iceland. During the early morning of the 28th the *Gneisenau*'s radar failed, and during 90 minutes of frantic repair work the *Scharnhorst* reported blips that gave every indication of being cruisers. Course was reversed and the two raced their way through an unnerving series of radar sightings and course changes until free. From there it was far to the north for a rendezvous with a tanker, which was 30 km from the expected position and very likely could not have been found during such long polar nights without radar.

With fuel tanks full they tried the Denmark Strait, a risky business in such weather with no lights on Iceland's shore nor stars to steer by. The Seetakt sets transformed it all into routine off-shore navigation by providing the range and direction to prominent features of the island, and the passage was almost routine with a couple of twists and turns to put distance between them and other vessels. By noon of 2 February they were clear.

On breaking into the open sea they separated during the day to enhance the chance of finding ships but rejoined again at night. On the 8th the *Scharnhorst* was sighted by the battleship *Ramillies*, which was escorting a convoy. Lütjens broke off well to the northwest for more oil, whence he began seeking convoys on the Halifax route and was rewarded by sinking five unescorted ships on the 22nd. Their wanderings continued until they lay off the coast of Africa, again sighted, then off to the north where their luck improved and they sank 16 unescorted ships. Lütjens was ordered to put into Brest, luring many of his pursuers after him and clearing the sea for the return of the *Hipper* and the *Scheer* to Germany by the northern route. The *Scharnhorst* and *Gneisenau* began preparing for expanded raiding with the new battleship *Bismarck* planned for the following month.

This cruise raised the mariners' opinion of radar to an unprecedented high. The troublesome early failures at sea were studied on this trip by an engineer from the Nachrichtenmittel-Versuchs-Kommando (NVK, the newly renamed NVA) aboard the *Gneisenau* who was able to isolate causes of many past and present troubles. They also found that the nearly identical wavelengths used by the two ships gave unmistakable indications on their respective sets when they approached one another after daytime separation, an unexpected sort of IFF that was much appreciated when extremely hostile ships were about [3].

3.3.3. *Admiral Hipper, 2–14 February, 16–31 March 1941*

The February sortie was made in accordance with the plan, once the two battle cruisers were observed in northern waters. During the night of 11/12th the *Hipper* observed a large convoy by radar, which they used to maneuver into the best position for attack at dawn. The convoy consisted of 18 unescorted vessels, seven of which were sunk by gunfire. The *Hipper* then returned to Brest to prepare for the return trip to Germany where her numerous mechanical problems were to be removed.

Thanks to Seetakt this return trip in March was uneventful, but even with radar *Hipper* required eleven hours to find the tanker needed for mid-trip refueling. In what had now become routine a heavy cruiser was successfully dodged in the Denmark Strait [4].

3.3.4. *Admiral Scheer, 27 October 1940–1 April 1941*

Of these warship cruises this one began first and ended last, ranged over North and South Atlantic and into the Indian Ocean, and accounted for

16 ships. The outbound passage of the Denmark Strait was typical of what was to follow weeks later by the other raiders: secure navigation and avoidance of unwelcome encounters. On 5 November she encountered the homeward-bound convoy HX84, escorted by the armed merchant ship *Jervis Bay* under Capt E S F Fegen, who drove straight at the raider while his charges made smoke and scattered. The *Jervis Bay* was, of course, destroyed but only five of the convoy were sunk. Capt Fegen received the Victoria Cross posthumously.

The *Scheer*'s long trip supplied her radar personnel with a generous supply of problems. They repaired an uncommon number of failures, generally caused by the shock of their own gunfire, and on leaving the Indian Ocean found themselves with only two spares of a critical tube. Radio provided for replacement at a supply-ship rendezvous on 11 March, only to have an oscillator crystal fail. Again replacement was sent, this time by submarine U-124. The weeks without radar left the command experiencing the sensations of the recently blinded.

They passed the Denmark Strait safely homeward bound, easily avoiding a heavy cruiser—and then had the set go out again, the consequence of a short circuit caused by condensed water that was removed the following morning. The remainder of the voyage to Kiel was uneventful both navigationally and electronically [5].

3.3.5. *Bismarck and Prinz Eugen, 18–27 May 1941*

Late in the evening of 18 May 1941 two new units of Raeder's surface fleet left their Baltic port after completing careful shakedown cruises and training. They were the 15 inch battleship *Bismarck* and her consort, the 8 inch heavy cruiser *Prinz Eugen*. The former was named for the chancellor whose foreign policies had made friendship with England a vital element, attained by avoiding naval and colonial rivalry. The latter was named for the comrade-in-arms of Winston Churchill's ancestor, John Churchill, 1st Duke of Marlborough, Eugen's partner in a long and successful struggle by the Germans and the British against Louis XIV's attempts to subjugate Europe. Both ships were the ultimate of naval architecture. Both were equipped with Seetakt; both had special radar rooms as a part of the original design. Their assignment was commerce raiding under the command of Admiral Lütjens. More was expected of them than of previous surface actions, for with their armor, speed and radar they would be difficult to stop, an opinion shared in Berlin and London.

Previous surface raiding had found the Royal Navy radar poor and the raiders making good use of their own. Now the balance was to swing in the other direction with the Royal Navy, Coastal Command and the Fleet Air Arm radar equipped to some extent. The exact time and route of the pair were not known to the Admiralty, but the break-out was no surprise, and a significant reception was prepared.

On the 21st the two were sighted at Bergen by air reconnaissance. That observation had been followed up on the following day in near-impossible flying weather, and the harbor was found to be empty. The two had sailed for the Denmark Strait, at the time about two-thirds blocked by ice and with most of the remainder the recent depository of 6100 mines. Retreating ice had left a safe passage that Seetakt easily traced, allowing them to avoid the floating bergs as well as the pack ice even in the deep fog that kept British non-radar air patrols from sighting them.

The cruiser *Suffolk* had received one of the first two 7.5 m type 79Zs in May 1939, later upgraded to type 279, and now was also equipped with the 50 cm type 284 radar for directing the fire of her main armament. She waited at her station at the exit of the mine field. The cruiser *Norfolk*, which patrolled 80 km to the west, had only the 1.5 m fixed-antenna type 286M, the one that required swinging ship for direction [6].

At 1920 hours on the 23rd the *Suffolk* and the *Bismarck* sighted one another visually as the latter broke briefly from a fog bank. The type 284 transmitter tubes were pushed to the limit to gain the needed power at such short wavelength; this normally allowed operation for only a couple of hours at a time, not too restrictive for gun-laying but hardly suitable for searching. The vertical lobe structure of the 7.5 m set precluded using it for surface search except at very close range. It was the intermittent use required to conserve the 284 that caused the British sighting to be visual. *Suffolk* scurried for fog before 15 inch shells could be sent her way, got off a sighting report and began tracking the big ship with the 50 cm type 284.

The *Bismarck*, whose two 80 cm sets were not restricted in duration of operation, had located the *Suffolk* both with radar and underwater sound before the visual sighting. Fortunately for the cruiser the Seetakt did not incorporate lobe switching and thus could not direct blind fire, having a directional accuracy of only 5° [7]. Because of iced insulators on the radio antenna the *Suffolk*'s first sighting report was received only by the *Norfolk* and the *Prinz Eugen*, where it was promptly decoded [8]. The *Norfolk* soon had a glimpse of the battleship and narrowly escaped a salvo of heavy shells. The shock of gunfire had the effect of knocking out the forward Seetakt [9] to Lütjens's great displeasure, so *Prinz Eugen* had to lead, as both her radars still functioned. The *Suffolk* managed to keep her quarry in optical or radar sight and hold the *Norfolk* close with radio. The Admiralty soon learned of the chase and dispatched the new battleship *Prince of Wales* and the flagship *Hood* to intercept. They met the enemy early in the morning of the 24th, despite the *Suffolk* having lost contact a few hours before. Vice Admiral L E Holland, commanding the squadron, ordered complete radio silence for his ships, including radar, until the German ships were sighted, his fear being that with their greater speed the Germans could escape if alerted.

The *Hood* was the finest of that most unfortunate kind of warship, the battle cruiser. As large as a battleship with guns as heavy, it sacrificed armor to gain speed. It was a stylish idea in naval circles before the demonstration that a 5 knot difference in speed did not matter to well aimed projectiles that easily penetrated thin steel. Three ships of this type had disappeared in the Battle of Jutland in catastrophic explosions. (The German battle cruisers *Scharnhorst* and *Gneisenau* traded gun power for speed rather than armor plate, having only 11 inch artillery, small for battleship-class vessels.)

The *Hood* had a type 279M air warning radar and a type 284 gun-laying set, but radar did not protect her from the first salvos of the two German ships, and she blew up in a mighty explosion, the presumed consequence of a heavy shell penetrating her thin deck armor and detonating the magazines. The German optical fire control was up to the same high standards it had so startlingly demonstrated in action in the North Sea in the previous war and the *Bismarck*'s defective radar was not missed.

The *Prince of Wales* had a 3.5 m type 281 air warning set and nine fire control radars, but the ship was so new that civilian workmen were still on board, as bad luck would have it, because of problems with the main armament. She was also so new that the gunnery officers had not incorporated radar into their procedures. The radar officer reported accurate ranges throughout the brief fight, but they were not used in calculating gun orders, and it was only the sixth salvo that had the correct range [10]. So it came to pass that in the first encounter of big-gun ships equipped with radar the use of the new technique is enveloped in fog: the forward German set on which the First Gunnery Officer would have relied was dead, and the British set was ignored [11]. What the *Hood* did will remain unknown, but her first salvo was not on target.

The *Prince of Wales* developed serious malfunctions in her artillery and sustained enough damage to cause her to withdraw behind a smoke screen. The *Bismarck* had unintentionally begun replacing fuel oil with seawater though retaining a speed of 28 knots. Why Lütjens did not pursue and very likely sink the *Prince of Wales* is a puzzle few have understood. At this point the *Bismarck* was sufficiently damaged that commerce raiding without repair was not possible, and sinking the two most powerful ships of the Royal Navy would have certainly justified the attempt. Lütjens detached the *Prinz Eugen* to proceed independently to the south and began a straight run for the safety in the Bay of Biscay.

Now the *Bismarck* was pursued by an ever growing assortment of very heavy ships with the *Suffolk* again doggedly tracking, but on the 25th she lost radar contact, the almost certain consequence of the intermittent use required of the 284. Lütjens was so impressed with the ability of the *Suffolk* to follow that he broke radio silence to inform his chief of the radar capability of which he had not been informed and the range capability of which he greatly overestimated. The overestimation probably resulted from navigational errors of one or both ships, as Lütjens compared his calculated position with the continual flow of messages that the *Suffolk* was transmitting [12]. Lütjens's message allowed British radio-direction finders to get a rough idea of his position, but at the time he incorrectly thought he was being held fast by British radar.

This incident is linked to reports [13] that the *Bismarck* had a passive radar receiver and had monitored the tracking. If so, it must have been an experimental set of which there is no other record [14], and the passive receivers that first came into use more than a year later would not have responded to 50 cm waves. It is plausible that the radar operators, presumably briefed on British use of long waves, picked up on communications receivers some of the abundant 7.5 m transmissions, which they would have recognized as radar. Given the circumstances it is unlikely that they would have realized that this equipment was incapable of observing them at the ranges involved.

A sighting through the swirling clouds over a rough sea by a Catalina flying boat equipped with ASV mark II [15] established the *Bismarck*'s position accurately enough for the cruiser *Sheffield* to be ordered to pick her up with the type 79Y radar, if possible. At this point aircraft from the carriers *Victorious* and *Ark Royal* were decisive. Both were equipped with the famous Swordfish biplanes, slow but very tough and possessed of a remarkably long range and a deceiving agility, if not encumbered with torpedo or bomb. They probably sank more tonnage than any other torpedo bomber during the war and were valuable participants until the very end. We shall return to them when describing action in the Mediterranean, the high point of the Swordfish's service.

One of the Swordfish from each carrier was equipped with ASV mark II, and green fliers from the *Victorious*, which had not had time to work up her crews, even to allow them to practice take-off and landing from the deck, found the target and got an ineffective hit on the armor belt. The first attack by 14 planes from the much more experienced *Ark Royal* went after the shadowing *Sheffield* instead, of whose presence they had not been informed, but their torpedoes missed. Their next attack of 15 planes found the *Bismarck* with radar in conditions of 'low rain cloud, strong wind, stormy seas, fading daylight and intense and accurate enemy gunfire'. One torpedo struck the armor belt, another jammed the steering gear, and with that the great ship was doomed. The radar that found the target also found the home ship, and all 15 aircraft returned, to be sure with wounded crewmen, perforated fabric and three crash landings [16].

With the stricken ship no longer able to reach the protective cover of land-based bombers, dawn came as a death sentence to be executed by the battleships *Rodney*—ordered to the spot with a deck cargo for installation in America and 300 passengers—and the *King George V*. Accurate fire, soon delivered at close range, destroyed the ship that refused to surrender. There are several accounts of this famous battle. The reader is advised to read the one by the *Bismarck*'s Adjutant and Fourth Gunnery Officer [17] and that of the under-water explorer who found the wreck in 1989 [18].

The sinking of the *Bismarck* put an end to German surface raiding with large ships. Even without that dramatic climax it was becoming increasingly obvious that it simply did not pay. The *Scharnhorst* cost as much to build as 100 submarines, required a huge crew and elaborate supply, and was not immune to sinking. There was an attempt by the pocket battleship *Lützow* to renew raiding, but her sortie of 10 June 1941 was countered by a torpedo-plane attack that sent her back to Kiel for months of repair. When Hitler attacked the Soviet Union, he required many of his surface units for the Baltic. The disguised raiders continued until the Royal Navy removed them, their tankers and supply ships from the seas. Commerce raiding would be left to the U-boats of Konteradmiral Karl Dönitz, and all traces of romance disappeared.

The use by the Kriegsmarine in 1939–1941 of Seetakt was a most impressive consequence of the power of pure radar, the result of a naked radar set mounted on a ship for which no thought had been given as to what its exact tactical function was to be. The naval personnel received little training, but the set was simply ideal for a commerce raider. It was the kind of thing that every alert officer recognized when he first encountered it—the torpedo officer of the *Hipper* a conspicuous exception. Application came immediately and instinctively. There is no evidence of captains considering radar as just 'an interesting device'; they regarded its malfunction to be a major problem for which they demanded the delivery of spare parts by special ship and submarine.

It had not been planned that way by Raeder. On first seeing a radar demonstration he was impressed enough not to interfere but cautioned Kühnhold that his primary research mission was under-water sound. It was the line officers who recognized the new weapon for its value, and their use of it in the few months of surface action was beyond criticism. Except for a technically dull-witted command they could have had blind-fire directed gunnery in 1938. German naval radar had a brilliant beginning that led nowhere.

Typical of the want of understanding at the top was the vacancy of the position of Chef der Abteilung Entwicklung der Nachrichtenmittel (Chief for Development of Signals) from November 1939 until April 1943 [19]! Moreover it was not until mid-1941 that the Marine-Nachrichtendienst (Navy Signals Service) was formed and with it a naval career specialty for radar, Seetaktischer Funkmessdienst (Tactical Radar Service) [20]. Progress remained slow, and Dönitz was to find his U-boats completely outclassed in either defensive or offensive radar techniques.

A comparison between the two navies offers instruction about their respective use of radar 21 months into the war. The Germans had mounted a prototype Seetakt in 1938, modified it in small ways, and haltingly made it reliable aboard a warship, the obvious responses of competent engineers; it was their only shipborne radar for months yet to come. Despite the Navy's introduction of the equally good air-warning Freya, it was never taken to sea except on vessels in the North Sea as part of the country's air warning system, nor was the excellent gun-laying Würzburg used aboard ship to improve AA fire, although GEMA soon adapted the Seetakt for dual purpose. The British by contrast had by May 1941 almost a dozen different kinds of shipborne radar installed [21], but it was not until the 10 cm type 271 appeared, with sea trials in March and April 1941, that they had a surface-search set competitive with Seetakt [22]. In their hunt for the *Bismarck* only one shipborne radar set of the entire pack of hounds was effective, and its inability to maintain continuous search caused it to lose the target vessel at a critical moment, saved by the splendid ASV mark II. It remains a puzzle that a naval command that gave high priority to radar placed so little importance on surface search equipment. The answer to the puzzle probably lies in Britain's approach to radar from the long-wave side.

3.4. FRIEND, FOE OR HOME?

As soon as radar sets began to move from plans to reality designers began to consider the matter of determining whether the blip observed on the cathode-ray screen came from a hostile or friendly ship or plane. It was a problem as old as warfare itself. Visual identification has never been certain, and in the confusion of war a significant fraction of losses has always come from 'friendly fire'. In unconstrained naval battles fought at long ranges in poor visibility it had become a pressing problem before the complication or help of radar. For the relevant case of war in the air there were many cases of attacks on the wrong aircraft through failure of visual recognition, and the new kind of vision offered by radar was to bring the same old problem back in a new form.

Watt discussed it in his original memorandum to the Tizard Committee [1]. The British name for it, IFF for identification friend or foe, never had any serious competition in English. The first idea was the same everywhere: equip radar targets from your own side with some structure or circuit that would resonate at the radar wavelength and that could be modulated and re-radiated. This would give an enhanced return signal, the amplitude of which would vary in a manner recognizable to the operator at the screen.

The people at Bawdsey elected a tuned dipole with a switch in the middle that allowed keying a signal recognizable at the radar set. Arnold Wilkins produced equipment of this kind for testing in the Home Defence Exercise of 1938. The return signal proved difficult to distinguish from the variation of intensity produced when the target was a flight of planes, so he decided on an active rather than a passive unit [2]. Here passive means that only a reflected wave is returned; active means energy is added to the response through a vacuum tube circuit. The Naval Research Laboratory tried similar tuned dipoles in testing XAF during the Caribbean exercise of 1939 and also decided to use an active system [3]. The opening of hostilities found neither side with operating IFF. Britain rightly expected to be under air attack soon and felt dire need for the new device. This led to the introduction of IFF mark I in time for the Battle of Britain.

Owing to the urgency of the times, the designers elected to have mark I respond only to CH, a band from 10 to 15 m, but not to MRU, CHL, GL or Navy wavelengths because of the complication of incorporating these bands. It consisted of an airborne receiver whose tuning was swept mechanically through the range in a few seconds. The receiver was a super-regenerative circuit that functioned as the responding transmitter as well; it was a receiver from the earliest days of vacuum-tube radio, but which was seldom encountered in 1940.

This kind of circuit came out of the first experiments by De Forest and Armstrong. A triode used for the amplification of radio waves had a tuned circuit in its output, comprised of a coil in parallel with a capacitor. Both investigators found that if one connected some fraction of the output to the input, a process called feedback, the circuit would go into oscillation. This was the first electronic oscillator, which transformed radio completely and soon eliminated spark and arc equipment. They also found that with feedback just below the threshold for oscillation the circuit would amplify with much higher gain than without it, and the regenerative receiver was born. One of the skills required of people listening to early broadcasts was that of adjusting this critical element. There was an intermediate region with even higher sensitivity in which the detector, once tickled by an input signal of the frequency to which it was tuned, had its output grow into sustained oscillation. This was clearly of no use for the reception of voice, but could be adapted to Morse code if the oscillations could be appropriately squelched.

IFF mark I functioned by being triggered by the CH signal when its tuning matched the frequency of the interrogating radar, which meant that for a given CH station the IFF would respond not for every radar pulse but only those for which its tuning corresponded. On being triggered the oscillation amplitude would grow until the detector became a small transmitter. The radar operator would recognize a friendly blip by its increasing in size in a repetitive manner determined by the speed at which the IFF transponder tuning was being swept. This had both simplicity and economy. Unfortunately, it required in-flight adjustment of the feedback by a flier whose mind was fixed on other matters and was not appreciated by the pilots and radio operators who had to use it. An improper adjustment resulted in no response or in radiating random pulses to the great annoyance of the ground stations [4]. Installation of the first 100 sets, made by Ferranti, began in November 1939 with 1000 eventually delivered. Operational experience was poor, as only about half of the interrogations with properly adjusted equipment yielded an identifying response [5].

Inadequacies of mark I were apparent before it went into service, development of mark II was under way in the spring of 1939 and even that design was seen to be only a provisional solution. Mark II responded to the 10–15 m band of CH, the 7 m band of the Mobile Radio Unit and the Navy, the 3–5 m band of the Army sets and the 1.5 m band of CHL and the Navy. It functioned the same way as mark I but had a 'complicated system of cams and cogs and Geneva mechanisms' [6]. Unfortunately, it retained the sensitive in-flight adjustments. Improved maintenance, supervision and training brought operational success up from the 50% of mark I but never to satisfaction.

After the disclosure of radar information to the United States by the Tizard Mission in September 1940[4] mark II was also adopted by American forces as SCR-535 and manufactured by Philco. By July 1942 the US Army Air Forces had 18 000 on order [7]. IFF equipment soon dominated the costs of radar.

When it became apparent that microwaves would have to be added to the spectrum for which IFF must respond, the basic design of mark II was recognized as unusable because of the disparity of electronic techniques that separate 15 m waves from 10 cm. Furthermore, mark II would respond just as well to German radar, if they placed equipment using Allied bands in service. F C Williams began to concern himself early with this problem, casting it in terms of a universal interrogating system, one independent of the primary radar frequency. IFF sets were not to reply directly to the radar signals but to a special interrogating transmitter. The interrogation took place in the range 1.6 to 1.9 m, which the transponder swept, with reply on the interrogation wavelength. It used the super-regenerative method but had automatic gain stablization to maintain the receiver at optimum sensitivity, eliminating in-flight adjustment. The reply was a pulse of coded length with an extremely long pulse signifying distress. The basic design was ready in early 1941 with manufacture begun by Ferranti Limited. American production was desired, and Hazeltine Corporation received a contract to design the set for American production with an acceptable prototype completed by mid-1942 [8].

The mark III became the Allied standard for the war but not without a small serving of acrimony, quickly settled by America's precipitate entrance into the war. The Naval Research Laboratory and the General Electric Company had already designed an IFF system not dependent on interrogation by the radar pulse, and the Signal Corps Aircraft Radio Laboratory had adopted it, the airborne component being SCR-515. The Americans did not like mark III because they saw the sweeping of a 30 MHz band an open invitation for enemy mischief. Their own system had passed preliminary tests and was scheduled to begin final trials on USS *Hornet* on 8 December 1941. The British did not like the NRL-GE design because the interrogation used a single frequency that overlapped the Würzburg band [9] and that would be easier to jam [10]. In retrospect these seem a curious

---

[4] See Chapter 4.2 (pp 159–66).

objections. The Würzburg would generally be encountered with IFF sets turned off, and mark III responded to a range not far removed from the Freya frequencies; jamming IFF never became an important part of countermeasures. The NRL-GE system also used a single but different response frequency.

That Britain would use mark III come what may presented the awkward situation that would have required American planes operating in Europe to have two transponders, impossible for fighters. This quickly forced the decision in favor of mark III as the Allied standard. The American system, of which several hundred had been manufactured, was subsequently referred to as mark IV and held in reserve in the event mark III was seriously compromised [11]. It was compromised as early as 1943 and used as range-enhancing secondary radar regularly by Germany, but these radar struggles come later in the story. Mark III was retained.

As use increased IFF problems grew apace. When large formations began taking to the air the interrogating equipment began to show masses of signals, quickly earning the name of 'IFF clutter' and easily obscuring the presence of enemy aircraft. Just as bad, the transponder of one aircraft could trigger the response of another, the consequence of using the same frequency for interrogation and response; the solution was to lower transponder gains [12]. At sea the vertical lobes of the interrogating radiation were much larger than those of the shorter-wavelength primary radar, which when coupled with low transponder gain, led to vessels being subject to radar-directed gunfire from their own side [13].

Few things illustrate the organizational problems that plagued German radar better than IFF. German electronics work was invariably marked with high degrees of competence at engineering levels but confusion and want of understanding at the upper levels of command with obtuseness increasing with altitude. It is obvious that successful IFF requires agreement by all branches of one's own and allied forces. On the American side a Joint Radio Committee had agreed on a national IFF standard early enough for a system to be starting production in late 1941. Even before America's formal entrance into the war discussions had begun with Britain over this matter, leading to a decision shortly after the Pearl Harbor attack. The German contrast is stark.

The German work started soon enough. In November 1938 a meeting took place about recognition. It was only then that the Naval Command learned of the separate radar work for the Luftwaffe by Telefunken and Lorenz. At that time GEMA offered an IFF, eventually to be called Erstling (first born), for Freya. The need to have such equipment respond to 80, 62 and 53 cm as well as Freya's 2.4 m was obvious, but no agreement came from the discussions. Not only was there no agreement then, there was still none by the end of the war. In the fall of 1939 after a GEMA demonstration the Chief of Luftwaffe Signals ordered 2000 to 3000 units of Erstling, still with response only to 2.4 m [14]. At about the same time Dr Hans Plendl had designed an IFF for use with the Würzburg called Stichling (a prickly fish). It gave identification only in direction, not range, at the time not thought important for Flak. Martini's demands that the functions of these two devices be combined were ignored [15].

The Technical Bureau of the Reichsluftfahrtmisisterium (National Air Ministry), the government agency resposible for aviation matters, brought forth a modification of Stichling called Zwilling (twin) for which production began in early 1941 with 10 000 eventually being installed, this despite a report from the Air Research Station at Rechlin that the device was completely unusable. Martini was not able to stop production of this unit until 10 January 1942 [16]. GEMA cannibalized these units for components used to make Erstling [17].

Zwilling was deficient in that it responded only to the Würzburg and did not even do that satisfactorily. If a Würzburg sighted an aircraft for which the identification was in question, it altered the pulse repetition rate from 3750 Hz to 5000 Hz. The receiver, a diode detector sensitive to a wide range of wavelengths, responded to this change of repetition rate by transmitting on an auxiliary 1.9 m wavelength Morse code formed by a notched rotating disc in an audible 800 Hz. The radar operators listened on a receiver tuned to the auxiliary frequency. The deficiencies of Zwilling quickly became apparent: there was no range information and the direction of the reply signal could not be determined from the auxiliary transmission, as only a simple receiver was used at the radar station. This meant if there were more than one plane in the few degrees of the Würzburg's beam, one did not know which was responding. Further, when the radar set interrogated a target its normal functions were completely disturbed because of the need to alter the pulse repetition rate [18].

Dissatisfaction with Zwilling led to a field modification called Zwilling J1, which dispensed with the change in pulse repetition rate and replaced the amplifier that had been sensitive to the 5000 Hz with an amplifier for the radar pulses. These were then transmitted to the radar set on the 1.9 m auxiliary wavelength. The 1.9 m receiver at the radar set was altered to broad band width in order to accept the pulses, and its output fed to the radar oscilloscope screen [19].

On 9 July 1942 Erstling was introduced for the Würzburg as well as Freya by simply equipping the radar set with an interrogation transmitter on 2.4 m [20]. The interrogating system received the name Kuckuck—the same cover name used early by the British Army for MRU.

German IFF problems grew seriously in 1943 as a result of their own confusion, Allied jamming and the fear among their own pilots that IFF was being used against them by the enemy [21]. Various designs, the most important being Neuling (newcomer), were never deployed. The Allies also planned more advanced IFF equipment, Hazeltine and the Naval Research Laboratory designing IFF mark V [22], also not deployed during the war. For both sides IFF had become such an enormous endeavor that change came to be looked on with horror, not only because of the magnitude but because of the complications of a changeover that would necessarily extend for an uncomfortably long period.

The problem of IFF between aircraft became critical for Germany as the great struggle between Bomber Command and the Luftwaffe night fighters became intense in 1942 and was a problem never satisfactorily solved for the defenders—or the attackers either for that matter. Night fighters on both sides were twin-engine aircraft because two-man crews were required, necessary not only because of the complications of radar operation but also because of the absolute requirement that the pilot retain his night vision, which was ruined by looking at an oscilloscope. The absence of German plane-to-plane IFF led to an order to pilots not to attack any two-motor aircraft [23].

IFF did not prevent losses to friendly fire. An air crew sometimes forgot to switch it on when returning from enemy territory, where it should have been off; sometimes it had been damaged in action or it occasionally malfunctioned as a result of postponed or deferred maintenance; it was, after all, a piece of equipment that did not call for attention as would a defective communications transmitter or receiver. There was always the problem presented by hastily trained crews, who were not alert to a device that was supposed to be automatic. There were cases of an insufficient number of sets available for an operation, with occasionally disastrous outcome, the paratroop invasion of Sicily being one [24]. Ill use of IFF reached such alarming levels for the US in 1944 that a 40 min joint Army–Navy training film was issued for the benefit of front-line units [25]. A P Rowe, Director of TRE, perhaps put it best: '. . . the problem of IFF, like the poor, was always with us' [26].

IFF was, of course, very secret yet it invariably fell into enemy hands from aircraft wreckage, despite the explosive charges and thermite incorporated to destroy it, and furnished the eagerly sought information about interrogating frequencies. The first certain evidence that General Martini had about CH wavelengths was from a reconstructed IFF mark I in April 1940 [27].

Radar beacons are technically members of the IFF family. The early application of ASV radar by Coastal Command in 1940 to this purpose led to its use growing enormously on the Allied side. Beacons quickly became important to British night fighters, who were just as desirous of knowing where they were and how to get home as were their comrades in Coastal Command. This enthusiasm led to much user individualism so that matters became and remained rather complicated and local [28]. It grew beyond these uses. Portable beacons were set up behind enemy lines to assist drops of agents, supplies or paratroops. For these the general response of IFF mark III was not desired. One wanted the beacon to respond only to a specific interrogating wavelength rather than a band of wavelengths and to respond with a coded identification, which required a special interrogating radar that the British called Rebecca and a beacon called Eureka. A prime virtue of Eureka was in Watson-Watt's words 'to speak only when spoken to' [29].

3.5. THE JAPANESE REALIZE THEY ARE BEHIND

Japan's first dealings with Nazi Germany aroused both esteem and suspicion. Hitler's ejection of constitutional government coincided with the Imperial Army's own achievement and fostered a moderate amount of political comfort. The Anti-Comintern Pact concluded between the two in 1936 and expanded to include Italy the following year was much to Tokyo's liking, being directed toward the ominous power that threatened the Manchurian border, where, in fact, two full-scale battles erupted in 1938 and 1939. But Japan was outraged when Germany concluded the Non-aggression Pact with that same Soviet Union, considered an enemy three years earlier, and renounced the 1936 treaty—not all of the outrage at the German–Soviet pact was found among the democracies. A year later, seduced by Germany's astounding defeat of France, Japan was persuaded to enter the Tripartite Pact, pledging unspecified help to Germany and Italy for equally equivocal aid in return. The Berlin–Rome–Tokyo Axis was thus formally presented to the world.

As a consequence of this renewed and presumably friendly relationship, Japan sent two missions to Germany to learn what they could of military value. An Imperial Army delegation of 20 left in December 1940, but it was the Navy group that most actively sought information touching on radar. This group of 22, the largest naval mission ever sent abroad, had Commander Yoji Ito handling electronic matters with two others from the Naval Technical Research Department (NTRD) accompanying him [1]. They decided against the Trans-Siberian Railway used by the Army and chose a sea route through the Panama Canal. The separate travel arrangements illustrate the serious rivalry of the two services. The naval delegation left Yokohama on 16 January 1941 aboard the *Asaka Maru*, the same day they learned that American armed guards would board all foreign flag ships during passage of the Canal. By 6 February the *Asaka Maru* received assurances from the United States that this would cause no affront to the prestige of the Imperial Navy and dropped plans for a voyage around one of the capes. Indeed the passage of the Canal proved to be quite friendly with the senior officers accepting a dinner invitation ashore from the Commanding Officer of the Canal Zone. They departed Port of Cristobal 9 February and arrived in Berlin on the 24th by way of Portugal [2].

Ito and a few others immediately met with the military attachés in Berlin and left to see the points of naval interest from the French campaign, boarding 13 new Packard sedans at the Düsseldorf railway station, which they noted was severely damaged by a recent visit of the Royal Air Force. The first sight of radar that Ito gained was British, not German, the remains of GL mark 1s and MRUs displayed in the wreckage at Dunkirk. At the submarine base of Lorient Admiral Dönitz greeted the delegation, which proceeded with their inspection after a reception. It was then that Ito saw a Würzburg; it took no effort on his part to ascertain that it was a gun-laying radar, and he was suitably impressed. To obtain a closer look at the Würzburg took a bit of persuading and was limited to half an hour for a few, including Ito and Rear Admiral Naosaburo Irifune, head of the Navy's Gunnery School. Attempts to obtain details came to nothing, as Göring had declared the Würzburg the most secret of Germany's weapons; not even their formal ally, Italy, had been given such information [3]. Ito noticed a cage-shaped antenna on a pole that he took to be for a very-high-frequency directed beam. It was probably a Freya, as they were being installed along the coast at that time, but his inquiries received no answer [4].

The delegation toured Germany for several weeks with numerous meetings and exchanges of information, but this was not an 'Axis Tizard Mission', the very open exchange of secrets between the British and Americans that took place in late 1940[5]. Ito did not learn many details of German radar, but neither did he let out anything about his magnetron. Given the continued bad performance of German torpedoes, one must assume that they told them nothing about their own designs. One suspects that the Germans did not expect to be taught anything by the Japanese, which may have restrained their probing.

While the Naval delegation was in Germany, alarming intelligence began to accumulate, both to the delegation and in Tokyo, about British and American radar. This was more often than not inaccurate in the direction of exaggerating the capabilities, but this only added to the rising concern that Japan was being left behind in the knowledge of what was finally realized to be an important new weapon.

While the delegation was in Germany the Axis suffered two serious naval defeats at the hands of the Royal Navy, both resulting from British radar superiority. During the night of 27/28 March 1941 Italy lost three cruisers and two destroyers in the night action off Cape Matapan[6]. Rumor of this filtered to Tokyo from the Italian Naval Attaché in Washington in the form of a report that British ships could deliver accurate blind fire at night without searchlights. As a consequence of this report Commander Iwao Arisaka, an ordnance officer then the Naval Inspector resident in New York City, was ordered to investigate discretely any radar capability that the US Navy might have. His report, based on what evidence is not clear, came back with exaggeration that matched the Italian: American battleships, carriers and cruisers all had antennas on their foremasts suspected of being this new method of night vision. Inasmuch as the Japanese considered themselves to be masters of naval night fighting, this was serious [5].

---

[5] See Chapter 4.2 (pp 159–66).
[6] See Chapter 5.1 (pp 208–9).

This was followed by news reports—attended by rumors of radar's involvement—of the sinking of the German battleship *Bismarck* on 27 May[7], which led to instructions to the Naval Attaché in London, Commander Ryo Hamazaki, to look into the matter. He came up with precious little about what radar had to do with the sinking of the *Bismarck* but did send a description of what must have been a GL mark II that he saw set up with an AA battery in Hyde Park [6].

Part of the delegation to Germany had planned to leave in June, leaving Ito and the others to remain a little longer to extract what more they could, but the news of the radar capabilities of Japan's prospective enemies brought a peremptory order on 19 June for all to return immediately. If information gained about radar had been meager, the overall results of the mission were considered to be extremely valuable for the details of industrial processes, submarines, high-speed torpedo boats and methods for manufacturing artificial rubber.

The delegation returned home in two different groups. The first, which included Ito, departed Rome on 15 August in Italian aircraft for Recife, Brazil by way of Villa Cisneros in western Sahara. German aircraft took them from Recife to Rio de Janeiro, whence they boarded a Japanese steamer for Yokohama. The remainder went through Switzerland and Vichy France to Spain where passage on a Spanish ship to Rio was arranged, thence by freighter to Japan [7]. Although the Army's interest in radar did not equal the Navy's, one officer of its delegation, Lieutenant Colonel Kinji Satake, remained in Germany and became well instructed about radar and the Würzburg in particular after Japan entered the war.

By summer of 1941 the reports from abroad had begun to alarm key members of the Naval General Staff (Kaigun Kansei Hombu) and caused the issuance of an order on 2 August for the expenditure of ¥11 million ($4.4 million) on radar. Rear Admiral Kiyoyasu Sasaki, head of the Electrical Engineering Research Department (Denki Kenkyu Bu), was quite eager to get started and called a meeting of his entire staff. Commander Chuji Hashimoto was placed temporarily in charge until Ito returned, and he consulted closely with Ito's staff. Industrial support was added from NHK Japan Broadcasting Corporation's Technical Research Laboratory, specifically and importantly by Kenjiro Takayanagi, who had developed Japan's television, and NEC (Nihon Denki). Most of the design elements needed for a meter-wave apparatus were at hand among these groups, and assembling them into a lashed-up pulse-modulated 4.2 m set of 5 kW proved remarkably easy.

A 3 m refined prototype was set up on the grounds of the Naval Mine School on the Miura Peninsula by 8 September. At this time they had no data on the reflectivity of aircraft nor any idea as to the best polarization to use, but they were soon able to detect a medium-sized bomber at 97 km

---

[7] See Chapter 3.3 (pp 124–9).

and a flight of three at 145 km.

From this prototype the mark 1 model 1 (Ichi-go Ichi-gata) was contracted to three firms for immediate production: NEC for transmitters,
Japan Victor (Nihon Victor) for receivers and Fuji Electrical Apparatus
Manufacturing Co. (Fuji Denki Seizo) for antennas. The first industrially
produced radar was placed in November 1941 at the Katsu-ura Lighthouse,
where it was used throughout the war [8].

The Army relied primarily upon NEC and Toshiba for its radar, the
key developmental elements of which were the NEC Ikuta Research Office
Branch (Ikuta Kenkyu Bunsho) and the Toshiba Research Institute (Toshiba
Kenkyusho) [9]. The Doppler interference between widely spaced transmitters and receivers, such as had been observed at the US Naval Research
Laboratory, in the Watt's Daventry experiment, by David in France and
by Oshchepkov in Russia, had not escaped Japanese observation, and the
Japanese Army put this equipment, designated as type A, operating on 4
to 7.5 m, to use in China in 1941. The longest such link was from Formosa
to Shanghai, a distance of 650 km [10]. In June 1942 NEC and Toshiba each
began developing 1.5 m searchlight radars, designated Tachi-1 and Tachi-2
respectively. Both proved too fragile for field service and only about 25 of
each were manufactured, although Tachi-2 proved a financial success to
the manufacturer at ¥200 000 ($80 000) each [11]. The Army's air-warning
radar was the 4 m type B, soon designated as Tachi-6. It used a broadcast transmitter reminiscent of CH but with some degree of electronically
adjustable direction and had up to four separately directional receivers
[12].

Trustworthy information about Allied radar came with the first Japanese conquests. A report by the military commander in Singapore described
what were thought to be electronic weapons captured from the British,
which led to a delegation that included Masatsugu Kobayashi of NEC
and Shigenori Hamada of Toshiba flying to inspect [13]. At Singapore
they obtained a GL mark II and a searchlight control radar (SLC), which
startled them in its use of Yagi antennas. This antenna, named for their own
illustrious high-frequency expert, had been used extensively in America,
Great Britain, Germany and Russia for experimental work and a few final
designs, but it had found very little use in Japan. Along with the SLC came
a nice extra. A Corporal Newmann had made—in violation of draconian
orders to the contrary—a complete set of notes that the Japanese had typed
in English and duplicated in a 22-page booklet [14]. In the Philippines they
obtained an SCR-270 and 268.

The British SLC had impressed Hamada with its compact simplicity
and its similarity to the design already employed in Tachi-1 and 2, and he
had its improvements copied into Tachi-4. It had a single Yagi for the transmitter and four Yagis positioned about it for the receiver and connected
through a rotating capacitor that generated a conical scan, a technique
already used in Tachi-1 and 2. It also worked on 1.5 m and had trailer-

mounted antennas that could be pointed in azimuth and elevation as had the two predecessors.

Kobayashi allowed Masanori Morita to make the improved copy of the GL mark II called Tachi-3 that had dominated his thoughts since he had learned of the British set [15]. Transmitter and receiver were mounted separately about 30 m apart over underground shelters that rotated in azimuth. The radiation pattern of the transmitter could be adjusted in elevation by altering the phase between pairs of dipoles. The receiver had five dipoles in a horizontal array, four forming a diamond that yielded lobe-switching for azimuth and elevation with the fifth used for determining range. Both Tachi-3 and 4 became widely used sets for searchlight and somewhat limited gun-laying use.

The Navy successfully developed its own version of SCR-268 as mark 4 model 2 and placed it in operation at the important base of Rabaul in 1943. It was widely used and one of the most produced Japanese radars, 2000 units having been manufactured. As the war progressed the nomenclature of Japanese radar becomes more confused, although generally with regard to equipment that was never deployed.

With Japan's entrance into the war against two of Germany's three principal enemies a change in attitude prevailed in Berlin—specifically Japan could have the Würzburg secrets. To obtain the details and an operating set the new and very large submarine I-30 was dispatched on a long, adventurous voyage to France with 120 men on board and arrived at the growing U-boat base of Lorient on 8 August 1942 to a rousing welcome. I-30 departed on 22 August with all that could be desired for building Japanese copies of the now famous gun-laying radar and arrived at the Penang Naval Base on the Malay Peninsula on 10 October. After refueling she continued this remarkable voyage only to strike a British mine seven days out of Penang. The sample Würzburg was damaged beyond repair and the drawings became a soggy mass. In summer 1943 Colonel Satake and a Telefunken engineer, Heinrich Foders, made a harrowing voyage on an Italian submarine to bring Würzburg capability to Japan; unfortunately, important components and data were on an accompanying submarine that was sunk [16]. The Navy re-engineered it as mark 2 model 3, and the Army produced its own version independently as Tachi-24, illustrating the fundamental problem the nation had in apportioning scarce resources. Neither advanced beyond prototypes completed in April 1945 nor affected the war in any consequential manner [17].

GEMA also sent an engineer, Dr Emil Brinker, to Japan by a submarine that landed in December 1943 with details about Freya, including anti-jamming circuits. The Japanese considered their own air-warning equipment adequate and were not interested in Freya, so Brinker spent his time developing radar test equipment. Although an expert in IFF, he was not allowed to enter Japanese radar research because his security clearance was only for work on a Freya [18].

By the end of the war Japan's radar had been completely outclassed by the Allies. This came about from having resources unequal to the task and a military divided from beginning to end as to radar's relative importance in the disputes about allocations. Until dramatic examples were presented in the late war years of the great power of this new method of waging war, this attitude held back Japanese radar. There were high levels in the government that appreciated the military need for scientific and technical research, so Prime Minister Fumimaro Konoe established the Cabinet Board of Technology (Gijutsuin) to organize such efforts and appointed Viscount Kyoshiro Inoue, Professor at Tokyo Imperial University and Minister of Railways, as president of the Board on 31 January 1942, but the Board was severely hampered by contemptible annual budgets and ignored by the feuding Army and Navy. Its effect on the course of the war is difficult to find [19].

Shigeru Nakajima headed research at Japan Radio, which conducted Japan's research in magnetrons, and saw his staff shrink from 800 at the beginning of the war to half that number at the end [20]. Technical specialists were simply drafted into the Army. It was a startling comparison to the huge growths of similar American, British and, after an initial pause, German groups.

Nevertheless, much was accomplished. Fielding an industrially produced 3 m air warning set in November 1941 in a program begun in August is rivaled in speed only by the South African production of the JB and the Australian of the LW / AW[8]. As we shall see, the Japanese 10 cm equipment was at sea only months behind the British and weeks behind the Americans. Yet the value of their sets was limited by retaining only the A-scope, the most primitive indicator, the display of signal size against range, something that immeasurably reduced the effectiveness of this equipment and that so remained until the end. Only 100 IFF sets were manufactured [21].

[8] See Chapter 5.2 (p 221).

PHOTOGRAPHS: AIR WARNING—PACIFIC

Australian–American cooperation. A 1.5 m LW/AW mark II antenna (light-weight/air-warning radar) at Rose Hill Racecourse in Sydney with US personnel in training. This set, which the outbreak of the Pacific War caused to go from design to production in a few months, was light enough to be transported in the ubiquitous DC-3 transport and was widely used by Allied forces in the South Pacific during the early years when American equipment was both rare and too heavy for some island operation. No landing was considered secure until an LW/AW was on the air. The frame below the antenna supported a tent. Australian knowledge of the jungle requirements of electronics proved valuable for the Allied forces operating in the South Pacific. Worledge Collection, Radar Research and Archive Collection, RAAF, Williamtown, NSW.

An American SCR-271 3 m air warning radar at New Caledonia in May 1943. The development of this set resulted from the Army Air Corps eagerness for a device to protect their bases from surprise attack. The 271 was a fixed-station mount of the electronically similar SCR-270, which was portable. The cabin at the left rear housed the engine-driven generator. The military unit was Company A, 579th Signal Air Warning Battalion. National Archives photograph 111-SC 245874.

British Light Warning radar LW. This 1.5 m set weighed only 1200 kg and was generally the first radar on the beaches. It was copied by the United States as the SCR-602. Troops in the Pacific found its electronics superior to the Australian LW/AW, which they used until mid-1944, but preferred the latter's dipole-array antenna to the four Yagis of the original design. Historical Radar Archives. Crown Copyright.

An American SCR-602 light-weight air-warning radar. This 1.5 m set was a copy of the British LW and went through a number of modifications. Here the original Yagi antennas have been superseded by a dipole array. It replaced the Australian LW/AW for American units as the war progressed. The device shown here was used by the 583rd Signal Air Warning Battalion on Tanahmerah Beach, New Guinea, 23 April 1944. This battalion placed detachments with even smaller radars, AN/TPS-2, among Filipino guerrillas to monitor Japanese air and sea traffic before the invasion of the Philippines. National Archives photograph 111-SC 254238.

Japanese Navy 3 m fixed air-warning radar mark 1 model 1. This set was one of two captured in August 1942 on Guadalcanal to the great surprise of American radar men. It had a peak power of 5 kW, pulse lengths of 3 to 20 μs, repetition frequency of 750 to 1500 Hz and a range of about 150 km. National Archives photograph 80-G 11293.

Japanese Army 4 m Tachi-6 air-warning radar transmitter. This used a broadcast transmission with a vertical dipole array configured to produce radiation lobes covering the region from which attack was expected. Three or four directional receivers completed the station. This design was the Army's standard air-warning radar. This well camouflaged antenna was captured on Noemfoor Island, Dutch New Guinea in July 1944. National Archives photograph 111-SC 267148.

Japanese Army 4 m Tachi-6 air-warning radar receiver. Rotatable horizontal dipoles with reflectors on two levels provided directional accuracy of 5° by maximizing. The receivers were located about 100 meters from the transmitter. The system has elements common to CH but was almost certainly designed without knowledge of the British equipment. This antenna was captured on Noemfoor Island, Dutch New Guinea in July 1944. National Archives photograph 111-SC 267161.

CHAPTER 4

NEW IDEAS

4.1. MICROWAVES

Most of the originators of radar wanted to use microwaves from the very start. The main reason for this was the desire to construct a radio searchlight, and the mirror for such a searchlight had to be several wavelengths in diameter. By forming a narrow beam one could also concentrate more power onto the target, technically called increasing the antenna gain, something of no mean importance considering the low power of the generators then available. A narrow beam also yielded much less reflected power from nearby ground or water, making the operator's task simpler. Not appreciated at the time was the decreased susceptibility of narrow beams to jamming or to the reflecting dipoles that attacking aircraft were to throw into the radio eyes of the defenders. That shorter wavelengths allow one to resolve smaller target structure did not figure into their thoughts either, although the importance would soon be felt.

Preliminary experiments, such as done in the United States, France and Germany, had shown clearly that microwaves yielded detectable reflections from ships, automobiles and aircraft and indicated that practical equipment could be made, had there been a generator of sufficient power. But there were arguments from theorists that microwaves would not be effective because of specular reflection, and an experiment in the summer of 1936 at the Nachrichtenmittel-Versuchs-Anstalt gave the opponents of microwaves some supportive data. A metal screen was suspended between the masts of a small vessel from which the reflection of very short waves was observed. Swinging the ship by 90° produced very large effects [1]. This experiment, which satisfied some investigators of the unsuitability of microwaves, was about forty years ahead of its time and was supported by inadequate theory and computing capability. An airframe whose surface is made up entirely of flat surfaces—an aerodynamic horror—can be made invisible to radar for any reasonable orientation of aircraft to radar beam. This is the basis of the Stealth technique [2].

For the first few decades of microwave research the only design element not common to radio in general was the parabolic mirror. This drew

145

on both optical and acoustical experience, the optical from astronomical telescopes and searchlights, the acoustical from the whispering chambers of renaissance architecture. The next truly microwave design element came from ancient acoustical experience. Almost every child has learned for himself or from his playmates that blowing across the mouth of a bottle or jug produces a low-pitched tone. This phenomenon comes about from the sound being restricted to certain discrete wavelengths because the air molecules cannot vibrate freely when in contact with the walls of the jug. Hermann Ludwig Ferdinand von Helmholtz, whose investigations probably encompassed more branches of learning than any other prominent scientist, used these simple devices in explaining the physics of sound and in formulating a physical theory of music. For our part he earned a place of honor for having guided his favorite pupil, Heinrich Hertz, toward electromagnetic waves.

Stimulated by von Helmholtz's 'On the Sensation of Tone', Lord Rayleigh undertook a mathematical analysis that ended in the publication in 1877 of *The Theory of Sound*, a book still in print for its tutorial as well as historical value. The analysis of acoustical resonators can be transferred to electromagnetic waves in cavities by requiring that the tangential component of the electric field be zero at the conducting surface of the cavity, essentially the same requirement as saying that air molecules cannot vibrate when touching a hard surface. This transfer was made by W W Hansen of Stanford University in the mid-1930s [3]. Hansen's analysis went beyond solving the partial differential equations that describe the waves with the boundary conditions imposed by the cavity. He reduced the results to a form that allowed a designer to consider them as equivalent circuit elements and described how one made the connection from a two-element conductor to a cavity. He predicted and verified that energy loss in such chambers was remarkably small.

Hansen's original goal was to accelerate electrons to produce high-energy x-rays for medical application. David Sloan had built an accelerator based on this principle at Lawrence's laboratory in 1934, but the frequencies available from the oscillators then available restricted its use to ions much heavier than the protons, deuterons and alpha particles that the nuclear physicists wanted to use. Electrons required techniques of much shorter wavelength applied over distances that precluded accelerating electrodes connected conventionally [4].

Closely related to a resonant cavity and just as vital for microwave work is the waveguide. Rayleigh had shown that electromagnetic waves could be propagated between two parallel conducting sheets whose separation was comparable to the wavelength, and this was seen to be the explanation of the worldwide propagation of very long wavelengths, the earth's surface and the ionosphere [5] forming the two conducting surfaces. He also showed that waves could be propagated in tubes of circular or rectangular cross section [6]. George C Southworth's thesis at Yale had left a residue of ideas that grew into waveguides some years later at Bell Telephone Laboratories [7]. At the same time at the University of Kiel an experimental thesis by O Schriever reported the propagation of radio waves in dielectric rods using the new Barkhausen–Kurz oscillator [8]. Southworth read and appreciated the paper's significance, but he remained unaware of Rayleigh's paper until years later when it was found during the research for the waveguide patent.

Southworth continued work in high frequencies at Bell Labs and as sources of 15 cm radiation became available took up again the dormant idea of propagation through a pipe. In August 1933 he succeeded in building a circular waveguide of 12.5 cm diameter. He had had to hide the early work as a form of 'test' rather than 'research' because of the restrictions on his duties. When the truth came out it attracted the attention of a theorist, who almost wrecked the project through a calculation error. The transmission of electric power without a return conductor was a bit too startling and evoked skepticism, but success brought approval and theoretical help. Soon Bell was full behind the project. Southworth used the term 'wave-guide' in an early memorandum [9].

At the Massachusetts Institute of Technology (MIT) Wilmer L Barrow was making waveguides independently. After earning a degree in electrical engineering at MIT, Barrow went to the Munich Technical Institute, completing a dissertation there in acoustics. He also received instruction in mathematical physics from Arnold Sommerfeld with important consequences for work he was to do. He returned to MIT and began research in very-high-frequency antennas and propagation. His approach to the waveguide, unsuccessful at first, was as a method of picking up signals reflected from a paraboloid mirror. By Christmas 1935 he understood the critical elements and by the following March he had worked out the nature of the vibration modes in a rectangular guide and derived the cutoff wavelength beyond which the guide would not transmit [10].

The two investigators learned of one another's work a few weeks before both were scheduled to present papers to a combined meeting of the American Physical Society and the Institute of Radio Engineers in May 1936. Patent problems were worked out amicably.

An open-ended waveguide squirts radiation much like water from a hose. Barrow determined the optimum shape for the horns that terminate the waveguide—the nozzle, as it were—so that a minimum of energy was reflected back down the pipe. There was soon great activity at Bell Labs and MIT on working out the details of waveguides and horns as circuit elements.

Microwaves can be transmitted by coaxial cables but reach practical limits beyond which waveguides have the advantage. As frequencies increase the rapidly changing electric field is able to penetrate ever shallower depths of conductors, the skin effect. For a coaxial cable this means that less and less of the center conductor is used, which results in ever increasing resistance. The same thing happens on the inner surface of the outer shield, but it has more surface area. For a coaxial cable this means that with microwaves only a small fraction of the center conductor is used and transmission becomes increasingly attenuated. For high power the cable will become hot, even with the low-loss insulator, polyethylene. In a waveguide all the necessary conduction is in the large periphery, so attenuation is much less.

Most of this waveguide research was carried out with the Barkhausen tube, later replaced by something far better—the klystron.

The idea of using the electron's relatively slow speed to effect the generation of very short wavelengths had been in the air since the invention of the Barkhausen–Kurz tube and the split-anode magnetron, both of which we have encountered. In 1935 a paper from Italy [11] presented a new idea, the Heil tube. A beam of electrons passes through an electrode that is connected to a resonant circuit; a gap precedes the electrode and another gap follows it; when the resonant circuit oscillates the electrons are bunched in the first gap (the buncher) and energy extracted from them at the second (the catcher); the oscillations build up, if the electron velocity is properly matched to the dimensions of the electrodes and the resonant frequency. Electron bunching is brought about by reducing the speeds of the first of a group of electrons and increasing those of later ones, thereby producing a grouping further down the beam. Some excellent Heil tubes were made, but the dependence on external circuit elements greatly reduced their suitability for microwaves [12].

A couple of years after the Heil tube Hansen built a microwave tube that combined one of his resonant cavities with an Acorn triode and an electron beam in a design he called the 'rhumbatron' [13], but it was quickly swept aside by the klystron. The klystron was the product of two brothers, Russell H and Sigurd F Varian, working with Hansen in the Stanford Physics Department. They dispensed with the Acorn and used an electron beam much as in the Heil tube (of which they were unaware) but used a resonant cavity, which Hansen designed specially for the tube [14], in such a way that the standing waves of the cavity acted directly on the electron beam instead of imposing resonance with an external resonant circuit. This proved to be a superb generator of microwave oscillations and went through several development generations in short order. Its name is derived from the Greek verb 'klyzein' that describes the breaking of waves on a beach.

The team of Hansen and the Varian brothers combined three remarkable men. Hansen was simply brilliant. He graduated from high school at the age of 14 and went immediately to Stanford where he mastered the disparate skills of master mathematician and machinist. Russell Varian, the older of the two brothers, had such difficulty in learning to read that he graduated from high school four years older than his classmates but with a solid grasp of the things he learned and a good knowledge of science. His perseverance earned him a BA in physics at Stanford. Sigurd wanted nothing to do with college, purchased a war surplus Curtiss Jenny and entered a career as pilot, eventually trying out new routes for Pan American Airlines. His knowledge and interest in aerial navigation later supplemented by concerns for air defense led him to work with Hansen and his brother on microwaves, contributing his enormous practical and organizational skills [15].

In frequency stability the klystron was all that the radar engineers could have wanted, but it did not have adequate power. The power level is proportional to the intensity of the electron beam, and as in the cathode ray oscilloscope[1] the electrons repel one another, so that as one attempts to increase intensity the beam diverges and the desired effects are soon dissipated. Unlike the transmitter triodes, no dramatic increase in power could be obtained by pulsing the tube because the limit was space charge, not the thermal characteristics of the anode. If not useful as a transmitter, although there were attempts, it proved to be invaluable for microwave radar in more delicate functions, such as amplification and as the local oscillator for producing intermediate frequencies.

The Stanford work attracted the attention of Professor Edward L Bowles of MIT and the Sperry Gyroscope Company, who were experimenting with methods for assisting pilots in making blind landings. As a consequence of Bowles's favorable recommendation Sperry gave financial assistance to Stanford and opened a shop in nearby San Carlos for designing a blind-landing system based on the klystron. The Varian brothers went to the new Sperry shop, and Hansen remained at Stanford to continue klystron development. In the fall of 1940 Sperry transferred these operations to Brooklyn [16].

Alfred Lee Loomis was perhaps the last of a long tradition of amateur scientists, amateur only in not earning his livelihood from science; his income was derived from investment banking, and his science done on the side. He was also a successful lawyer and Army officer in World War I. It was while serving as a major at Aberdeen Proving Grounds that Loomis became friends with the Johns Hopkins physicist R W Wood. After the war Wood helped Loomis set up a private laboratory at Tuxedo Park, New York from which an impressive list of publications arose [17]. Conversations with Bowles convinced Loomis of the value of the new microwave equipment, and he agreed to set up a program to be conducted by J A Stratton to study propagation [18]. They soon had a continuous-wave radar measuring automobile speeds at Tuxedo Park.

The collapse of France in June 1940 altered entirely Americans' view of the war. It placed the heads of four leading scientific organizations in much the same position that Tizard and his associates had found themselves five years earlier: Vannevar Bush of the Carnegie Institution, James

---

[1] See Chapter 2.1 (p 33-5).

Bryant Conant of Harvard University, Karl T Compton of MIT and Frank B Jewett of the National Academy of Sciences. They had discussed informally in the past the ways that the country might organize its scientific and technical strengths for defense, but the defeat of France and Britain's extremity demanded immediate action. Bush incorporated their ideas into a memorandum for President Roosevelt that he was able to present in person in early June, receiving the famous 'OK FDR' initials. The result was the creation of the National Defense Research Committee, NDRC.

Bush began to form sections for the varied forms of research to be pursued and chose Loomis to head Section D-1, the Microwave Committee. Loomis selected Bowles as Executive Secretary, included representatives from Bell Labs and Sperry Gyroscope and called a meeting at Tuxedo Park on 14 July 1940. They quickly added the heads of General Electric, Westinghouse and RCA along with nuclear physicist Ernest O Lawrence. The Committee took as its first business the survey of what was being done by the armed forces in radar and within days visited the Naval Research and the Signal Corps Labs. One of the impressions that they gained, and which was to remain preserved in the memory of the NDRC people, was the conviction that the two services kept their radar work so completely covered in secrecy that neither knew what the other was doing [19]. This bit of misinformation was soon expanded to include the belief that the service laboratories had never consulted the electronic industry.

In surveying the work that had been done in microwaves the Committee learned of another very-high-frequency tube, a demountable, continuously pumped beam tetrode having mechanically tunable cavities; it delivered 1 kW continuous and 5 kW peak power at 50 cm. Generally referred to as the Sloan–Marshall tube or the Resnatron, it had been developed at Lawrence's Berkeley Laboratory and reported at the June meeting of the American Physical Society. Fortunately for the security that now began to cover microwaves, the abstract [20] disclosed no details in its record briefness of two sentences and no publication followed. The Committee wanted a shorter wavelength but nevertheless thought the work of value and secured $20 560 to further it. This tube would be swept aside within weeks, when the long-sought generator for 10 cm waves arrived, but was to have a curious part to play by and by.

Britain had placed far and away more resources into the radar program prior to the outbreak of war than any other nation but nothing into microwaves, although the Royal Navy was developing decimeter equipment. The need for very short waves became pressing as the shortcomings in the equipment for airborne interception (AI) became apparent. This concern was not really so serious as thought, as the horrendous losses that Bomber Command were to suffer from German night fighters equipped with 3 m radar were to demonstrate, but radar AI had not been used in combat in early 1940, and the designers were rightly concerned with what they perceived as a serious defect of their product.

Bowen had proposed very early pushing microwaves but found 'no interest within Bawdsey itself and anyone talking about centimetre waves was thought of as some kind of crank', although he did find interest with Sir Charles Wright, Director of Scientific Research for the Admiralty [21]. Decimeter waves had already engaged the interest of the defense establishment, and the Admiralty awarded a development contract to the General Electric Company, Ltd (GEC) in November 1938. GEC entered into the problem of AI a year later and had a complete 25 cm pulsed radar lashed up at their Wembley laboratory in April 1940. It used high-power triodes and attained an output of 2 kW. Of particular note was its use of crystal diodes for the mixer. The set had, in fact, exceeded the goals set for it four months before [22].

Two eminent nuclear physicists recently brought into radar, John D Cockcroft and M L Oliphant, became concerned about the AI problem and about the general absence of research in very short waves. Oliphant secured a development contract for microwaves in September 1939 for the Physics Department of the University of Birmingham from the Admiralty Department of Scientific Research and Experiment [23]. Work began immediately on constructing klystrons, and Oliphant soon had James Sayers design one yielding impressive continuous-wave power but of temperamental performance: 'ten minutes once a fortnight' some claimed [24]. Such results were hardly encouraging, especially the inability to compress the impressive power into pulses [25]. One of the Birmingham staff, J T Randall, known at that time for having been the inventor of the phosphor used in fluorescent lamps, was assigned to microwave generation with a lecturer in radiophysics, H A H Boot. They saw that a resonant component had to be incorporated within the vacuum tube and that Hansen's cavity offered no solution beyond the klystron. This conclusion pointed them toward the magnetron.

The demands for speed fortunately prevented them from making the usual survey of the magnetron literature, much of it not in English, which by that time had grown to impressive dimensions. Hoping to find a new approach, Randall read a translation of Hertz's work and recognized in the loop-wire resonator, used in the very first radio experiments, the design element they sought. They thus dropped Hansen's three-dimensional cavity for Hertz's two-dimensional loop [26].

The result was a success on the first trial that fixed the fundamentals of this device. A plurality of loops was connected in a circular pattern around the cathode to increase the output power; the magnetic field was oriented perpendicular to the plane of the loops. This had two important advantages over any of the magnetrons previously made: the Hertzian loops were made out of copper as thick as practical and maintained at ground potential. This allowed the anodes, which must take up the energy of the electrons that had not been expended in radiation, to be much larger for a given magnet size than the split-anode magnetron and to be water cooled, as in the first model, although generally replaced by air cooling as a consequence of the high efficiency of the device in converting direct-current power into microwave power. The difficulties with previous magnetrons—soon to disappear from the radio world—were the requirement of an external resonant circuit, the inability to dissipate the heat generated on the anode except by thermal radiation, and an inherent frequency instability.

With the goal of generating 10 cm waves they made their first anode block according to the formula for the Hertzian loop, which specified 1.2 cm diameter; the thickness was arbitrarily chosen to be 4 cm. The first model was pumped by a diffusion pump and sealed with wax in the tradition of the golden age of physics. The first trial on 21 February 1940 was an immediate success. The continuous wave output at 9.4 cm was 400 W and, unlike the early klystrons, pulsed operation would yield greatly enhanced power. The powerful continuous output allowed dramatic demonstrations of glowing lamp bulbs and heated flesh. Worries about serious frequency instability, a continuing problem with split-anode magnetrons, were soon set to rest.

In April a contract was awarded GEC for the design of a manufacturable, sealed-off tube. E C S Megaw incorporated a large, cylindrical oxide cathode, drawing on the experience of Henri Gutton of the Société Française Radioélectrique, who had demonstrated the high currents attainable and the ability of such cathodes to work at high voltage [27]. Pulsed operation followed quickly because of Blumlein's recent contribution of the high-voltage modulator. Despite the importance of the work, Randall and Boot worked alone on the project until June, a measure of the shortage of qualified personnel. Within a few months tubes delivered peak powers of 10 to 50 kW with 10 to 20% of the power supplied converted to microwaves [28].

In June E G Bowen was relieved of his duties at St Athan in developing meter-wave AI and ASV, which was turned over to his assistant, A G Touch, who continued in much more suitable quarters at the Royal Aircraft Establishment at Farnborough. He and Brown soon completed development of ASV mark II. Bowen was re-assigned to ill defined duties at the principal Air Ministry Research Establishment (AMRE) laboratories, which by then had left their unsuitable quarters in Dundee for Swanage, located on the Channel coast 35 km due west of the Isle of Wight. It was suitably placed to participate first hand in the invasion Hitler would soon be planning. There Bowen learned about the capabilities of 10 cm radiation and lectured informally on the general problems of airborne interception.

The resonant magnetron naturally changed the direction of the 25 cm work being so successfully carried forward at GEC, and their research was combined with the work being done at Swanage at a meeting held on 29 July 1940. It was not to prove a 'happy marriage', primarily because the GEC engineers saw their new colleagues as being considerably behind in radio engineering. Dr C C Patterson, Director of GEC's Research, wanted

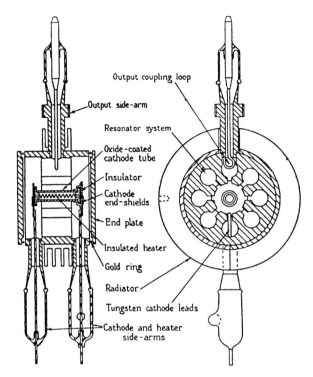

Output coupling loop

Output side-arm

Resonator system

Oxide-coated
cathode tube

Insulator

Cathode
end-shields

End plate

Insulated heater

Gold ring

Radiator

Tungsten cathode leads

Cathode and heater
side-arms

*Plan and section views of the resonant magnetron E1189. E C S Megaw of GEC
designed this tube from the laboratory model of Randall and Boot. It used a cylindrical oxide cathode and produced in pulsed operation 12 kW at 9.5 cm wavelength.
Number 12 of this series went to America with the Tizard Mission. E B Callick,*
Metres to Microwaves.

AI development done at their well equipped and well staffed laboratories,
whereas Rowe wanted the work done at AMRE under the supervision of
P I Dee, a nuclear physicist with little radio experience. That the working
conditions at Swanage were primitive added more to Patterson's displeasure. Rowe won out [29].

The most effective receiver for microwaves would use the heterodyne
principle, by then almost universal for communication work. This mixes
the receiver input with a tunable local oscillator to form a much lower
and constant beat frequency, called the intermediate frequency (IF) that is
more conveniently amplified. To do this requires a tunable and stable local
oscillator of frequency comparable to the signal frequency. The klystron
came immediately to mind. It was stable enough but tuning was done by
mechanically altering the shape and hence volume of the resonant cavity,
a clumsy way. Robert W Sutton, a master tube designer at His Majesty's
Signal School, was presented the problem and soon returned with a very

valuable microwave component based on principles outlined earlier by the Stanford group—the reflex klystron [30].

As the reader now knows, the klystron generates microwaves by exciting one of Hansen's cavities to a resonance that matches the velocity of the electron beam. If the electron velocity is not perfectly matched to the cavity, oscillations will still take place but at a smaller amplitude; there is a range of frequencies centered on the cavity resonance for which oscillations with amplitudes usable for purposes of a local oscillator will occur. Sutton placed an electrode to reflect the electrons back into the tube. By varying the voltage on the reflector, he could alter the frequency slightly, which enabled the development of circuitry to perform that function. In this way the intermediate frequency could be held constant, a requirement for constant amplifier gain, when the magnetron frequency drifted slightly. A protype tube operated in September 1940.

The mixing of frequencies for an IF amplifier requires a non-linear circuit element. A diode rectifier, a tube or circuit element that permits current to flow predominantly in one direction, is certainly non-linear, but for wavelengths of 10 cm the vacuum tube diodes had too much inter-electrode capacitance, the same thing that made trouble for some uses of triodes at very high frequencies and prevented use of the usual heterodyne mixers. The solution lay in resurrecting another circuit component from early radio to join the super-regenerative receiver in doing modern service—the crystal diode. Many a home wireless enthusiast in the 1920s and as many children in the 1930s had delicately positioned a needle on the surface of a galena crystal to make their crystal sets work. The delicacy of this operation saw its replacement by vacuum tubes occur without a great deal of nostalgia, but the crystal diode had for the radar engineer the important property of having much lower capacitance than one formed by the electrodes of a vacuum tube.

The utility of the crystal diode had not escaped the attention of early experimenters with microwave devices. H E Hollmann described their incorporation in regenerative receivers [31], and George Southworth reported their routine use at Bell Labs as detectors [32].

Silicon crystals proved to have better properties than galena, and a plentiful supply of metallic silicon was found in the Birmingham department. H W B Skinner at TRE designed a wax-encapsulated diode suitable for production but for a time was the sole manufacturer [33]. The main problems in constructing a microwave receiver had been solved: high-power generator, local oscillator, mixer.

Bernard Lovell went into radar during the summer of 1939, one of the university physicists assigned by the expedient of professors putting together duty rosters much as sergeants assign men to guard or kitchen duty. After his apprenticeship at a CH station he began to work with Bowen and Brown on airborne radar and by the summer of 1940 he found himself making his first and fateful encounter with microwaves. He first set up shop at Worth Matravers but soon moved to Leeson House nearby in the collection of TRE laboratories [34].

When the first of the General Electric Company's sealed magnetrons came to TRE on 19 July, Lovell and his colleagues set about to make a crude pulsed system as soon as possible and by 12 August they had observed reflections from aircraft [35]. Within a few weeks they had demonstrated its utility for detecting ships, including a surfaced submarine. This was a simpler design problem than AI, and the Admiralty type 271 was rushed to production at HM Signal School—the first operational 10 cm radar [36]. It used separate transmitting and receiving antennas, which were enclosed in a lantern-like structure of teak and plastic at the top of a mast. Its function was to scan about the ship, but the waveguide technology was not far advanced nor was the technique of making simple rotating connections. Also missing was the duplexing technique for microwaves that would have allowed a common antenna to be used. Polyethylene coaxial cables were available, and they allowed rotation of the antennas for two revolutions, after which the cables had to be unwound. The plan position indicator, which gives the operator a maplike presentation of the targets as the antenna continually rotates, was not yet available, so this restriction was not particularly onerous. The maximum range for detecting a surfaced submarine was 5 to 7 km [37].

Having something good invariably seems to goad humans to want something better. The stability both in amplitude and frequency of the cavity magnetron was so much better than that of the split anode variety that it was hailed as a triumph, but it was soon noted that there were different oscillation modes for which the device showed somewhat random preference and that required an adjustment of the local oscillator of the receiver in order to return the target blips to the scope. This had been an incentive for the invention of the reflex klystron, but Sayers sought a more direct correction and determined that the problem lay in maintaining the same phase between non-adjacent poles of the oscillating Hertzian loops. He forced the phase with by-passing anode conductors, called straps, which stabilized the frequency sufficiently that one had to wait another generation for new complaints to arise, and power miraculously grew in strapped magnetrons [38].

There are no simple explanations of the cavity magnetron, only complicated ones, none satisfying. The best approach is to describe the forces acting on the electrons as they attempt to reach the anode. The cathode has a potential some kilovolts negative relative to the anode, so electric forces push the electrons toward the anode. Perpendicular to the plane of their motion is a magnetic field that pushes them at right angles to the direction of their velocities. The magnitude of this magnetic force is proportional to the electron velocity but at right angles to its ever changing direction. Left to itself this situation would admit analysis, but a further complication enters—a complication on which the whole matter depends. All of these changes of electron velocities, both of their directions and speeds, require accelerations, and Maxwell's equations entail radiation whenever there is acceleration, just the process that produces the microwaves. So as the electrons dance their way to the anode, they lose energy and in so doing produce the microwaves as well as introducing severe difficulties into the equations of motion.

The design elements of the magnetron were known 20 years before Randall and Boot put them to use: (1) vacuum good enough for electron tubes, (2) the diode, (3) the law governing the motion of electrons in electric and magnetic fields and (4) the Hertzian loop resonator. Thus it is not really surprising to learn that it had already been invented, not once but three times.

That one of these should have come from Japan can hardly be a surprise. Yoji Ito, the Imperial Navy's leading electronics expert, had been greatly impressed by a magnetron based on the split-anode configuration invented in 1927 by Professor Kinjiro Okabe of Tohoku University, but Okabe left magnetron research in 1936. In 1937 Ito was rewarded by two members of his group, Tsuneo Ito (unrelated to Yoji Ito) and Kanjiro Takahashi, producing a tube with eight segments that excited an internal resonance in the manner of a cavity magnetron and that produced centimeter waves. The concept soon yielded a 6 cm magnetron that produced 30 W, a 3 cm tube of 3 W and a 1.5 cm tube of 1 W. The value of industrial electron-tube laboratories led to the decision in early 1939 to reveal the secret aspects of the magnetron to Japan Radio where Shigeru Nakajima soon operated a 10 cm eight-sector tube that delivered 500 W of continuous power [39].

Some Japanese have subsequently credited Arthur L Samuel of the Bell Telephone Laboratories for the invention of the cavity magnetron [40]. This is an interesting credit because Samuel was never able to extract enough power to warrant describing the device in a publication, and the Bell Labs—certainly interested in microwaves—did not carry the idea further. An examination of Samuel's patent [41] shows one of the three electrode shapes to be highly suggestive of the cavity magnetron, and Okabe was sufficiently impressed that he showed it in his 1937 book [42]. A critical difference lies in the wide gap Samuel placed at the mouth of the loops; the Japanese design differs primarily in tightening this parameter and thereby forming a Hertzian resonator. The arrangement Samuel chose for the electrodes forced the use of a coil without an iron core to produce the magnetic field, which very likely kept operation below the minimum field required for magnetron oscillation, were it possible. Both the Samuel and the Japanese designs extracted microwave power through two balanced leads attached to cavities of opposite polarity. Samuel had seen the need to place the resonant elements within the tube but thought in terms of discrete components having inductance and capacitance.

The 1939 Japanese tube was water cooled and had an anode 1.2 cm thick. Continuing work naturally brought forth a variety of anode designs.

The delegation to Germany in 1941 had learned the advantages of pulse modulation, which was applied immediately on their return to Japan and put an end to continuous-wave methods.

These magnetrons had glass envelopes and used electromagnets throughout the war, owing to a shortage of materials for permanent magnets. The first sets used a super-regenerative receiver that was extremely difficult to tune and that was replaced by super-heterodyne equipment. Each type required a stable oscillator for the detector for which a very low-power magnetron, the M-60, with the anode made from a thin molybdenum plate rather than massive copper was designed. It operated at much lower voltages and currents than one normally encounters in cavity magnetrons [43]. Mechanical deformation of the thin anode of the M-60 altered the frequency slightly, an effect utilized for tuning [44]. The Japanese did not use klystrons even though descriptions of them were in the open literature and never learned of the reflex klystron as a microwave local oscillator. The M-60 was paired with the M-312, which had the high emission needed for high-power pulse work; it also had cavities of alternating size, which suppressed unwanted modes. The 10 cm prototype radar was quickly transformed into the mark 2, model 2 Shipborne Surface Search Radar and mounted on the battleship *Hyuga* just before the Battle of Midway. Maximum ranges were 55 km for aircraft at 3000 m and 20 km for a battleship [45]. This radar had separate horns for transmission and reception, a configuration that marked Japanese microwave sets throughout the war.

The Japanese Navy excelled in night actions and wanted to add this new technique to their ways of penetrating the darkness, although the microwave project did not gain really strong support among line officers until events in the Solomon Island later drove the point home. When the Navy was first introduced to the possibilities inherent in microwave techniques they saw it as an improved way of station keeping at night that would prevent collisions, a use that came to other navies only after having had experience with it at sea [46].

As already discussed, Soviet radar development suffered from lack of interest in the high command, confusion as to its mission and the dispatch of excellent radar engineers to the Gulag during the purges of the late 1930s. That anything at all came out is remarkable. Because of or in spite of these extraordinary circumstances there occurred what must be one of the most baffling incidents in the history of radar. In April 1940 when the cavity magnetron was Britain's most precious military secret, when it traveled under armed guard, when its use was discussed at cabinet level, when it was described as the most valuable cargo ever to arrive in America [47], when the United States was preparing to open a special laboratory just to exploit its properties, when all these circumstances applied, two Soviet engineers published a complete description of it in the open scientific literature [48].

During 1936 and 1937 N F Alekseev and D D Malairov produced a series of cavity magnetrons as part of a project for building an anti-aircraft (AA) gun-laying radar at Scientific Research Institute 9 (NII-9) from proposals by its director, radio-eminence Mikhail Alexandrovich Bonch-Bruyevich [49]. The magnetrons were discarded in favor of a pulsed trans-mitter that used very-high-frequency triodes that worked on 64 cm, had 12 kW peak power, was called Zenith and was abandoned in 1940 [50].

One can presume—and little other evidence is at hand—that the lack of success of magnetrons in this work, for whatever reason, taken together with Professor Joffe's long-standing opposition to microwaves for radar allowed the publication of the paper. Irrespective of the reasons, the paper is a complete disclosure of the elements of the cavity magnetron. One does not even need to know Russian. It suffices to see the tables giving wavelengths and powers and to think a bit about the drawings of the characteristic electrode shapes. That the drawings showed water-cooled anodes tells one a lot. It was all there. There is a report of the independent invention of the klystron at NII-9 by N D Devyatkov during those same years, but even less was made of it than the magnetron. It was quickly followed by a reflex klystron [51].

At the Swiss electro-technical company of Brown, Boveri & Cie, F Lüdi began experimenting with a magnetron for generating centimeter waves in 1936. By 1939 he had developed an oscillator that would have delighted radio engineers in Britain or America, a delight they could have enjoyed, had they followed foreign technical publications and studied them with a bit more attention than usual, for the basic elements were described in an article published in 1937 [52]. Lüdi continued to work on the device, eventually called the 'Turbator', but without incorporating it into a radar set. It did not have as much power as the Randall–Boot magnetron but had good frequency stability, certainly better than the un-strapped magnetron. A high power level could have been attained, had that been a design goal, but it was unnecessary for the microwave communication links that were intended. With a large oxide cathode it attained 10 kW peak power in pulsed operation.

Lüdi's magnetron resembled the Randall–Boot device except that the resonating cavity was in fact a three-dimensional toroidal cavity with a rectangular cross section, which was maintained at positive potential rela-tive to the axially located cathode by 0.5 to 2 kV. Alternating slits coupled the cavity to the rotating electron stream that excited various modes of oscillation when this 'electron wheel' was synchronized with one of the cavity modes. The wheel's rotation velocity depended on the anode volt-age and the magnetic field. The anode cavity was supended in a glass envelope and cooled by thermal radiation, thereby limiting power. Lüdi published another paper in 1942 [53] that expanded on the first. It in-cluded a photograph of the device that clarified ideas until then generally obscured by pages of not particularly enlightening mathematics, although still a little coy about the details of the resonator. Experimental results for 10 cm removed it yet another step from the abstract. The Turbator is best described in a postwar paper [54].

The Swiss were keenly interested in devising modern weapons, both for the defense of their homeland and for export, but neither Brown Boveri nor the Swiss Army initiated a program to provide methods of radio location. Army intelligence became aware of radar early in the war and set up high-altitude monitoring stations using equipment of their own design and from belligerent aircraft that landed in Switzerland [55], but for enigmatic reasons, possibly related to the sensitive nature of Swiss neutrality or to a negative evaluation of its tactical significance for the defense of the mountain réduit, the Army did not tell Brown Boveri about their knowledge of radar.

I leave it to the reader to ponder the strange meanings found in these last two tales.

4.2. THE TIZARD MISSION

It was to Professor A V Hill that Wimperis had turned in 1934 when considering the formation of what became known as the Tizard Committee, and it was to Hill that Tizard turned in November 1939 with the idea of opening a scientific liaison with the United States. Tizard, ever thinking well in advance of present problems, feared a long war for which Britain's technical capacities were inadequate and saw America as an important resource. Hill, whose scientific stature rested on medicine and physiology, had made important contributions to air defense in the 1914–1918 war and had become thoroughly acquainted with radar in serving on the Tizard Committee. As a consequence of his reputation and position as Secretary of the Royal Society he was well known to many American scientists. All these reasons spoke for his visit to the United States and Canada in the spring of 1940 to investigate the possibilities of scientific liaison and to test the Americans' attitude toward the war.

Although Americans were overwhelmingly pro-Allied they were also isolationist, the consequence of two decades of disillusionment over the causes and outcome of the previous war and from their interpretation of the manner in which their entrance into it had been secured: a naive, unsophisticated people deceived by cunning Europeans. The case made by Walter Millis in *Road to War* [1] caused reflection by many a serious reader. But Hill found American scientists so repelled by the Nazi movement that these concerns had been swept aside. Some were openly interventionist and many wanted to co-operate with their British counterparts. Hill noted that Canadian scientists were prevented by the requirements of secrecy from collaborating with their natural and willing American colleagues.

On returning to England Hill proposed that official exchanges of secret scientific material be opened between Britain, Canada and the United

States. There was a meeting about the proposal on 2 May 1940 made up of the Tizard Committee, considerably expanded and including Watson Watt, who maintained 'that the Americans could not teach us anything, and that we should get much the worst of the bargain', and Admiral Sir James Sommerville, who believed 'that anything told to the American Navy went straight to Germany' [2]. The German offensive a few days later brought a high degree of urgency and reality to the deliberations and an agreement by the top military and civilian leaders for a unilateral disclosure of scientific and technical material to the United States. It was hoped for, but not placed as a condition, that the Americans would respond in kind. The most important gain would be access to the full resources of the American radio industry, which necessarily required some disclosure. There were opponents to the idea on the western side of the Atlantic as discussion of this remarkable proposal spread, but it had the support of Lord Lothian, Ambassador to the United States, and gained the full support of President and Prime Minister. Churchill gave Tizard a free hand [3]. Arrangements were embodied in a memorandum between President Roosevelt and Lord Lothian dated 8 July 1940.

The mission, officially the British Technical and Scientific Mission, consisted of Tizard as leader with Brigadier F C Wallace (Army), Captain H W Faulkner (Navy), Group Captain F L Pearce (Air Force), Professor John Cockcroft (Army Research), Dr E G Bowen (radar) and Mr A E Woodward-Nutt as Secretary [4]. Tizard insisted that the military officers had to have served recently in combat and have personal knowledge of the technical equipment they described. He flew across in the miserable conditions that often served even the mighty as they crossed the Atlantic in wartime; the remainder came on the *Duchess of Richmond*, named by them the *Drunken Duchess* for her behavior in the seaway encountered. They landed in Halifax and arrived in Washington on 8 September with the famous black deed box containing documents describing Britain's technical secrets and one of the 12 magnetrons that had been produced by General Electric Company.

Tizard and Lord Lothian set up a series of meetings with the radar discussions arranged through Major General Joseph O Mauborgne, Chief Signal Officer since 1937, and Rear Admiral Harold Bowen, Director of the Naval Research Laboratory (NRL). Equally important openings to the recently established National Defense Research Committee (NDRC) came easily through Vannevar Bush, one of which led to the Microwave Committee.

Tizard had planned to demonstrate two of his mighty gifts: the magnetron and ASV mark I. The magnetron was ready for display, having come over in the black box, but an ASV mark I set to be mounted on arrival on a suitable airplane was not sent, someone in London having decided to send mark II instead, consequently not demonstrated until after the meetings were over. The demonstration was left for Bowen to make after Tizard left, but the magnetron demonstration went as planned.

The meetings started on 12 September in the Army and Navy buildings located on the Mall, temporaries from 1918 and now mercifully dismantled. First the British described their radar equipment but held back for the moment the details of how they actually produced the prodigious power for centimeter waves with which they were so pleased. They were then given tours of NRL in Anacostia, the Signal Corps Lab at Fort Monmouth and Loomis's Lab at Tuxedo Park [5].

The Americans representing radar interests were of three general groups: the first group were Signal Corps and NRL men, who had struggled intensely with the practical problems of radar for half a decade, the second were the microwave men, who knew more about that technology than any other group in the world—except how to make a high-power generator—but had yet to make a useful radar set and the third were the Army Air Corps officers who knew little about the techniques of radar but had already realized its value.

The first group held the view prevalent among the American military establishment that, if left to fight alone, Britain's position was hopeless and that their duties called for them to prepare the United States for the consequences of Britain's defeat. They saw little hope that America would enter of its own volition and more or less assumed the Soviet Union was Germany's ally. They were also acutely aware of the incredibly bad state of the Army, one of the smallest in the world, and of its Air Corps. On the naval side of things there were still traces of a long-standing rivalry between the British and American fleets.

This group was favorably impressed by only two British sets: CD/CHL, which had strong similarities to American equipment, and ASV of which they had no counterpart, although they did have a good radio altimeter. They were unable to separate what the British had done from what they were planning to do. The NRL people thought the Royal Navy's radar second rate. CH came across as crude, expensive and immobile, and the extravagant claims made by Brigadier Wallace [6] for GL mark I must have brought serious doubts to the Americans about the veracity of their informants once they had examined the specifications of that set. They feared that giving any secrets to Britain would bring them to German hands through capture. They also found the British 'snooty, crusty, scornful and antagonizing' [7], but after their long series of disappointments in microwaves these thoughts did not prevent a full appreciation of the importance of the cavity magnetron.

The second group held the view dominant among American scientists, deeply stirred by Churchill's speeches and the ferocious air battle then being fought, that Britain had set her face against a monstrous tyranny and must be given every possible support. And this new magnetron was the circuit element they had dreamed about. The excitement of learning how radar was saving the island from invasion carried them to a high degree of enthusiasm for the work they saw cut out for them. The military's cold analysis of Britain's situation, if anything, heightened their zeal. They wanted to work with their colleagues across the sea and start immediately! In this they were to succeed in an unparalleled manner of international wartime technical collaboration.

The third group, the US Army Air Corps, formed the opinion that American radar was significantly inferior to the British [8]. This came about in part because of the British mixing of plans and achievements coupled with the Army and Navy's slowness in telling what they had, the result of personal reluctance on the part of those attending and a slowness of higher commands to free them. From this unfortunate muddle came the demand by the Air Corps that the United States begin producing British sets for American forces, a demand bitterly resented and resisted by the Signal Corps.

Signal Corps and Air Corps officers soon visited Britain, but each group saw what it had been led to expect. The former found CH poorer than advertized, especially in its height-finding capability; the latter found British air defense superb and reasoned it came about because they had the best radar. Both were right in part but were unable to reconcile their points of view [9]. These biases stuck for most of the war. The microwave group, soon to become the MIT Radiation Laboratory, retained a very fruitful and even loving relationship with the British, eventually opening a branch in England for closer liaison. E G Bowen found the working atmosphere so congenial and his memories of Rowe so raw, especially after having been relieved of his airborne development duties, that he arranged to remain at Rad Lab.

The NDRC people formed a relationship to the Signal Corps that was at least correct but with NRL it became destructively poor with Admiral Bowen singled out by NDRC as the cause. He had been strongly opposed to the exchange of information and at times boorish about it [10]. The atmosphere did not improve until Rear Admiral Van Keuren replaced Bowen as Director of NRL. Individuals learned to cross these hindrances, but they did not disappear.

E G Bowen's demonstration of ASV mark II mounted on a PBY flying boat in December did much to dispel lingering American doubts of British competence and led to the Navy ordering 7000 from Philco [11]. The lesson was reinforced by a quick, unsuccessful attempt to put the SCR-268 transmitter and receiver with small external dipoles in a B-18 bomber [12].

For their part the British 'were agreeably surprised to find that both the US Army and Navy had progressed a good deal further in radar than we had been led to expect' [13]. They found the reception by the Microwave Committee exhilarating. Here were men that understood all and could hardly wait to start. The enormous success attributed to the Tizard Mission rests almost entirely on microwaves, which changed warfare during the next few years more than any single weapon. The enmities engendered between the British and the American services labs were sufficiently well suppressed to prevent open hostility—until Watson Watt's visit in early 1942, but the Alliance was strong enough by then to sustain even that.

An indirect result of the Tizard Mission was the Signal Corps Electronics Training Group, the suggestion of James B Conant with implementation through President Roosevelt. It was composed of men with sufficient technical training and education to be commissioned directly as reserve officers and sent after a minimum of preliminary training for a radar apprenticeship with the British. The demand for persons with such qualifications was high both by the armed services and industry, a circumstance that only became worse with time, but on 12 September 1941 the first group of 35 left Fort Monmouth for their assignment. Seldom in the course of human events have soldiers served in such a strange organization. They were publicized yet secret, observers yet combatants (two killed before America was officially in the war), receiving and giving orders to British. The locations of many were lost to American authorities for months [14]. They were remembered in Antiaircraft Command by its commander: 'From America we got a wonderful bunch of recruits. America was not yet in the War, so these scientists operated on our gun-sites, in the guise of civilians' [15].

The item brought in the black box was a naked magnetron, no magnet, no power supply, no modulator, so demonstration had to wait until suitable auxiliary equipment was available. That this heightened the dramatic impact did not trouble Tizard and Bowen. The test came on Sunday, 6 October at Bell's Whippany Laboratories where the necessary magnet and a modulated power supply for 10 kV were present. Bowen's fears that something might have happened to the tube in its travels were quickly forgotten as the air around the output terminal glowed for an inch as a result of the radiation [16]. Power and wavelength were quickly measured with the power exceeding the advertized amount and the wavelength was right on, 9.8 cm. The Bell Lab's tube department was to manufacture 30 for NDRC immediately, which were ready in a month.

Microwave equipment quickly moved from the United States to the United Kingdom: klystrons, Acorn tubes, lighthouse tubes, waveguides and horns, all essential elements in building microwave radar.

What to do now? On the weekend following the dramatic test of the new tube, a meeting took place at Tuxedo Park. Included among others were Loomis, Bowen, Edward Bowles from MIT, Carrol Wilson representing NDRC and Ernest Lawrence, Loomis's friend and a figure of eminence in American science. There was immediate agreement that a new laboratory modeled on the Air Ministry Research Establishment, soon to be re-named yet again as the Telecommunications Research Establishment (TRE), had to be founded. They undertook to discuss locations for the imagined laboratory without deciding where and to discuss its staffing. The British example of drawing on university research groups, not just physicists and engineers but from all disciplines, brought immediate agreement, and Lawrence was ready to start recruiting. That out of the way, the objectives of the laboratory were fixed: (1) a 10 cm airborne interceptor (AI) set, (2) a 10 cm gun-laying (GL) set and (3) an as yet unspecified long-range radio navigation system. It was an exhilarating weekend to say the least, and it all came to pass [17].

On 18 October a meeting with Bush at the Carnegie Institution in Washington saw the organization of a Laboratory Management Committee, and the MIT Radiation Laboratory (Rad Lab) was about to be born, the location of which had only just been selected [18]. Bush saw to it that orders were placed for magnetrons from Bell Labs, permanent magnets for them from General Electric, modulated power supplies from Westinghouse and RCA, 12 inch cathode-ray tubes and IF amplifiers from RCA, parabolic reflectors from Sperry and various other components [19].

A decision was made to have the service labs continue their meter-wave work, leaving microwaves to Rad Lab and Bell. The who, how or when of this decision remains unclear. It just seems to have happened. It may have been a collective decision compelled by its obvious wisdom. Meter-wave radar was now a proven and very valuable weapon just entering production in the United States. Indeed it was to be decisive during the next two years, and there would be an abundance of tasks for its designers during these months. Microwave radar showed great promise but some risk and would require the invention of a whole new series of techniques. How much better to leave it to a fresh crew, unencumbered with concerns about production, mobilization and design modification.

For whatever reason and however justified, the decision to entrust the magnetron to a new, as yet non-existent laboratory did nothing to remove the vexation of the service lab men. After delivering the country first-class radar prototypes ready for production, done with trivial support but much personal effort during the preceding years, they were now considered country cousins. That these numerous frictions were not allowed to develop into nasty jurisdictional fights with attendant administrative obstructions is a credit to those involved.

While all these events were unfolding the engineers at Whippany thought about the wonder they had just seen but from a more practical point of view. They were preparing the production of the CXAS fire direction set that they had demonstrated to both Army and Navy about a year earlier. There had been two versions, one for 40 cm, the other for 60. Power and with it range had dropped sufficiently in going from 60 to 40 that the Navy had elected the longer wavelength equipment for its FA or mark 1 fire direction radar. The prototypes were modular in design with a self-contained transmitter unit, which used special high-frequency triodes. The way to improve this set was almost immediately apparent. Re-design the transmitter unit for a magnetron at 40 cm and obtain better directional resolution and increased range, not to mention a longer life for the output tube, as the special triodes were used up fast. The magnetron scales linearly with wavelength and made a manageable package for 40 cm operation. All the other modules of the set would remain the same. As FC or mark 3 it would be the first American magnetron radar [20]. It quickly evolved into FD or mark 4, which had vertical as well as horizontal lobe switching that made it the Navy's standard AA fire control radar for the entire war. And thus it came to pass that the first American application of the device capable of making the long sought 10 cm waves required its modification to produce waves four times as long.

Canada was included in the Tizard Mission, but almost as an afterthought. During the early discussions about whether to share technical secrets with the United States, Canada did not fare well. Reluctance came about not from fear that secrets would be lost to Germany, as some had expressed as a concern for America, but that it simply was not worth the effort.

Dr John T Henderson, Chief of the Radio Section of the National Research Council, visited Britain on invitation in early 1939 to be given 'information respecting a most secret device which they have adopted for the detection of aircraft'. Henderson's background prepared him for the principles of radar but not for the extent to which it had been developed as a weapon—without previous hints to Canada. During the winter of 1939–1940 a group of Canadian radio scientists began to approach this new technique, and by March their number had grown to 22. Design details and prototype sets from Britain that would have been extremely useful were not forthcoming, so the Canadians began building from their own designs using commercially available components. By June 1940 a 1.5 m set for ship location proved successful in trials at Halifax and was given the name of Night Watchman.

The Tizard Mission entered through Canada on its way to Washington but primarily as a courtesy. It was only after they were actively engaged in discussions south of the border that they learned of the extent of Canadian progress and decided to deal with them as intellectual equals and gave them the same information imparted to the Americans. The Canadians set about designing a 10 cm gun-laying radar as their first priority, a device destined to become GL mark 3C. (Mark 3B would designate the British set and mark 3A the American, better known as SCR-584 [21].)

The Tizard Mission opened up liaison with scientists in more fields than just radar, but it will be remembered more for the radar than for any of the other matters. It was an overwhelming success for both sides. Britain gained the much needed additional electronic manufacturing capacity along with a big jump in the knowledge of microwave techniques, and the United States gained the magnetron in time to make good use of it. The discovery might have been made in America independently, just as it was made in Russia, Japan and Switzerland, but at the time of the Tizard Mission none of the American microwave people seem to have been thinking in terms of magnetrons because the klystron was exerting a strong influence on their thoughts. One might rather expect that they would have added an axial magnetic field to that device to allow much larger electron beams, but who knows?

It came as a somewhat rude surprise for both parties to learn that someone else had operating radar. In this encounter the British representatives were better prepared mentally. They had at least come to terms with the necessity of disclosing their secrets weeks earlier and had their stories ready. The Americans had learned of this high-level deal only shortly before the meetings, did not have immediate clearance to tell all and were personally reluctant to let things out, so they were psychologically unprepared for what was to take place. The result was an unfortunate split of the American service-lab people not only from the British but also from the Air Corps and the future Radiation Laboratory. It also gave rise to a reputation they did not deserve of having produced inferior equipment. From the *Administrative History of the Office of Scientific Research and Development*: 'Historians may differ as to the reasons why with all of its remarkable scientific advances the United States lagged so dangerously in the development of weapons, but none will deny the fact' [22]. It is hoped that the fallacy of this statement is apparent to readers who have come this far.

4.3. THE RADIATION LABORATORY

Within most physicists there lies an engineer eager to design something. The two branches of learning have always had a thin boundary separating them, one frequently crossed or dismissed. In principle the distinction is simple enough: physicists discover nature's laws, engineers synthesize these laws into apparatus for good or evil. But in the day to day performance of their trades the two are at times indistinguishable. An experimentalist designing equipment with which he plans to measure some atomic quantity looks for all the world like an engineer designing some part of an engine or radio; a theorist striving to describe a nuclear scattering process with quantum mechanics looks like an engineer analyzing the stress patterns of a bridge design. Names often confuse more than enlighten. Designers of lasers manipulate atomic properties so as to produce some form of radiation, operations no different from an electrical engineer designing yet another form of high-frequency oscillator, but the former seems invariably referred to as a laser scientist whereas the latter is an engineer.

Beauty is a common attribute for which both strive. There is never a more satisfying achievement for a physicist than the reduction of some complicated phenomenon to description by a simple, closed mathematical statement. It is one of the misfortunes of our civilization that the beauty of an equation cannot be shared with so many who specialize in the appreciation of beauty. It is in beauty that engineer and scientist differ: for the scientist it is nature's beauty, for the engineer it is the beauty of creation. Engineering is a form of art and has filled the world with things of obvious visual beauty but also with subtle forms, much like a theorist's equations.

A poet may enjoy the grace of the Brooklyn Bridge but will never appreciate the charm of a well designed electronic circuit, but the beauty is there for all that, and an electronics man will recognize grace, symmetry and style in a design without ever having had a course in electronic art appreciation. If ever there was a place where the latent engineer in the scientist burst out, it was at the MIT Radiation Laboratory.

Lee DuBridge was Lawrence's choice for Director of the new laboratory even before the formation of the Laboratory Management Committee in Bush's office on 18 October, and Lawrence's choice met no opposition. DuBridge had built a cyclotron at Rochester University that had delivered its first beam in 1938 and formed the basis of a very active group in nuclear physics. Recognition of DuBridge's scientific and administrative skills was apparent in his being Chairman of the Physics Department and Dean of the Faculty of Arts and Sciences. Loomis and Lawrence persuaded him to accept the post. DuBridge quickly selected another unrelated Loomis, F Wheeler Loomis of the University of Illinois as his executive officer. Alfred Loomis was never a member of the Rad Lab staff, but he was always in the background and made significant contributions.

Physicists not in the Microwave Committee were quickly added to the roster. Kenneth T Bainbridge of Harvard was probably the first, and recruiting was substantially helped by a meeting on applied physics at MIT during 28–31 October attended by 600. A luncheon meeting at the Algonquin Club in Boston introduced I I Rabi of Columbia, Edward U Condon of Westinghouse, John C Slater and John G Trump of MIT and others to the opportunities of working at the Radiation Laboratory, and acceptances came quickly. DuBridge presided over an organizational staff meeting on 11 November, attended ex-officio by Lawrence and Alfred Loomis, in which eight sections were created: (1) pulse modulators, (2) transmitter tubes, (3) antennas, (4) receivers, (5) theory, (6) cathode-ray tubes, (7) klystrons, (8) integration. They chose sections like children picking sides for ball games. Bainbridge wanted modulators, Rabi the magnetron until all sections were taken [1]. The various design projects were to draw on these sections for support.

Recruits began arriving from far and wide: Luis Alvarez from Berkeley, Ivan Getting from Harvard, Norman Ramsey from MIT. Most were nuclear physicists. As Alvarez remembered: 'The first weeks at the laboratory were like a family reunion' [2]. The atmosphere was distinctly that of Bawdsey with Bowen transferring tradition and spirit. Real work started at once.

Project I, airborne interception, held E G Bowen's attention fast. Britain was experiencing the Blitz, the night attacks to which the Luftwaffe had been forced after defeat during the day, and the bombers were delivering their loads, inaccurately but with great indiscriminate damage, and departing with few losses either to fighters or guns. A 10 cm AI set seemed the most vital contribution toward stopping this. By 16 December

Rad Lab was beginning to take on some semblance of being a laboratory, so a schedule was drawn up with AI the first goal. On 6 January 1941 a 10 cm set was to be operating on the roof; on 1 February it was to be installed in a B-18, the bomber version of the famous DC-3 transport, and a month later in an A-20 attack bomber, the most likely candidate on hand for a night fighter. It was, as intended to be, a hard-driving program.

The crucial components that Vanevar Bush had ordered on the basis of the Microwave Committee's suggestions allowed the roof-top equipment to transmit its first beam in late December. Duplexing, or TR (transmit–receive) switching as it became known among the microwave men, was not available for this first unit, so two antennas were used, and by 4 January there were reflections from buildings in Boston. But if aircraft were to be tracked from another aircraft, it was imperative that a single antenna be used. They found that using a klystron as a pre-amplifier for the diode detector prevented the transmitter pulse from burning out the diode; an overloaded klystron did not transmit a damaging signal and it recovered fast enough to function for the reflected pulses. This temporary expedient allowed a common antenna but introduced tube noise at the input. Finally by pointing the parabolic dish visually the hard-pressed crew observed reflections from an aircraft at 3 km on 7 February 1941, an event so important that the information was telephoned to DuBridge and Lawrence at a meeting in Washington [3].

The problems with the roof-top unit were just about everything. Frequencies were misaligned because of the lack of adequate test equipment, there were no polyethylene cables and mismatched waveguides were producing standing waves. A new engineering art had to be mastered, and the most important teachers of this new art proved to be the radio amateurs, with James Lawson especially remembered. By 5 March the roof-top unit was delivering a creditable performance and was transferred to a B-18 with a plastic nose transparent to the radiation. On 27 March Edwin M McMillan, E G Bowen and Luis Alvarez took off with their creation and were rewarded by observing Cape Cod and numerous ships on their screen [4]. After this success Rad Lab could proceed, if not in a relaxed manner at least with assurance, toward other projects; by then the rolls listed 140 employees, six of them Canadian. There were to be many more roof-top units, as all new equipment started by examining the Boston skyline.

A solution to the microwave TR switch problem came from Oxford's Clarendon Laboratories in spring 1941, the invention of A H Cooke. One of Sutton's klystrons was filled with water vapor at a low pressure so that the electrodes of the resonant cavity, which were close together, would pass a low-level signal picked up by the antenna but would form an immediate plasma discharge from the high power of the transmitter, creating a short circuit that reflected the high power back, thus preventing it from destroying the detector diode. It was Page's duplexer transferred from two-conductor transmission lines to waveguides. The results were sufficiently encouraging to warrant making a special tube, and several kinds of gas and of electrode configuration were tried. One problem was to have the discharge begin quickly enough, because the diode was unforgiving of even the briefest application of high power. To do this a small number of ions had to be present always. Radioactivity was tried, but a small ambient current from a filament proved best. An acceptable design, the CV43, took lots of painstaking trials, but other versions were to follow as transmitter power levels kept pushing the TR switch designers to new models [5].

Project I then began to undergo changes of direction and emphasis. The Air Corps doubted whether the A-20 would make a suitable night fighter and preferred designing the equipment for the P-61, an aircraft specially designed for such service. A set was flown to Britain in June 1941 for comparison with their 10 cm AI equipment. The American set had greater transmitter power, the British a more sensitive receiver, and the return flight brought Rad Lab the much needed TR tube. Western Electric made a few of these sets, designated SCR-520, by the end of 1942 but concentrated on designing a lighter version, SCR-720, for the P-61 [6].

By June there was much less emphasis on airborne interception. Germany had attacked the Soviet Union, greatly lessening the night attacks on Britain, and the attacking planes had begun to take severe losses from meter-wave AI during the month before the greater part of the Luftwaffe had been sent east. On the other hand U-boats were proving increasingly effective as Doenitz's ideas began to be taken seriously. The SCR-520 was modified for this use and designated SCR-517. This set had only forward scanning, and it was obvious that ASV radar should scan a full circle and present the observations on a PPI scope. The set to accomplish this, the Navy ASG and called 'George' by its users, was installed in blimps in June 1942 and shortly thereafter in airplanes. It was to become the Navy's favorite ASV equipment [7].

Project II, gun-laying, was not nearly so high on the British list as AI but struck a resonance among the Americans. First, it followed a direction that the Coast Artillery Corps had pushed for over a decade; second, it was a seed planted in rich soil for the development of such equipment and was a project that quickly became self-generating. Out of it came the finest fire-control radar set of World War II—SCR-584. The engineering was done by Ivan Getting and Lee Davenport.

Getting was born in New York City into a family with strong Slovak ties, his father being deeply involved in the interests of his countrymen. With the formation of Czechoslovakia in 1919 his father accepted a position in the new government, and the family moved to Bratislava only to return when the senior Getting received a position in the Czechoslovak Embassy. After seeing the postwar turmoil of Central Europe the mother decided that the children were to remain in America. It was for Ivan the beginning of an ever expanding range of experience that led him to nuclear physics at Harvard by way of a Rhodes Scholarship. A useful skill learned along the way came with a reserve commission in the Coast Artillery, giving him first-hand knowledge of AA artillery. As one of Bainbridge's students he was one of the first recruits to Rad Lab [8].

Davenport's background was quite different from Getting's. It was typical American training for a career in physics in the 1930s, one shared by many of Rad Lab's younger staff. He was born in Schenectady, New York and grew up with a self-generated interest in all matters technical with encouragement from his father, who taught high-school mathematics. On graduating from high school at the bottom of the depression he got a Federal Youth Administration job drawing the illustrations for a revised edition of Kimball's *College Physics* and then entered the local Union College to study physics, which allowed him to live at home. Summer work at General Electric paid his way. A teaching assistantship at the University of Pittsburgh had brought him almost to a PhD when he was called to the Rad Lab, where he immediately found himself on Project II [9].

In the first discussions the novice designers decided almost immediately that the new equipment would incorporate automatic tracking, a course neither expected nor desired by the higher ups, certainly not by the members of the Tizard Mission. Reasons for their decision are not hard to find. In 1940 MIT led engineering in servomechanism design, and these ideas had been incorporated into the microwave equipment tried out for Bowles's blind-landing experiments. Barrow had even built a microwave horn system used to track students moving about the campus [10]. The opposition thought this approach would be too time consuming and preferred human interpretation of the signals. Automatic tracking proved neither difficult nor time consuming and was a most fortunate choice, doubly fortunate because the Ordnance Department had incorporated automatic tracking in the 90 mm AA gun that was replacing the 3 inch and because Bell Labs was designing an electronic analog computer for predicting AA fire that was to be much better than the mechanical computer. These were decisions for which London would later be thankful.

A roof-top unit was the first step with the dish mounted on a servo-controlled machine gun turret of the kind being manufactured for the then developing B-29. Automatic tracking required something akin to lobe switching, allowing the amplitude of two signals to be compared. The solution was the conical scan generated by a rotating feed as used on Telefunken's Würzburg. The roof-top unit was tracking aircraft on 31 May 1941 with remarkable skill, the principal uncertainty being the defect inherent in lobe switching or conical scanning caused by the ever changing aspect of the target that yielded reflections of varying amplitude, giving the tracking a nervous twitch. This was quickly removed by circuits that produced running averages of the signals [11]. It was time to show off the results.

A visit by Alfred Loomis soon brought the by-then Brigadier General Colton to see the replacement for his beloved SCR-268. The effect on Colton was dramatic. He was to provide the group with all the support it needed from the Signal Corps, and Getting was just the kind of man to work effectively with Colton. The General was so impressed that he decided immediately that the Army had to have such equipment, but he was not certain that Rad Lab would be able to deliver, so he immediately placed an order with Bell Labs to design something similar as a back-up [12].

The next radical design step was to put everything in a closed truck, designated XT-1 for experimental truck 1, with the dish lifted to the top when in action and carried within for transport. This provided not only mobility but gave the operators comfortable working conditions. They would not be exposed to the weather as on SCR-268 or forced to keep their head pressed against a light shield in order to see the oscilloscopes traces during the day. In the production model a trailer replaced the truck. Transmitter power and receiver sensitivity made important advances, and the maximum range kept increasing until it reached 90 km. This allowed it to function as a search radar so a PPI scope was added. On 6 February 1942 XT-1 was tested at Fort Monroe with the Bell Labs T-10 director and a 90 mm gun firing on towed sleeve targets. They 'shot down targets with as few as eight rounds, all without human intervention or visual contact with the target' [13]. The Coast Artillery Board recommended the XT-1 be procured as the standard gun-laying set with an allocation of one to each AA-gun battery. On 2 April the Chief Signal Officer ordered 1256 units [14]. XT-1 became SCR-584 and the Bell Labs director became the M-9.

Industrial production had problems of its own that even had Getting confronting the president of the Chrysler Corporation about fabrication of dishes and their high-precision drives. Production did not begin until a year after the first order, and it was not until early 1944 that the first set reached combat, delays in part attributed to a misunderstanding of material priorities in the War Department that had their ultimate origin in secrecy [15].

Even before reaching the battlefields SCR-584 began to attract a lot of attention. Luis Alvarez saw it as the solution to the problem of blind landing. It was not the solution, as we shall see, but its superior properties put Alvarez onto the path that led to it. Its ability to track shells accurately uncovered errors in the firing tables for the 90 mm gun, which detracted from accuracy and were corrected for M-9 director, and the ability was extended to determining bomb trajectories. At the front uses were found for it never dreamed of during design: location of enemy artillery and mortar positions by tracking projectiles, surveying islands by tracking a photographing aircraft, accurately directing bomb release for a plane above the clouds on near-front targets, detecting the movement of enemy vehicles at night. A plan for countering the V-2 rockets was based on the 584, but the launch sites were taken before it could be tried [16]. It was a masterly design [17].

AI and gun-laying were the two microwave designs for which the laboratory had been established, but if one were to ask what Rad Lab set had the greatest effect on the course of the war and certainly if one asked what set had had the greatest effect on the immediate postwar world, the answer would have to be one that was not discussed in the excited meetings that followed the Tizard disclosures and the founding of the Laboratory. This was the SG, the naval set that Samuel Eliot Morison, historian but no stranger to the bridges of warships, called the 'greatest boon of scientist to sailorman since the chronometer' [18]. The antenna of its modern descendant can now be seen rotating on the mast of almost any vessel traveling the great and small waters.

It grew directly out of the first roof-top set for the same reason that the Admiralty Type 271 went forward faster at HM Signals School than airborne interception at Telecommunications Research Establishment—it seemed easy and sensible. The B-18 equipment lacked two vital components necessary for the SG to be an improvement over the 271: (1) the TR switch to allow, without the degradation caused by a klystron preamplifier, a common, rotating antenna and (2) the PPI scope. When these elements were added the resulting equipment had much the same electrifying effect on seamen as the installation of the radar beacon for ASV mark I had had on airmen of Coastal Command. It presented to the watch officer a map of the area surrounding his ship showing coastlines, harbor markers and other ships. It greatly simplified the difficult problem of keeping station for convoys and fleets running in blackout and foul weather. It became the indispensable navigation aid for landing troops in an invasion where the approach had to be made under cover of darkness. It allowed a surfaced submarine to be picked out in the clutter of a convoy far more easily than with the type 271. It allowed vessels to run for the first time (legally) at full speed in fog. In the tangled naval actions of the southwest Pacific it removed the confusion the meter-wave sets had when operating where ships and islands appeared much the same on their A-scopes. All this became apparent when the SG was first used. It was never planned to be such; this set was an afterthought.

The first model of what was to become SG was mounted in May 1941 on USS *Semmes*, a four-stacker destroyer working out of New London. The first sea trials of the device on 5 June caught the vessel in fog on her way back to port, but a sea-sick Ernest Pollard picked out buoys with his radar and piloted for the skipper, Lieutenant Commander W L Pryor, who made an excited telephone call to Washington on docking [19]. One of the first things learned was that the dish needed to be stabilized because the roll of the ship—and the four-stackers were unsurpassed in their ability to roll—caused the relatively narrow radar beam to rise above the surface or to examine the near ocean too carefully, depending on the orientation of the mast. Stable verticals use servomechanisms to maintain a direction established by a gyroscope, and this solution was tried but rejected in favor of the cut paraboloid, which was wider than tall and formed a beam 5° wide and 15° tall, a vertically extended fan, and so the design has remained [20].

The *Semmes* gained a quasi-permanent laboratory crew and cruised about the eastern seaboard for much of the year as successive alterations were made in the design. At the insistence of Commander Pryor the final form of the set remained aboard the *Semmes* until replaced in March 1943 by a production model of SF, the destroyer model of SG. The Naval Research Laboratory made the design sea and battle worthy and Raytheon produced it. The first production set of SG was installed aboard USS *Augusta* on 5 April 1942. It reached the Solomon Islands, where it became an instant favorite, in October.

These three sets, ASG, SCR-584 and SG, and their descendants form a solid basis for Rad Lab's high regard. They were created during the heroic period in ways similar to the heroic years at Bawdsey. Just as Bawdsey grew into the large, well run engineering laboratory at Malvern, so Rad Lab became a large, well run engineering laboratory at Cambridge. Its appropriations for 1941 were about five times what the service laboratories had spent on radar before 1940, and by 1945 it spent as much in a day as the service labs had spent annually during those austere years. The total cost for the war was $142 million. Personnel reached a peak only slightly less than 4000, and building floor space grew accordingly.

An Army Air Corps detachment stationed at East Boston Airport began flying for the lab in July 1941. They outgrew this base and a new one was created from scratch at Bedford, opening in May 1944. By the end of the war 95 aircraft made up Rad Lab's private air force. In addition to the shops for building experimental equipment there was the Model Shop operated under contract by the Research Construction Company, Inc. that made small production runs for immediate military needs that could not wait on industrial contract. There were field stations far and wide, and Rad Lab men gave expert help to those at the front. To strengthen the ties with their colleagues overseas the British Branch of the Radiation Laboratory was established in September 1943 at Malvern where the TRE had taken up final residence, ensuring a steady flow of Americans to the places where the equipment was being used to advise and learn.

The Radiation Laboratory was a big business within a bigger business. By the end of the war the United States had spent $2819 million on radar, 48% being for equipment of Rad Lab design. The number and kind of radar that had appeared on the military scene by 1945 was staggering to anyone who had known radar in 1940. Few of those interesting sets can be treated in a book this restricted and many came too late to affect the outcome of the war in any important degree [21].

DuBridge directed the laboratory in a relaxed manner, but it was a relaxed dictatorship, for it was he who made the decisions. He formed a Steering Committee for guidance, eventually made up of about 20 persons he thought best capable of helping him. He relied heavily on them and seldom introduced his own technical ideas. It generally met at monthly intervals on Saturdays at 1 pm, just when the normal work week concluded. It was in the Steering Committee where policy, jurisdictions and contentions were hammered out and where long-range plans were made [22]. Positions on the Committee were much sought after and filled by competitive men. Pollard remembered the sessions as very rough afternoons. 'I was worn down more by the Saturday afternoons, which used to go from one until six, than the whole rest of the week' [23]. The Steering Committee differed from Rowe's Sunday Soviets in having as its objective the making of decisions rather than the exchange of ideas. Only rarely were outsiders present.

By spring of 1943 Robert Oppenheimer began approaching nuclear physicists at Rad Lab to join his Los Alamos staff. The attraction of a new, mysterious project having closer ties to their scientific lives coincided with a feeling that the Laboratory had fulfilled its most important work and the details could now be entrusted to other, respected members of the staff. Bainbridge, Alvarez and many others thus experienced two astounding engineering laboratories within five years. Few of their coworkers knew what was going on, but they noticed that people started to disappear.

Unlike Los Alamos the Radiation Laboratory shut down soon after the end of the war, officially terminated on 31 December 1945. As a final task Rabi had insisted that they make a record of the technology they had learned, in part out of concern about possible future Congressional investigations as to what the country had obtained for the money spent. Writing began in fall 1944 under the editorship of Louis N Ridenour and produced 28 volumes [24] from 49 authors (not including Rabi, who did not write books), most of them protesting strongly about this waste of time when men were dying at the front [25]. But it was everything but a waste of time. Volumes from this electronics encyclopedia would be found on the bookshelves of almost every electronics engineer and experimental physicist for more than a generation, indeed some are found there today and not just as mementoes. Adding to the impact of this injection of knowledge into the postwar world was the return of the staff to civilian tasks, either the academic ones they left or the industrial ones many had learned to like. The effect on America of this influx into physics departments and electronics companies is difficult to judge, but it was extremely large.

4.4. THE PROXIMITY FUZE—THE SMALLEST RADAR[2]

By 1938 Britain's air defense problem was acute. Almost all of her fighters were biplanes that were slower than the monoplane bombers that Germany was building, an impossible situation for the defense. Hurricanes and Spitfires were entering production and CH was advancing rapidly, but air defense cried out for new ideas. One such idea was the 'bomb the bombers'

[2] Based on 'The Proximity Fuze' by Louis Brown that appeared in *IEEE Aerospace and Electronics Systems Magazine* Vol 8, pp 3-10, 1993. Copyright IEEE.

scheme. A number of relatively light bombs dropped from above might have hope of doing damage, if a direct hit were not required. A nearby burst could damage the bomber much as an AA artillery shell might, but time fuzes such as used in AA shells required rapid measurements and calculations impossible for air crew members; a fuze was required that sensed the presence of the bomber [1].

Professor P M S Blackett, a distinguished nuclear physicist from Rutherford's Laboratory and veteran of the Royal Navy in World War I, proposed a fuze based on the photoelectric cell in a memorandum to the Tizard Committee on 7 July 1937. The idea was discussed at a meeting of the Royal Aircraft Establishment on 22 October in which the idea of acoustical triggering was injected. By May of the next year ground test results were encouraging, but premature detonations could be caused by the fuze looking at the sun or clouds. Tests made by dropping fuzed bombs on balloons in March 1939 were at best a modest success, and the Tizard Committee recommended 500 be manufactured for service trials. Disagreements and misunderstandings marked the next few months with service tests called off at the outbreak of war, it being proposed to substitute trials against the enemy. Matters continued without anyone insisting on a fixed goal. Use of the fuze against ground and sea targets was also pushed; using rockets instead of bombs was tried. By mid–1940 there was little hope for this approach.

The Air Defense Experimental Establishment experimented with an acoustical fuze and went through a similar series of tests. It had a Rochelle-salt crystal microphone incorporated in specially shaped tail fins. Tests in August 1939 showed directive response to sounds of frequency above 5 kHz, but only three out of 18 detonated correctly. Rockets were tried with no mentionable success.

Into this stepped W A S Butement, designer of radar sets CD/CHL and GL, with a proposal on 30 October 1939 for two kinds of radio fuze: (1) a radar set would track the projectile, and the operator would transmit a signal to a radio receiver in the fuze when the range, the difficult quantity for the gunners to determine, was the same as that of the target and (2) a fuze would emit high-frequency radio waves that would interact with the target and produce, as a consequence of the high relative speed of target and projectile, a Doppler-frequency signal sensed in the oscillator. Discussions with E S Shire and A F H Thomson yielded a simple design of a continuous-wave oscillator capable of responding as desired [2].

William Alan Stewart Butement came from a pioneer New Zealand family. He was born in 1904, educated in Australia and England, graduated from London University with a Bachelor of Science Degree, and joined the Signals Experimental Establishment of the War Office. His early experiments with P E Pollard in radio location, which resulted in a design having all the elements of an elemental radar, are mentioned in Chapter 2.3. That he did not continue the bent so clearly disclosed was the immediate result of the lack of a transmitter of sufficient power at the 50 cm used and the more enduring lack of any real interest at the War Office. One of many idle speculations is what the history of radar might have been had an enlightened attitude toward scientific research given Butement support [3].

One of the disclosures of the Oslo Report[3] was the knowledge that the Germans were working on a proximity fuze; a tiny vacuum tube from the project was even packed in the envelope containing the report. The design attempted to utilize the change in electric capacitance between a nose electrode and the shell body when some object came near. The German work did not proceed to any useful result but made a small ripple in England. The author of the Oslo Report was a Siemens und Halske technical expert, Hans Mayer, who was a friend of an English instrument maker and entrepreneur, Cobden Turner, owner of Salford Electrical Instrument Company. In the summer of 1939 he and a few of his engineers visited Siemens und Halske on business and got a hint of the fuze work. On returning home they designed a radio-influence fuze and even tried it in some bombs [4].

Unfortunately, Britain had too many serious problems to deal with. Butement was heavily involved in designing radar for the Army. There was no time to be spent on a device that showed so little promise. The proximity fuze needed development by forced march and was proceeding at a stroll. Fortunately, by the summer of 1940 others in America, less pressed, began considering the problem.

While working for the newly formed National Defense Research Committee (NDRC) C C Lauritsen, a nuclear physicist from Caltech, noted in July 1940 that the Western Electric Company and RCA were manufacturing 20 000 thyratron and photoelectric tubes for the British Army [5]. A thyratron is a vacuum tube filled with low-pressure gas and has an electrode configuration like a triode. Unlike a triode it conducts only negligible current for low-level signals, but goes into a plasma discharge once the grid voltage exceeds a certain threshold value, thereby passing a large anode current. It is an electronic switch. The combination of the two types and the specifications for them brought a quick guess that something in the nature of a proximity fuze was being made.

This information soon became an item of discussion between Vannevar Bush, President of the Carnegie Institution and Chairman of the NDRC, and Merle Tuve, a physicist at Carnegie's Department of Terrestrial Magnetism (DTM) located in Washington. Tuve was already well known to people at the Naval Research Laboratory through his invention with Gregory Breit of ionosphere sounding with radio waves. Since 1927 he had worked to build a particle accelerator for nuclear physics, succeeded in adapting the Van de Graaff generator to that end, and by 1940 had created one of the major centers of experimental nuclear physics in the

---

[3] See Chapter 3.1 (p 104).

United States with three Van de Graaff accelerators operating and a 60 inch cyclotron under construction. Tuve and his colleagues at DTM were very concerned about the war and wanted to get into war work immediately. Bush had taken a quick liking to Tuve and brought him into discussions of defense matters from the start. After discussions with naval ordnance Bush formed Section T (for Tuve) on 17 August to work on a proximity fuze at DTM [6].

Thus in mid-August 1940 Tuve asked Richard Roberts whether he thought a vacuum tube could stand an acceleration of 20 000 g, and received a tentative answer of yes the next day. Roberts mounted an obsolete tube, a number 38, on a lead brick that he suspended from the ceiling and then fired a bullet at the brick, the oft repeated experiment that demonstrates the conservation of momentum to students in introductory physics and that had completely altered the scientific study of guns two centuries earlier. The tube still worked, and calculation showed it had briefly sustained an acceleration of 5000 g. The next day Roberts mounted a tube on a hemisphere of lead and dropped it from the roof of a three-story building onto a steel plate. The indentation of the lead allowed an estimate of the acceleration, which was even higher than before, and the tube still worked. The fuze project was under way [7].

In fact, Tuve and Roberts were already on a war project, for both were on President Roosevelt's Advisory Committee on Uranium. In January of the year before, Roberts had demonstrated fission in a startlingly simple experiment to Niels Bohr, Enrico Fermi, Edward Teller and Gregory Breit, who were attending a scientific meeting (on low-temperature physics!) in Washington at which the knowledge of this new nuclear process had got out. Roberts continued to work on fission and subsequently discovered delayed neutrons, which allow fission to be controlled in a reactor, but the events of the spring and summer of 1940 brought the men at DTM to the viewpoint that an atomic bomb would come too late to affect the outcome of the war. One of the DTM staff, Norman Heydenburg, continued making measurements for the uranium project until all such work was transferred to Los Alamos and construction of the cyclotron continued, but most of DTM went to work on the fuze. Other thoughts may have been in Tuve's mind. When asked about leaving the bomb project years later he said: '... and I didn't want to make an atomic bomb' [8].

Tuve and Roberts made interesting contrasts. Tuve was the son of Norwegian immigrant grandparents who had settled in a small town of South Dakota. He and his childhood friend Ernest Lawrence had linked their houses with a telegraph line, replaced with wireless sets when Ernie's family moved. Both went on to build pioneer nuclear physics laboratories. Roberts traced his lineage to colonial roots, had financial independence with origins in Pennsylvania oil and had gone to the best schools. The two of them guided DTM for four decades with a scientific leadership that kept them active laboratory partners of their colleagues. They were implacable

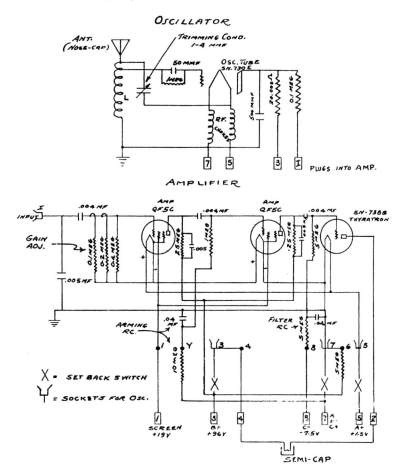

Circuit for the proximity fuze. This is a slight modification (prescription E) of the design of W A S Butement. The dc component of the plate current of the free running Hartley oscillator was altered by a change in radiation resistance when a conducting object came within a few wavelengths, modulating it by the motion of the projectile and forming a signal that passes through a low-pass filter into a two-stage audio amplifier. When the output of the amplifier exceeds a fixed level, the thyratron conducts, discharging the semi-cap. Archives of the Department of Terrestrial Magnetism, Carnegie Institution.

enemies of big science.

Roberts's first experiments obviously called for shooting vacuum tubes out of a gun, so the machine shop made a small muzzle-loading smooth bore, which was taken to a farm owned by a friend of Tuve's in what is now the Virginia suburb of Vienna. The gun was pointed straight up, a projectile with a small tube potted in wax loaded, and the gun fired.

And failure! Although the glass envelope had survived, the electrodes collapsed completely. Navy ordnance experts suggested that they try again using smokeless instead of black powder, which explodes instead of burns and gives much higher initial acceleration than smokeless. A 37 mm gun of 1916 vintage was procured, and tubes began to survive. For the next few months projectiles were fired, sometimes hundreds a day, testing tubes and other components. Initial nervousness of the experimenters about where the shots would land was soon replaced by confidence on learning that they could predict the point of impact within less than 100 m.

While the tests to determine whether electronic components could be fired were being conducted, the Tizard Mission arrived in Washington, and on 14 September R H Fowler and John Cockcroft had dinner at Tuve's home, open exchanges of information about fuzes soon following. The Americans had not settled on the method of influence yet and were examining the same methods the British had. Lawrence Hafstad worked on a photoelectric method, and G.K. Green on an acoustic. The electronic circuit designed by Butement, Shire and Thomson for a radio proximity fuze was extracted from Tizard's famous 'black box', and Roberts, who brought the additional skills of an electronics enthusiast as well as a reserve officer of Field Artillery to the project, had the circuit working in the lab in a couple of days. The basic circuit remained unchanged throughout the project [9]. The anode resistor of a Hartley oscillator was connected through a low-pass filter to a two-stage audio amplifier connected to a thyratron, which passed current through a detonator when its grid voltage exceeded a given threshold. Just four tubes!

The laboratory circuit, tuned to 100 MHz, worked beautifully. The thyratron output responded sensitively to the motions of a half-wave dipole anywhere in the room. Roberts's brother, Walter, a radio engineer who had helped design the oscillator for the DTM cyclotron, worked out the theory of the thing and found that if the target came within a few wavelengths of the oscillator, it altered the loading of the antenna, thereby changing the direct component of the anode current, which varied at a rate determined by the relative motion of target and projectile. This amplified and filtered signal triggered the thyratron. Doppler was not really needed. It was an elegant design.

With evidence that vacuum tubes could be fired from guns and that a simple electronic circuit could be made to trigger the explosion it was obvious that a greatly expanded project was needed. Vacuum tube manufacturers had to begin furnishing prototype rugged tubes while preparing for mass production. Batteries presented particular problems. Circuit and mechanical design had to proceed toward a usable device, and a greatly expanded testing program undertaken. All this required an increased staff, which quickly had over a hundred persons working in a building that had housed only a dozen a few weeks earlier. Tuve put out a set of rules, the first of which was: 'I don't want any damn fool in this laboratory to save money. I only want him to save time'. For those who had experienced Tuve's frugality before or after the war this was a startling rule.

In October fuzes made of non-rugged components in non-miniature circuits for both the radio and photoelectric fuzes detonated 100 lb bombs dropped at the Naval Proving Ground, Dahlgren, Virginia. At about this time Tuve decided that fuzes for non-rotating projectiles presented different kinds of developmental problem and turned the work on bombs and rockets over to the Bureau of Standards under the direction of Harry Diamond, who continued to work on the photoelectric method but dropped the acoustic as impractical. The DTM group dropped all methods except the radio fuze.

In February 1941 tubes were fired in 5 inch star shells with the parachute intended to lower the flare being used to bring down the components tested. On 20 April 1941 an oscillator was shot from the 37 mm and observed to function, and about two weeks later seven oscillators were fired from a 5 inch gun at Dahlgren, four being heard in flight. An oscillator with a modulator to calibrate microphonics generated in flight disclosed no such problem. It was time to make complete fuzes.

The small size of the 37 did not allow the firing of complete fuzes, so the vertical firing was transferred to a 57 mm at Dahlgren. This gun had not only a larger shell but a higher muzzle velocity. The Dahlgren firings were enlivened by the caretaker's dog, who raced into the river with each shot, expecting that such a powerful gun would bring down plenty of ducks, and who needed weeks of duckless firing to learn that the hunters were incomparably bad shots. Firing became routine for testing prototype industrial tubes as well as production lots. A more exciting aspect of the Dahlgren firings was a poorer ability to predict where the shots would land, the consequence of them ascending to much higher altitudes through more complicated wind patterns. One landed completely out of bounds, a mile from the gun.

Numerous tube manufacturers entered the competition, but Sylvania proved most successful. Its T-3 tube weighed less than three grams. One must remember that small-sized electronic components so common today were not so much admired in 1940. It is also worth noting that the entire US production of vacuum tubes in the last peacetime year was 600 000 per day. By 1945 the production of tubes for proximity fuzes was 400 000 per day with 95% from Sylvania.

The first batteries were specially adapted dry cells furnished by National Carbon, but they soon showed serious shelf-life problems and were replaced by wet batteries that had indefinite life with the added advantage of being activated only at the firing of the gun. A sealed glass ampule containing acid was placed within a stack of annular discs. One side of each disc was zinc, the other carbon. On firing the glass ampule shattered and the acid was flung into the plates by centrifugal force. If its shelf life was long, its active life was short, about two minutes, just long enough for the flight of any proximity-fuzed shell.

A fuze was mounted in a 5 inch shell with a hole cut in the side for an ammeter in the anode current circuit. With this the radiation pattern of the little transmitter driving a dipole formed by a cone-shaped electrode at the nose and the shell body, was measured. The next step was to fire pilot production fuzes in 5 inch guns at Dahlgren, which took place in August 1941. On 29 January 1942 the success rate at Dahlgren exceeded 50%, and full production started while the bugs were still being removed. Unfortunately, removing bugs did not mean they would stay removed. One of the greatest problems in producing fuzes proved to be quality control at all levels. It was a never ending problem, and there was no let up. A dud rate no greater than 5% was sought, but it was hard to attain.

The introduction of the high-explosive shell at the turn of the century brought an awkward period during which guns exploded on firing from time to time, owing to imperfections in the fuzes. Improvement in design soon made the simple impact and time fuzes bore safe, but the proximity fuze obviously had many more ways to fail. Tuve was determined that his race against time was not going to result in dead gunners, so a major effort went into safety devices. The explosion was initiated by a detonator that was activated by some tens of milliamperes, so the first line of safety was to keep it shorted until the projectile was clear of the muzzle. A clockwork located in the base of the fuze and actuated by projectile spin removed a short circuit and a mechanical gate in the powder train half a second after firing; it was eliminated in later models. The wet-cell battery also helped by requiring a tenth of second to come up to voltage. A mercury switch functioned in two ways. Before firing the mercury resided at the center of a porous cylinder located slightly off the projectile axis where it effected a second short. On firing, centrifugal force spun the mercury through the porous material thereby opening the short and closing a switch that activated the electrical components with a delay determined by the diffusion time through the diaphragm. The thyratron and the last stage of the audio amplifier, which operated in the range from 30 to 300 Hz, were initially biased to cutoff and became active only after a time delay determined by a capacitor charging time. Finally, the presence of the gun tube so loaded the antenna that the oscillator would be quenched while within the gun. Thousands of rounds with only one operable safety and a reduced charge of black powder were fired to evaluate each separately.

Firing at air frames suspended from balloons and from towers at the New Mexico Proving Ground by H R Crane and David M Dennison measured the burst patterns, which could then be adjusted with the only available parameter, the sensitivity. If the sensitivity were too great, the shell would burst too far from the target; if it were too small, the shell would burst close enough to assure the target's destruction but allow many possibly damaging rounds to pass by. These tests were all made with explosive charges just great enough to permit photography, otherwise target replacement would have become a major waste of time. After these successes it was time for the critical test: firing from a ship at radio controlled targets, called drones, under routine service conditions. The tests were made on the shake-down of the new cruiser USS *Cleveland* in the Chesapeake Bay on 12 August 1942. Roberts was aboard and later recorded the event.

> The next day all was ready off Tangier Island and a drone approached on a torpedo run. At about 5000 yards the ship opened fire with all its 5 inch guns. Immediately there were two hits and the drone plunged into the water. Commander Parsons called for another drone and out it came on a run at about 10,000 ft altitude. Once again it came down promptly. Parsons called for another and then raised hell when the drone people said there were no more ready for use. He enjoyed this very much as he had been on the receiving end of a lot of comments by the drone people in other firing trials. The drone operators had one back-up drone ready in case of troubles but they never expected to have one shot down. In fact the Navy photographic crew who took pictures of all the firing trials of the fleet had never seen a drone shot down before. The ship was ordered to the Pacific with no stops, as the crew had seen too much [10].

As the *Cleveland* was not to dock on her outbound voyage the technical personnel were loaded into a launch to take them ashore. In a somewhat humorous gesture the skipper gave his evaluation of them when he presented each a life preserver as they descended to the small boat, which naturally had a normal supply of such articles.

It is ironic that this test, which showed that a warship could defend itself very well against air attack, took place less than 100 miles from the location where, some 20 years before, Mitchell thought he had proved that surface ships were obsolete as a result of air power.

By mid-November 1942 about 5000 rounds were on the way to Pearl Harbor of which 4500 were sent to the South Pacific on USS *Wright*. At Noumea they were distributed by Vice Admiral Halsey to the ships considered most likely to see action. On 5 January 1943 USS *Helena*, on her way back with two other cruisers and two destroyers from an attack on an airstrip on New Georgia the day before, shot down a Japanese plane with a shell equipped with an industrially produced fuze [11], less than 30 months after the first discussions at the newly formed NDRC about the need for such a device.

The security surrounding the device was extreme. Early models were called 'T3G Device' and all shipments were guarded by Marines and signed for by a commanding officer. Afloat and ashore they were kept under lock and key, and on arrival at port no one was allowed to leave the vessel until the fuzes were accounted for [12]. In production it became 'mark 32', and in the summer of 1943 called the 'VT' after British suggestions meaning 'variable time' or 'velocity triggered'. The proximity fuze may have been enveloped in extreme secrecy, but it was the subject of enough rumor by the time of the Battle of the Eastern Solomons to be mentioned in the after-battle reports, the same month as the *Cleveland* trials in the Chesapeake Bay [13].

The supreme driving force behind fuze development was its use against aircraft, but once this problem was solved thoughts naturally proceeded to an older problem of the artillerist: air bursts against ground targets. The first explosive artillery shells, which introduced the term 'bomb shell' into the language, had used powder-train fuzes. If this fuze were cut short, it led to 'bombs bursting in air'. With better fuzes and more accurate guns this had been refined by General Henry Shrapnel of the British Army into a shell filled with lead balls and that burst in the air with devastating effect on exposed infantry. After World War I shrapnel had been replaced by the high-explosive shell that did its killing with jagged shell fragments instead of lead bullets, but the time fuze remained. Up to 15 seconds flight time could be obtained with a powder-train fuze, 25 with a clockwork fuze. With flat trajectory guns at moderate ranges and observed fire these could be effective. At long range, at night or in fog, or unobserved, time fire was almost useless. Use of the proximity fuze was obvious.

The Field Artillery had gone over to howitzers to a large degree, and they presented a few problems. They never had the high muzzle velocities of the AA guns and even had a variety of velocities from which to choose, determined by the amount of propelling charge loaded. Varying muzzle velocities meant varying spins, and spin operated the safeties. Thus high acceleration, that horrible problem in the summer of 1940, became a necessity. It was soon decided that only the top three powder charges for howitzers would be considered. Fuzes were soon ready.

The army equivalent to the *Cleveland* firings was a demonstration to the Field Artillery Board at Ft Bragg on 24 and 25 September 1943 with Lieutenant General Leslie McNair, Chief of Army Ground Forces in attendance. It was fouled up, yet a stunning success. The fuzes for different caliber weapons were mixed up at the gun positions causing up to 30% duds and bursts at the wrong heights. The Section T men were frantic, and it showed. The Board was so startled to see air bursts at extreme ranges, air bursts unobserved, air bursts with high-angle fire (shells descending almost vertically), air bursts at night that its excitement was almost uncontrolled. When the fuze men went on about the performance, McNair answered: 'Gentlemen, you want all this and the moon too?' [14].

The account of the story at this point does not convey a proper picture of what had been going on. Tuve's objective was a weapon to be placed in the hands of the warriors—and soon! This meant that production had to be brought in early, well before designs were final, and the entire project grew at an incredible pace. The early fuze work had more than 40 industrial and academic contractors, and Canadians helped with battery design. The year 1940 was a good time to place orders because industrial mobilization had just started and there was plenty of slack yet to be taken up. In April 1942 Section T had outgrown the space at the Carnegie department and moved to a large building on Georgia Avenue in nearby Silver Spring, Maryland. At that time the Carnegie Institution transferred administrative control to Johns Hopkins University, and the newly established unit was named the Applied Physics Laboratory. By the time of the *Cleveland* firings production of fuzes was already beginning. Needless to say, this gamble brought on no small number of emergencies. Strange infirmities would appear, in a product that had a built-in bias against diagnosis, yet diagnosis was demanded immediately. US and British forces had between them 40 different kinds of shell for which the fuze was required, and each had to be individually fitted.

Secrecy had adverse effects in complicating procurement, and curious ways were found to conceal the true function of various components. The plastic noses were ordered through Johns Hopkins Medical School under the name of 'rectal spreaders'. Worse, because they were not told what they were making, workers came to believe it was not important, and to keep from arousing curiosity, fuze plants were never given the Army-Navy 'E' for excellence flag. In a product requiring high quality control, this was a definite embarrassment [15].

By the end of the war 112 companies were engaged in production work on fuzes and more than 22 000 000 had been manufactured with the price eventually falling to $18. As a wartime project it was exceeded in magnitude only by the bomb and what we might call 'large-set' radar. Yet the entire project was directed to the end by Tuve, who controlled both the technical and the business aspects and who before 1940 had never supervised more than half a dozen persons.

The first wide-scale employment was in the Pacific, in part because it was more the Navy's weapon than anyone else's but mostly because fleet use gave the smallest probability of one being captured. Section T was well aware that the first danger from a fuze falling into enemy hands was jamming, and recovery of just one of the all too many duds could give the whole thing away. Jamming really meant causing premature bursts and could be effected by sweeping a high-frequency oscillator through the frequency band of the fuzes. When the frequency of some electronic device interfered with that of the Hartley oscillator of the fuze prematures did occur. On Okinawa 105 mm howitzers using the fuze had to stop using them because of bursts all along the trajectories, bringing severe protests from the infantry. The cause was determined to have been the meter-wave radars of nearby destroyers [16]. (This explanation is questionable. There are no Navy reports of fuzes being set off at sea, where the radars in question would have had excellent opportunity. If radar was the cause of prematures, the source is more likely the SCR-270, which was present on land, as its frequency band of 100 MHz was that of the fuze, and its 100 kW peak power and pulse repetition rates of 200 to 400 Hz were well suited to trigger the fuze.)

The effect on naval action was immediate. Each naval air engagement saw the new weapon playing an ever greater role, culminating at the Battle of the Philippine Sea on 19 June 1944 and in the defense against the suicide pilots, principally at Okinawa. The first use in the European theater was again naval, during the invasion of Sicily.

The most spectacular triumph of the fuze was in the defense against the flying bombs, but it was a triumph shared with the gun-laying radar SCR-584 and the electronic director M-9. It was crucial in the Battle of the Bulge, where it was used to devastating effect against infantry advancing in fog.

The German proximity fuze work continued in fits and starts. Siemens und Halske dropped work on the original balanced-capacitance fuze, but others took up the task later in the war when large AA rockets were being designed at Peenemünde, the location of the V-2 rocket-bomb development. A proximity fuze was necessary for these, but it was not subject to the severe constraints of space and shock resistance imposed on the fuze of an artillery shell. The work was directed from Peenemünde West and was both of local and contracted origin. Four fuzes were under simultaneous development: Kranich, Kakadu, Marabu and Fox.

Kranich was a purely acoustical fuze that had a resonant cavity dimensioned to the principal frequency of heavy-bomber motors. A wire whip was fastened to a diaphragm that formed one wall of the cavity. Vibrations set the whip in motion, causing it to touch a ring electrode and close the firing circuit. Kakadu and Marabu were 50 and 70 cm continuous-wave transmitter–receiver pairs with separate tuned-dipole antennas. Kakadu made use of the Doppler shift in the reflected wave, Marabu a frequency-modulation effect. Fox operated on 3 m and was similar to the Butement design in using the alteration of antenna loading. It was not small, having a dipole antenna.

These competing designs were tested at Peenemünde by fastening them to long poles mounted on a wooden tower that held the necessary test equipment. The individual designs caused lights to flash when actuated, and their spatial relationship to the aircraft that flew over the poles was recorded on film. The end of the war prevented the work from proceeding further [17].

Britain also continued fuze work. By November 1940 GEC furnished miniature pentodes that withstood the shock of firing, and in August 1941 a shell fired from a gun was detonated in the air by a radio pulse from the ground. This approach continued, and in February 1942 yielded a test in which 75% of the shells were burst by signals from the ground. By October 1942 a report to the Prime Minister placed the American efforts well ahead; work continued nevertheless, as it was not clear that US fuze production would suffice for Britain as well as for the extraordinarily hungry Pacific fleet. When it became clear that Britain would receive an ample supply of the new devices, work lagged and no satisfactory design emerged from the war effort [18].

4.5. GREATER AND LESSER MICROWAVE SETS

By 1942 radar engineers had begun to think of their powerful new microwave generator as a standard element of design rather than as a sensational invention. It was still unknown to the Germans, ignored by the Soviets and faltering in application by the Japanese. It brought with it a completely new kind of electronics, one that used waveguides—pipes!— for transmitting power, a concept that had caused a small amount of consternation when first demonstrated, as electricians had long thought conductor pairs necessary for this function. It required technicians to master an assemblage of new crafts for the workbench, and antenna designers found themselves pulling down textbooks on physical optics. These first microwave years saw a truly astounding growth in the number of practitioners of the new art, many quite new to electronics, and to an overwhelming degree they found themselves fairly taken with it. The predictable result was a harvest of designs for equipment both to excel in the tasks that radar had been doing and to enter unsuspected fields of usage.

In Britain radar was being developed by three service laboratories. For the Air Force it was done by the Telecommunications Research Establishment (TRE), the lineal descendant of the Air Ministry Research Establishment (AMRE), in turn the descendant of Bawdsey Research Station. For the Army it was done by a group that started as the Army cell at Bawdsey and which had had three name changes, first the Air Defence Experimental Establishment (ADEE), next the Radar Research and Development Establishment (RRDE) and finally the Air Defence Research and Development Establishment (ADRDE). Naval radar began at His Majesty's Signal School (HMSS) but conformed to the prevailing administrative atmosphere by becoming the Admiralty Signal Establishment (ASE).

On the American side of the Atlantic four laboratories, all retaining their original names, provided designs: the Naval Research Laboratory (NRL), the Signal Corps Laboratory, the Bell Telephone Laboratories (sometimes BTL) and the new Radiation Laboratory (Rad Lab). The two service laboratories had extensive commitments to improving existing sets and to supervising the general supply of radar and communication equipment to combat and training units. They also had the important responsibility of insuring that the designs from Bell and Rad Lab were satisfactory for the harsh conditions under which they had to function.

The constraints on the two American service labs applied even more strenuously to the three British service labs, which were soon in the middle of a radar war with the Luftwaffe that had designers and technicians being rushed from one emergency to another and that frequently had them building special sets for combat one day from a prototype of the day before. They also found working in a war zone disrupting of the calm conditions generally sought for research. As a consequence the clever new designs using microwaves came more often from Rad Lab and Bell.

P I Dee came from the heroic age of nuclear physics at Rutherford's Cambridge laboratory, as did others who occupied themselves with radar, such as P M S Blackett, John Cockcroft, C W Gilbert, W B Lewis and M L E Oliphant, all of whom acquired honors as a consequence of their scientific and technical contributions. Dee arrived at Swanage in May 1940, the time when the cavity magnetron had reached the stage of an industrial prototype. His intense nature quickly led to his assuming control of this new work. At the onset airborne interception was to be its obvious purpose. Defense against night bombing during the Blitz was bringing down few attackers and airborne radar was the only hope. Meter-wave AI mark IV working with GCI finally became effective in early 1941 and the shift of the Luftwaffe to the east took much of the pressure off obtaining 10 cm AI, or AIS as the project was called. The S in AIS was derived from the code designation for the 10 cm band; X was soon added for 3 cm and K for 1 cm. These designations have remained despite modern attempts to rationalize the nomenclature of microwave bands.

Those were years during which new disasters were ever ready to replace any whose virulence had receded, and the struggle with the U-boats replaced the Blitz. Meter-wave ASV mark II proved to be one of the best designs of the war and was able to go into significant production once priority went to ASV.

Microwaves were better than meter waves for both AI and ASV, so these projects continued, but there was quite a bit of uncertainty during the summer of 1940 about how best to use microwaves. To the general surprise of the designers, Air Chief Marshal Sir Philip Joubert de la Ferté, Assistant Chief of Air Staff announced on 22 September that AI mark IV seemed satisfactory and 10 cm gun-laying equipment for AA should be the goal rather than AIS, which involved Dee in extensive discussions with those of higher authority [1].

When the GCI-AI skills began to mature toward the end of 1941, the Luftwaffe began changing their tactics. Noting that the aircraft assigned to planting mines in British waterways were seldom intercepted, the bombers became fond of low altitudes too, where the ground return on AI mark IV swallowed up all but very near targets [2]. The situation was not desperate because the raids were much fewer since Hitler had attacked the Soviet Union, but they required improvisation by the air crews and placed renewed pressure on TRE to produce 10 cm AI.

Independently, Oliphant's group at Birmingham was working with the firm British Thompson–Houston on a 10 cm GL set using the powerful klystrons they had designed. Oliphant had pushed to completion a working, two-paraboloid set that he called for some unrecorded reason the

'Dog's Breakfast' [3], and the Navy offered its own diversion, as we learned earlier, in making the first operational 10 cm radar in its type 271. A new, unsuspected purpose was to grow out of these contradictory tasks—guiding the night bombing of Germany—and it was soon to dominate TRE.

4.5.1. AI marks VII, VIII and IX

A critical element for AIS was the microwave TR switch that came from Oxford's Clarendon Laboratories in spring 1941 (the invention of A H Cooke) because it removed the need for separate transmitting and receiving dishes. Double dishes required too much space and for aircraft even a single dish had to be made as small as possible. By March 1941 the first AIS flew successfully and was compared to the Rad Lab's AI-10, which they sent over in June[4].

Accuracy of direction was attained for meter-wave equipment through lobe switching, a method neither practical nor desirable for microwaves, which allowed the more elegant solution of changing the direction of the beam. This was done first for a 50 cm beam in the Würzburg by off-setting the dipole at the focus of the antenna dish and rotating it, a method invented independently at Rad Lab for SCR-584. The solution at TRE was to move the dish rather than the feed, the ingenious design of Alan Hodgkin working with the firm of Nash & Thompson. Dee pushed the project through the General Electric Company, which required convincing them that 10 cm was better than the 25 cm equipment they favored [4].

In March 1942 100 sets for operational use were designated AI mark VII, and installation in Beaufighters and Mosquitoes began. Within a few months an improved version, mark VIII, that corrected a few faults and included IFF, had replaced mark VII in production. Fifteen hundred sets were ordered.

The next step was to have the AI set lock onto or track the target and allow blind firing, which was a more difficult problem than simply tracking the target. The project took a bad turn on 23 December 1942 when A C Downing, who had taken Lovell's place in the work, was shot down and killed by a Spitfire that misidentified the Beaufighter in which he was testing the only mark IX prototype for a Ju-88. The arrival of the successful SCR-720 that Bell Labs had developed out of the original Rad Lab AI-10 effectively put an end to the mark IX [5].

4.5.2. H2S and ASV mark III

Hitler's attack on the Soviet Union changed Britain's strategic situation to a remarkable degree and with it the tasks facing TRE. On one day defense was paramount, defense against air attack and against the U-boat. On the next day air defense was much less important—so long as Russia held out.

---

[4] See Chapter 4.3 (p 168–9).

It was not in Churchill's nature to use this relief to rest; he wanted to use the resources freed to attack Germany, and the only way Britain could attack was from the air with Bomber Command. As the night raids intensified it became ever more apparent that the navigation of the bombers was wildly inaccurate. Thus when Lord Cherwell, whose primary interest—at times it seemed his only interest—had become the bombing of Germany, reported in September 1941 a study proving that two-thirds of British bombs were falling at distances greater than 8 km from the target, it was clear that a first-class crisis was at hand.

At Rowe's Sunday Soviet of 26 October Cherwell insisted that radar was going to have to furnish the navigational aid to guide the bombers into Germany beyond the 500 km that current radio-navigational methods were planning to attain[5]. All present knew that Cherwell spoke with the voice of Churchill. No one had a solution, but Bowen had noted observing towns in experiments with airborne radar as early as 1938 [6], so Dee had an AIS set modified to aim downward at 10 degrees with some lateral scanning. In flying toward Southampton echoes from the city were evident. On 29 December 1941 Rowe placed Bernard Lovell in charge of making a device to be mounted in the four-engine bombers for guiding them deep into Germany by looking at the ground with 10 cm radar.

Lovell quickly enjoyed the sensation of having high priority and soon had good co-workers joining him along with cooperation from A D Blumlein at EMI. On 23 April 1942 they tested an experimental system mounted in the belly of the microwave section's own Halifax. It had a rotating antenna that pointed slightly downward and presented the observations of a map-like display on a PPI scope. It received the code name H2S, meaning 'home sweet home' or 'it stinks', depending on the raconteur.

The prospect of flying their most secret of secrets, the magnetron, over Germany proved exceedingly troubling. All were certain that it would soon be in German hands, and they were right. This led to a second Halifax being equipped with a similar system using klystrons instead of magnetrons, but their pulsed power was 20 to 30 times less than magnetrons produced, and the klystron project quickly died.

The project was delayed by two severe setbacks. First, in March 1942 TRE moved abruptly yet again and again with brief planning, but more about that later. Second, and more tragic, Blumlein and two of his EMI co-workers died when their Halifax crashed in an accident on 7 June 1942. That evening Lovell recognized only the magnetron in the smoldering wreckage of their single prototype.

On 3 July Churchill summoned the Secretary of State for Air, the Minister of Aircraft Production, the leader of Bomber Command to meet with Dee and Lovell. He demanded 200 sets of H2S by 15 October, although not a single aircraft had flown with H2S since the crash that killed Blumlein.

---

[5] See Chapter 6.2 (pp 301–4.

This unrealistic deadline was not met, but two squadrons were ready by the end of December.

What had been built as a device for navigating—most imperfectly—deep into Germany was an almost perfect ASV set, something not lost on those seriously concerned about U-boats. Lovell's group made the minor modifications that created ASV mark III out of H2S and mounted three of them in Coastal Command Wellingtons by the end of January 1943 [7].

Microwave antennas have the capability of having the beam shaped for specific purposes by the forms given the reflector and the feed. This came about first for the Royal Navy's type 271 that produced a vertical fan-shaped beam for 10 cm surface search radar that would yield a good horizontal resolution and not be affected by the ship's roll. With H2S a more sophisticated approach began that has seen significant later evolution. In sweeping the ground from a bomber one does not need or even want as much antenna gain for close-range points as for far range, but one does want to observe close-in reflections that are nearly directly below. The fulfillment of these requirements gave the PPI display a more maplike appearance and resulted in a reflector for H2S that produced a reflected signal at the receiver with an approximately constant amplitude independent of range for the same size target. Its dependence on vertical angle caused it to be called a 'cosecant-squared' beam pattern.

The maplike display for H2S had two inherent and troubling problems. The intensity of the display depended on the amount of reflected signal, which depended on the composition of the target and the angle at which it was irradiated. Of the various target materials, only a surface of water could be relied on to give specular reflection with essentially nothing returned to the receiver; this presented a dark surface that was easily identified on the scope. But even here the edges of a body of water were affected by landforms, vegetation or structures that lined the banks, preventing an unambiguous definition of the shore, and the picture changed, depending on the distance and direction of approach. Even worse was the consequence of the radar necessarily measuring slant range from altitudes comparable to the ground ranges, not only distorting the depiction but having the distortion change as the aircraft moved [8].

4.5.3. *Gun-laying radar*

Joubert's surprising announcement in September 1940 of the high priority to be given the design of a 10 cm radar for AA artillery had been carried over with the Tizard Mission as one of the three high-priority projects suggested for the new Radiation Laboratory, the other two being airborne interception and an unspecified long-range long-wave navigation system using ground stations, which became Loran[6]. The end of the Blitz changed all priorities, and neither of the two microwave projects received the priority in Britain

---

[6] See Chapter 10.1 (pp 430–1).

that it did in the United States. When one combined the general lack of belief in AA artillery in Britain with Cherwell's disdain of it and favor for any method of improving the means for bombing Germany, the result was low priority indeed for GL mark III [9]. That this lessened zeal did not visit the United States resulted from the happy combination of Ivan Getting's drive and the Coast Artillery's long-standing objective of radar-controlled AA fire.

Yet when SCR-584 made its spectacular appearance in early 1944 four competing 10 cm GL sets were also in production, pointing either to a very high priority that demanded insurance of success by duplicating design efforts or to confusion in inter-Allied planning. As we have seen, Col Colton saw the roof-top demonstration of automatic tracking at Rad Lab as so important that he immediately covered his bets by ordering Bell Labs to design a 10 cm GL set, which became SCR-545. This equipment added a 1.5 m array to the 10 cm dish for increasing the field of view and had automatic tracking for both wavelengths [10]. It was an excellent set but was outclassed by the 584.

Although aware of the American work, ADRDE continued the design that had begun as the 'Dog's Breakfast' and that became GL mark 3, or more correctly GL mark 3B, designating the British set. The obvious superiority of SCR-584 led to British purchases for which the designation GL mark 3A was applied. There was also a GL mark 3C, the Canadian design. Neither 3B nor 3C had automatic tracking, and both used separate transmitter and receiver dishes [11]. They were not in the same class as the two American sets and had deficiencies that became all too apparent when the robot bombs headed for London in 1944, deficiencies that could be traced to Lord Cherwell and the prevailing British attitude about AA artillery.

4.5.4. The Alvarez antenna

Of the 16 chapters of Luis Alvarez's autobiography [12] only one is devoted to radar, the record of two and a half years, for he went to Los Alamos in July 1943. In that brief period he initiated two remarkable radar sets, Eagle and Ground Controlled Approach (GCA), and had an important influence on a third, Microwave Early Warning (MEW). The brilliance of his style makes an interesting contrast with that of Getting. Alvarez would perceive a completely new concept and follow its design through to the point where the principles were safely established, and then rush on to the next idea. Getting appreciated the dire need for directing AA fire and pushed through the fundamental design, prototype, industrial production, battle-field deployment and final modification based on experience to make SCR-584 the finest radar set of the war. For Alvarez the basic idea was everything. For Getting the final device was everything. Both were true to their styles in their equally brilliant postwar careers.

The three radars that interested Alvarez all required higher resolution of targets than any previous sets. Resolution, the ability to distinguish between adjacent targets, comes with narrow beams, and narrow beams come with increased antenna size for a given wavelength. To obtain the beams he wanted with reflecting dishes required impossible dimensions. The lobe needed to be narrow in only one dimension, which allowed the antenna width to be a linear array of dipole elements. Alvarez questioned whether a waveguide with slots, a 'leaky pipe', could not be configured so that each slot radiated as a dipole. This proved possible but invariably yielded unwanted side lobes, which allow targets several degrees off the beam direction to be observed as if they were in the beam. The next approach reduced side lobes to tolerable levels by replacing the slots with dipoles of alternating polarity connected to the waveguide [13], mimicking the technique of meter-wave arrays.

In the first application of this antenna it was not possible to scan by rotating the structure, so Alvarez devised an electrical technique capable of scanning 30° to the right and left. The wavelength in a waveguide is not the same as in free space and depends on the size of the guide. The direction of the emitted beam is altered if the phase of the wave differs by some amount from dipole to dipole, and phase shift is just what happens if one changes wavelength. The wavelength in the waveguide can be altered by changing a resonant dimension mechanically, specifically its width, and this became the basis for the electrical scanning method, frequently referred to as the Eagle system.

4.5.5. *Eagle (AN/APQ-7)*

The US Army Air Force faced the world conflict in 1939 with a confidence in their ability to do accurate high-level bombing based on their regard for the Norden bombsight, a mechanical analogue computer that incorporated data entered by the bombardier for altitude, ground speed and wind. It was so secret that the United States refused to share it with Great Britain, even though the enemy would inevitably gain samples soon after American bombers appeared over their territory. The RAF had, in fact, a very good, very practical bombsight, and the accuracy of both air forces was affected in a trivial way by the shortcomings of these instruments.

There were two serious problems with its application: (1) accurate bombing required a carefully adjusted approach at constant altitude, deviating from a straight line only to the degree necessary to follow the bombardier's corrections, which made it ideal for a defending AA gunner; (2) it depended on being able to see the target for nearly all of the entire run, something that was routine in the American deserts but often a rarity in the skies of Europe and, of course, it depended on doing the bombing during normal business hours.

Alvarez became interested in this bombsight and was permitted to visit Carl L Norden at his Brooklyn shop where he personally made all the drawings. Alvarez recognized the application weaknesses but saw that if radar replaced optical vision, darkness and clouds became advantageous for the attacker rather than the defender, as optically controlled gun fire would be much less dangerous and with proper jamming, then actively being planned, even radar-controlled fire would be nullified.

Thus Alvarez approached the same problem of radar navigation and blind bombing as Lovell but found an entirely different solution than H2S. First, he insisted on having much finer resolution on the ground. To obtain this he used the 3 cm radiation that Rad Lab was quickly perfecting and a wide antenna capable of projecting a much narrower vertical fan than the 10 cm with a small dish. Second, he saw no reason to scan the entire horizon, just a 60° swath ahead of the plane, which he could do with electrical scanning. The wide antenna, a linear waveguide array of 250 dipoles, was first intended for the leading edge of a wing, although this idea was dropped in favor of a separate airfoil vane mounted beneath the plane. These elements became Eagle or AN/APQ-7 as it was entered in the catalogues [14].

The novel antenna awoke a great deal of skepticism among those not directly involved with it, and the Eagle group received low priorities and numerous challenges to have it replaced by some other system. On becoming project engineer E A Luebke had to defend it against cancellation and to decide between five competing computers: Norden mark 15, General Electric, Librascope, Bell Labs's BTO and UBS, some weighing half a ton [15]. A design emerged that was independent of computer choice [16], but a Norden design was used. Eagle had a moderately successful prototype flight test on 16 June 1943, and a very successful production model from Western Electric was tested on 16 May 1944. It was capable of picking up cities at ranges of 250 km and had a beam width of 0.4° [17].

Because of the delays resulting from its poor support Eagle did not reach the European theater in time for useful deployment, although it was used effectively against Japan. Its importance lies in the new concepts of antenna design that it introduced, concepts that were to have a profound effect on later radar. It was renamed Eagle by DuBridge, who found the acronym in use by the group, EHIB, too flippant for transmission to outsiders. It stood for Every House in Berlin [18].

4.5.6. *Ground-controlled approach (GCA, AN/MPN-1)*

It was the flier in Luis Alvarez who saw in Getting's XT-1, the prototype of SCR-584 a method for landing an airplane under conditions of near blindness. There had been 'blind-landing' methods since the early 1930s, the Lorenz system being the most widely used [19]. The early microwave work of Edward Bowles at MIT was aimed at this problem, but all these schemes only got the pilot to the point where the landing field or its lights could be seen. If things were really 'souped in', they were not much help, and all required that the aircraft carry special equipment that the pilot had to master. Blind landing was no small affair during the war. A large fraction of aircraft losses were through accident, and poor landing conditions led the list of causes.

The XT-1 could locate an aircraft accurately in three coordinates and had been observed a couple of times to follow the plane all the way to the ground during a landing. Alvarez saw this as a way of 'talking' a flier down simply by comparing his measured location with a desired glide path. He secured the use of XT-1 in April 1942 [20] with Lee Davenport testing the idea by following a number of aircraft onto the runway. The result was a failure.

For reasons within the electronic soul of XT-1 it would sometimes and without warning break away from the line of sight and locate the plane below the runway! The radar beam saw the reflection of the target on the ground. No amount of adjustment or operator skill altered this uncomfortable fact. When Alvarez finally conceded that the problem lay in deficiencies of the method rather than in the immediately assumed inadequacies of Davenport's operator skill, he consulted with Alfred Loomis. The two of them worked out the solution [21] that was incorporated in mark I GCA, for which L H Johnston became project engineer in July 1942 with the construction of ten units contracted to the Gilfillan Brothers Company, an organization that contributed much to the final design [22]. The design used three separate radars: a 10 cm search set, which scanned the horizon and fed its information to a PPI, and two 3 cm sets, one having a narrow horizontal fan beam for elevation, the other a narrow vertical fan beam for direction. The antennas used the electrically scanned dipole arrays from Eagle. The vertical positioning antenna scanned up and down, the horizontal from side to side, thereby fixing the three coordinates of the unseeing pilot. The shorter wavelength of the two fan-shaped beams provided accurate positioning through beam narrowness, which eliminated the possibility of a target being observed through its reflection on the ground. An ideal glide path was measured and reproduced as an electrical analog signal. Deviations of the aircraft's path from the ideal were presented to the controller by instrument deflections by which the controller could tell the pilot the corrections necessary to position himself about 40 m above the runway, although on occasion completely blind landings were made. The search radar with PPI allowed planes to be stacked while awaiting landing [23].

A single mark I, which had mechanical scanning, was tried under field conditions in England in 1943. The demonstration was so successful that demands for the equipment began to come from all theaters. Five models of mark II, which had advanced to electromechanical scanning (Eagle), were quickly produced, and by the time mark III appeared there had been more than 2000 successful blind landings by crews new to the equipment [24]. Many bombers and their crews returning from Germany were saved by these sets, but its greatest triumph was to come during the Berlin blockade in 1948–1949, as the weather would have never permitted the continual flights needed to supply the city without GCA.

Curiously, GCA was not applied to commercial aviation after the war, and airports continued to cease operation when obscured by fog. The reason for this was the refusal of civilian pilots to relinquish responsibility for the aircraft during such critical moments. During blind landing the GCA controller effectively flew the plane. A few civilian airports had GCA sets for military aircraft, and there were times when civilian airliners were brought home with them. The very high installation, operation and maintenance costs did not appeal to commercial aviation.

4.5.7. *Microwave early warning (MEW, AN/CPS-1)*

The great problem of air warning with radar using wavelengths of meters and tens of meters was the vertical lobe structure that allowed aircraft to approach undetected at low altitudes. This was recognized for CH and resulted in CHL. The difference was the use of wavelengths one-tenth as long for CHL as for CH. But 1.5 m sets still could not detect low fliers until they were closer than fighter controllers would like to have things. A further shift to microwaves was the obvious solution. Poor target resolution resulting from the large beam widths was also a problem that severely reduced traffic handling capacity and microwaves would help there equally well. The gaps formed by the vertical lobes were used for height determination, but they had as a consequence gaps in coverage at moments when things became hectic and when no one had time for height determination; microwaves would not eliminate the gaps but formed so many that the transitions were indistinguishable from normal target fluctuations. There was, of course, no height finding capability.

Morton H Kanner of Rad Lab went to England in January 1942 to study air warning requirements with the defense of the American west coast uppermost in his mind. On returning he discussed the problem with Alvarez, whose use of the 'leaky pipe' feed then being considered for Eagle might be used for producing the desired narrow vertical beam. Kanner became project engineer in June 1942.

Rough calculations indicated an S-band transmitter of the order of a megawatt would be required, a power level comparable to those of the big meter-wave sets. The reduction in size would require insulators to hold much higher voltages than for the same power at longer wavelengths. The polyethylene coaxial cables that had so pleased designers only months before for their splendid electrical characteristics were inadequate. All radio-frequency power would have to go through waveguides. The work encountered arcing of magnetrons and waveguides in forms never before encountered in the long two and half year history of microwave design and met equally exasperating problems with the transmit–receive (TR) junctions that allowed crystal mixers to burn out. All had to be solved [25].

The antenna design that would secure good low-level coverage at long ranges would not be tall enough vertically for near-in targets that were high, so a second antenna was planned for high coverage. Beam widths of 0.8° required a horizontal parabolic-cylinder reflector 7.6 m wide; the low-coverage reflector was 2.4 m high and the high coverage 1.5 m. Each antenna was fed through a linear array of 106 dipoles mounted on a waveguide similar in concept to that used in Eagle but the scanning was by rotating the antenna, as 360° coverage was required. These dimensions and the component sizes needed for the peak high power, eventually to be 700 kW, began to give microwave early warning, MEW as it was called in the field and lab, and AN/CPS-1 in the office, a tremendous size. The two reflectors were mounted back to back. The large number of targets that the system could track required five 30 cm faced oscilloscopes with assorted auxiliaries, all of which went into a small house. When it was all put together MEW weighed 66 tons and consumed 23 kW from an engine-driven generator. Transport required eight trucks [26].

Only a few MEWs were manufactured, all hand crafted at Rad Lab. Set number one operated in England in January 1944 and quickly established its value to both air forces, and two were ready in time to help control air traffic on D-Day and locate the V-1s, something for which the absence of height finding capability was of no consequence [27]. For fighter control over France it was necessary to add separately a set with a very narrow, horizontal, fan-shaped beam, a modification of a Bell Labs design for the Navy, much like similar equipment that had come into use to determine heights better at British GCI stations. A veteran British fighter control officer gave MEW the highest marks for its ability to deal with large numbers of planes [28].

In the Pacific where the density of aircraft was not so great as in Europe and a general satisfaction with meter-wave equipment for air warning prevailed, the result of island siting, MEW was not initially greeted with enthusiasm—66 tons to be moved under the worst transport conditions and no height-finding capability! MEW Number 4 arrived at Saipan, the base for the B-29 attacks on Japan, on 21 September 1944. On 27 November it was still not in operation, and the base was surprised by fighters from Iwo Jima that came in just above the ocean and left burning and damaged bombers behind them. Repeat performances brought telegrams from Washington ordering immediate installation of MEW. After a sizeable engineering effort the set was established atop Mount Tapochau by New Year's eve. A raid on 3 January was picked up at 200 km and intercepted. This naturally changed attitudes, and fondness for the monster grew when it began to give remarkably accurate positions for downed air crews in need of rescue [29].

4.5.8. *Airborne early warning, Project Cadillac*

By the end of the war one-fifth of all the Rad Lab personnel together with 160 officers and men from the Navy were working on a single project, one that became responsible for 12% of all outside purchases for the entire five years of the Laboratory's existence. It was Project Cadillac, named for the mountain in Maine where sunlight first touches the United States each morning, not for the luxury automobile with the same name, not that that would have been inappropriate. It was a project that was not finished in time to have any effect on the outcome of the struggle, but it pointed to the future for radar and weapons systems in a manner unlike anything previously designed.

Cadillac was to overcome some of the restrictions placed on radar by the Earth's curvature. It was a 10 cm search radar mounted in an aircraft flying so high as to extend the range, especially for low-flying planes and ships well beyond what a ship's search radar could detect. The need had been recognized early. An inter-service committee had recommended in June 1942 such a system—and here 'system' begins to take on a modern meaning—and requested Rad Lab to investigate. In August RCA loaned a television link with which an oscilloscope trace was transmitted between an airplane and a ground station. The amplitude-modulated equipment showed serious problems as the distance separating transmitter and receiver continually changed, but Zenith Radio had developed a frequency-modulated television of the kind that came into use after the war, and a PPI display was satisfactorily transmitted in May 1943. An air–ship link distance of 160 km was attained.

This part of the task was, in fact, the easy part. The amount of information about the location of ships and aircraft, with the perennial problem of IFF, would be simply enormous for the area observed and over which a fleet might be maneuvering. To be of any value this information had to be communicated rapidly to all of the fleet's CICs. It also required for command communication many separate channels of very high-frequency radio that had to be established through the aircraft in order to avoid the same line-of-sight restrictions as had the radar. Methods had to be incorporated using radar beacons for the accurate determination of the observing aircraft's position.

J B Wiesner as project engineer presided over a true crash program. There were five divisions of labor: shipboard system, airborne system, relay radar, relay radar transmitter, and beacons and IFF. Two complete systems were to be delivered from a complex assortment of suppliers by February 1945, and one did become available in March. USS Ranger entered tests in April 1945 that extended two months and established Cadillac's value beyond doubt, but the end of hostilities removed the dreaded need for it in the invasion of Japan [30].

PHOTOGRAPHS: LAND AND NAVAL FIGHTER CONTROL

British AMES type 16 fighter direction station. This 50 cm equipment had a pencil beam that allowed ground controlled interception to be made with a single radar. Its normal mode was a panoramic scan at about 2 per minute as it oscillated vertically between 0 and 10°. Historical Radar Archives, RAF photograph. Crown Copyright.

British AMES type 13 height finder. This 10 cm set generated with its truncated paraboloid antenna a fanlike beam narrow in vertical extent but wide in horizontal, which allowed the height of a target to be determined. It worked in ground-controlled interception, generally with the AMES type 15 1.5 m CGI set, which could accurately determine horizontal but not vertical coordinates. It became the congenial companion of the Rad Lab MEW (microwave early warning) radar. The design was the basis of the American SCR-615. Historical Radar Archives, RAF photograph. Crown Copyright.

Fighter control with radar: Britain. The indicator of a CHL radar. This 1.5 m equipment could observe low-flying attackers for which the large vertical lobe structure of CH allowed evasion until too late. The coordinate grid on the left oscilloscope indicates it was an A-scope display yielding range; the right scope had a maplike PPI. Historical Radar Archives, RAF photograph, Crown Copyright.

German Würzburg-Riese (FuMG 39T-R). This radar used the same wavelength and electronics as the small Würzburg but attained greater range by increasing the size of the reflecting paraboloid dish from 3.0 to 7.4 m in diameter. Its function was to guide night fighters close enough to the attacking bomber for the pilot to see it visually or with his airborne Lichtenstein radar. Photograph courtesy of Bernd Röde.

A Würzburg-Riese used as radio telescope. The excellent structural integrity and availability of these dishes led to them to be used to observe the 21 cm radiation of atomic hydrogen in The Netherlands, England and America. One is shown here attached to an equatorial mount at the Department of Terrestrial Magnetism, Carnegie Institution in April 1953. DTM Archives, photograph 18167.

Fighter control with radar Germany: a Jagdschloss (FuMG 404) panoramic set based on the electronics of the Freya. It had 18 dipoles arranged on a 20 m supporting structure to produce a narrow vertical fan-shaped beam. Jagdschloss incorporated a PPI display, which was transmitted from the station to command headquarters by high-frequency cable or the widely used 50 cm directed-beam communication links. It had a broad-band antenna surmounted by an IFF antenna that allowed modest changes of frequency and specifically that allowed a change to excite the fighters' IFF set, which then served as secondary radars that enhanced the positions of the German aircraft at the closing of a switch. The scaffolding to the left is not part of the antenna. National Archives photograph 111-SC 269084.

Air defense center for Berlin. This huge concrete building located in the Tiergarten (Zoo) coordinated fighter and AA defense for the region around the city. Mounted on the top are a small and a giant Würzburg. Similar towers were used for gun batteries to allow unrestricted fields of fire in cities. The military unit was Turmflak-abteilung 123 (tower flak battalion). Photograph courtesy of Fritz Trenkle and Werner Müller.

Fighter control with radar: Germany. The women shown here received filtered information from radar stations and projected it in coded form onto a large map so that the squadron-control officers could make use of it. This photograph presents the unusual circumstance of having officers of both the Royal Air Force and the Luftwaffe observing the tracking of British aircraft during Operation Post Mortem. Imperial War Museum photograph CL 3317. Crown Copyright.

American microwave early warning (MEW, AN/CPS-1) on Okinawa. This 10 cm set removed two serious problems of meter-wave air-warning sets: their poor angular resolution of targets and their blindness for low-flying aircraft. In providing excellent 0.8° horizontal location it relinquished any knowledge of height, which proved to be no disadvantage in guiding fighters and alerting gun crews against the uniformly low-flying V-1 flying bombs. A very narrow vertical fan-shaped beam was attained through a linear array of 106 dipoles cut into a waveguide across the focus of a cylindrical parabolic reflector. A reflector that would secure good low-level coverage at long ranges would not form a beam tall enough vertically for near-in targets that were high, so there was a second antenna for high coverage, the two mounted back to back. Only six hand-crafted sets were made, none of which were available at the outbreak of the Korean War, leading to the use of SCR-270 for air warning for Japan. National Archives photograph 111-SC 238090.

Fighter control with radar: Germany. Filtered data from radar stations were projected onto the large map located in front of the desks of the squadron-control officers. Officers of both the Royal Air Force and the Luftwaffe are shown discussing the tactics of their recent encounters during Operation Post Mortem. Imperial War Museum photograph CL 3316. Crown Copyright.

American Little Abner height finder (AN/TPS-10) on Okinawa. This 3.3 cm set was used to make good the inability of MEW to determine heights. The very thin horizontal fan-shaped, or beaver-tail beam, allowed it to observe close to the ground without excessive ground returns. The design was essentially a copy of the US Navy mark 22. National Archives photograph 111-SC 238093.

American MEW Operations Room on Okinawa. The combination of an MEW and a Little Abner could provide extraordinary amounts of data that had to be converted rapidly into information useful combat units. There were five 30 cm PPI scopes and 5 18 cm B-scopes (plotting range against azimuth in rectangular coordinates) for the MEW alone. Two men are seen plotting on the reverse side of the plastic map display. National Archives photograph 111-SC 238089.

CHAPTER 5

YEARS OF ALLIED DESPAIR AND HOPE

5.1. THE MEDITERRANEAN, 1940–1942

Neither Germany nor Britain had wanted war in the Mediterranean. Geography should have given Italy a dominant role, and pre-war estimates of the strength of Mussolini's navy and air force suggested that he might have been able to assert power successfully once France was out of the way, if her fleet were not to come into British hands. With the vital passageway to Middle East oil, India and the Far East passing through Suez Britain would have certainly preferred quiet in the waters north of Africa. But Mussolini, Il Duce, did not want to miss 'the march of history'.

In entering the war against the Allies when France was near surrender he gave the first of many demonstrations of policy determined by delusion. Mussolini saw Italy's entrance into the war as heroic; others, even the new German allies, saw it as despicable. Furthermore, for all the show-window displays in the 1930s of the power of Fascism, for all the military might exhibited in Ethiopia and Spain, Italy was woefully unprepared even for a minor war, something nearly everyone at high level in Italy seemed to know but Mussolini. The German command saw the Italian army as third rate in equipment, logistics, training and spirit, and acting on their and his own evaluations Hitler insisted that Italy not begin land actions that could require German assistance. Harass the British in the air and on the sea but do not try to take on even the meager British forces in Egypt.

The German high command did have one immediate military request for Mussolini: seize Malta at once! World War II was to rage over the Mediterranean from Gibraltar to Suez, on every shoreline except those of Turkey and Spain, and everything would turn on Malta.

It required only a map and certainly no great training in naval strategy to see that this island—the unsinkable aircraft carrier—was the key to the Mediterranean. The German request was not only strategically sound, it was at the time of Italy's entrance into the war tactically achievable. The defenses of the island were extremely weak. Fear of air attack had caused the fleet to depart for the safer harbor in Alexandria, and the garrison was but a handful with air defense no better. In the Mediterranean the Italian navy greatly outnumbered the British in every class except aircraft carrier, a deficiency countered by the large number of planes based on Sicily. Mussolini disregarded the German request while Churchill put every possible reinforcement into the fortress that could be sent from a homeland expecting invasion, slowly and painfully correcting the island's defenselessness. The need to take Malta never left the minds of Axis strategists, but when another opportunity presented itself Hitler faltered.

The issue in the Mediterranean was supply, supplies for two armies contending for the restricted space between an ocean of sand and an ocean of water. These supplies had to reach them across the sea—generally on, but sometimes beneath or above the surface. Supplies were dispatched or intercepted with whatever cunning man could devise, precipitating some of the most vicious convoy battles of the war. The struggle assumed the aspect of a double siege: Malta and a few North African ports.

For two years navies struggled continually, augmenting the traditional fire of guns and the half-century old torpedo with the attack of airplanes in a completely new kind of warfare, a kind of warfare soon to take on increased dimension in the inappropriately named Pacific. Naval and merchant seamen were pushed to exhaustion and death in a struggle that wavered from month to month, favoring first one, then the other. Britain, stretched to the limit, quickly proved capable of dealing with Italy alone. Germany's contributions would be momentarily decisive but not consistent, as they were provided by a command riveted to the great struggle in the east where first opportunity and then danger dominated the thoughts of the dictator.

In the first months of the war the Royal Navy sent four radar-equipped vessels to the Mediterranean: the battleship *Valiant*, the cruisers *Orion* and *Ajax* and the carrier *Illustrious*. All except the *Orion* had type 279, a 7.5 m set with separate, rotatable antennas for transmission and reception, the small antenna size and long wavelength resulting in a very broad beam. It was capable of detecting planes 3 km high at a range of 80 km. Owing to a large vertical lobe pattern it did much worse with low-flying aircraft or ships. The *Orion* had type 286, the 1.5 m set intended for small ships that was an adaptation of ASV mark II. Despite the shorter wavelength the early models did not have rotatable antennas, requiring that even very rough directional information be obtained by swinging ship. The shorter wavelength and smaller vertical lobes allowed it to do a better job in detecting ships and low-flying planes [1].

Because she was commissioned in April 1940 HMS *Illustrious* was the first Royal Navy carrier to be radar equipped, as the older carriers were considered to be too busy for the installation, but the initial absence of radar aboard flagships points to some high-rank coolness toward this new method. Early tactical doctrine also displayed a little aversion. Information as to the whereabouts of enemy aircraft obtained by radar in capital ships was to be shared with carriers by flag signals. The Navy's strict rules for radio silence, following a lesson learned from the ill effects of indiscriminate German use of wireless in 1916, were applied to radar as well—one radar sweep every five minutes, which was at least an improvement on the earlier restriction of one minute every hour. Serving officers quickly learned the power of their new weapon and disposed, at first unofficially, later officially, of such restrictions. Aboard the *Illustrious* fighter direction evolved out of combat at the same time American sailor–aviators were developing their combat information centers in fleet exercises. Officers aboard non-radar carriers were quick in picking up the new tricks. Lieutenant Commander Charles Coke of the *Ark Royal* gained renown in the technique despite the clumsy communication imposed [2].

As experience accumulated with long-wave radar the operators encountered properties that diminished the initial euphoria for the miraculous device. When operating near land the A-scope indicators presented confusing echoes from the extended land masses illuminated by wide beams. There was a clear need for panoramic scanning and screens with plan position indicators, the manner of presentation that draws the operator a map of his surroundings and is now the form universally associated with radar. In principal the height of aircraft could be determined using a knowledge of the vertical lobes, but in practice this was difficult or impossible in a rolling ship or if multiple targets had to be tracked. The lobe structure yielded such weak reflections for low-flying aircraft that they were often detected visually first. A related electronics problem caused the fighter control officers as much grief as the deficiencies of the radar. The intelligence gathered by radar was difficult to communicate rapidly to the fliers, both because of restrictions on radio transmissions and deficiencies in the frequency band used—a deficiency removed for Fighter Command in England with very-high-frequency (VHF) sets. The HF bands then in use were noisy and allowed relatively few channels. The US Navy was to encounter many of these problems a year later in the Pacific.

Two weeks after Italy entered the war her bombers attacked Malta, for which the air defenses were a few RAF Gladiators, biplane fighters of a type that saw wide use during the first year in the Mediterranean and throughout Africa. Coupled with the 7.5 m Mobile Radio Unit (MRU) radar installed on the island at Fort Dingli in April 1939 [3] they proved a match for the tri-motor Savoia bombers and their protecting fighters, but by the end of July Malta had only three Gladiators left, the legendary 'Faith', 'Hope' and 'Charity', and five Hurricanes. Air raids continued and battles over convoys to the island began, the first ending in a fleet action off Calabria fought without radar [4]. In September Mussolini blithely ordered his forces in Libya to invade Egypt and followed a month later with an invasion of Greece from Albania, specifically the kinds of action that his German allies did not want. He soon needed German support.

The Italian navy outnumbered the British to such an extent in the early stages that Admiral Sir Andrew Cunningham knew he must do something to improve the odds and to gain the moral upper hand. He did this on 11 November 1940, and radar helped. On the 7th a force left Alexandria escorting convoys bound for Malta and Greece but with other things planned than the safe conduct of merchant vessels. HMS *Illustrious* and escorts headed for Taranto where important elements of the Italian fleet were based. The harbor, alerted to the movements of the British, was guarded by reconnaissance planes, but the *Illustrious* located them with her radar and vectored ship-board fighters to eliminate them [5]. Twenty-one Swordfish torpedo bombers from the carrier removed three battleships and various lesser vessels from the enemy lists but did not gain this through the hoped-for surprise, for the entire harbor erupted in a 'volcano' of anti-aircraft (AA) fire well in advance of their arrival. Instead they gained it through their famous skills in night action [6]. The success of this venture was not lost on the Japanese military attaches.

This daring attack, made with a realized expectation of a strong defense, was carried out in machines that were to gain a mythical reputation. The Swordfish was a sturdy, open-cockpit biplane of 1933 design, affectionately known as the 'Stringbag', and not fast even by World War I standards. It was the Royal Navy's air striking arm from land and carrier through most of the war [7]. It also became in July 1941 the first aircraft in the Mediterranean to carry 1.5 m ASV radar with a transmitter dipole on the upper wing and receiver Yagis fixed to the wing struts. If Luftwaffe planes protected Axis convoys during the day, the Swordfish could find and destroy them at night. An aircraft with ASV would locate the ships and drop flares to illuminate them for the planes with torpedoes. In late summer 1941 a few ASV-equipped Wellingtons began performing a similar function for Malta's Force K of cruisers and destroyers by guiding them at night towards a convoy and illuminating them with flares to permit sinking by gunfire. In both cases arrangements were made for surviving vessels to be met by submarines during the day [8].

Four months later Cunningham dealt another crushing blow to the Italian fleet and this time radar was decisive. British supplies to the Greeks forced the Italians to attempt a surface interdiction that led to action off Greece's Cape Matapan, the southernmost tip of mainland Greece. During daylight of 27 March 1941 a Swordfish had disabled the cruiser *Pola*, and Admiral Angelo Iachino decided to send back two cruisers and four destroyers to aid the stricken ship under cover of darkness. An alert radar operator on the cruiser *Orion* insisted that an insignificant blip on his oscilloscope was a ship and soon had the squadron heading toward it. Two other ships had radar, the battleship *Valiant* and the cruiser *Ajax*, and their sets soon showed a number of Italian ships. Approaching unseen and unsuspected they closed to comfortable and accurately known range and illuminated their targets with searchlights. The result was destruction from which only two of the destroyers survived [9].

The Italian navy had relied on the 'fleet-in-being' for much of its effect. That is, its mere existence should tie up enemy units as a guard against sorties. It was a strategy attributable to Mahan that contrasted notably with the British doctrine of 'bring the enemy to action'. Whatever value Italy's naval force had ever had as a 'fleet-in-being' was lost after Matapan.

From the disastrous outcome of this battle the Italians realized that the suspicion of Britain having radar must be true and re-initiated the program they had abandoned. By the end of 1941 they were able to install a set aboard the battleship *Littoria*, although without significant effect. Mussolini had consistently opposed requests by his navy for aircraft carriers, arguing that land-based planes could satisfy the need. With the power of British carriers demonstrated beyond doubt, he also gave orders to convert two liners into carriers, not completed before Italy left the war [10]. The Italian Navy suffered from having no aircraft under its own control, being dependent on requests of Regia Aeronautica and the Luftwaffe, both considering requests as just that, not orders.

South African forces had joined the British in stopping the Italian drive against British and French territories and then in clearing them out of Ethiopia, the first liberation of a nation conquered by the Axis. Moreover they had brought their own, home-made radar with them, called JB (for Johannesburg) and provided air warning for Mombasa and Nairobi. After the Italians had been defeated they were assigned positions in the Sinai desert for the defense of the RAF installations covering the Suez Canal. There they had the satisfaction of seeing their equipment frequently outperform the huge factory-produced MRUs. When passage of the Mediterranean became blocked, greatly increased amounts of shipping, vital to Allied forces in Egypt, had to pass the Cape, so the South African radar personnel were withdrawn to take up the duties originally intended for them and their JBs: protection and navigation of ships off the storm-lashed southern coast. The COL sets (an overseas version of CD/CHL) were slow in coming, but twelve JB-3s together with a few ASV sets modified for ground use were erected in the vicinity of Cape Town, Port Elizabeth, East London and Durban.

The story of the JBs is another example of radar being a straightforward, not particularly mysterious device. Dr Basil Schonland, Director of the Bernard Price Institute, had learned about it from Dr Ernest Marsden of New Zealand, who imparted what he could of his recent instruction in England while his ship made the short passage from Cape Town to Durban, weeks after the outbreak of war. Schonland learned enough to start radar development at his institute with a small group that kept a copy of *The Radio Amateur's Handbook* nearby. They recorded their first echo on 16 December 1939; by March 1940 they had detected aircraft at ranges of 80 km. It worked on 3.5 m, radiated 5 kW peak power from a steerable array of dipoles and was built from components obtained locally [11].

Radar was not the only secret weapon Britain had in the field. The Mediterranean was the training ground for Ultra as an important, later the most important segment of military intelligence. Its use in the Battles of Britain and of the Atlantic was relatively simple, but operations in North Africa necessitated heavy traffic by radio, furnishing ample feed stuff for Bletchley Park but little of it simple. Just as practice and experience had been needed to transform radar equipment into a weapon, practice and experience were needed to turn the English translations of Enigma signals into useful information. The first problem was speed. The Axis transmissions had to be recorded and transmitted to England. Decryption there was not necessarily immediate, depending on the key used and the care with which the messages had been encrypted. The vital intelligence had then to be made available to the commanders without disclosing the source; the secrecy surrounding Ultra was necessarily severe—and successful. This presented numerous problems that had to be worked out with intelligence personnel, the greatest being the general distrust of information thought to be obtained from agents, which bitter experience had long taught generals to suspect.

Information from Ultra about Axis convoy sailings became reliable in June 1941 but had to be used with care for fear of compromising the source. The policy imposed required that vessels be spotted by aircraft apparently on normal patrol before being attacked. As sinkings increased the Germans suspected Italian treason, which created a few side benefits for the British [12]. Information from Ultra about strategic and tactical affairs was much more confused than details of ship sailings, as it described only part of a complicated situation. When Ultra was integrated into a comprehensive intelligence organization that gave each part thoughtful evaluation, significant benefits accrued. Ultra's part in stopping Rommel's advance at El Alamein demonstrated the maturity of such application, but this maturity had been reached in a tough school that had begun with the confusion Rommel caused by disobeying orders in starting his first offensive in April 1941 [13]. After El Alamein Ultra became thoroughly integrated in command thought.

The war in the Mediterranean was brutally continuous. Behind the complex military struggle between Italy, Germany and Britain was an even more complex political struggle, often lapsing into violence, involving the three principals with various combinations of Arabs, Jews, Iraqis, Iranians, Turks, Vichy French, de Gaulle French, Greeks, Spanish, Balkans and minor groups. To describe it in detail here would serve no useful purpose, for the story is well told in many books. Let it suffice to describe the principal acts of this drama, pointing out the parts that radar played.

In early 1941 the British army with strong Commonwealth components under the inspired leadership of Major General R N O'Connor drove the Italians out of Egypt and back into Cyrenaica. This predictably brought

German aid, which took the form of Generalmajor Erwin Rommel with the Afrika Korps and Fliegerkorps X, an air unit to protect Axis and attack British convoys. Fliegerkorps X had specialized in operation against shipping when based in Norway. This initiated the 'First Malta Blitz', often called the '*Illustrious* Blitz' because of the attractive presence of the damaged carrier in the harbor following a convoy action.

The air defense was stout but technically deficient. By the end of January 1941 an additional MRU was operating at Fort Dingli, the two working alternately to ensure continuous coverage, and low cover had been secured by COLs at Forts Madalena, Ta Silch and Dingli. (Overseas versions of CH and CHL were designated CO and COL and had minor technical differences.) Probably because of bad siting and the lack of enough aircraft flying time for the extensive calibration required, the Malta radar gave very poor indication of altitude; it also had blind spots, and the small size of the island precluded effective use of ground observers. The intense activity can be appreciated by noting that there were 20 000 plots in the Filter Room in 34 days [14]. When coupled with poor communication this made the basic tactic one of sending up fighters to wait near the island rather than meeting the enemy far out [15]. Fliegerkorps X caused near strangulation of the island, and soon Axis convoys were reaching their destinations with few losses.

With their supply line secure German and Italian forces attacked British ground forces in Libya on 31 March 1941 and scored substantial advances because British troops had been sent to help Greece and because Rommel proved to be a master of desert warfare. Fliegerkorps X shifted much of its strength to support the Afrika Korps, giving the island a slight breather but adding to the British Army's troubles. It was a difficult spring for Britain, to say the least, for in April the Germans invaded Greece and Yugoslavia, subsequently driving the British from the mainland. A contingent remaining to defend Crete, which had some of the characteristics of an eastern Malta, was defeated by paratroop assault. The British radar deployed in Greece and Crete had little effect on the melancholy outcome [16].

The Crete victory was extremely costly for Göring's airborne troops, something not forgotten in Berlin during the ensuing months. The losses to the Royal Navy in its generally successful withdrawal of the Army were so great that surface forces could no longer be based at Malta. The Navy's losses came about by operating beyond the range of land-based air cover and because their AA fire was inadequate to protect them alone. It was a lesson that they had, in fact, already learned; the decision to proceed was made in cold blood as the Army had to be supported. Rommel soon stretched his supply line to the limit and Fliegerkorps X was sent to help support the invasion of the Soviet Union, which began on 22 June 1941.

A year of desert warfare had caused the RAF to forget their prewar reluctance for army support duties. When General Wavell attempted to dislodge Rommel from the frontier of Egypt they demonstrated themselves increasingly skillful under Air Marshal Tedder in attacking ground targets. These skills made little use of help by ground radar to keep fliers informed of the approach of enemy planes by GCI stations, although they were pleased enough to learn the location of a downed aircraft for rescue. Secrecy may have played a part, as pilots did not know the basis for the instructions they received over the radio. A ground radar officer of the time did not remember having had discussions with the Air Officer Commanding's staff as to how radar would fit into air defense [17]. A suggestion of the reason can be found in Tedder's memoirs, which mention radar only three times in 417 pages. It was a problem that did not improve much with time so far as army support went. A ground radar officer in the Normandy invasion found the attitude of the fliers toward radar control to be 'do not interfere' [18]. The unyielding belief in the absolute necessity of preventing radar equipment from being captured, coupled with its poor mobility, kept sets sufficiently far from the front to reduce their effectiveness. The best use of radar in the desert campaign was in the long defense of Tobruk by the Australians.

Although outnumbered in men and material the Italian and German forces held. With O'Connor a prisoner of war the deficiencies in Army leadership, so obvious in Belgium the year before, were there for all to see.

The Germans introduced their first radar into North Africa, a Freya to Tripoli, in January 1942, with no small degree of apprehension. This was still the time that everyone thought himself master of a very secret thing, but tactics won out as they built an effective air warning system—with local sharp-eyed Arabs added as irregular observers. The first Würzburg made its appearance at Bomba Bay in April, and by June there were three Freyas and two Würzburgs in service [19]. The Germans made good use of their radar in air defense, probably because the Luftnachrichtentrupp had been so closely integrated with the flying units. An attempt by commandos to take a Freya on 14 September failed [20].

With Hitler's attention fixed in the east, Malta quickly recovered and was soon dealing harshly with the Italian merchantmen bound for North Africa. The combination of Britain being able to read relevant Axis signals and searching with airborne radar at night resulted in 63% of the vessels dispatched in November 1941 being sunk, the same month the British opened a successful land offensive and drove back into Libya. In an example of desperation the Italians loaded two cruisers with drum-loaded fuel for transport. They were discovered by airborne radar and transformed into gigantic conflagrations by destroyers. By the end of the year Axis forces were reeling.

The successful attacks on Axis convoys caused Hitler to order many of his U-boats to the Mediterranean in August 1941 against the strong objections of Admiral Dönitz. The damage they did there was more than balanced for Britain by the damage they did not do in the Atlantic. Furthermore, the Mediterranean was a submarine trap because the currents at Gibraltar made entrance relatively easy, exit nearly impossible, but they long remained a problem there, sinking carriers *Ark Royal* and *Eagle*, battleship *Barham* and 95 merchantmen at a cost of 68 of their own number [21]. Coastal Command four-engine Sunderland flying boats equipped with ASV began seeking them in the eastern reaches, where radar confusion of the numerous fishing boats with submarine conning towers gave the chase its own flavor [22].

With the eastern front in winter stalemate Hitler sent Field Marshal Albert Kesselring as Commander in Chief South with a contingent of his Luftflotte 2 to turn matters around. Malta had to endure the 'Second Malta Blitz', and it proved to be worse than the first with starvation seriously threatening garrison and population. The nadir of the island's fortunes was marked by the failure in February 1942 of a convoy to deliver even a single shipload of cargo. Other convoys that had attempted to run the gauntlet had been able to bring at least a token cargo through. Malta's air defenses were reinforced by US and British carriers ferrying Spitfires to within flying distance of the island, a technique developed during the first Blitz.

During this Blitz the Germans set up radar on Sicily to monitor British air activity and used it to observe the transfers of fighters from carrier to land, attacking them on the ground shortly after their arrival [23], which motivated extreme measures to get them airborne again as quickly as possible. They also started electronic countermeasures. Martini supervised the installation of jammers for the Malta radars, which effectively put them out of action. This was reported to R V Jones in London, who told them to continue operation as if nothing had happened [24]. Those on the scene took more direct action. An Air Force officer struggling with an MRU and an Army engineer sent out to rescue the GL sets independently discovered that the jamming, either by accident or design, was modulated at the 50 Hz power frequency. They synchronized with the peaks of the jamming and operated where the disturbing noise was least, restoring much capability [25]. For whatever cause German electronic intelligence concluded the jamming was ineffective, and it was stopped. When Martini met Jones after the war his first question was how the jamming had been circumvented. Whether he or Jones ever learned what happened is not clear.

For some reason Kesselring did not attempt to destroy Malta's radar sets, although his photo intelligence had picked 14 positions that they marked as radar stations in June 1942 [26]. This came as a pleasant surprise to the operators, who grimly expected a full-scale attack [27].

Despite clever radar tactics by the defenders Malta was neutralized and Axis shipping again went through. If the Malta Blitz succeeded in opening the supply route to Rommel again, it was not able to subdue the air defense of the island completely and came to be like the labors of Sisyphus to the attacking fliers, who took heavy losses [28]. Kesselring and Rommel insisted that Malta must be invaded and began planning an airborne invasion, measures that found little enthusiasm in Berlin because of the losses in taking Crete. By May Axis supplies on the desert had reached a level sufficient to sustain an offensive, and British strength had shrunk as a result of urgent demands for help from Burma and India. To no-one's surprise Rommel attacked but with an effect that surprised both sides—the heavily fortified port of Tobruk fell. This disaster saved Malta, curiously enough, because within the Tobruk fortress were immense quantities of supplies that fell to the attackers, Rommel thought enough to sustain him to Suez, if the air forces attacking Malta were transferred to him. The invasion of Malta was forgotten, to the relief of Hitler and Göring. But the gamble failed; the British Eighth Army, its left flank secure on the Qattara Depression and its right on the sea, stopped Rommel at El Alamein in early July 1942, and he was never to go beyond it. The air and naval forces of Malta soon began again to take their toll of Axis shipping.

On 10 August a large, heavily escorted convoy departed Gibraltar and soon found itself the center of a bitterly fought convoy action, attacks coming from aircraft, surface ships and submarines. The type 279 radars were unable to sort out the many targets, resulting in poor fighter direction [29]. Several ships were lost, including the Mediterranean veteran carrier *Eagle*, but the sacrifices were not in vain, as a significant cargo was landed, effectively ending the siege. On 30 August Rommel opened a desperate attack on the British positions with inadequate fuel and ammunition. Its failure was a direct result of the decision not to take Malta. On 23 November the British forces opened a long-prepared attack on Rommel's lines and succeeded in forcing him to retreat after ten days of heavy fighting. Less than a week later American and British forces landed in Morocco and Algeria. The nature of the struggle altered significantly after that as did the employment of radar, which concluded a phase one might well call 'heroic'. It entered a new phase in which it was one of many weapons. Unnoticed by nearly all of the participants, high and low, was the capture of German radar equipment at El Alamein, in a variety of smashed conditions, to be sure, but Allied electronic intelligence soon had an operating Würzburg [30]. Those German units that retained their radars did so with determination, holding their tractors at gun point from others during the retreat [31]. Of course the capture of radar did not work only in one direction. The Afrika Korps captured a Wellington equipped with ASV mark II [32].

The question now arises. What effect did radar have on this part of the War? Did it turn the outcome in favor of the Allies? The radar enthusiast will be quick to point to Matapan and sometimes to Taranto as important victories in which radar was a major contributor. He will also point to the steady use of the technique in interdicting Axis and protecting British convoys and in defending Malta. That it was an important weapon for Britain—it also served Germany but more as a technician than as a warrior—is beyond question, but did it prevent Axis victory?

For the Axis only one victory counted: to seize Alexandria and the canal. The consequences for Britain would have been severe. Their loss would have almost certainly meant the loss of the vital oil of Iraq and Iran and its gain for Germany. Lost too would have been the reliable supply line to the Soviet Union, and all this when Singapore had surrendered, Burma had been given up and India was threatened. There were limits on what America's growing power could do. They were terrifying times.

Would Rommel have reached Suez, had Britain not had radar? Is it really possible to decide? Indeed one can just as well pose the question in terms of Ultra, and will be no closer to an answer for having done so. There must be a dozen instances in which the outcome seems to have turned on a single engagement. But there were for both sides defeats that were made good, and much of what took place resulted from the variation of Hitler's attention. Both Axis failure and Axis involvement in the Mediterranean had a common origin: the near absence of strategic planning in Berlin and Rome. It is possible that had some of Britain's early successes, obtained with the help of radar and Ultra, not been made Germany might have sent a smaller force to aid Italy with a correspondingly smaller threat to Suez. Action in the Mediterranean until the end of 1942 was unquestionably the most tangled of the War. Change any part and one is lost in a morass of conjecture. One thing is clear. Britain needed every source of strength available just to stay in the war, and in such circumstances any new weapon, like any new ally, must be remembered as vital to victory.

5.2. WAR IN THE PACIFIC

Harold Zahl's first thoughts when he learned of the Japanese attack on Pearl Harbor were about SCR-270 [1]. He knew that this early warning set into which he and many others at the Signal Corps Laboratory had put so much labor and ingenuity was deployed in Hawaii. Had it failed? It was a question that stood out above all others as the engineers at Fort Monmouth discussed the news, especially as the extent of the disaster became known to them. For a decade their professional lives had been spent in devising the means to prevent just this, and now it had all come to nought.

The defeat suffered by American arms on 7 December 1941 was remarkably similar to the one suffered by the British a year and a half earlier on the fields of Belgium and France and had the same roots—commanding officers lacking comprehension of the importance of air power compounded by their inadequate imaginations and poor communication with their subordinates. For the British it was accepted as impossible that an armored force could come through the Ardennes; for the Americans it was accepted that Hawaii would not be attacked by carriers at such a great distance from their home base. Both should have been in high states of readiness: the British were at war, and the Americans had received a war warning on 27 November, but both had made the ancient military blunder of basing action, more properly inaction, on what was expected of the enemy.

Like many catastrophes Pearl Harbor has compelled extensive studies to determine its causes and to extract the lessons that it has to teach, and the literature matches the magnitude of the event, demonstrated by the publication of a bibliography of 1514 relevant publications and documents [2] and by the 110 cm of shelf space occupied by the Congressional and various service investigational reports. Not surprisingly it has generated controversies, two of which remain. One is whether a misguided foreign policy of President Roosevelt and Secretary of State Cordell Hull unnecessarily forced Japan to such desperate action. This clearly lies beyond the limits of this book. The other is whether Roosevelt and Army Chief of Staff George Marshall withheld specific knowledge from the local commander of the impending attack because they wished to bring the United States into the war on the side of Britain. This stretches the limits of common sense, being founded on the incredible assumption that an attack resolutely defended would have left America at peace!

But what about radar? Had it failed? Here the answer is as unambiguous as are the consequences of the attack. No, the instrument had functioned perfectly but to no avail. The responsibility for failure to act on the warning that radar provided can be fixed uncluttered by issues that complicate determining overall responsibility. The object of the Japanese attack was the destruction or severe injury of the Pacific Fleet commanded by Admiral Husband E Kimmel. When in port it was the responsibility of the Hawaiian Department of the Army under Lieutenant General Walter C Short to protect it against air or ground attack. The key to air defense was early warning, something that the Air Ministry in Britain had understood from the first and that had given the proposals for radio location such power. Similar thoughts had guided the prewar work at the Naval Research and Signal Corps Laboratories. What was lacking in the Hawaiian Department was the realization that radar gave warnings of only 20 to 50 minutes and that the utilization of this precious intelligence required an organization whose complexity equaled that of a radar set. It was Dowding's clear grasp of this elemental fact that was decisive in 1940.

Of the officers who reported for duty with the Pacific Fleet in 1941 few had had the varied service of Lieutenant Commander William E G Taylor. Commissioned as a reserve officer in 1927 he had transferred to Marine Aviation after a year as a carrier pilot aboard USS *Lexington*. Ordered to inactive duty during the depression he had worked a variety of flying jobs until going to England in the summer of 1939 in anticipation of war. The Royal Navy accepted him as a Sub-Lieutenant, and he soon had duties as a fighter pilot both from carriers and land. When the RAF began forming squadrons of Americans he resigned from the Navy and became the Squadron Leader of 71 Eagle Squadron. Intimate contact made him extremely interested in British fighter direction with radar, and he secured a rather extensive knowledge of it. At the instigation of the American Naval Attache he was offered, with British good will, a commission in the American Navy that he accepted and returned to the United States in August 1941 where he lectured on fighter control [3].

On arriving in Hawaii he was sent to work with Major Kenneth P Bergquist of the Hawaiian Department Air Force [4], who was building an air warning system based on radar equipment that began arriving during the summer of 1941. Taylor liked what Bergquist was doing, the two worked well together and both were very impatient with progress. Bergquist had been in the first class of ten officers and 40 enlisted men who attended the new Air Defense School at Mitchel Field (named for John P Mitchel, former Mayor of New York City) in March and April to learn the new technique. His teachers, Air Forces [5] and Signal Corps officers just returned from inspection trips to Britain, had quickly put together demonstrations of radar air warning and fighter control, and Bergquist returned to Hawaii filled with evangelical zeal.

On return to his duty station he was assigned to his old interceptor unit until the end of May, when he was told to establish an air warning system, something he had been pushing since his return. But official consent was not official drive, and equipment and personnel were difficult to obtain.

The Signal Corps began setting up the SCR-270s as they arrived and training operators and maintenance men. That went well, the main deficiency being that the sets required siting where no electric power was available, and no amount of talking to Chiefs of Staff could bring in electric lines, making it necessary to run with engine-driven generators, which were having enough breakdowns that practice was severely restricted. Finally they were able to build a small, wooden information center at Fort Shafter with tactical and administrative telephone lines to the five radars in operation. By December the radars functioned and the plotters moved the markers about the big map board during the few hours a day that they could practice, but in the balcony sat neither fighter control nor liaison officers. This left the 'Information Center' devoid of information about their own interceptors, bombers, naval air and AA guns on ship or ashore. Equipment might be acquired in devious ways but officers were in much too short supply to be obtained for assignment to this work. They could never be spared; there was too much to do. Majors have difficulty instructing flag ranks in their duties.

Thus by 7 December Bergquist and Taylor's little center was able to track what the SCR-270s reported but had nothing to help identify the various plots; IFF lay months in the future for Hawaii. If through some occult process they decided a particular radar target was hostile, they could do nothing more than call some commanding officer at his office. There were no tactical lines out of the Information Center to any fighting unit capable of making use of the information. There were no serviced fighters ready for take-off, no rooms filled with pilots waiting on call [6].

On the fateful Sunday morning two privates were operating an SCR-270 alone on Opana Mountain, the northernmost radar on the island of Oahu. They were restricted by General Short to operate between 4 and 7 AM and had reported distances and directions of the planes that made up a quiet morning's traffic. Joseph Lockard was experienced with the set and was teaching George E Elliott its use. A truck was to arrive at 7 to take them to breakfast and leave replacement guards, but the truck was late and Elliott asked whether they could continue with his practice, to which Lockard agreed, although in fact a violation of orders. At 7:02 Elliott saw a blip on the cathode-ray tube at a range of 220 km, which Lockard said was the biggest reflection he had ever seen at such a range. It was a large formation of planes nearly due north and flying straight at them.

They decided to call the Information Center but were unable to obtain an answer on the tactical line, as the exercise was over. It seemed important enough, although they were not sure why, to try again by calling on the administrative line where they eventually spoke to First Lieutenant Kermit A Tyler, who had been ordered to observe the morning exercise to become familiar with the procedures as the first step in becoming a fighter controller. He had no other operational function and had no way of knowing what the formation of planes was. He did know informally from a friend in the bomber command that flights of B-17s were passing through on way to the Philippines but knew nothing officially. He had also been told the flights used a broadcast station playing Hawaiian music uninterruptedly all night as a radio homing beacon and had noted that the station had been playing just that as he drove to his 4 AM duty station. He reasoned that this was what the Opana station was seeing and told the two not to give it more thought. There was indeed a flight of B-17s bound for the island, but that was not what made the large reflections, they were made by Japanese carrier planes [7], who listened to the music too.

After the attack there were plenty of investigations into the causes, all neatly collected in the congressional publication, to keep people busy for decades. Some of the recorded interviews wander in the direction of a disordered intellect: two privates and a lieutenant being interrogated with an underlying implication that they might have saved the naval base.

Needless to say Hawaii's air warning system received ample support in equipment and personnel during the days following the attack. There was a valiant but generally ineffective effort to make use of it during the hours of and after the attack, but it was unable to prevent aircraft of USS *Enterprise* from being shot down by Navy guns as they flew in from the carrier that evening. The Opana station had followed the departing Japanese aircraft back to the north, but this information was lost in the confusion of the day and searches for the Japanese fleet went south, all considered probably just as well. Radar was now the favored instrument for which an almost insatiable appetite grew. The CXAM from the seriously damaged

USS *California* found its duty station on land.

The garrison in the Philippines knew of the outbreak of war hours before any Japanese action against them, but General MacArthur used this advantage to little effect. As in Hawaii the failure lay at the top, not at the bottom, and radar had brief but honorable service. A Signal Corps Aircraft Warning Company had arrived in the islands in August 1941, and their first SCR-270 followed on 1 October. They set about unpacking and assembling the device and learning to use it. It worked well from the start, which was particularly helpful as no test equipment came with it.

Lt C J Wimer and a detachment of 30 men positioned this first set at Iba on the coast about 150 km northwest of Clark Field, the base for the ever increasing number of B-17s, the Flying Fortresses that were the hope for defending the islands, as the Air Force was confident that these bombers were the match for any invading fleet. The Iba station was operational by the end of October. More equipment arrived, and by the outbreak of war there were additional sets positioned, though hardly fully operational, on the coast of the Philippine Sea about 200 km southeast of Manila, at Tagaytay Ridge 70 km south of Manila, and at Burgos Point at the extreme northern tip of Luzon. In late November the Marines called the Signal Corps to report the arrival of an SCR-268 for which no one was trained, and a detachment was sent to get them started.

In the early morning hours of 8 December (7 December on the other side of the International Date Line) and before they had learned of war the Iba station picked up intruders at 180 km and a squadron of fighters was scrambled to intercept; they failed, probably because of an inaccurate height datum. From then on Japanese planes made many attacks that kept the radar station busy. It reported a large formation of bombers at 11:45 AM headed for Clark Field. The responsibility for allowing the destruction on the ground of half of the B-17s from this attack, hours after the start of war, has been a matter of continuing dispute, but the transmission of the Iba station report by teletype to Clark Field is not contested, although its reception is. The Japanese bombed the Iba airstrip and destroyed the radar at 12:20 PM.

Radar played no further role in the defense of the Philippines. What reports came from the remaining sets went to combat air units unable to utilize them [8].

The Philippine radar was not caught off guard, sending fighters to investigate intruders even before they knew of the outbreak of war. Their disposition of the initial attacks was as good as their equipment allowed.

The United States was not alone in experiencing disasters. The Japanese struck south not only toward the Philippines but toward Singapore, approaching through landings on the Malay Peninsula to the north and proceeding through the jungle, a path that the defenders had ruled out and against which they had prepared no fortifications. To put an end to this the Royal Navy sent the new battleship HMS *Prince of Wales* and the old battle cruiser HMS *Repulse* to destroy the landing parties. They moved without air cover, and were sunk by land-based torpedo planes on 10 December 1941 in the Gulf of Siam.

Radar may have played an unfortunate part in this. Admiral Sir Tom Phillips, who commanded and died with the ill fated squadron, had from his first day on board the *Prince of Wales* shown a keen interest in her radar, to the point of discussing in detail the use of each set with the operators. He was aware that this vessel was the best equipped in the world with this new technique. It had a type 281 3.5 m air warning set, a type 284 50 cm main-armament gun-laying set, four type 285 50 cm AA gun-laying sets, four type 282 50 cm close-fire AA gun-laying sets and one type 273 10 cm close-surface search set. Types 284 and 285 both employed lobe switching. It was an impressive suite of equipment to say the least. The *Repulse* had one type 286P 1.5 m air-warning set, much improved over the original 286 in having a rotating antenna, and one type 284 [9].

Phillips had held strongly for the invulnerability of modern capital ships against air attack at a meeting of the Joint Planning Staff just before outbreak of war [10] but presumably understood by late 1941 the dangers so recently emphasized at Pearl Harbor. He had no carriers but could call on land-based planes from Singapore. One is led to suspect that he may have relied too heavily on radar's ability to protect his ships. Two elements may have sealed the fate of the two great ships: first, maintenance at Singapore was interrupted by the outbreak of war, leaving inoperable three of the four type 282s, needed for the very important 40 mm AA fire; second, he delayed breaking radio silence to call for air support until well after the first radar sighting of impending trouble. The ships fought well but were overwhelmed. Air support arrived but too late.

Beyond the Philippines and Singapore lay Australia, clearly in the path of Japanese aggression. As loyal members of the British Commonwealth, Australia and New Zealand had declared war on Germany promptly and both had sent sizeable contingents to the deserts of North Africa, contributing mightily to General R N O'Connor's devastating disposal of the Italian army's invasion of Egypt.

Dangers of Japanese attack, especially from the air, were never far from mind and had led to the organization of the Radiophysics Laboratory on 29 November 1939, a name intended to mask its radar function. Sir John Madsen, Professor of Electrical Engineering at Sydney, went to England for discussions with Tizard and Watt among others. On his return the Radiophysics Advisory Board proceeded on a course of radar training, research and construction. In September 1940 Wing Commander A G Pither of the Royal Australian Air Force (RAAF) left for training in England, whence he returned in May 1941 to take over Section 7 of the Directorate of Signals with responsibility for radar. On 15 September 1941 training began with 6-month courses for officers in radiophysics at Sydney University and for mechanics in maintenance at Melbourne Technical College. Instruction of women members of the Air Force to become operators commenced in June 1942.

In October 1941 Australia had three radar sets, one of their own experimental construction and two British. Additional imports would certainly be far too few for the wide expanse of the continent's northern coast and island responsibilities. Concern about air defense became acute after Pearl Harbor and brought forth from the Radiophysics Lab an air warning set, designated AW, working on 1.5 m that was successfully placed in operation for Sydney's defense only five days later.

Australia had no large electronic manufacturers, but this proved to be no hindrance in making either the AW or its successor, a light-weight air warning set, referred to only by the initials LW/AW, and surely one of the most remarkable pieces of radar equipment to emerge from World War II. Its critical light-weight antenna—making it the first radar set transportable in a single airplane—was designed and built by the New South Wales Government Railways in conjunction with the RAAF. The electronics for both were designed by the Radiophysics Lab, who also made the first six production sets during the early weeks of 1942, after which production was by HMV Gramophone. There is nothing in radar history to compare with this feat for speed linking development to full production and then into action. Quite obviously the circumstances had 'concentrated their minds wonderfully'. LW/AW, which came off the production lines in the last half of 1942, proved to be perfect for the fighting that soon enveloped the islands of the southwest Pacific and equipped numerous American units during the early years. The Australians also coined (of unknown origin) the name 'doover' that held its own against 'radar'.

Components for a prototype AW set were flown to Darwin on 5 February 1941; it was not operational in time to help counter the first destructive raid of 19 February but did detect the arrival of the fifth, becoming the first RAAF station to detect an enemy raid [11]. After the first radar-detected raid on Darwin the Commander of the American fighter Group asked that someone from the RAAF be stationed at Cape Fourcroy on Bathurst Island to extend the range of air warning. The first radio report by Corporal Bill Woodnutt gave an extra 20 minutes warning that enabled the P-40s to gain height and intercept with excellent results. One observer and one radar blunted further Japanese raids [12].

In February 1942 the first American troops, intended originally to reinforce Philippine or Dutch East Indies garrisons but unable to reach them, arrived in Australia. They brought with them a few SCR-268s, which they turned over to their hosts, as no AA units came with them. There being no directors to link them with Australian AA guns, they were modified for air warning and called Modified Air Warning Devices (MAWD). Obtaining ranges as great as 160 km, MAWDs became highly esteemed doovers, particularly in the Darwin area.

New Zealand was no less interested in radar than Australia and sent

Dr E Marsden to the same conference in Britain as attended by Madsen in 1939. Both Auckland and Canterbury Universities trained scientists in radiophysics, whom the Department of Scientific and Industrial Research employed to design some experimental air warning equipment. New Zealand elected not to produce their own equipment but to import British equipment, something their less vulnerable strategic position allowed. The radar program yielded a large number of trained personnel who, in addition to serving with their own Royal New Zealand Air Force in the Solomons and beyond, could be found wherever the RAF went, from Greenland to Burma [13].

The attack on Pearl Harbor, followed a couple of days later by Hitler's declaration of war, brought the United States, both people and government, to a state of high consternation and replaced feelings of security with those of apprehension. AA guns appeared in New York's Central Park and the Greenpoint section of Brooklyn, and of course, there were false alarms. Such events took place more frequently on the west coast with radio broadcasting abruptly stopped and an impromptu blackout imposed on the San Francisco region, accented by a submarine putting a few shells on the sacred soil of California. In February the air over Los Angeles was filled with bursting AA shells attempting to destroy an unidentified balloon.

No branch of the Army could have been more harassed during those first few months following the outbreak of war than the Signal Corps. It was deeply involved in expanding, purchasing equipment for the combat arms as well as itself, instituting training for its extremely technical functions, not to forget manning the electronic defenses needed to protect the continental nation and its outlying territories. In such trying times it is considered wise to call in experts to help officers and officials to do what they have been wanting to do anyway, and such support was found for those working in radar in the person of Robert Watson Watt.

Immediately after the Pearl Harbor attack the US military mission in London suggested that Watt visit the United States to advise on radar and associated matters. He accepted with alacrity and was on his way to Lisbon for the first leg of his flight to the United States by 9 December. On arrival in Washington he had discussions with Secretary of War Henry Stimson followed by meetings with lesser dignitaries [14]. Stimson had seen the great value of radar quite early, was naturally distressed about the Hawaiian turn of events and wanted a report from Watt [15]. Secretary of the Navy Frank Knox made no similar request.

After a visit to the Radiation Laboratory and sometimes in the company of Alfred Loomis (Stimson's cousin and an accomplished radarman in his own right), Watt examined the air defenses on the west coast, at Panama and in the Caribbean. He found nothing but praise for what he found at the Rad Lab and nothing to praise in the Signal Corps. Much of what he criticized was the result of finding inadequately trained (or even untrained) personnel in an army mobilizing at an unprecedented rate, a problem about which the Corps considered itself adequately informed, but he also demonstrated a narrow point of view that had been hardened by four years away from the laboratory in his capacity of Director of Communications Development. He dismissed all Signal Corps radar as nearly worthless and pronounced that long-range early-warning radar must work in the 10 m band and hence could not be truly mobile, the obvious success of the Opana set notwithstanding. Consequently, he insisted that Chain Home stations be erected at many crucial locations, relying on the purchase of 100 MRUs temporarily until the required CH towers could be erected. The deficiencies he found at Panama, primarily the inability of following aircraft inland as a result of the mountainous terrain and of detecting low-flying attack, were inherently those of meter-wave equipment, but he ascribed them to the infamous design of SCR-270 [16]. To cover the dangers of low-flying attack the Corps modified the 1.5 m SCR-268 to serve for air warning as the Australians did in making MAWD, designating it SCR-516 and incorporating a PPI display.

The Army Air Forces had sought earlier the procurement of Ground Controlled Interception (GCI) equipment, which became SCR-527, but none was available at this critical time. Watt insisted that the west coast had to have GCI, but here reality took over. GCI was inherently a British, in fact an English set. Its function required siting in a shallow bowl with a minimum of fixed targets visible outside the bowl. This requirement is easily satisfied by England's gently rolling countryside but not on the American west coast, where reconnaissance had failed to find even an approximation of the right surface conditions in the Pacific northwest with things not much better to the south [17]. Everyone agreed there was a crying need for ASV mark II for patrolling both coasts, but there were none.

Watt's report to the Secretary of War was received with satisfaction in Washington, especially by Colonel Gordon P Saville, Director of Air Defense, whose use of it did not improve relations between Air and Signals [18]. Watt was thanked profusely and awarded the US Medal of Merit. He left Signal Corps officers writing detailed rebuttals and striving to get orders for CH and MRU equipment canceled. Within weeks the visit had been forgotten, swallowed by extraordinary amounts of work. For whatever value Watt's report had to the Signal Corps it left the unmistakable understanding in the War Department that Stimson considered radar important. On 1 April 1942 he added Edward Bowles of MIT as 'his special consultant for the purpose of getting radar upon a thoroughly sound and competent basis' [19], allowing him to apply not too subtle incentives when he and Stimson thought it expedient, something which worked strongly to the Corps's benefit.

5.3. THE CHANNEL, 1942

5.3.1. *The dash of the warships*

Students of comparative naval theory had assembled for them toward the end of 1941 at Brest a conundrum worthy of their best analytical minds, and because the solution was more than just an exercise for the faculty of some war college it engaged an increasingly large circle of thinkers, who in turn activated an ever increasing array of military force. In March the battle cruisers *Scharnhorst* and *Gneisenau* had ended a successful cruise of commerce raiding and went to Brest to prepare for their next sortie. Initially Raeder planned to have them join with the *Bismarck* and *Prinz Eugen* in a progressive advancement of the technique, but problems with their boilers could not be resolved in time for the action, and a change in the fortunes of war had brought the *Prinz Eugen* fleeing to Brest without having sunk a single merchantman and having left the *Bismarck* resting on the ocean floor.

Thus one naval theory, the value of warships for commerce raiding, had just lost a degree of credibility, although its proponents were by no means inclined to admit a general failure of doctrine. The presence of three large-gun warships in Brest brought into play another naval theory, one guarded by the sacred writings of Alfred Thayer Mahan: the value of a fleet-in-being. So would the matter have remained a generation earlier, but by 1942 there was another theory with which the contending admirals had to deal, one that came from the pen of Billy Mitchell, a prophet as highly honored by airmen as Mahan by seamen. Succinctly put, it stated that surface ships, however well armed and armored, could be destroyed with comparative ease by land-based heavy bombers. As so oft in affairs of abstract thought, the situation at Brest by the end of 1941 was concluded with little regard for theory.

A fleet-in-being is one harbored safely behind coast defense guns, mine fields, breakwaters, nets, booms and torpedo boats, and by 1941 it was one also guarded by components of air power. By its mere existence it is a threat to an enemy who wishes to travel the adjacent waters. This requires that forces must be at sea or in the air to prevent it from departing unexpectedly. The presence of these ships at Brest meant that convoys would have to have battleship escorts as well as the lighter vessels for defense against submarines; three heavy ships at Brest skewed completely the dispositions of the entire Royal Navy.

Brest lay within easy striking range of British heavy bombers, and although the RAF had never placed the abolition of warships as part of their mission, as had their sister service across the ocean, they were assumed to be capable of sinking ships with bombs. Almost automatically Bomber Command reluctantly found itself expected to destroy the ships at Brest—reluctantly, because its avowed mission was to win the war alone by destroying Germany, and every digression from this was resisted. This in turn required the Luftwaffe to defend the ships with large numbers of fighters and AA guns at a time when these resources were much needed on the Russian front and for defending the Reich against ever growing air attacks. So the three ships became a distracting evil for the Kriegsmarine, the Royal Navy, Bomber Command and the Luftwaffe.

Into this tangle of conflicting theories stepped Hitler. He had decided that the British would soon invade Norway in order to secure the northern route to Murmansk and Archangel and to rid themselves of some troublesome U-boat bases, which he rightly saw as a greater threat to his enemy than surface ships. This meant that these three ships had to be posted to Norway to aid in its defense, and if the experts were of the opinion that it was impossible—because of radar among other things—to return to Germany by the old northern route, then they must return through the English Channel. This brought on a naval chorus crying that it could not be done and that another Atlantic sortie was the better use of the ships. Hitler disregarded all this and ordered them to procede through the Channel, threatening to remove the heavy guns for coast defense and pay off the ships, if the admirals could not bring themselves to the task. This certainly fixed the attention of the naval staffs, who, on examining the matter in detail, began to realize that ocean raiding was probably finished and that a dash up the channel might not be as foolhardy as their first reactions had given voice [1].

On the English side of the Channel, naval thought began coming to similar conclusions, for although little damage had been done by the thousands of bombs dropped where the ships were thought to be, eventually they would be destroyed or at least damaged beyond the repair capabilities of Brest. A dash up the Channel was the only way of saving the ships. That they were worth saving was a thought as natural to British minds as to those of the Kriegsmarine. Only Hitler seemed to have realized that events had reduced them to three nuisances.

Thus two sides began to plan for the dash up the Channel that began in the evening darkness of 11 February 1942, ten months after the two battle cruisers had taken refuge at Brest. The Germans produced an extremely detailed plan that provided for scheduling the ships' movement to the minute, for sweeping mines from the route, for using radar for navigation, for jamming the British radar, for preparing a vast cover of fighter planes and for central control of all these activities. The British plan consisted of alerting high-level commanders that the ships were expected to attempt the passage. There were separate commanders for the Fleet, Fleet Air Arm, Army Coast Defense, Coastal Command and Bomber Command units, all of whom had responsibilities in combating the vessels, and no direct communication links between them. Secrecy allowed no planning at the command levels where the details of a response would have to be carried out, and the news of the breakout caught them by surprise. It was assumed that the Germans would elect to pass the Straits of Dover at night, which meant they would have to depart Brest during daylight, which the

Germans inconveniently and prudently decided not to do.

Radar featured prominently for both sides in this action. Gen Martini personally supervised this aspect on the German side. The Kriegsmarine insisted on complete radio silence to include the ships' radar. This prevented them from using their Seetakt sets to obtain their positions from prominent land features or radar beacons. Given that the antenna arrangements of the Seetakt allowed almost no back-leakage of radiation and that their direct beams could hardly have been received on the English side until near Dover, this was probably an unnecessary and not very helpful restriction, but the squadron was able to navigate successfully nevertheless. As a substitute the shore stations were to locate the ships, which had IFF, and radio their positions to them [2]. In practice this proved difficult, owing to the inherently poor directional accuracy of the shore-based Seetakt sets and an unreliable IFF performance that made distinguishing between the ships difficult [3]. On top of this, communication failures at times failed to get the information assembled and transferred [4], so the navigation officers had to pay close attention to their fathometers and the depths indicated on the charts. There were also a series of marker boats at various points.

The British radar of which Martini was aware was the 1.5 m CD/CHL sets all along the coast. No shore-based radar would be able to observe the squadron except in the vicinity of the Dover Straits, so Martini set up 1.5 m jamming devices there, starting with very little interference but increasing it slightly from day to day until the level was thought adequate to mask reflections from the ships. This tactic did not fool Lieutenant Colonel B E Wallace, who quickly called attention to it but was ignored. In desperation he went to R V Jones, Head of Scientific Intelligence, and implored to be taken seriously. Jones had a top officer from the Telecommunications Research Establishment on the spot the next day—just in time for the passage of the ships [5].

Unknown to Martini, however, there were new 10 cm NT 271 coast defense sets (also called CD mark IV) installed on the southern coast. These were essentially the same as the Navy type 271 but with a 2 m paraboloid. Four that were within range of the squadron's path were located at Ventnor, Beachy Head, Fairlight and Dover. A fire had put the Ventnor station out of operation the night before, and it may have been out of range anyway, but the others worked just as intended. During mid-morning of 12 February 1942 the Beachy Head and Fairlight stations reported a force of large vessels at the far side of the Strait moving much faster than a normal convoy. These two reports then began winding their way to the Dover Coast Defence Operations Room. At the same time aircraft observed the ships visually but made no report until after landing, as strict radio silence prevailed, and their report also began its way toward Dover. Well over an hour later, Dover observed the ships with their own NT 271. It was their first news of the affair, as word of the other sightings had still not arrived [6].

The most important radar sighting was the one that was not made. Patrols had been set up for Coastal Command in three regions in the vicinity of Brest by Hudson bombers equipped with ASV mark II. These planes had been watching the harbor for seven months and their procedures had become slack. The patrols did not overlap and equipment failures were not allowed to disturb routine. ASV mark II was capable of spotting surfaced submarines, and a capital ship would have been the largest target the operators had ever encountered, but they were not there. In a maddening series of equipment failures—both in aircraft and in radar—the German ships passed the patrol regions when they were not covered [7].

Martini made no provision for jamming these sets, which is strange given the demand that there must be no discovery during the dark part of the passage [8]. It is especially puzzling because British ASV capability had been known since the preceding May [9], yet the detailed plans saw no problem in the possibility that air surveillance might have radar. One can only conclude that this was another of the many examples of something kept secret from those who needed to know.

The feeble attempts to sink the escaping ships concern this account only peripherally. All attacks were made piecemeal. Some were marked by great courage enhanced by knowledge of the small chance of either success or survival. The heavy guns at Dover, which initiated British action, seemed impressive until one inquired about their rates of fire and ability to follow moving targets; none of their 33 rounds hit. Motor torpedo boats failed to penetrate the protective cover of the German destroyers and E-boats to a range that would have allowed hope of a successful launch. Next came the pathetic attack of six Swordfish torpedo bombers, all of which were shot down. Had they been able to attack as planned at night, the specialty in which they excelled, the result might have been otherwise, but they attacked during the day with one-tenth the fighter cover intended and suffered the fate of all torpedo bombers that had to face an overwhelming fighter and AA defense. A flotilla of six 20-year-old destroyers went straight after the big ships accompanied by Beaufort torpedo bombers properly escorted by Spitfires. In the waning daylight and bad weather the mixture of destroyers and aircraft from both sides gave generous examples of mistaken identity with fights between enemies at times appearing to be the exception, because visual IFF had problems as severe as radar IFF. Bomber Command's high-level attacks at the end of the day seldom found the target, let alone hit it.

As night closed the squadron had to thread its way—not too successfully—through mine fields. This would have been an excellent time to use Seetakt with radar beacons on shore to navigate, but strict radio silence was still enforced. The two battle cruisers were damaged by encounters with three mines. The *Gneisenau* received a blast followed by subsequent bombing from which she was never to recover; the *Scharnhorst* went on to meet a futile but heroic end in the waters off northern Norway; the *Prinz*

Eugen served in the Baltic and ended her career as a test ship for an atomic bomb at Bikini Atoll.

The removal of the fleet-in-being from Brest greatly simplified the Royal Navy's arduous Atlantic duties, so Britain emerged from this in a better position strategically but having suffered a major defeat psychologically. It happened at the time of the sinking of the *Prince of Wales* and the *Repulse* and the loss of Singapore. The reaction of the public and the lower levels of military command was one of fury, as it was seen to be incompetence of the kind demonstrated at Pearl Harbor. The Prime Minister's investigation found everything in order, and no senior officers were disciplined or replaced. Disparaging remarks in the report about the capabilities of radar triggered an extensive response in Watson-Watt's memoirs [10].

5.3.2. The Channel Convoys

Despite its proximity to the enemy, the Channel was an important supply line for both sides. Britain's western harbors could not handle all of the island's cargo, and distributing it inland strained an already overloaded rail network. This meant shipping had to face the dangers of attacks by German torpedo boats, long-range guns and aircraft. The coin had two sides, of course; transport shortages required the Germans to make use of their own convoys bound to and from the French and Lowland ports. Each side kept its ships close to a friendly shore, but neither shore was friendly around Dover and Calais.

The Germans had set up long-range guns shortly after occupying that part of the coast, and on 11 November 1940 scientists at His Majesty's Signal School were informed that a convoy had been subjected to accurate gunfire at night. Suspicion of radar brought N E Davis, an experienced Marconi television engineer, to the scene with a receiver capable of a wide range of wavelength. (Unknown to the Navy, technicians from TRE had observed the Seetakt radiation the month before and identified its origin as radar [11].) Davis quickly determined the radiative characteristics of Seetakt, providing the certain evidence of the existence of German radar. He then set out to provide a jammer.

Using a high-power decimeter-wave triode Davis built an oscillator tunable with Lecher wires over the Seetakt frequency band. Restrictions on the form of jamming modulation came from fear that an effective noise might be used against CH, forcing him to use a sinusoidal form. By February 1941 he had six experimental jammers in the Dover–Folkstone area, which greatly reduced the accuracy of the gunfire. It was the beginning the 'radio war' at Dover.

The experimental sets were replaced by engineered versions, the Navy type 91, which had large waveguides fitted to flared apertures fed with wide-band dipoles. The presence of Würzburgs, which could be used for coast watching, was known by then, so the frequency range was extended to encompass them as well. The Germans countered by altering frequency, and so it went. As the radio war expanded an inter-service organization was set up to collect relevant intelligence. Secrecy deemed it be called the Noise Investigation Bureau [12].

5.3.3. *The Bruneval Raid*

With the defeat of France the English Channel became the no-man's land of the Second World War. As on the shell-scarred ground between the trenches of 1914–1918, patrols felt out the enemy but by boat or aircraft rather than by crawling in the mud. Lives disappeared in clashes forgotten except in the most detailed of official histories and in the memories of loved ones and survivors. Both shores saw the placement of the tangible elements of the earlier war of attrition—barbed wire, mines, machine-gun posts, trenches and boredom. And as before, on both sides radio technicians set up receivers to record any transmission the enemy might make, while others set about to decode them, after which staff officers tried to fit the information so obtained into a coherent picture of the enemy activities or plans.

A new component of this electronic warfare was the signals intelligence units that specialized in radar. The Germans had begun this before the Battle of Britain and had quickly identified the CH and CHL/CD chains. There remained for some months afterwards the general British belief, unquestioned at upper levels, that the Germans had no radar. It was a belief not shared by TRE, where radar had lost its mystique, nor the Royal Navy, nor by R V Jones, where the combined results of the Oslo Report, the interrogation of prisoners and a number of other pieces of intelligence pointed strongly toward German radar [13]. As noted earlier, both H M Signal School and TRE had observed Seetakt emission in late 1940. The same techniques had seen signals from Freya but had confused them with emission from British equipment of similar frequencies, and their certain identification as radar had to wait until February 1941 [14].

The conclusive piece of evidence for Jones had to be a photograph, and for this he was in luck because the Air Force had had by the end of 1940 the remarkable Photographic Reconnaissance Unit under Wing Commander Geoffrey W Tuttle. This group of pilots flying unarmed and very smooth-surfaced Spitfires could sneak into and out of the most heavily guarded region to take pictures, which they then turned over to equally skillful interpreters. This capability came into being at the beginning of the war through the irrepressible efforts of F Sidney Cotton, wealthy Australian business tycoon, flier with experience in the previous war and photographic expert of the highest order. With private money he began doing his own freelance photo-reconnaissance of German military positions in January 1939. His annoying methods—the worst being that he did a much better job than the old RAF photographic units—alienated nearly all in the Air Force except Dowding, whose keen technical understanding convinced him to give Cotton two of his precious Spitfires at a time when Fighter Command needed them desperately [15].

Once the identification was certain that the emissions from Freya were radar, Jones quickly succeeded in obtaining a close-up view of one of the stations [16], which finally convinced the top levels of British command on 24 February 1941. The success of electronic intelligence by ground stations naturally called for equipping aircraft with suitable receivers, and a flight of 109 Squadron was outfitted with receivers in Wellington bombers, which gained the name of Ferret for this kind of aircraft, and began searching for emissions over a wide spectrum. By October electronic intelligence had located 27 Freya stations [17]. On 7 May they made the acquaintance of the Würzburg, observed as pulsed 50 cm transmissions from nine locations [18], but no suggestive antennas showed up on the kind of photographic coverage that had given the first hints of Freya. The information gained from the Ferrets indicated that the Würzburg had a much narrower beam than the Freya, and Jones suspected that this was superior equipment already widely deployed, possibly the cause of the unnerving speed with which searchlights were being brought to bear on bombers. The characteristics of Freya were pretty easily understood from the emissions, but this new set was surprising, and Jones wanted details. The need for a picture was obvious.

In late 1941 a careful study of a Freya station revealed an object close by and apparently associated with it, small but curious enough for Jones to insist on a close-up shot. This required two flights for success, but the second gave Jones the picture he wanted. This first view of a Würzburg disclosed two intriguing elements: the device was indeed small and located close to the beach. Could it or some of its important components be taken in a commando raid?

Jones was reluctant to recommend action that put the lives of many men at risk to secure this information, but a consultation with W B Lewis, by then Deputy Superintendent of TRE, bolstered his belief in the value of learning the details of the Würzburg. A raid also fitted with Churchill's wishes to agitate the German shore defenses as much as possible, so the request went to Lord Louis Mountbatten, Chief of Combined Operations, who favored the operation enthusiastically and ordered preparations. A parachute company underwent the detailed training for a night landing near the village of Bruneval where the radar station was located, and crews of the landing craft that were to remove the raiders and their precious loot from the beach practiced on a similar coastline.

The raid took place on the night of 27/28 February 1942 and was an unqualified success. Its professionalism, greatly admired by the Germans, made up for the bewilderment that had marked British response to the battleship dash of two weeks earlier. The disposition of the garrison was known to the attackers from aerial photographs and the reports of

French resistance agents, and it had been taken completely by surprise. The entire Würzburg was too large to be taken, but transmitter and receiver modules were easily removed and the antenna feed, which had the dipole, was sawed off. The radar man made understanding sketches and took a couple of photos until the camera's flash drew fire. They loaded the captured equipment onto a cart brought for that purpose and led a captured Würzburg operator with it onto the landing craft on which they departed. Loss of life on both sides amounted to two British and five German [19].

Four important pieces of information resulted from what was brought back. First and most important, the Würzburg was a fixed frequency set with a very narrow band over which it could be tuned. Second, it had no circuitry designed to deal with countermeasures. Third, it was an extremely well engineered piece of equipment, having modular design that made isolating faults and repairing them extremely easy. Fourth, the operator prisoner, though extremely cooperative, had poor technical competence. These last two characteristics were inversions of British procedure, which was to move prototype designs to production as fast as possible and have them operated by personnel trained almost to the level of electrical engineers.

Another benefit accrued from the raid. German radar stations became heavily entrenched, making them easy to spot from the air [20]—but also difficult to take. For whatever it was worth, the raid, combined with strict secrecy, gave rise to extravagant rumors within the Wehrmacht.

5.3.4. *The move of TRE to Malvern*

If the raid had convinced the Germans to fortify their radar stations, it also made the British apprehensive about the exposed location they had selected for TRE at Swanage. The move there in the summer of 1940 from the admittedly unsuitable quarters at Dundee must remain a puzzle to an outside observer. Fear of German air attack had caused the rapid departure from Bawdsey to Dundee, but Swanage was a terrible location after the defeat of France, yet the laboratory grew during the weeks when invasion was by no means a trifling danger. After the Bruneval raid, high command became concerned about a possible retaliation, and the knowledge that a German parachute company had been stationed near Cherbourg began to make a number of people uneasy.

Superintendent A P Rowe, who had been greatly alarmed by the position at Bawdsey, had become quite comfortable at Swanage. The laboratory had spread about the region and numbered 1000 employees. The method used to procure his cooperation was to provide rumors of German activity. The parachute company quickly grew in the telling to be 'seventeen train-loads', and the region around the laboratory was fortified. Rowe was convinced the Prime Minister had ordered the evacuation. For whatever reason, a new site was soon found in the buildings of a college at Malvern overlooking the Severn Valley in Worcestershire. With the emphasis on microwaves there was much less need for a location near the ocean. The spa town of 15 000 had to take up the addition of the 1000—soon to grow to 3000—employees and their families, and found this decidedly less pleasant than furnishing hotel space for the annual festivals of plays by Bernard Shaw that had marked the pre-war decade.

So the last and final move of TRE began on 25 May 1942 [21]. It was not the final name, however. It remains in Malvern today, although not in the college. Five more name changes have left it the Defense Research Agency.

5.3.5. *The Dieppe Raid*

At the first light of day on 19 August 1942 landing craft touched the beach at the channel port of Dieppe at about the time of similar landings to the east and to the west of the town. It was an action similar to others on the channel during preceding months, but this one was conducted at division strength, almost entirely by the Canadian 2nd Division with some British Commandos and a token force of American Rangers. The interrogation of an Allied officer prisoner posed a question: 'What was it? It was too big for a raid and too small for an invasion'. The answer speaks across the decades: 'When you find out, tell me'.

The first wave was met by infantry in their combat stations who delivered a withering fire, and when tanks followed—the first landed in a raid—they either fouled on the obstacles that the sappers had been unable to remove or were knocked out by antitank guns already in position. A western flank landing did better, but not even a dramatic success there could have altered the enormous measure of the disaster of the center, a disaster for Canada—of the Canadians engaged 68% became casualties— about as great in proportion to the nation's population as the Viet Nam War was for the United States and one that left as great a psychological scar. Controversy has naturally followed Dieppe with books and magazine articles in abundance. As the years pass an uncountable number of relevant documents have been located as German files have been sorted out and Allied files released from classification. Many have been lost or were destroyed, and what remains is a great mix—thoroughly stirred by the Official Secrets Act.

For those alert to the events of the time, Dieppe is well remembered, in part as something for which rumor augmented the news reports and led to their distrust. For those who came later the name does not have the same impact, or has none at all. The work by Campbell in the general references is an invaluable aid in sorting out the confusion that lies behind the written record but is not suitable as an introduction. Although written too soon for access to important material, Robertson's book [22] presents the important facts in a gripping way.

The radar story of Dieppe is a curious one. Succinctly put, one might say that nothing happened, seemingly belying the extent of what follows, but there are occasions where 'nothing' requires a bit of explanation. There are three parts to the radar component of Dieppe: (1) the operational use of radar by British and German forces, (2) the technical intelligence gained by the raiders and (3) the effect on subsequent use of radar, specifically for the invasion.

German coast watching was the responsibility of the Kriegsmarine, which began to set up Seetakt equipment on the French coast as it became available, the first going to the batteries of heavy guns at Cap Gris Nez and Calais, and the first detected by British signals intelligence. A Seetakt had been placed at Pointe d'Ally, a few kilometers west of Dieppe, quite capable of observing the attacking flotilla. Its existence was unknown to the British, but it was removed a week before the landing, bringing British intelligence up to date. The Kriegsmarine made the change without troubling to tell the Army, telling something about interservice cooperation.

The British did know of a Luftwaffe Freya–Würzburg pair at Pourville to the west of Dieppe, the location of the Green Beach landing. Although a 2.4 m air-warning set, this Freya was capable of observing large surface craft at 30 km as a consequence of its splendid cliff location [23], more or less the capability of the Würzburg, but the Allied operations order had no provision for any kind of radar countermeasures. Air Vice-Marshal Trafford Leigh-Mallory, commander of air support, had discussed various means of jamming or deception, but the Pourville Freya was beyond the horizon of ground jammers on the English coast, so only deception was considered, such as flying aircraft to distract the operators or having motor boats tow balloons carrying dipole reflectors. Neither was done [24].

This becomes a particularly puzzling attitude because complete surprise was crucial to the operation, as there was to be no preliminary bombardment. The Royal Navy refused to commit any vessel larger than a destroyer to the operation for fear of losing it to dive bombers, and Bomber Command initially refused, owing to a policy of not bombing French cities at night. When the policy was relaxed by Churchill for the occasion, the bombardment was rejected by the planners as unnecessary [25].

What would seem to have been an excellent capability of radar warning for the defenders was spoiled by poor interservice cooperation. The Freya stations reported air and the Seetakt stations reported surface movements, although after the Bruneval Raid local arrangements were made to have Luftwaffe stations report maritime activity to the Kreigsmarine plotting centers [26]. The Freya station, under the command of Oberleutnant Willi Weber, sighted the flotilla at 0232 hours and decided by 0330 that it was a raid of substantial size. His report to the Navy plotting center was brushed aside because they thought it was of a coastal convoy headed for Dieppe from Boulogne. Deduced from the nature of the movements Weber was sure it was not the convoy and called the 302nd Infantry Division, where he thought the warning caused the troops to go into a condition of enhanced alert. The outcome of the landing naturally reinforced in his mind the opinion that he had given the alarm [27], but the enhanced alert had had other origins. Since February Hitler had transferred his invasion fears from Norway to France with the result that all lower levels of command were thoroughly exercised. On top of that, British propaganda had been trumpeting a second front for 1942, amplified by all communists or communist sympathizers in the west. Troops went into alert positions whenever moon, tide and weather were suitable, as they were that night.

The operation did have one radar objective in putting a technician, Flight Sergeant Jack Nissenthal, ashore on the Green Beach to examine and, if possible, remove components of the Freya. Unlike the Bruneval Raid, Jones had not requested this as he considered his knowledge of Freya adequate; the task had been tacked on because the planners of every raid now wanted something about radar. Nissenthal went ashore with Company A of the South Saskatchewan Regiment, who were unable to take the well fortified radar station, but he was able to observe the movements of the antenna, which told him the set had lobe switching [28]. Some extravagant claims have been made by Nissenthal [29] and others about the accomplishments of this exploit. A big point was made and has been disputed of his having cut telephone lines out of the Freya station [30], forcing them to use radio to transmit their findings, but these stations had already used radio for this purpose to the extent that there was little more intelligence to be gleaned from listening to such transmissions [31]. However that may be, his adventures on the beach had a distinction denied any of the others, for he had had a personal guard assigned to protect him or kill him (he learned later) rather than allow his capture. This resulted from an order originally intended to apply to a TRE scientist and not altered when the assignment was given to a technician [32].

The German convoy bound for Dieppe was an unforeseen complication for the attacking fleet, as the two collided in the darkness with a resulting exchange of gunfire. This was interpreted by the Kriegsmarine to be British torpedo boats attacking the ships. British shore radar had tracked the convoy with their NT 271 equipment since 2140 hours of the 18th, even achieving remarkable long-range plots as a result of anomalous propagation [33], but this information did not reach Captain John Hughes-Hallett, the Naval Force Commander because of a communications failure [34]. Two destroyers in the force had 10 cm type 272 (a modification of the 271) radar but failed to detect the convoy [35]. The reason for this probably lies in the absence of a PPI indicator in this set [36], which made it difficult for the operators to untangle the confusion of so many ships. The resulting melée scarcely helped get things off to the right start for landings dependent on surprise, although the short fight actually raised no alarm. Subsequent studies show that part of the confusion arose because some officers thought plots beyond normal radar range, the result of anomalous propagation, were not trustworthy data [37].

Thus the extensive radar activities add up to very little, if somewhat more than nothing. German radar contributed nothing to warning the garrison despite the alert radar commander's report and insistence. No jamming or deception was attempted by the British despite the importance of surprise. Information about the German convoy learned from shore stations was not passed on to the responsible commander, and his own ships did not spot it. The one piece of radar intelligence that Nissenthal learned, that the Freya had lobe switching, was not deemed worth preparing a special report [38].

The small thought given to radar in preparing for Dieppe was not repeated in planning for the invasion of Normandy, somewhat less than two years later. The invasion had a detailed plan for radar at every level and used jamming and deception in the most advanced forms as well as radar in its primary mission. The debacle at Dieppe had many lessons to teach for the Normandy and Mediterranean invasions. That they had to result from such a bloody defeat does not necessarily follow.

5.4. CARRIER WARFARE DEFINED

In one day the entire foundation of naval thought had been turned upside down. On 6 December 1941 the battleship was the principal weapon, by 11 December it was the aircraft carrier. The supremacy of the former had been accepted doctrine by all navies, but it was replaced in one day's action, with the lesson hammered home three days later in the Gulf of Siam for any who still needed it. The air power demonstrated by six fleet carriers at Pearl Harbor was simply enormous. There remained the concern that the fragile nature of carriers, the result of thin decks covering large quantities of gasoline and explosives, would make their careers short, if exciting, in the initial phases of a great sea battle, but the dive bomber and the torpedo plane had demonstrated that the battleship had pronounced weaknesses too. The coming year was to see the rapid evolution of a new kind of maritime warfare, one increasingly controlled by radar, something that had been unknown when most carriers had been laid down and absent from tactical thought during the design period of all then in existence.

Vice Admiral Chuichi Nagumo, who led the Japanese fleet that had attacked Hawaii, was concerned for the vulnerability of his carriers and had opposed the mission as too risky. He was to use these carriers for the next four months in one of the most spectacular naval exploits of all time, sinking five battleships, one carrier, two cruisers, seven destroyers and large numbers of lesser craft without losing a single ship. He was to drive the British Eastern Fleet away from the water it had to defend, a humiliation roughly equal to that suffered earlier by the US Pacific Fleet.

The Pearl Harbor Striking Force returned directly to Japan, though two carriers were dispatched temporarily to support the second and successful invasion attempt of Wake Island on 21–23 December. Nagumo then supported the invasions in the southwest Pacific where lay the oil fields and various minerals and produce of Borneo and Java, vital to sustain Japan's war effort. The hastily organized ABDA Command united the disparate American, British, Dutch and Australian forces but had neither the material nor the essential common training to deal with Japan.

Darwin, on the northwest coast of Australia, was the supply port for the defense of the Malay Barrier, and Nagumo sent a powerful carrier attack against it on 19 February 1942 that reduced its ability to function and destroyed many of the vessels needed. A week later Vice Admiral Takeo Takagi disposed of the ABDA fleet under the Dutch Rear Admiral K W F M Doorman without carriers. The newly formed allies fought bravely and desperately but to no avail, and Doorman went down with his ship. USS *Langley*, the Navy's first carrier, was lost in one of the actions fought around Java, while serving as an aircraft ferry.

Nagumo's next target was Ceylon and control of the Bay of Bengal to secure the naval flank for the invasion of Burma and, for all anyone on the Allied side knew, to open the way to India and a link with the Germans in the Middle East. To counter this the Admiralty scraped together a force that seemed strong enough on paper; three carriers, five battleships and assorted cruisers; but the carriers had too few aircraft and those inferior to the Japanese, and the battleships were old and slow. Action ended when the Japanese left with other business in mind after sinking a carrier and two heavy cruisers. Nagumo had lost no ships but had to note attrition of his superb air crews that would prove much more difficult to replace than the aircraft.

In none of these actions was radar used.

From the safe historical perspective that time allows, one can say that Japan began the Pacific War unfavorably. The success Japan achieved at the beginning masked far more enduring errors. Their concentration on battleships at Pearl Harbor was in part a necessary consequence of absent US carriers, but their neglect of fuel tank farms and repair yards resulted from failure to appreciate fully the logistics of modern naval warfare. In fact they left the American Navy with most of its fighting ability, although this was hardly appreciated at the time. That the US Navy was outnumbered in important units did not result from the losses at Pearl Harbor, and for whatever losses in ships and territory America suffered, it gained a public united to a degree that would have been thought impossible the day before the attack.

Battleships would be important, better said useful in the coming struggle, but the encounters long planned by all naval tacticians of great masses of capital ships were not to be. In their place came task forces made up of one or more carriers as the principal striking element with protection and support from cruisers, destroyers, oilers, supply ships and—if they were available and fast enough to keep up—battleships. The enemies of the task force were aircraft and submarines. The new battleships, such as USS *Washington*, would be highly valued not for their main batteries but for the enormous number of 5 inch dual-purpose guns they carried, especially after the Bell Labs FD fire control radar and the proximity fuze became standard in 1943. It was a different world from the time of HMS *Dreadnought*, which had had initially no secondary batteries at all.

American naval officers had recognized the value of the radar demonstrated to them in the Caribbean exercises of 1939 and were quick to invent the Combat Information Center (CIC) to evaluate and act on radar intelligence. At first put into whatever cramped cabin could be found on carriers, not known for unused space, the CIC evolved over the months and years into an elaborate battle station for the commander.

When the Pacific War began all American carriers had CXAM air warning equipment, the RCA production model of the XAF tested in 1939. On active duty it was to do all things expected of it, if not all things desired. As an early warning set it was as valued in 1945 in its updated version, the SK, as in 1939. (Later, when radar seemed a province only for microwaves, the meter-wave air-warning set remained as early-warning champion, when it was observed that the reflectivity of a jet, unencumbered with propellers, radial engines or radiators, was substantially less for 10 cm than for 1.5 m.)

Use of the vertical lobes for height determination was not considered in the design of the CXAM as it had been in British sets, but by late 1941 it was well understood by the Navy radarmen, who could make a fair determination on a single aircraft approaching at great distance provided that the vessel experienced very little roll. A gyroscopically controlled stable-vertical mounting might have helped, but in practice height determination was and continues to be a difficult art, making the complications of a stable vertical for such a heavy antenna of questionable value. Good height determination also requires following the target for a substantial part of its path with frequent referrals to the fade charts, which is not the kind of thing one did well when the scope became crowded with targets and action was imminent.

Perhaps the worst aspects of CXAM were its dependence on the A-scope display (the plot of blip size against range) and on the operator having to point the antenna. If there were many planes in the air at various directions, the radar officer became a very harassed man. Each sighting had to be hand plotted. The SK, which began to see service in early 1943, had a rotating antenna and a PPI display [1]. Without computers good use of these data relied on almost intuitive skills of the radarmen, and careless preoccupation with one part of the sky could mean being taken by surprise from another quarter. Small wonder the results were less than perfect in the confusion of battle. One must bear in mind that fighter control for a carrier was a different thing than for southern England during the Battle of Britain, as all activity concentrated within a few km of the ship rather than over hundreds of km of land; reaction times were correspondingly short and the resolution of target blips more easily confused.

One thing radar was not able to do was replace or even greatly assist the all-important task of reconnaissance. Early location of enemy ships was the key to victory in this new kind of battle, and it needed to be done at ranges greater than the striking range of the bombers. The maximum range for the detection of surface ships by radar was generally determined by the curvature of the earth, although atmospheric conditions occasionally produced anomalously long ranges by forming propagation ducts. (In some parts of the Indian Ocean anomalous propagation is often the rule.) Moreover whereas aircraft might be detected 200 km distant, if suitably high, surface ships could be difficult to observe at 15 km with meter-wave equipment, the consequence of the vertical lobe structure and the curvature of the earth. All carrier planes had much greater ranges than radar. Reconnaissance was accomplished by the American PBY Catalina and the Japanese Kawanishi 97 flying boats from island stations, by carrier planes with fuel substituted for bombs and torpedoes, and by catapulted float planes. The difficulties surrounding naval reconnaissance exceeded those of radar. The observing aircraft often did not know their own position nor consequently that of the ships sighted; for understandable reasons they were shy about approaching closely, which caused serious misidentifications, frequently transforming whatever they saw into capital ships.

5.4.1. The raids

The immediate concern of Admiral Chester Nimitz, newly appointed commander of the Pacific Fleet or CinCPac as the Navy chose to call him, was the protection of his carriers and the initiation of action against the enemy. During the first months of 1942 he did this skillfully by task-force raids on Japanese island bases and finally on the home islands. At first glance these appear to have been merely a method of striking back at the hated foe, who was then inflicting slow defeat on American and Philippine armies, but they yielded a variety of benefits. They kept the carriers at sea at unknown locations—until, of course, a task force struck—and offered excellent training and experience against the Japanese before having to meet them in a full-fledged carrier battle; they did much for fleet morale and kept the enemy off balance.

On 1 February 1942 two task forces, one under Vice Admiral William F Halsey with USS *Enterprise* and the other under Rear Admiral Frank Jack Fletcher with USS *Yorktown*, raided the Marshall and Gilbert Islands respectively. Japan had seized the Marshalls from Germany in 1914 and had retained them as a League of Nations mandate and had taken the Gilberts, a British Crown Colony, at the outbreak of war. On these raids American radar began its apprenticeship in war, and Americans began to add some exotic names to the list that had Shiloh and the Argonne.

Both forces caught the enemy by surprise. The *Enterprise's* radar opened this chapter of naval history by spotting a snooper, which was determined to be on a course that would not allow it to see the ships, so it was not molested [2]. The absence of IFF made itself acutely felt once enemy planes were in the air, leading to less than desirable fighter control, but in alerting the task forces to enemy reconnaissance aircraft, to be disposed of or not as circumstances directed, the CXAM was superb. The absence of the FD (mark 4) fire-control radar resulted in poor, sometimes wild AA fire from the 5 inch guns. The Marshall–Gilbert raids did not do a lot of damage, but American losses were light.

These raids were followed a few days later by a third under Vice Admiral Wilson Brown with the *Lexington*. The objective was the recently captured harbor of Rabaul in New Britain, a port quickly becoming a major Japanese base. A Kawanishi discovered the force uncomfortably early but radar made Brown aware of having been seen [3]; two of the big flying boats were downed by radar vectoring. The discovery quite predictably brought an attack by land-based bombers on the morning of 20 February, but radar gave alert for a timely interception with a repeat performance for a second attack in the afternoon. The defending fighters broke up the attacking formations before they could reach the fleet, which prevented damage to the ships and allowed few of the bombers to escape. Because of the extreme range the bombers had had to fly, the defenders did not have to contend with a fighter escort and as a result suffered few losses [4]. Brown called off the attack on Rabaul as not worth the risk to his carrier, but his presence troubled the Japanese and delayed the advance on Port Moresby.

At the end of February the *Enterprise* raided Wake Island and added two more firsts for its radar. It tracked the attacking planes and corrected their course through the YE homing radio. The YE was a meter-wave directional beam used to guide planes back to the ship. It normally rotated automatically, sending a coded signal that gave the direction to the ship. Pilots listened as its beam swung by them and thereby learned the course they were to take to return. Its directionality made it a relatively safe way of communicating with the attackers. After the successful raid a plane with a non-functioning homing device was located by radar and guided back by risking a short normal radio communication [5].

The next raid had significant strategic impact. Brown led the *Lexington* and the *Yorktown* through the Coral Sea to attack from the south Lae and Salamaua on the north side of New Guinea. Surprise was complete and the raid gave Japan the greatest losses of the war to date, and their advance on Port Moresby was delayed yet again. Radar was by then just part of the varied techniques of a technical force.

The last raid of this preliminary period appeared to be one for the public, the Tokyo Raid, but it so unnerved the Japanese that they made overly hasty plans for what was to lead to the Battle of Midway. Radar's observation of picket boats [6] caused Halsey to launch the Army B-25s earlier on 18 April than planned, but the raid was a spectacular success and no ships were lost; the bombers were lost, but most of the crews were saved by the Chinese amongst whom they landed. Japan exacted a terrible retribution on the populations of the regions that had helped them.

5.4.2. *The Battle of the Coral Sea*

Japan's strategic planning was marked from the beginning by disagreements between the Army and the Navy. The overwhelming strength of the Army was tied up in China and Manchuria where the intentions and strength of the Soviet Union were matters of great and varying uncertainty. They refused to provide the troops necessary to invade Ceylon and India, and their refusal put an end to that Navy plan.

The Tokyo Raid brought these disputes to a head. Admiral Isoroku Yamamoto realized the American Navy had to be dealt a crucial blow before the great industrial power behind it could make it too strong. This he planned to do by initiating a decisive battle. This fitted the Japanese warrior tradition and had produced the desired effect in their famous defeat of the Russian Baltic Fleet in the Strait of Tsushima in 1905. Yamamoto elected to bring the Americans to battle by attacking Hawaii with the seizure of Midway as the first step, reasoning correctly that this would make the Pacific Fleet fight. Preparations for this grand battle began immediately after the Tokyo Raid. There were to be no more such insults.

A faction in the Imperial Naval General Staff had been favoring the isolation of Australia from American help with the drive on Port Moresby, and the weakness of Australia's defense made them disinclined to drop it. The General Staff decided that Port Moresby was to be taken, as it could be accomplished with relatively small naval and military forces. The recently evacuated Australian base on Tulagi Island in the Solomons was to be occupied and an air field built on the larger, neighboring island, Guadalcanal, while another force would take Port Moresby, on the south side of New Guinea. These two bases would permit a thrust to the south that would control the approaches to Australia. Yamamoto's Hawaiian strategy was not immediately apparent, but this move was; following President Roosevelt's order that Australia and New Zealand were to be defended, Nimitz moved to counter the threat by sending two carriers under Fletcher to the Coral Sea to stop the invasion of Port Moresby.

The Battle of the Coral Sea was the first important naval battle fought during which the surface vessels never saw one another. It had everything needed to try the nerves of the participants: confused reconnaissance, garbled and failed radio communications, complicated refueling of ships and aircraft, weather manifestations to drive everyone mad, difficulties in coordinating air strike units that were straining to hit the fragile but dangerous enemy first and worries about whether aircraft would be able to find their way home in the dark—and whether home would be waiting for them.

The degree to which radar had taken hold of the naval airmen was illustrated in an order by the commander of the *Yorktown* assigning his most experienced squadron commander as Fighter Direction Officer, which required him to take his duty in the radar room, not in the cockpit of a fighter [7]. Given the responsibilities and the fallibility of his equipment it was not an enviable job and certainly not the one sought by an accomplished flier. For the aircraft of both carriers there were only six IFF sets, and two were lost with their Wildcat fighters in a strike against Tulagi that opened the battle.

Radar picked up a snooper on the morning of 5 May with the now well exercised CXAM, but it being too dangerous to guide the pilots to the encounter by radio they had to head out on a compass bearing and depend on skill and luck, which prevailed [8].

On 7 May the carriers *Shokaku* and *Zuikaku* under Rear Admiral Chuichi Hara (the whole Port Moresby force was commanded by Vice Admiral Takeo Takagi) launched an early strike force that destroyed a fleet oiler and a destroyer, misidentified as a carrier and a cruiser. At about the same time an American reconnaissance plane reported surface vessels that were taken to be carriers through a coding error, and they launched their 'knock out blow', not at the *Shokaku* and the *Zuikaku* but at the covering force for the invasion, which did have the light carrier *Shoho*. The *Shoho* and her consorts were caught completely by surprise and the Americans showed the good effects of their on-the-job training. Their attack on the carrier was nearly perfect—except that American torpedoes were slow and often did not run true or explode—and demonstrated the fragility of carriers by quickly converting the vessel into an exploding inferno.

The remainder of the day was spent by each side unsuccessfully seeking to find the other. A Japanese strike went ineffectually after another group of the wrong Allied ships, which were bombed equally ineffectually by US Army Air Forces planes from Australia. Some of the returning Japanese aircraft were ambushed by Fletcher's fighters on the basis of a radar sighting [9], some were lost trying to find their way home in the dark, all significant losses of excellent flight crews. That night found both Fletcher and Hara intently examining every piece of what was by then a huge amount of confusing and often contradictory evidence as to the location of the other.

At first light of 8 May reconnaissance planes from both task forces sought and found their adversaries with strikes launched immediately. The American attack, which obviously came as no surprise, did not have the sting as the one on the *Shoho* but damaged the *Shokaku* seriously enough to put her in repair yards for the coming decisive Battle of Midway. *Zuikaku* hid in a convenient rain squall. Both *Lexington* and *Yorktown* were damaged, only the former seriously, but inexperience in damage control resulted in gasoline fumes accumulating unnoticed in an electrical room where a spark set off a tremendous explosion followed by uncontrolled fires that destroyed the *Lexington*.

It certainly had not been the *Lexington*'s day. Her radar had picked up the approach of the attackers, but the height determination was sufficiently in error to cause half the interceptors to miss entirely and the rest to engage too near the ship for good defense [10].

As the battle ended, warning signs began coming from a very secret source. The US was reading the Japanese wireless signals and piecing together their strategy from an abundance of clues. The picture that came from the decoding rooms at times bordered on the incoherent, obtained as it was from the messages that could not go by wire and underwater cable, but Nimitz had compared the cryptographers' earlier handiwork with subsequent events and had come to trust it. A recent change in code had made life difficult for the decoding personnel, but by March 1942 they had broken the new version of JM25. Yamamoto's plans for a great decisive battle became, if not clear, strongly suggested.

Fletcher was ordered back to Hawaii where maintenance crews in a record three days of hectic work were able to patch up the *Yorktown* well enough to fight again. American losses of flight crews were much smaller than the Japanese. The invasion of Port Moresby was called off, and the two Japanese carriers would not be available for Midway. The *Zuikaku* had lost too many of her flight crews; in the Japanese Navy fliers were part of the ship's crew and there was not enough time to work up replacements. In the American Navy air squadrons were assigned to carriers, which were simply their current floating air fields, and replacement squadrons could relieve the battle weary. It was an administrative difference but an important one.

In preparing for Midway Nimitz added a ruse that radar allowed. Halsey with *Enterprise* and *Hornet* was ordered back to Hawaii from a canceled assignment but only after he was certain his force had been sighted. A snooper, for whom the whole task force steamed, was duly picked up by CXAM and determined to be on a course that assured he had seen them. The snooper was allowed to depart unharmed at which moment the task force made a rapid change of course [11].

5.4.3. The Battle of Midway

In American historical memory two battles will stand on equal footing: Gettysburg and Midway. Both were fierce, dramatic and decisive, even though the Civil War would continue nearly two years and the Pacific War more than three; both left the defeated fighting a hopeless cause. For either battle to have gone the other way would have had consequences one does not like to consider. A victory by Lee in 1863 might well have brought about the severing of the Union; a victory by Yamamoto in 1942 would have almost certainly resulted in the United States abandoning the 'Germany first' policy, a somber thought when Stalingrad and El Alamein lay six months into the future.

Midway's epic nature has naturally resulted in careful study and the retelling of its story. Here only the barest essentials needed to link the events with radar will be recounted. The reader who does not know it would do well to start with Morison [12], a description almost fresh from the scene. For accounts drawing on a greater number of sources Lord [13] and Prange [14] are excellent. For those interested in minute details about the air battles there is Lundstrom [15].

Yamamoto's plan, which came into Nimitz's hands from the Pearl Harbor decoding rooms of Lieutenant Commander Joseph Rochefort, had three parts: the main carrier force under veteran Nagumo approaching Midway from the northwest, a diversionary attack on the Aleutians, and the Midway occupation force approaching the island from the west. Yamamoto committed six carriers against which Nimitz could oppose only three. The recently repaired *Saratoga* was on the West Coast, but had neither escort cruisers and destroyers nor fully trained flight crews to man her. Yamamoto conveniently attached two carriers to the Alaskan diversion that he could have well used in the main force and that accomplished nothing in the northern waters; this left the ratio four to three. Yamamoto followed his carriers with a mighty fleet of battleships; Nimitz had six but left them at San Francisco, considering them too slow and not yet converted into the floating AA batteries that would really help. The Americans had more IFF sets than a month earlier in the Coral Sea, but many fliers had none.

Early on 3 June a Catalina spotted the Midway occupation force, and nine Army B-17s dropped bombs moderately near it. Nimitz was not deceived by this or the Aleutian task force and placed the two task forces that made up his strength north of Midway. Halsey was in the hospital and his force, made up of the *Enterprise* and the *Hornet*, was entrusted to Rear Admiral Raymond Spruance, subordinate to Fletcher, who commanded the *Yorktown* force.

Unaware of the location, or even the presence for that matter, of Nimitz's carriers Nagumo launched a very strong strike at Midway on the morning of 4 June. This attack was sighted by a Catalina and shortly thereafter by the island's SCR-270s [16]. Strike planes immediately left the island to hit the carriers, including the B-17s that were to drop bombs all over the ocean in three different sorties that day; Marine fighters in obsolete Buffalos and a few Wildcats knocked down a few bombers but were badly mauled by the escorting Zeroes. These attacks on Nagumo's carriers did no harm but did contribute to a hectic command atmosphere, which was not helped by the leader of the Midway strike force calling for a repeat attack and by the first faltering evidence from a reconnaissance plane of the position of American carriers to the northeast. The difficulty of CXAM in dealing with the many targets provided by the task force's patrolling aircraft, which were incompletely equipped with IFF, delayed the identification of this observer for the better part of an hour, allowing him to report the presence of a carrier.

Fletcher and Spruance launched their attack at the earliest possible moment and provided one of the most dramatic episodes in naval history. The first to sight and attack were the ineptly named Devastator torpedo planes. Given the circumstances of their attack it is doubtful whether the use of the new and much improved Avengers would have made any difference for they were set upon by fighters with no more protection than their rear cockpit machine guns. Of the 41 that attacked only six survived and not a single torpedo went home. Their sacrifice very likely provided victory, for while the Japanese combat air patrol and AA guns were destroying the Devastators and their crews, dive bombers caught Nagumo's carriers by complete surprise.

Radar's only part in this was in giving certain warning to the Midway garrison, important but hardly decisive. More important was radar's absence. Nagumo had no idea that dive bombers were so near and consequently did not make the critical preparations for receiving an attack, and the decks were full of armed aircraft with filled tanks and gasoline hoses lying about. *Akagi*, *Soryu* and *Kaga* found the consequences of this in a matter of minutes. Yamamoto had radar but it was on the battleships Ise and *Hyuga*, some hundreds of kilometers behind where it served no function [17]. Had the US Navy been using electronic intelligence receivers they could have received a shock. The *Ise* had a 1.5 m air-warning radar of mark 2 model 1, and the *Hyuga* had a 10 cm surface-search mark 2 model 2, which was given credit for preventing a collision in the bad weather following the battle [18]. When the one remaining carrier *Hiryu* launched a strike later in the day on the *Yorktown*, it was seen by the CXAM in plenty of time to make preparations that prevented her from being turned into an inferno [19]. It was to no avail, as two torpedoes from a submarine (the Japanese variety worked consistently and well) caused her eventual loss. *Hiryu* found the same fate as the other three carriers later in the day.

An evaluation of radar for the early stages of the Pacific War has many similarities with the evaluation of it for the Mediterranean. America's great disadvantage initially was in being outnumbered in carriers and in having inferior aircraft and less experienced air crews. After Midway the inequality began to be reversed, but it was a change that had had to be earned, fate had not foreordained it. The advantages that the US Navy had during the time before the country's industrial might came to the fore were (1) a well trained navy, if not at the peak that Japan had in December 1941, (2) an excellent understanding of the logistic support that a modern navy requires, (3) the ability to read some enemy signals and (4) radar. The question turns on whether the Navy would have emerged victorious had any one of these elements been absent; more specifically, could it have won without radar?

The prelude to the victory at Midway was the carrier raids. Given the extreme value of an individual carrier, these raids were risky affairs, but could be undertaken with moderate assurance that the task forces could not be surprised from the air, and this proved to be the case. The Japanese examples during those years demonstrated the fragility of these vessels. Without radar a carrier was like a battleship without armor plate. In effect radar gave carriers armor. One can easily imagine—and here one enters the great swamp of conjecture—the aborted Rabaul raid taking a bad end for the US had the land-based planes not been intercepted before reaching the *Lexington*.

The confidence radar built up among its users was obvious by the Battle of the Coral Sea. Just its use in ambushing the returning strike force planes on the evening of 7 May may have spelled the difference the next day by adding to Hara's serious attrition of air crews and planes. The presence of those lost planes might have led to the sinking of the *Yorktown* too. But this is all idle. There is no end of possible outcomes. There is no denial that radar was highly valued by those whose lives depended on it. It was not a perfect instrument but it was something no one wanted to sail without. CXAM had been designed as an air warning set and had functioned as planned. Its performance drove all users to make their CICs into good fighter direction devices. What was sorely needed was IFF, PPI, very-high-frequency radio and CICs properly set up and manned. Radar's value at the tactical level was comparable to the ability to read Japanese signals at the strategic level.

One tenet of air power doctrine was quietly laid aside as a result of the experiences of these few months. The Army's four-engine heavy bombers proved useless against surface ships. It was one thing to put a bomb into a pickle barrel from 6 km high, but something else again to put it onto a fast moving ship. Army B-17s from Midway made a total of four squadron sorties without hitting a single ship, although they certainly frightened a destroyer on their last attempt. Such gross inaccuracy was welcomed when their comrades in Australia had mistakenly targeted American ships in the Coral Sea but not when they missed the *Shoho*, which they had attacked before the carrier planes arrived to show how these things were done. The Army had obtained four-engine bombers on the basis of their ability to protect the nation's coasts; they had been the great hope for protecting the Philippines. Billy Mitchell's thesis had agitated defense circles for two decades and should have been forgotten, but airpower advocates transferred with an adroit inversion of the scriptures the prediction of carrier power to their prophet and the myth endures.

5.5. THE SOUTH PACIFIC, 1942

Admiral King decided after the Midway victory that the Allies must take the strategic initiative even though their resources were not adequate, especially in view of the Germany-first decision. This put him at cross purposes with the Army General Staff, which was entertaining fantasies of an early invasion across the Channel and was far more concerned with the slaughter of Allied merchant shipping in the Atlantic, something for which it and the British saw the US Navy as showing no interest. There was certainly an element of truth in the accusation that the Navy was only interested in the Pacific War. It was the war they had been preparing for, thinking about, and training for for two decades. It had started for them as a humiliation and they wanted desperately to redeem themselves. Despite their half-year apprenticeship as *de facto* belligerent on the North Atlantic, keeping U-boats off the convoys by their mere presence, they were ill prepared for the submarine enemy, ill prepared in equipment, planning, organization and training, and it was going to take time for those elements to be ready. In the meantime they were prepared for action against Japan.

It was apparent to the Allies that the Japanese drive toward Port Moresby, which Tokyo had not written off, had to be con- tained. The discovery on 4 July 1942 that Japan was building an airfield on Guadalcanal provided King with support for the strategic initiative he wanted and insured that the Army would agree, for such an airfield was almost as great a threat to Australia as one at Port Moresby, which justified seizing it, but to do so required attack, not defense. Furthermore the attack had better be made before the airfield became operational. The only ground troops available were the 1st Marine Division, formed of units that had undergone multiple fissions to provide cadre for new units being organized. It not only was filled to the brim with officers and enlisted men having less than one year's service but had had no unit training beyond battalion. They had expected six months training on arriving in New Zealand but only saw New Zealand briefly on their way to 'that stinking island'. King named Vice Admiral Robert L Ghormley Commander South Pacific and as such, leader of the combined forces Operation Watchtower, laconically renamed Operation Shoestring by those selected to execute it. The Allies assembled a major naval force with three carriers to invade an island of no possible significance except for war, which must have given Ghormley long thoughts about the lessons of Midway. The Japanese intelligence had not discerned the plan, making it importantly different from Midway.

The two powers had their logistic lines stretched to the limit, leading to major ground actions being fought by companies and attempts to bring reinforcements of regimental or even smaller size precipitating great naval battles. By the time the issue had been settled five major surface engagements had been fought—without the meddling interference of air power—along with two carrier battles in which the men of the surface vessels never saw their adversaries. In between major actions there were almost daily fights between ships and ships, ships and shore batteries and ships and aircraft. A restricted part of the ocean north of Guadalcanal became the resting place of more naval tonnage than any comparable place in the world and was named Ironbottom Sound by the survivors. The air field on the island, the possession of which hungry, often starving ground troops contended, was continually subjected to air attack and defended by gunner and pilot.

Radar assumed an ever increasing importance, becoming by the end of 1942 a significant weapon, but the infantry noted it not. Their war had few refinements: rifles, pistols, bayonets, machine guns, explosives, mud, jungle, insects, disease, boredom, terror.

The reader who does not know the details of this campaign should correct this fault. The best source for completeness, accuracy and readability is Richard Frank.

At the beginning of 1942 the American Navy had CXAM radar mounted only on a few capital ships. As the year progressed Bell Labs FC (later mark 3) sets began to be mounted on battleships and cruisers for main-battery fire direction, and by March FD (later mark 4) sets for dual-purpose batteries began to appear on all capital ships with some destroyers outfitted toward the end of the year. These radars operated at 40 cm and used lobe switching for accurate direction, FC only for horizontal, FD for horizontal and vertical. Originally designed for the very-high-frequency triode invented at Bell Labs, their design had been altered immediately after the disclosure of the magnetron, and they became the first American sets deployed using this invention. The basic FD design remained the Navy's AA fire-direction radar until the end of the war with a 33 cm automatic tracking version, mark 12, introduced in late 1943 [1].

The FD had, of course, larger lobes than 10 cm equipment, but skilled crews made good use of the size by tracking one air target while keeping others on their scopes, ready to be picked up when the first departed the place of honor. As it was, the beam of the FD was already so narrow that picking up the target initially required good teamwork between search and fire-direction radars. On the other hand the large lobes made it nearly impossible to track very-low-flying planes, something not long kept secret from Japanese fliers.

For surface targets the FC and FD were less satisfactory, for here the large lobes did not have the usefulness apparent in AA fire direction. Sea clutter, the reflection from surface waves, was a nuisance but could be worked through in firing on single isolated targets, but multiple targets, especially with similar range, caused the operators attempting to get correct bearing with lobe switching to aim at the center of gravity [2]. This defect led to Bell Labs making the mark 8 fire control radar, about which more in a later chapter. At about the same time the FCs and FDs were being distributed, cruisers and destroyers began to receive the type SC radar, a 1.5 m set similar to the CXAM but with a smaller antenna that generated larger lobes, of course.

More important, however, than any of these was the SG, the 10 cm set that resulted from the first Rad Lab experiments that had been begun with other purposes in mind. After its success on the *Semmes*, the prototype was rushed to the Naval Research Lab for alteration that made it agreeable to the harsh conditions of a warship; it then went to the Raytheon Corporation for production, starting that company in a field where it would long remain. It was mounted on the cruiser *Augusta* (which does not figure in these events) in April 1942 and on the *San Juan* and the carrier *Saratoga* shortly thereafter [3]. USS *Helena* had one when she came to the Solomons in August [4], and when fighting began in the waters among the islands the SG was the only radar that distinguished clearly ship from shore. What was confusion on the A-scopes of the meter-wave sets became clarity on the PPIs of SG. It became a navigation instrument mariners did not want to forego. By the beginning of 1943 cruisers and destroyers began to carry routinely an SG for surface search, an SC for air warning and one or more FDs for fire control. It was their basic radar equipment for the remainder of the war.

The US Navy was well served by Nimitz's comprehension of radar's strengths and limitations. He actively supported the establishment at Pearl Harbor of the Radar Center comprised of schools for maintenance, operation, fighter direction and tactics, established in that order. The last, the Radar Tactical School, gave instruction to commanding and flag officers and was to prove its worth in overcoming two misconceptions among that naval population: (1) radar was not used in the Battle of Jutland and was therefore unimportant; (2) radar was a magic box on which one need only press the button and the battle was won. Both of these onerous attitudes were prevalent during 1942 [5].

American Marines had to fight briefly for the tiny island of Tulagi, but they went ashore unopposed across the channel onto Guadalcanal after a short bombardment on 7 August 1942, the garrison having been caught most puzzlingly by surprise and prevailed upon to take up quarters in the jungle. Even more surprising than the absence of resistance was the discovery of two radar sets [6].

During the first months of the Pacific War no thought had been given to the possibility that Japan might have mastered this technique. This was partly the result of an affliction common to the British and Germans at the beginning—not believing the enemy was smart enough to duplicate, let alone exceed one's own achievements—and partly because the first few months of the war were occupied with thoughts about what the Japanese obviously did have rather than what they might have, so no special receivers were available to search for the tell-tale signals that would have disclosed radar. Had such Ferret aircraft been in operation earlier they might have located a similar radar that had begun operation at Rabaul in the spring.

The sets were Navy type mark 1 model 1 for air warning. One set with its boxlike array of dipoles was restored to service shortly after its arrival at the Naval Research Laboratory. It operated on the 3 m band with a peak power of 5 kW, with pulse lengths of 10 to 30 μs [7]. It seemed a poor thing compared with SCR-270, but that judgement was pronounced without appreciating the short time that had elapsed between perceived need and production. Furthermore, the Allies were soon to make good use of the Australian LW/AW radars of similar power. The greater range of SCR-270 had to be bought at great price in power and weight; the latter quantity was at times uncomfortable in the islands of the southwest Pacific.

An SCR-270 was on one of the transports unloading Marines and their supplies but did not get ashore before the ships were pulled back to Nouméa; a night surface fight had suddenly left them with little naval protection, and they departed. In desperation the landed radar technicians tried to put the Japanese set into operation [8].

From the time of the American landings until the last Japanese withdrew from Guadalcanal on 8 February 1943 almost continuous fighting on land, sea and air ensued. The goal of the antagonists was quite simple: control of the airfield. The methods of attainment would prove to be extremely varied. For the Americans the object of the ground fighting was to hold the airfield, named Henderson Field in honor of one of the Marine aviators who died at Midway. That done they had to expel the Japanese infantry from the island to insure a more secure operation of the air base.

Naval forces on both sides had to bring reinforcements and prevent the enemy from doing the same. A key element in this was control of the air during the day—lack of sophisticated radar prevented any serious night air activities—but control of the air turned on whether Henderson Field could function. For the first two days carriers provided the airplanes overhead, but Fletcher withdrew them as too precious to risk long in such a tough neighborhood. Marine fighters did not arrive at Henderson Field until 20 August when the engineers with their single bulldozer had made it capable of getting planes aloft. Guadalcanal's air defense came to be called Cactus after the island's radio identification.

Attacking aircraft came from Rabaul on New Britain and from Buka and Buin on Bougainville near the north end of islands so arranged that the path to Guadalcanal gained quite naturally the name of 'the Slot'. Early warning was crucial and came from two sources: radio messages from the coast watchers, Australians with native assistance positioned on the islands past which the raiders flew and Cactus radar. The first SCR-270 reached Guadalcanal on 20 September, quickly followed by two more 270s and two 268s. A few days later Lieutenant (USNR) Lewis C Mattison and three other officers arrived from the Fleet Fighter Director School at Pearl Harbor. Radar established itself as so important that Lieutenant Colonel Walter L Bayler, an experienced flier, took over fighter direction. The Japanese were unable to locate and eliminate the coast watchers, so the pilots began taking routes that avoided observation, but by then radar was able to give reliable warnings.

An effective air defense grew out of the strengths and weaknesses of the equipment. The SCR-270s detected the raiders, depending on altitude, at ranges as great as 200 km. The operators became skilled at determining both number and type of the attacking aircraft but had little success in determining altitude. The warning time so received sufficed for the defending fighters to spiral directly over Henderson Field to an altitude that gave them advantage. Their radio equipment was too weak to direct an interception in advance of the island, and the pilots did not want to fly without control, something that produced excellent radio discipline. When the attackers came within range of the 268s a reliable determination of their altitude could be obtained and transmitted to the fighters overhead. An extremely valuable aspect of the radar warning was that after its installation no air patrols were needed. This conserved fuel, aircraft and pilots, all in short supply and pushed to the limit [9].

So long as Henderson Field functioned, American surface ships were relatively secure during the day, but when night blinded the fliers Japanese destroyers ventured into those waters working as transports, attacking any American ships they found, and putting a few shells into the American camps. The rapid arrival and equally rapid departure of these ships caused them to acquire the name Tokyo Express. This pattern of Japanese supply was maintained throughout the struggle. The larger naval surface actions, which were fought almost entirely by gunfire and torpedo, had the same form, just heavier ships and bigger fights.

Five major surface actions were fought over Ironbottom Sound, and two carrier battles were fought to the north and east of the islands. At other times the carriers moved nervously but not idly about the Solomons. They were fundamentally aggressive units, yet fearful because of their few numbers and strategic importance. For details of their activities between and during major battles read John Lundstrom.

5.5.1. *The Battle of Savo Island*

In 1898 the US Navy fought two very one-sided surface battles with the outclassed Spanish Navy. Except for the encounter of a few of its ships as part of the ABDA fleet in the Battle of Java Sea it had fought no gunfire actions since that time: the main engagements of the war it was then fighting had employed its air arm. During the night of 8/9 August 1941 it came to know how the Spanish felt.

Landing the Marines on the 7th had necessarily concentrated a number of transport and cargo ships near the beach, offering a target of which the Japanese were aware and to which they responded with alacrity. American carrier planes made an attack by surface ships during the day much too risky, but the Japanese Navy excelled in night action and sent a sizeable force of cruisers and a destroyer to destroy the supply ships at the beach.

To intercept any raiding force were five American and three Australian cruisers with about as many destroyers under Rear Admiral Richmond Turner. Two American destroyers with SC radars, *Blue* and *Ralph Talbot*, were placed as pickets beyond Savo Island. Shortly after 0100 hour Rear Admiral Gunichi Mikawa led a somewhat smaller but much better controlled and undivided force from Rabaul out of the Slot around the south of Savo Island. It should have been detected by *Blue* but was not. Japanese lookouts—unaided by radar—saw *Blue*, but *Blue* saw nothing with either optical or radio eyes, and Mikawa left her to continue her patrol. The reason for this serious failure seems to have been a mixture of poor equipment performance, inadequate training and the confusion for meter-wave sets caused by the presence of nearby land. The SC displayed its observations on an A-scope, target amplitude against range with direction selected by the operator. Large nearby land masses, even if many degrees off the antenna axis, could form echoes capable of hiding targets at greater range, very much as reflections from the ground obscured targets for meter-wave AI equipment. The SC, like nearly all meter-wave sets, also leaked radiation out the back to a small degree, and when this was reflected off a land mass it gave the operator the appearance of something small to the front. PPI and microwaves were the only sure way in such circumstances. Nevertheless, the SC should have disclosed the Japanese squadron and the reason for failure is not clear.

The cruisers *San Juan* and *Quincy* and the destroyer *Patterson* had radar that should have been useful, but the *San Juan*, which had the only SG, was placed such that she was never engaged. The *Quincy* used her SC to note but disregard a Japanese float plane [10], the *Patterson* her FC to fire briefly on the *Chicago* [11]. The Japanese squadron retired after receiving only minor hurts but in fear of the planes that dawn would bring and without executing the orders to put the beach and supply ships under fire. The Allies had four cruisers sunk, including the Australian *Canberra*, and one badly damaged, a thorough whipping. The beachhead, however, remained.

Naval tacticians have examined this battle in great detail, as the reader can well imagine. Such study lies outside our purpose, but it is obvious that radar had been more a handicap than an advantage because of the false confidence it gave, reducing the emphasis on alert lookouts. The reasons for radar's failure were equipment unsuitable for close-in surface action (the SG was not committed), inexperienced operating personnel, the absence of IFF on ships and a want of understanding by commanders of radar's limitations as well as its capabilities. Fortunately, the Americans recognized these reasons as the cause of grief with the means of correction evident; there was no loss of confidence in radar.

The Japanese immediately followed this battle with attempts to bring in reinforcements by destroyers that raced down the Slot under cover of darkness packed with men but little heavy equipment. Not wanting to rely on this tactic, Admiral Yamamoto set much larger forces in motion, and for two weeks the two opposing navies prepared for a major encounter.

5.5.2. *The Battle of the Eastern Solomons*

This classic carrier battle resulted from the attempt to land 1500 men on Guadalcanal. To accomplish this Yamamoto sent two carriers, two battleships, nine cruisers and many destroyers under veteran Nagumo south from Truk to counter any American fleet opposition, and a lesser invasion force under Rear Admiral Raizo Tanaka destined for Guadalcanal. The Americans were not well served by the intelligence service that had stood them so well at Midway, but it was not particularly difficult to know that something of the sort was afoot. Fletcher had three carriers, a battleship and seven cruisers to counter Nagumo but had sent the carrier *Wasp* to the rear for refueling.

The use of the air warning radar was very effective in picking up snoopers and attacking squadrons. There was some improvement in determining target altitude but just as much confusion for the air controllers when aircraft began to tangle close in. IFF mark II gained few admirers. Worst of all, however, was the over-crowded high-frequency radio channel used between pilots and controllers. The excitement of combat overcame communications discipline and filled the air with unnecessary talk, greatly hampering controllers in their duties. Furthermore, that frequency band was noisy and many transmissions simply did not get through, especially distressing when sightings of the enemy failed to be received or were garbled. The new, very-high-frequency frequency-modulated radios were needed almost as much as radar.

The Japanese carrier *Shokaku* and the battleship *Kirishima* received air-warning sets, mark 2 model 1, just before sailing [12], and the operators and technicians had just begun learning how to use them. On the carrier they noted the approach of bombers from the *Enterprise*, but the report was lost to the ship's command; it was alarm from the lookouts that set off defensive action [13]. Allied commanders were not the only ones who had to learn this new way of a ship.

Action began near noon of 24 August with an attack on the island launched by the *Ryujo*. The flight was sighted by the radar of the *Saratoga* at a range of 150 km headed for Guadalcanal. Poor communications kept the warning from being received on the island, but this sighting indicated where the home carrier could be found, and it soon became a blazing inferno. Guadalcanal did not yet have operating radar, but a patrol plane alerted their newly arrived Marine fighter group, who managed to keep damage and losses to a minimum and from the outcome to lose their fear of the Zero [14].

The troop convoy for the invasion was effectively attacked with the largest vessel set afire and sunk and the other two forced back. It was here that the US Army Air Force finally sank a ship with a B-17: a hove-to destroyer, so unconcerned at the approach of the heavy bombers that she did not even get under way when the bombers appeared. Attacks by them earlier on all three carriers were as ineffective as usual.

The Japanese lost a light carrier and a destroyer. The Americans lost no ships but had to send the *Enterprise* to Pearl Harbor for repair. It was by any measure an American victory. The CXAM radar of the *Enterprise* gave ample warning of the attack that damaged her, allowing a strong defense and the prompt clearing of combustibles and explosives, but it failed for most fighter control, although as much from communication breakdown as saturation of the radar scopes.

In the interlude between this and the next important naval battle something happened of illustrative importance to radar. On 15 September during relatively quiet and routine operation the Japanese submarine I-19 passed undetected through the destroyer screen surrounding the carrier *Wasp* and hit her a fatal blast. Earlier that morning radar had vectored the combat air patrol to dispose of a snooper, but the destroyers' sonar had not been up to the same standard. Without a warning *Wasp* was fragile indeed and became after three major carrier battles the first American carrier to become an inferno, the fate that had by then befallen six Japanese carriers, all from surprise air attack. No amount of damage control skills sufficed when plentiful quantities of gasoline were about. It left the fleet in a critical condition because submarine I-26 had sent *Saratoga* off for three months at repair yards two weeks earlier in a similar attack. Both submarines escaped. The Navy's anti-submarine ability was consistently the same in both Atlantic and Pacific Oceans in 1942—poor.

5.5.3. *The Battle of Cape Esperance*

The piecemeal reinforcement of Guadalcanal by both Japan and the US began to favor the Japanese, leading to a decision by Ghormley to send the 164th Infantry Regiment to the island escorted by a strong naval force much as Yamamoto had attempted. Reconnaissance aircraft let both sides know what was under way, and Japanese cruisers raced down the Slot to be met by an American covering force under Rear Admiral Norman Scott at Savo Island near midnight of 11 October 1942.

Scott was not well informed about radar and specifically made the heavy cruiser *San Francisco*, which did not have SG, his flagship. He was also erroneously informed that the Japanese had radar-detecting receivers for SC so he ordered them shut down, although this probably was not of consequence, aircraft proving to have had no importance and the meter-wave sets not having shown themselves particularly useful around Savo Island. All cruisers had fire-control radar and the light cruisers *Boise* and *Helena* had the vital SG search equipment that functioned throughout the fight and were to prove crucial to the victory that emerged. Captain Edward J Moran, commander of the *Boise*, was well informed about radar and saw the struggle clearly on his SG when Scott was essentially blind. Moran applied a bit of judicious disobedience of orders to make Scott's simple and effective plan—crossing Rear Admiral Arimoto Goto's T—a success; however, he failed to communicate clearly what he saw to Scott, with substantial confusion arising. Moran probably did not realize the extent of Scott's blindness.

When the cruisers opened fire in darkness their first rounds were hits, and the Japanese squadron was caught completely by surprise and severely damaged in the first few minutes [15]. The Japanese failed to execute their mission and suffered the loss of a heavy cruiser, three destroyers and the life of Admiral Goto; the Americans lost one destroyer. The *Boise* was seriously damaged as a result of fire received when she began to illuminate the enemy with searchlights rather than maintaining reliance on radar, which became useless with so many ships and shell splashes about. In one last contribution the *Boise*'s SG prevented her from grounding while withdrawing [16]. The accurate opening fire produced effects far away in Japan where elements of the Navy had resisted further development of microwave equipment; the *Boise*'s salvos decided the matter: microwave radar was to go into all kinds of warship [17].

The next night two Japanese battleships pounded Henderson Field and vicinity in what was thereafter referred to as 'The Bombardment'. It left the air defense a shambles and gave the 164th Infantry more than a taste of how life on the island was going to be. It also served as a cover for the arrival—at no small cost—of the 'Fast Convoy' that unloaded several thousand men who, when added to those already ashore, formed the Japanese 17th Army. Thus, despite the victory of the cruiser action and the American reinforcement, matters at Guadalcanal stood at their lowest. The garrison on Guadalcanal was particularly displeased with the Navy's inability to cut off the steady flow of nightly reinforcements down the Slot. Nimitz chose to replace Ghormley with Halsey, and whether justified by events or not, it raised the spirits of the command measurably.

The 17th Army was to take Henderson Field and force the Marines and soldiers to the beach; with the air base safely out of the way Yamamoto's ships and planes would destroy the American fleet that would predictably come to the rescue. Acting on cryptanalysis that revealed the key elements of the plan [18] Halsey moved his carriers from their usual position south of Guadalcanal, where their function was defensive, to the north to make a spoiling attack, and the Battle of Santa Cruz Islands began.

5.5.4. *The Battle of Santa Cruz Islands*
This battle was Japan's last chance to regain naval superiority. Yamamoto sent four carriers against Nimitz's two—but for an engine room fire it would have been five; he still had a small core of his excellent fliers, and

a large fraction of the Americans had only just completed air training. Japan's shipyards and training schools were being completely outclassed, so it was now or never, although not fully appreciated at the time. For the Americans the odds were not as unfavorable as the carrier ratio indicated because Halsey had a new battleship, *South Dakota*, to help guard the *Enterprise*. The basis for this apparently incongruous statement was the transformation of this vessel into an floating AA regiment. It had an amazing number of 5 inch semi-automatic dual-purpose guns directed with FD radar and had just been equipped, as had the *Enterprise*, with the new 40 mm Bofors automatics to replace the weaker and jam-prone 1.1 inchers. Not only was the 40 a better gun for close-in defense, it had the excellent mark 14 optical computing sight designed by Stark Draper of MIT. The cruisers were not yet equipped with 40s, although all had the new 20 mm machine guns. They did have FD radars, but heavy AA fire despite radar control accounted for a much smaller fraction of kills than did the close-in defense by 40s, 1.1s and 20s [19]. It was the beginning of a new phase of carrier warfare, one that would change even more dramatically during the subsequent year when the 5 inch shells would carry proximity fuzes.

Radar added another new element in the form of ASV mark II sets mounted in the long-range Catalina observation planes, which spotted the enemy shortly after midnight on 26 October 1942, formally initiating the battle. They even attacked these ships with bombs and torpedoes after reporting the sightings [20].

The carrier battle had the usual confusions in communications and reconnaissance. It began with what had now become a tradition: B-17s dropping bombs into the ocean. Both sides used air-warning radar with attendant strengths and difficulties. The *Shokaku*'s radar returned the best range of the day with 155 km, but American radar had an especially bad day. On assuming command Halsey took the well proved Fighter Direction Officer of the *Enterprise* to be his Communications Officer at Nouméa to clean up problems that had plagued that command's signals, and unfortunately some excellent radar men from the carrier's radar plot went with him. The new Fighter Direction Officer was hardly able to find the china-marking pencils before the battle began [21], and by coincidence the operation also began with the maximum ranges of the CXAMs of both *Enterprise* and *Hornet* down to about half. Other vessels were not so handicapped but did not transfer the information to where it was needed [22]. Fighter pilots may have once resented control by a voice on the radio, but by October 1942 they depended on it and were vocal in their criticism of its failures at Santa Cruz.

The air combat can best be described as wild and confusing. Some Japanese pilots, either wounded or with damaged aircraft crashed or attempted to crash their targets, a morbid hint of things to come. American fliers severely damaged carriers *Zuiho* and *Shokaku*; their enemy counterparts crippled the *Hornet*, which the defending airmen blamed on the poor fighter control. Japan lost 99 of the 203 aircraft engaged, 48 of them to the vicious new AA fire, especially in the *Enterprise*'s task force; the US lost 80 of the 175 engaged [23].

The Japanese Army failed to take Henderson Field as a result of the difficulty of maneuvering in the jungle, the tenacity of the defenders and the use of infantry tactics that went out of style for other armies in 1914. Severe losses of air crews forced the Japanese to withdraw even though two carriers were still operational. American ships withdrew to the south leaving the stricken *Hornet* to be sunk by Japanese destroyers. The Japanese did not choose to pursue, having injuries of their own, especially the loss of their best aviators, and no wish to tangle with island-based bombers. The Battle of Santa Cruz Islands was very likely unique in the balanced importance of struggles on land, sea and in the air.

5.5.5. The Naval Battles of Guadalcanal

Just after midnight on Friday 13 November 1942 Rear Admiral Daniel J Callaghan led a column of five cruisers and eight destroyers northwest from Guadalcanal to protect the transports and supply ships he had just left and collided with Rear Admiral Hiroaki Abe leading two battleships, a cruiser and 11 destroyers intent on giving Henderson Field another taste of 14 inch shells. Callaghan had been Ghormley's Chief of Staff and had displaced Scott as Commander owing to 15 days' seniority. Scott remained as second in command that night.

This First Naval Battle of Guadalcanal is the most difficult to follow and the most contentious; much criticism has been directed at Callaghan, generally centered on his use or misuse of radar. Callaghan had little experience with radar and had not exercised command in a night action. Of the five cruisers three had SG radars; Callaghan took the 8 inch gun *San Francisco*, which had none, as his flagship and had not used his first two weeks of command to alter that arrangement. The SG-equipped cruisers were all 6 inch gun vessels; in expecting a rough fight a commander prefers the most powerful ship for his flag. Scott, with the experience of Cape Esperance behind him, elected to place his flag in the *Atlanta*, the only remaining cruiser without an SG. Two destroyers had SGs, the fourth in column and the last. Abe had an air warning radar on *Kirishima*.

That the value of radar had become common knowledge to American seamen is attested in a bit of doggerel that circulated after the Battle of Cape Esperance:

> Yes, we're heading for hell in column,
> Scott is as proud as can be.
> Only one thing he is lacking,
> A brand new, working, SG! [24].

The SG was unquestionably the best sea-surface radar of the time but it was not without faults, some of which negated its usefulness. The PPI indicator was mounted vertically as part of the control panel. With time a PPI repeater would become one of the vital instruments on the bridge, but such was not the case in 1942, and a crowded radar room was hardly the place for a commander. The intelligence gained by the SG was transmitted by telephone to the bridge whence it went by the single voice-radio communications channel (Talk Between Ships) to the two admirals. Under the best conditions this was a terrible way to transmit the complicated, rapidly varying information that appeared on the SG screens, something overlooked by the radar poet just cited.

The battle quickly became one of complete confusion—'a barroom brawl after the lights had been shot out'. Fire direction was generally optical because of ranges so close that machine-gun fire was exchanged. There were a lot of star shells, searchlights, fires and gun flashes [25] to aim with and at. Once this close-range battle began it is questionable whether central command could have been exercised regardless of the use of SG. Whether the battle might have developed more satisfactorily had Callaghan watched the PPI in the approach phases is clearly beyond the ability of later judgement to decide.

Callaghan and *Scott* died within minutes of the crash; both were awarded posthumous Medals of Honor. Henderson Field and the transports were saved from bombardment at the cost of two cruisers and four destroyers, but the battleship *Hiei* was so crippled from the many 8 inch shells fired at close range that she could not escape the air attacks that began the next morning. It took direct hits by four 1000 and one 500 lb bombs and 11 torpedoes (some of which actually exploded) to cause her to sink the following night, abandoned and unobserved [26]; she was a tribute to her 1910 builders and designer, Sir George Thurston [27]; she was also Japan's first battleship loss and as such was the first of two terrific psychological shocks to the Imperial Japanese Navy. Two of her accompanying destroyers ended with the accumulating collection of wrecks in Ironbottom Sound.

During the night of 13/14 November cruisers, destroyers and destroyer transports of the Tokyo Express landed reinforcements and bombarded Henderson Field with 8 inch shells, but most of a transport convoy was blown apart by aircraft next morning. Halsey did not want to submit his one remaining carrier, *Enterprise*, although he had her air group operate out of Henderson Field and ordered his two battleships to stop the next attempt at bombardment. Naval gunfire was very much more damaging both to material and to the spirit than any of the air attacks Guadalcanal suffered during that campaign.

Thus in the late hours of the 14th began round two. A battleship, four cruisers, 18 destroyers and four transports under Vice Admiral Nobutake Kondo and Tanaka came down the Slot to be met by two battleships and four destroyers under Rear Admiral Wilis A Lee. Lee had a flagship with an SG and an understanding of what it could do. None of his destroyers had an SG and only one had an FD fire direction set. These ships had not worked together before; the destroyers were selected at the last minute according to how full their fuel tanks were.

Lee kept control throughout and sank a destroyer and the battleship *Kirishima*, which ended in defiance of all hydrodynamic knowledge bottom up in the mud [28]. The transports were run aground to discharge the troops, but few escaped the morning air attack. Lee emerged with his flagship, USS *Washington*, unscathed and at the end fighting alone but with the loss of three destroyers and quite a bit of damage to the other battleship, *South Dakota*. He attributed his success to radar in turning back this large force and protecting the troops on Guadalcanal from another version of 'The Bombardment' [29].

5.5.6. *The Battle of Tassafaronga*

This was the last of the major naval actions that determined the fate of Guadalcanal and took place during the night of 30 November/1 December 1942. It was fought primarily with long-range torpedoes, which almost guaranteed defeat for the US Navy.

The difficulty of landing food and ammunition for the Japanese garrison, which by this time had been reduced to a condition of near starvation, had brought the Imperial Navy to loading provisions in steel drums, transporting them as deck cargo aboard destroyers, and pushing them into off-shore waters to be retrieved with small boats by troops from land. It was a desperate method of supply, but the Americans were determined to cut it off and put an end to the terrible bombardments and jungle fighting.

American supplies to Guadalcanal came from Nouméa and Espiritu Santo over a stretch of ocean beyond the range of bombers from Rabaul, and unloading was hampered primarily by relatively infrequent naval bombardments and air attacks. Japanese supplies to the island had to follow paths that were within range of Henderson Field, resulting in slow convoys being attacked in daylight, which forced the use of destroyers capable of moving supplies much faster at night. The result was a serious loss of ships, whether the slow or fast option was taken. This lack of symmetry caused Japan to lose despite winning most of the naval engagements. Whoever held Henderson Field controlled supply. The object of the battle was also the key.

Rear Admiral Thomas Kinkaid, Halsey's immediate subordinate at the Battle of Santa Cruz, took command of the cruiser squadron being assembled to counter the next Tokyo Express and quickly formulated a plan that incorporated lessons learned in the water about Savo Island with special emphasis on the use of radar. Two days after he had assumed command he was ordered by Admiral King to duty in the north Pacific, and the cruisers were given to Rear Admiral Carlton H Wright, who the next night led five cruisers and six destroyers to meet Tanaka with eight destroyers.

Wright's flagship, USS *Minneapolis*, had SG radar and led the column of cruisers with four destroyers ahead but none far enough to serve as pickets. At 2306 the flagship's radar picked up the first evidence of intruders, who were seen by the radars of the van destroyers shortly thereafter. Radarless Japanese observers made out the presence of the Americans six minutes later. Both sides launched torpedo attacks at long range. Wright's force delayed a few minutes and missed the optimum position; it also had inferior torpedoes. Tanaka launched at optimum position and had superior torpedoes. Wright lost the *Northampton* and had three other cruisers sent to repair yards for many months. Tanaka lost a destroyer to cruiser gunfire. It was another Japanese naval victory that went unnoticed by their troops on Guadalcanal for it gained them no vital supplies.

Ground action pressed the Japanese ever harder which, combined with the inability to supply these men, led to a decision on 25 December to evacuate, completed 8 February. It was a decision to fall back, not abandon the whole region, and they began construction of an air base further up the Slot—about a third of the way from Guadalcanal to Rabaul—at Munda on the Island of New Georgia. It was on the return from a bombardment of this construction during the night of 4/5 February 1943 that the cruiser *Helena* shot down a dive bomber using an FD and proximity fuzes. In 1943 there were nine more night battles in the Solomons, one of which claimed the *Helena*, but the story of Guadalcanal makes a closed chapter, one rich with lessons about radar.

Radar's contribution to this campaign is mixed. Taken all together this was one of the great naval actions of all time, but radar was seldom decisive, although it certainly contributed to the American victories at the Battle of Cape Esperance, the Battle of the Eastern Solomons and the second part of the Naval Battle of Guadalcanal. Its worst failure was at Tassafaronga in which the Americans had a strong force with the best surface-search and fire-direction radars; the Japanese had a weaker force with no radar.

The naval radar failures can generally be attributed to the ever changing upper levels of command not understanding that this new technique had to be studied and exercised. The aviation commanders had begun designing their operations around radar as soon as they had encountered it and had employed embryo combat information centers to direct their first engagements. The surface commanders did nothing comparable, which was understandable so long as they were constrained by the deficiencies of meter-wave equipment for locating ships, but they showed inattention when the extreme advantages of the microwave equipment became known, advantages strangled by the means through which the information had to reach commanders. As the problems of fighter direction illustrated all too clearly, one had to practice seeing with the new eyes.

But the same can be said about learning to fight at night. Every engagement had had a different American commander, frequently scarcely able to orient himself before action began. Retrospective criticism is cheap and often unjust, but one wonders why a single commander was not chosen, at least for the waters around Guadalcanal, and allowed to develop his skills at night fighting with the excellent equipment the inventors had rushed to completion. As the verses composed after the Battle of Cape Esperance witness, the power of radar was understood in the lower decks.

Failure of command was not restricted to the American side. The Imperial Navy was skillful but timid. They allowed the Americans to extract strategic victories from tactical defeats. Fear of seeing their ships sink prevented them from disposing of the beachhead immediately after the Battle of Savo Island, and this pattern repeated itself. For whatever faults the American commanders had, they protected the troops on the island, cost what it might. The Navy's casualties were three times those of the men on Guadalcanal.

There is instruction to be gained by comparing the use of radar in naval surface action by the Americans in the Solomon Islands in 1942 and the Germans in the Atlantic the two years before. In every one of the five night actions fought around Savo Island the Americans had the benefit of the superb 10 cm surface radar, SG, as well as the FC and FD gunnery sets. One may disregard the 1.5 m SCs as being unsuitable for surface work in such tight quarters, and there is little evidence of their utility in this function. The German surface raiders had only the 80 cm Seetakt, but it proved to be a good surface-search and ranging set, although not in the same class as the SG. It was also never put to the test of multiple targets in confined waters. The American use of this new technique was on the average poor; the German use was excellent.

Modest reflection explains the difference and provides a lesson in how to introduce radically new weapons, if time allows. The upper command levels of the Kriegsmarine were much less interested in radar than the American. Furthermore, they introduced the new equipment with much secrecy and little technical support aboard ship. The German officers were given little or no training before embarking but had ample time during their month-long cruises to encounter and learn about radar. Duty aboard a surface raider was perhaps the most relaxed of any kind of wartime service. There was time for examining and trying out Seetakt. A few times in which an enemy cruiser was effortlessly avoided at night was enough to make true believers of the most technically recalcitrant. By the time of the *Bismarck* actions in May Admiral Lütjens had come to depend on radar superiority—real or imagined—to an extreme, perhaps crippling degree.

By comparison the American commanders at ship or squadron level in the Solomons were not given the luxury of studying this new equipment in the detail that builds understanding and confidence. Each day brought a new crisis to which they had to react. When not in action, they were deeply involved in planning and in the myriad details of a rapidly changing tactical situation. They were briefed about the new radar sets, probably even shown them, but this is no comparison to devoting days to learning what the equipment could do. One may object that the comparison between German and American surface actions is an unfair one because the battle conditions were so different and that it would be better to compare German surface action with the American carrier raids of early 1942. Here the fighter controllers had been allowed to master the new methods in much the same way as the Germans and with an equally professional result. The Solomon Islands were the US Navy's surface-action radar school. The teacher was strict, the radar conditions the worst, but the lessons were learned.

The story of the ground radar on Guadalcanal itself is not so checkered. Without it the Cactus Air Force would have simply been worn down. The reason why ground radar functioned well and the Navy surface radar did not is obvious at this stage: the Cactus radar men worked their job continuously and became proficient. The significance is stated without qualification by Richard Frank: 'Without these warnings, Henderson Field, and ultimately Guadalcanal, could not have been defended' [30].

In 1942 there were two threats to the Australian line of communication with the United States, and the descriptions of the fighting in the Solomon Islands tend to obscure the one farther west. Where the struggle for the control of Guadalcanal brought on air and naval actions that excite one's sense of drama and used the latest methods of technical warfare, the defense of Port Moresby and the eventual ejection of the Japanese ground forces from Papua offer the reader the grim story of emaciated, rag-clad soldiers of both sides disputing a pass through the Owen Stanley range, but it ended just as effectively the threat to Australia and was the beginning of the Australian–American push north under General Douglas MacArthur and Major General Robert L Eichelberger. After occupation of Port Moresby from the sea had been thwarted by the Battle of the Coral Sea the Japanese elected to push across the mountains from Buna and to land at Milne Bay, the eastern-most part of Papua.

In preparation for invasion an Australian radar unit set up a CHL at Milne Bay. The size and weight of this equipment, completely out of place in this kind of warfare, drew heavily on the ingenuity of 37 Radio Station, which was furnished with irregular transport to say the least. The jungle was deleterious to man and English electronics, but the station was on the air by 8 August and greatly assisted in fending off air attacks preceding and during the invasion attempt of 25 August to 5 September 1942. In this they were aided by Japanese pilots uninstructed in radar who flew perfect paths for detection [31].

The Papua offensive began about the same time as the American invasion of Guadalcanal. It was contained by the Allies by the middle of September, and the Japanese had been forced back to the beachhead at Buna by late November [32].

Radar played a small part in all this compared with its use in the

Solomons. It did not yet count as equipment useful to the infantry in its inventory, and the naval action was one of supply conducted with fleets of trawlers and landing barges, whose attack and defense were neither the stuff to delight naval historians nor to make use of the radar of the time. But the Milne Bay example was continuously repeated—although not with CHL—on all the beachheads of the South Pacific. No landing was considered secure until the LW/AW was on the air.

In 1944 the American-made SCR-602, a copy of the British 1.5 m lightweight set, began arriving in the Pacific Theater. Later it was followed by the highly admired SCR-602 type 8. This 50 cm set made use of a high-frequency triode, VT-158, designed by Harold Zahl at the Signal Corps Laboratory and manufactured by Eitel-McCullough, the source of the fabled Eimac tubes [33]. It incorporated lines of thought followed independently by the makers of the Bell Labs door knob, the British micropup and the German LS180, TS6 and RD12Tf.

As American ground troops and air units began to assemble in northern Queensland in preparation for the coming push north, the Army Air Forces set up five SCR-270s in the vicinity of Townsville to reinforce the Australians, who had by July deployed ten of their Air Warning (AW) sets and eight of the converted SCR-268s (MAWDs); by the end of the year they had added another 27 pieces, two of them special rigs set up at Milne Bay for sea search made out of ASV mark II equipment for which the originally intended aircraft had been destroyed [34]. This was the beginning of a growing Australian–American radar effort that was to see a mixing of personnel and equipment as the Allies fought their way across New Guinea, the Halmaheras and Borneo.

5.6. THE EASTERN FRONT

The paths of civilization are marked by milestones set in place by wars. By the strict meaning of the word these wars are countless, having varied in size from clan feuds to strifes that envelop entire continents. They have been fought in every portion of the globe to which man has had access except the polar regions. They have invariably seen the use of all the technical aids to killing and destruction that the levels of culture the warring groups had attained. They have ranged from gentlemanly encounters defined by strict ritual in which prisoners were treated as guests to clashes of mutual extermination in which enemy individuals counted as nothing. The Second World War offers extremes of every one of these manifestations, but the Eastern Front had only the loathsome ones—in magnitude of suffering and bestiality to the helpless. Its nature was foreordained. Nazi Germany and Communist Russia were both ruled by criminal bands who condoned, in fact relished the murder of opponents as acceptable elements of their political systems. Their clash in June 1941 secured for the 20th century the distinction of being history's most bloody century, bloodier perhaps than all previous centuries combined.

Many people of eastern Europe, wearied by interminable suffering in war, revolution, civil war, collectivization and purges, and divided by hatreds extending beyond memory, had recollections of the disciplined armies of the Kaiser and looked on the Germans as liberators and welcomed them. With the arrival of the Wehrmacht their hopes remained alive, but with the passing of the first wave came the SS and SD, and the true face of the enemy was quickly revealed. The struggle soon became one of unfailing cruelty. Russian prisoners were so maltreated that even the Poles pitied them, and those few who returned to their homeland were received as traitors and sent to the Gulag. German prisoners received all the harshness of which Siberia was capable, and those who returned to their homeland years after the end of hostilities often came as broken men.

One horror was spared the people of the East—there was no significant strategic bombing.

When the Luftwaffe opened the Great Patriotic War by destroying a substantial fraction of Stalin's air force, it was unhindered by Soviet radar. There was only one kind of radar in use when the conflict began: RUS-2, the pulse-type air-warning equipment working on 4 m. That statement discounts completely the few units of the radio screen, RUS-1, the production version of Oshchepkov's Rapid of 1934, for which the war found no use; yet despite its complete failure during the Finnish War of 1939–1940, 13 RUS-1 sets were manufactured in 1941. RUS-2 proved of value as an air-warning set but suffered from the need to have transmitter and receiver separated by about a kilometer, the antennas of which had to move synchronously. There were six sets in existence when war broke out, but they had no effect on events. The Scientific Research Institute of the Radio Industry (SRI) had devised how to use a common antenna before the war began and had incorporated it into the modification, RUS-2S, but production had not yet begun [1].

The radar groups in Leningrad (LFTI, NII-9) and Kharkov (UFTI) soon found their principal problem was evacuation to the east, both of development laboratories and production plants, a process that removed five months of any useful activity. The death in March 1940 of Professor M A Bonch-Bruyevich, who had taken over the leadership of NII-9 after the purges added to the turmoil with which that group had had to deal [2]. The production of only 53 RUS-2S sets during 1942 tells the story more eloquently with numbers than is possible with words [3].

The Soviet dismissal of radar at the beginning of the war was not reflected in their other attitudes concerning AA defense. Large cities had hundreds of guns, although their accuracy was poor [4]; the fighter squadrons were based at all-weather fields, much superior to the usual Soviet bases. Moscow was the best defended city in the world and, despite its proximity to the ground fighting, did not suffer serious damage from bombing. Besieged Leningrad suffered in every possible way, but it too put up a very strong air defense [5]. Not surprisingly, the first effective Soviet use of radar was in augmenting the defenses of Moscow and Leningrad.

An experimental station at Toksovo near Leningrad, used before the war by the Physico-technical Institute (LIPT), assumed immediate tactical functions and was manned by members of its technical staff. Its equipment was RUS-2 but with more power for greater range. Transmitter and receiver were mounted on separate 20 m steel towers; antenna movement allowed a 270° sector of observation. Operation was turned over to military personnel once they had been trained.

The Research Institute of the Red Army (NIIIS KA), which had overall responsibility for radar, built in the first months of the war a large station for the Moscow air defense, which also used the RUS-2 principle. Specifications differ enough from those of other air-warning sets to be of interest: pulse duration 50–60 μs, which allowed a receiver pass-band of only 40 kHz and a repetition rate 50 Hz. It mimicked CH in more than pulse rate, for it too used special demountable vacuum-pumped transmitter tubes; they were designated type IG-8 and made by the Svetlana tube plant [6].

Leningrad received numerous air attacks, generally by formations of about 100 aircraft. In 1942 there were 38 such bombings, all of which were stoutly resisted. Radar's performance opened the eyes of theretofore uninterested military leaders, as 20 000 targets were picked up that year. Attacking squadrons showed up on oscilloscope screens in plenty of time to alarm the city and scramble fighters [7]. On a small scale the air defense of Leningrad was similar to the Battle of Britain, and by the end of 1942 the radar men did not have to beg for attention even though they had to beg for production. RUS-2 and RUS-2S gained reputations as simple, reliable pieces of equipment—if only they could have given height information.

Finland made common cause with Germany and attacked the Soviet Union in 1941 to regain the territory lost the year before in a defense that had amazed and thrilled the world and did succeed in keeping Stalin's armies out of most of the land. Their capital, Helsinki, and the nearby city of Kotka came under air attack during both wars. As the Russian air force began to recover from its disastrous 1941 defeat it renewed raids on Finland, causing Germany to furnish the Finns with two Freyas and four Würzburgs in the spring of 1943. Helsinki had four batteries of 88 mm guns with schoolboy volunteers operating the radar. The data from the Freyas were connected by telemetry to a control room that directed fighters and alerted the gunners. As in 1939 and 1940, defense was stout and attacks indolent. The fliers generally elected to drop their bombs on locations other than the targets, the ocean being found most convenient. Helsinki suffered less than any city that was subjected to major air attacks [8].

The air defense of Helsinki had an unusual element. A German freighter, *Togo*, had been made into a night-fighter-control ship through an uncommon cooperation between the Luftwaffe and the Kriegsmarine. Equipped with a Freya, a Würzburg Giant and a Y-Gerät she had a capability similar to one of Kammhuber's Himmelbett cells[1]. During the months of March, April and May of 1944 she was stationed at Tallinn (Estonian SSR) and would take position with the onset of darkness in the path taken by Soviet bombers headed for the Finnish capital. When the Finns began serious negotiations with Stalin in June 1944 to leave the war, the vessel was withdrawn to Liepaja (Latvian SSR) and provided similar protection for German capital ships [9].

The poor showing of early Soviet gun-laying radars did not eliminate this type from the minds of designers, and NII-9 organized an experimental battalion-sized AA unit in October 1941, employed in the defense of Moscow while trying out its new equipment. Initially the battalion had four 75 mm, six 105 mm (German guns obtained during the time of the non-aggression pact) and six 37 mm automatic guns. A team of engineers headed by M L Sliozberg worked directly with the unit. They introduced some experimental sets, Sleep, B-2 and B-3 that worked on 15 cm using cavity magnetrons [10]. The possession of the cavity magnetron, viewed in Britain and America as the ultimate microwave transmitter and the basis for uncounted radar successes in the coming years, seemed to hold no advantages for the Soviets. They were unable to produce a transmitter or local oscillator with sufficiently stable frequency to allow the construction of a heterodyne receiver, which was presumably attempted without a crystal-diode mixer. These gun-laying sets were failures and soon disappeared from the experimental battalion's gun positions.

The arrival of British GL mark IIs produced much more interest than the experimental microwave sets [11]. It was not much of a gun-laying set, to be sure, but it was a robust, reliable and well engineered piece that found use for searchlights, fighter direction and even air warning. The British technicians who had been sent to instruct the Russians, and who had been led to believe that radar was unknown to the recipients, encountered personnel who mastered the equipment rapidly despite a significant language barrier [12]. Sliozberg's people soon made a copy of it, called SON-2 [13]. It proved the favorite Soviet radar, but British imports of GL mark II (generally called SON-2) overwhelmed native production, which produced only 124 during the entire war [14].

Later Britain sent 44 microwave GL mark IIIs and America sent 25 SCR-268s, 15 SCR-545s and 49 of the superb SCR-584. A copy of GL mark III appeared as Neptune, and the 584 was copied after the war as SON-4 [15].

In the summer and fall of 1941 Stalin's gigantic army and air force suffered a defeat coupled with losses of men and material of magnitude unparalleled in history, but as Hitler's forces stood before Moscow, everything changed in an almost miraculous manner: (1) Japan was suddenly found to be completely occupied with America and Britain, thereby freeing many fresh Siberian divisions to board trains headed west; (2) the Russian

---

[1] See Chapter 6.1 (pp 283–4).

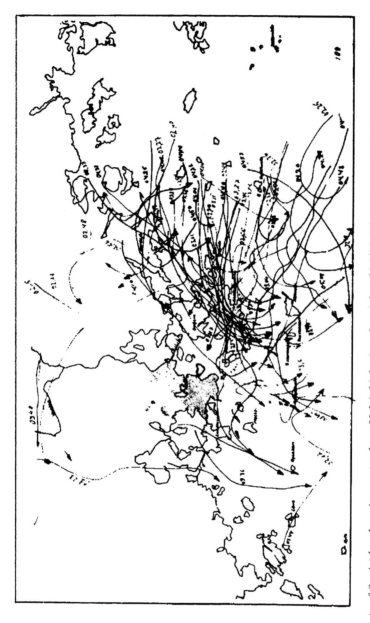

Radar plots of Soviet bombers in an attack on Helsinki during the night of 26/27 February 1944. The large number of plots came about because the Soviets did not attack in formations but in a long series of sorties, allowing the paths of individual aircraft to be determined. The city is shown by the shaded area at the left center. As can be seen, the attack was not pressed home and few bombs fell on the city. The scale can be ascertained by noting that the bottom edge is approximately 35 km. Archives of the Finnish Anti-Aircraft Museum.

people, who may have been originally indifferent to the downfall of the communist state, had come to realize that the war was against them, not just Stalin; (3) Hitler had also conveniently declared war on the United States, which was to prove a serious distraction for the Nazi state; (4) a Wehrmacht without winter clothing or equipment had been assailed by a winter as deadly as the enemy. In January 1942 Germany found herself in total war, a discomfort theretofore left to her adversaries. Only then did German total mobilization begin.

During the intoxicating summer of 1941 radar had been, if anything, even less important to the Germans than the Russians. The new weapon, so important in the west, was ignored in the east. The Luftwaffe dominated the air and found little need for equipment in short supply and required for the defense of the Reich against Bomber Command. There had been use of Freya sets before the surprise attack of 22 June to insure that no Soviet observation planes discovered the large assembly of forces [16], but few of the clumsy Freyas followed the Blitzkrieg.

The Soviet air force had to make its recovery in the face of German air superiority, but its slow progress called for correspondingly increased vigilance by the Luftwaffe. A measure of Russian progress can be found in the extent of German radar deployment. As Leningrad became besieged, the air struggle there became more advanced, and Luftwaffe Signals set up Freyas on the islands of Hiiumaa and Saaremann, located to the west of Estonia, to protect German shipping from air raids [17]. A Freya unit covering the south approaches of Leningrad found movement of the set in the terrible winter retreat of January 1942 so difficult that it had to be destroyed [18].

In 1941 radar was closely associated with strategic bombing, and its use with and against tactical ground forces lay a few years in the future. The steady growth of Soviet power came from factories beyond the range of German bombers, and the railway network continued to distribute supplies and troops, hindered but not brought to collapse by air raids. The four-engine bomber that General Walter Wever had favored to attack these resources was absent and not going to appear.

An early German use of radar came from an unexpected quarter—partisan warfare. When it became clear that Germany was the enemy of all those Soviet peoples that did not have some ethnic status that made them acceptable to the Nazis, partisan groups began to make no small amount of trouble behind the German lines. Made up of soldiers cut off but not taken prisoner and civilians escaping and fighting SS terror, these groups were organized and maintained by the Soviet command. Night flying served as the means of bringing vital supplies and officers to these units and the carrying out of information and wounded. Soon an elaborate air transport was established at night. Combating these infiltrating flights proved difficult, in great part because of the primitive Russian equipment used. The most important aircraft was a biplane, paradoxically designated the U-2. Flying slow and low and necessarily observing strict radio silence it was difficult to detect. Radar was obviously called for, and it came as railway radar trains, but an effective counter to the U-2 was never found [19].

Russia's notoriously muddy roads made movement by rail essential for heavy equipment—a Freya required 28 horses for movement by typical road [20]—and radar trains were the obvious answer, first placed in service in October 1942. They were portable fighter control units that consisted of a Freya for early warning and two Würzburg giants, one to track the enemy and the other to track the interceptor so the controller could bring the two together [21]. Some trains made good use of searchlights. It was the system called Himmelbett in the west[2]. As the air situation deteriorated for the Luftwaffe, the radar trains became more numerous and more important. In 1943 a radar train in the Orel-Bryansk sector took credit for bringing down about 30 planes [22].

(The first radar trains may have been placed in service somewhat earlier in France during the summer of 1942. By that time the activities of the underground were beginning to be troublesome, and light aircraft transported agents and supplies between the continent and England. Finding the resistance personnel was more important than bringing down the airplanes, so railway-mounted equipment that could be moved to suspected places of operation in order to observe where they landed was an obvious answer. It is reported to have led to several arrests [23].)

Growing Soviet air power began forcing the Luftwaffe to bomb at night, and their efforts had grown to such an extent that the Soviets began organizing night fighter units in late 1943. These units were not particularly effective because they lacked both airborne and ground radar capable of bringing about interception [24].

The absence of strategic bombing in the east meant there was no centralized air defense, so radar use on both sides tended to take on local character and ingenuity. A German bomber group at Shitomir (near Kiev) used two Freyas for night bombing Russian concentrations at locations beyond artillery range. One Freya directed a bomber by radio so as to follow an arc of constant radius while the second controlled the release of bombs. The attacks were not only complete surprises but remarkably accurate [25]. The reader will encounter a similar but more elaborate method of blind bombing, called Oboe, used against Germany in a later chapter[3].

By the time of the great tank–air battle at Kursk during 5–11 July 1943 the Soviet air force was something that had to be dealt with, and the Germans assigned five of the nine then existent radar trains to the sector. The Wehrmacht lost decisively. The wreckage of hundreds of aircraft and tanks littered the field, but one Freya was credited with saving Fliegerkorps VIII from complete destruction [26].

Any Soviet use of radar at Kursk has escaped mention in the sources

[2] See Chapter 6.1 (pp 283–4).
[3] See Chapter 6.3 (pp 301–3).

available. Indeed, Soviet use of radar in general was hardly noticed by the Luftwaffe until 1944 and never reached the stage where countermeasures were employed. They seem to have thought all of the enemy's radar was of British or American manufacture and been unaware that any of the Russian equipment was of indigenous manufacture [27].

German radar found ever wider use on the Eastern Front as ever more equipment became available and the pressure of Soviet air power increased—and not just Soviet. The oil fields of Rumania received a generous allotment of AA and fighter units and with them came Freyas and Würzburgs for Flak and fighter control. Their effectiveness is attested by the heavy losses of the American bombers that attacked Ploesti. The saving or at least preventing the capture of the extensive radar deployment in Rumania became a matter of serious concern when Russian forces secured that nation in August and September 1944 [28].

Such was radar in the east. Compared with the use in the west and at sea it was small indeed, being a mere perturbation on the cataclysmic battles that were fought there. Germany's deployment was, until near the end, trivial when measured against the air defense system facing the Allies. Russia used it first only in defense of her two largest cities, to what effect it is difficult to say. In the east huge ground forces struggled with air power restricted to army support. It was not until the appearance of remarkably accurate 10 cm equipment, such as SCR-584, that radar showed real value for this kind of warfare. In the hands of ingenious officers the equipment could be of benefit, especially to local fighter squadrons, but these contributions were never decisive. The actions in the deserts of North Africa are apt illustrations of this. Given this tactical background it is difficult to fault the Soviet command for not giving radar a greater priority. Were it not for their demonstrated capacity for confused and self-destructive administration, one might be tempted to attribute wisdom to the Soviet leaders for the low priority given radar. But whether from wisdom or folly, there is little reason to fault the result. The critical industrial strengths required for the manufacture and operation of radar were put to better purpose in communication equipment vital to mobile ground warfare.

The quality of Soviet radar development before and during the war must be evaluated by what was accomplished against what was attempted. Here is a bewildering confusion of competence at its highest and lowest. Soviet engineers invented the cavity magnetron, a device for which praise in Britain and America exceeds that for any comparable device. That not being enough they invented the klystron independently of the Varians and Hansen. But their attempts at putting them to use failed, owing to an inability to master the lesser arts of microwaves, and resulted in an especially bad gun-laying set that was never produced. The klystron does not seem to have entered a serious Soviet radar design. In meter-wave equipment the advantage of an early entrance was lost. Postwar design started from Allied and captured German sets.

PHOTOGRAPHS: NAVAL RADAR

German Seetakt (FuMG 40) surface-search radar captured at Toulon during the invasion of southern France where it was used with a coast-artillery battery. When used in this service it was referred to by British intelligence as 'Coast Watcher'. The same basic 80 cm set, known originally as DeTe-I, was used by the Kriegsmarine on ship and on shore. The pocket battleship Admiral Graf Spee received a Seetakt (60 cm version) in January 1938, the first warship to have a tactical radar. The Kriegsmarine rejected lobe switching that the manufacturer, GEMA, offered and thereby lost the capability of blind fire. National Archives photograph 111-SC 246248.

US Navy FC, later mark 3, fire-direction radar mounted on the forward main battery director of USS New Mexico in December 1941. Optical range finders can be seen ahead and behind the FC antenna. The mark 33 director could use data from either source. Its blind-fire capability gave a severe shock to the Japanese Navy at the Battle of Cape Esperance in October 1942. US Naval Historical Center photograph NH 84811.

US Navy FD, later mark 4, 40 cm fire-direction radar mounted on a mark 37 director of the destroyer USS Nicholas in January 1944. This set was developed by Bell Laboratories from a prototype demonstrated in July 1939 as the CXAS and used horizontal and vertical lobe switching that allowed it to direct 5 inch dual-purpose guns at surface or air targets; it was a three-dimensional version of the FC. Originally designed for the Bell Labs high-frequency triode 316A, its modular construction allowed the transmitter unit to be replaced by a 40 cm cavity magnetron and as such became the first American production unit to use the British invention. The similarity of the antenna to Hertz's original is striking. It was the standard fire-direction radar for most of the Navy throughout the war. US Naval Historical Center photograph NH 84804.

Radar of the carrier USS Yorktown on 5 April 1945. This replacement for the earlier Yorktown was commissioned 15 April 1943 and carried a full complement of the latest radar. At the left and right are mark 12 fire-direction radars with IFF antennas in their centers and mark 22 height finders at their right. The mark 12 was an improved version of the FD; the mark 22 was an elevation-only 3 cm set used to thwart the attacks of very-low-flying aircraft. Second from the right and insignificant in appearance but mighty in operation is the SG, the most valuable aid to mariners since the invention of the chronometer. Next is the flying bedspring of the SK, the improved version of the 1.5 m air-warning XAF and CXAM surmounted by its IFF antenna. Next and in the background is an SC, used as a back-up set for the SK. One step farther to the left finds in close proximity another SG and a YE aircraft homing system; the latter was not a radar but a very-high-frequency directed beam with a signal that gave the direction of propagation, thereby allowing fliers to set a homeward-bound course. Finally just abaft the forward mark 12 is an SM, a 10 cm radar with a parabolic reflector that is mounted on a gyroscopically stabilized platform; its function was fighter control close to the ship where the poor resolution and lack of height data made the SK nearly useless. National Archives photograph 80-G-376152.

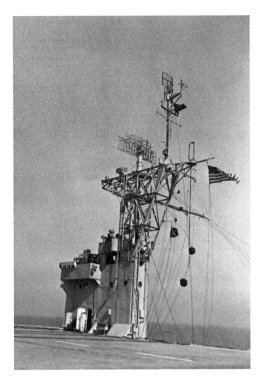

The radar on an escort carrier. These ships were intended primarily for operations against submarines and air support during landings and, as a consequence of the restricted space, carried a smaller amount of radar. The radar-like antenna at the top is the YE homing system, a very-high-frequency directed beam with a signal that gave the direction of propagation, thereby allowing fliers to set a homeward course; below is the 1.5 m SC air warning and the 10 cm SG surface search radar. National Archives photograph 80-G 214980.

American microwave phased-array radar, the Bell Labs FH (later mark 8) fire-direction set. Early use of the FC showed that the 40 cm wavelength and the dependence on an A-scope display resulted in confusion in a complicated surface action. The 10 cm mark 8 with its improved resolution and maplike display allowed multiple targets to be kept in a ±15° sector view without confusing the direction of fire on the one selected as target. US Naval Historical Center photograph NH 84813.

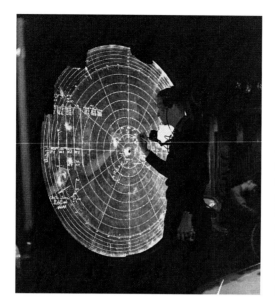

Fighter control with radar: an American Essex-class carrier. The Combat Information Centers of large carriers had elaborate means of displaying information. Here is a plastic plotting screen with a seaman plotting from behind. National Archives photograph 80-G-326751.

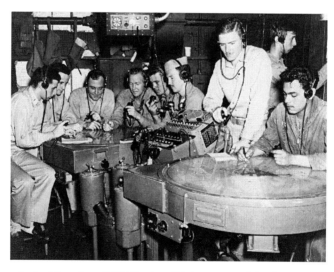

Fighter control with radar: USS Santee. Fighter control in the US Navy evolved the Combat Information Center (CIC) that grew into large specialized rooms that became the nerve center of the vessel in combat. The CIC shown is on an escort carrier in which space is at a premium, as these ships were intended primarily for operations against submarines and air support during landings. The plastic-covered table at the right serves as the plotting board from which the fighter directors extract information to vector onto the enemy. The carrier depended on the SC radar for data. National Archives photograph 80-G-342577.

Front view of Royal Navy type 271, 10 cm surface-search radar. This was the first operational microwave radar. Sea trials of a production model were made on 25 March 1941 aboard the Flower-class corvette Orchis. Transmitter and receiver had separate identical antennas, the transmitter at the top. A dipole backed by a rod reflector is located at the focus of a cylindrical parabolic mirror. The top and bottom plates earned this style the name of 'cheese'. Aboard ship this antenna was enclosed in a plastic cylinder for protection. By September 32 corvettes mounted type 271s. Churchill Archives Centre, Royal Navy photograph. Crown Copyright.

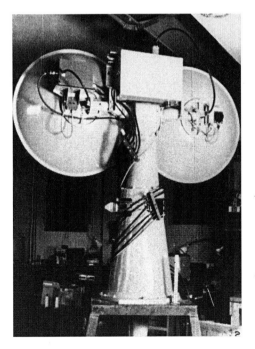

Rear view of Royal Navy type 273, 10 cm surface-search radar. This was the microwave radar intended for capital ships. It differed from the type 271 in using 90 cm paraboloid reflectors that provided higher antenna gain than the cheeses. The pencil beam required a stable vertical axis, the first radar so mounted. Sea trials were made on the cruiser Nigeria in August 1941. Churchill Archives Centre, Royal Navy photograph. Crown Copyright.

275

Fighter control with radar: fleet carrier HMS Venerable. The Aircraft Direction Room of British ships differed from the Combat Information Centers of US carriers in little other than name. Here a projection system, called the Skiatron, allowed control officers to survey the air situation. HMS Dryad archives, Royal Navy photograph. Crown Copyright.

Japanese 1.5 m air-warning radar mounted on the aircraft carrier Junyo. This mark 2 model 1 entered service in time for Midway. Had a carrier had it there instead of a battleship, which was far from the action, the battle might well have ended differently. National Archives photograph 80-G 264924.

Japanese 10 cm surface-search radar. This mark 2 model 2 also entered service in time to be used at Midway, but only on a battleship. It was used effectively throughout the Pacific War and gave Japanese submariners a weapon denied to their German allies. Japanese microwave equipment always used horns and circular waveguides. National Archives photograph 111- SC 290054.

Japanese submarine RO-58 with the mark 2 model 2 radar easily recognized on the conning tower. This 10 cm radar was widely used by Japanese submarines. It lacked a PPI display and could not be used with the vessel submerged, as could American boats, but was a technique that German U-boats lacked and could have used effectively. Photographed at Yokosuka Naval Base, 7 September 1945. National Archives photograph 80-G 339842.

Japanese Navy 10 cm surface fire-direction radar, mark 3 model 2. This set used lobe switching by a receiver having two horns below the single transmitter horn on top. By the time it had been developed the Japanese Navy was no longer capable of fighting surface actions. The set shown here is mounted for coast artillery but was never used for that either. National Archives photograph 111-SC 290052.

CHAPTER 6

THE GREAT RADAR WAR

6.1. THE DESTRUCTION OF GERMAN CITIES INITIATED

When Hitler turned his magnificent war machine toward the Soviet Union in June 1941 England found the air attacks they had borne for a year decreased to a much smaller intensity. But the war was expanded, not over, and Churchill, ancient enemy of communism, embraced his new ally with startling alacrity. Moreover Britain was wounded, not dead, and the nature of her race was to strike back at Germany with all she had, and Bomber Command was the only arm capable of hitting directly. The doctrines of the Royal Air Force during the between-war years had seen in bombing squadrons the vehicles for future victory, but the first 20 months of combat had left these doctrines devoid of reality. Somehow everything had got turned upside down. No 'knock-out blow' had been placed by either side, and the mission of army support, so decried before, was the Air Force's only effective activity against the enemy. Victory through air power was proving very elusive.

That the object of Britain's air war against Germany became the destruction of her cities did not come about as the result of a carefully thought out plan; it came about in the same way that the Battle of Britain and later the Blitz had been forced on the Luftwaffe. Both came as the result of an unforeseen military and technical situation. Without long-range fighters—and the delay in bringing them into the picture is its own story—massive daylight attacks could not be sustained and the precision of aim that had been assumed would prevail during the day was never reached. Without navigational techniques that permitted blind bombing with accuracies comparable to the size of a factory only area bombing could be done at night. Translated into practical terms this meant that the targets would have to be of city size. By mid-1941 this simple fact had put an end to the initial strategy of selectively destroying Germany's synthetic oil production [1]. Thus in phase one of the Great Radar War defensive radar had eliminated by its mere existence the possible selective destruction of German industrial targets, leaving cities the only thing the Air Force could hit—and all too often the bombers even missed them. Strategic doctrine was made to conform with reality in a directive of 14 February 1942, made in consultation with the Ministry of Economic Warfare, that designated the primary objective as 'the morale of the enemy civil population and in particular, of the industrial worker' [2].

While Bomber Command was trying to learn what it was to do and how it was to do it, the Luftwaffe began evaluating its unexpected role in defense. By the middle of 1940 it was obvious to them that Bomber Command was going to attack at night and that the raids were already showing the defense to be ill prepared. It was all well and good to say that the British were not capable of doing any serious military damage; when bombs fell on cities action was required, irrespective of the military significance. Night air defense had three parts: early warning, fighter direction and Flak (AA). The engineers built a new generation of air warning radar, the fliers organized to make best use of it, and Flak learned the value of the Würzburg.

Freya had demonstrated the ability to detect aircraft at distances of 100 km, given objects at sufficiently high altitude. The Freya-Fahrstuhl, a set wherein the antenna could be raised 20 m to alter the vertical lobe pattern, had emerged out of Diehl's experimentation at Wangerooge and had brought home the importance of controlling the vertical lobe structure in air warning equipment. Freya pointed the direction for improving early warning, and the resulting GEMA designs yielded the best equipment of that kind to be used until the appearance of MEW, the Rad Lab's microwave early warning in 1944. For the tasks at hand it was better than MEW, as it had good height-finding capability.

The deficiencies of Freya were a range shorter than the curvature of the earth imposed and a poor resolution of targets both in direction and in height, very poor for the latter. Both of these defects could be reduced by increasing the antenna size while retaining the basic 2.4 m wavelength and the circuitry of Freya. Increasing the number of dipoles caused a narrowing of the beam, which led to obvious improvement in the angular resolution but also concentrated the power, thereby leading to an increase in range. Directional accuracy was obtained through lobe switching (in German, AN-Verfahren or Leitlinienpeilung [3]), which Erbslöh and von Willisen had introduced in their early equipment only to see it dismissed as too complicated. Out of these considerations two complementary designs, Wassermann and Mammut, materialized from GEMA during the summer of 1940 at a test site at Jüterborg, about 60 km southwest of Berlin [4]. Both designs required little new in electronics because the improved performance came about from adding dipoles. It economized by relying on well tested basic circuitry. More powerful transmitters were to be added in time.

As time passed a number of different models appeared on the coasts of Europe, the details of which need not concern us here. Wassermann S (schwer or heavy) had an array of 188 dipoles in a vertical pattern mounted on a 4 m diameter steel pipe 60 m high, causing British intelligence people to name it Chimney. (Only Wassermann S used the pipe; others were constructed of structural steel braced with guy wires but retained both the German and British names.) This immense array of dipoles formed a thin, horizontal fan-shaped beam that could be moved up and down by altering the phase shifts in the transmission lines feeding the dipoles. This allowed a direct and rapid way of determining elevation to an accuracy of 0.75° for aircraft lying between 3 and 8°; lobe switching gave an azimuthal accuracy of 0.25°. At 100 km this localized the target in a box-shaped region 1200 m high, 300 m long and 435 m wide for planes above 5200 m. Aircraft 8000 m high could be detected at 210 km, and very high fliers at 300 km. Mechanical rotation about the vertical axis allowed a 360° field of observation [5]. For ease of repair a crane surmounted the tower except for Wassermann L.

The exceptional long-range accuracy of Wassermann was used to determine the positions of individual formations of attackers, a process that took minutes of valuable time. The designers assumed that they would have to contend with multiple formations, and Mammut allowed these to be tracked rapidly and accurately in the horizontal plane with no height information. Mammut had 192 dipoles arranged in a fixed array 16 m high and 30 m wide mounted on four vertical steel beams. Its appearance secured for it the British intelligence name of Hoarding (in American, Billboard). Its wide array produced a narrow, vertical fan-shaped beam that could be pointed in a horizontal direction by altering the phase shifts in the transmission lines feeding the dipoles, just as done for Wassermann in the other direction. This allowed 100° of coverage; lobe switching yielded a directional accuracy of 0.5° [6].

These two basic designs formed most of the Luftwaffe's early warning throughout the war. Together they proved capable of dissecting the immense numbers of bombers that were to attack Germany as the air war proceeded, numbers that would have saturated CH. Although these radars operated on a wavelength one-fifth that of CH, they shared its unsuspected characteristic of being difficult to put out of action. The tall, spindly towers proved resistant to explosives, and the electronics and operation room was in a bunker [7]. Later they proved vulnerable to rocket-firing aircraft. Like CH they were rushed to completion by a few forward-looking individuals in anticipation of the terrible ordeal that still lay in the future and was unsuspected by most.

Wassermann and Mammut are the first examples of what has become common in modern radar: securing beam direction through a phased array. The Bell Telephone Laboratory's 10 cm FH (mark 8) main-battery naval fire-control radar [8] and Alvarez's 'leaky pipe' were developed at about the same time, close enough and independent enough to make priority unimportant [9].

The air-warning difficulty for night attacks was straightforward. The basic ideas came to mind almost as soon as they were posed. Destroying the attackers was something else again. Hermann Diehl's invention of daylight fighter direction at Wangerooge in 1939 was an obvious beginning, but difficulties were easy to list. Freya was excellent for picking up an attacker 75 to 100 km away but gave no height data useful for night work. If a fighter pilot was to fire on a bomber, he had to be guided to within a couple of hundred meters of his adversary in order for it to become visible even to dark-adapted eyes.

Hauptmann Wolfgang Falck had commanded the Me-110s that Diehl had vectored so successfully to the attacking bombers on 18 December 1939. This had impressed him so much that he began practicing the technique using the new lobe switching for Freya with which Diehl had modified his equipment with help from GEMA. They began using it over the Zuider Zee in poor visibility and eventually in darkness, where local tactical conditions of predictable altitude allowed them to down several aircraft. Thus when Colonel Josef Kammhuber, just released from a French prison, was assigned the task of organizing night fighter defense on 17 July 1940, Diehl and Falck were incorporated into his new organization [10].

Telefunken had demonstrated in July 1939 the ability of the Würzburg to determine the three-dimensional coordinates of the target, and in October it was decided to use this device as an adjunct to Freya. Owing to the Würzburg's small size and portability, the first plan had been to sprinkle them liberally over the countryside in order to follow the incoming aircraft, making up for the short range of 25 to 30 km by fielding a large number. To this end 4000 were ordered, even though the design was not yet final. In the final acceptance tests on 9 April 1940 at the Erprobungsstelle (test station) der Luftwaffe Rechlin the performance was spectacular, and the sets that were soon to be in production were taken away from air warning and given to Flak, as they were obviously able to point guns or searchlights better than any other technique [11].

Flak invented an ingenious test firing for the new equipment at shoots in Kühlungsborn on the Baltic Sea, which they called 'mirror shooting' and which allowed trials on targets of higher performance than towed sleeves. For an aircraft passing to the west of the battery, the bearing angle calculated from the radar data was given a negative sign, resulting in the guns pointing to the east. A vertical glass plate lying in the north-south plane at the battery allowed the reflection of the target to be seen superimposed on the sight of the shell bursts. Theodolite data of target and bursts allowed numerical analysis [12]. In September 1940 a battery in the vicinity of Essen brought down the first bomber using gun-laying radar [13], and the future of the Würzburg was settled. It was also to be the heart of night-fighter control.

The night-fighter control evolved through a number of stages, the first two of which concern us now. Helle Nachtjagd (illuminated night fighting) came first. In it a Freya picked up the enemy at extreme range and provided a companion Würzburg with an accurate azimuth, which sufficed for the Würzburg to find it and point searchlights for almost immediate illumination. The Flak batteries also used radar to direct searchlights, and the speed with which the light caught a plane remained an unnerving experience for the fliers. The limited range of the little radar and the searchlight combined with the speed of the target gave the night fighter only a few minutes to intercept. Naturally this method failed if there was cloud cover.

Dunkle Nachtjagd (dark night fighting) was the next stage and required the cooperation of a Freya and two Würzburgs, one to track the bomber, the other the fighter. The fighter director would note the relative positions of the two aircraft in a three-dimensional display called the See-burg Plotting Table and guide the fighter to an interception close enough for the pilot to see the bomber. The Würzburg was sufficiently accurate to guide the fighter successfully often enough to make the method useful, although it needed airborne radar to become truly effective.

Generaloberst Ernst Udet, a distinguished pilot from World War I, greeted this idea with skepticism but found that it worked after trying it personally [14]. He has been credited with remarking that radar 'took all the fun out of flying', which disclosed an attitude toward technical matters that brings into serious question his fitness as a Chief of Air Armament. It can explain much of the backwardness that marked the end of the Luft-waffe. His suicide on 17 November 1941 certainly had part of its origin in his sense of failure [15].

Kammhuber recognized immediately that the range of the Würzburg, although adequate for Flak, was too short for night interception. At Telefunken Leo Brandt, who had taken over further development of the Würzburg, suggested that the quickest way to extend the range was to increase the antenna from 3 m to 7.5 m diameter just as GEMA had boosted the range of Freya by increasing the number of dipoles to make Wassermann and Mammut; therewith the Würzburg-Riese (giant) enters the story. It proved just right for night-fighter direction and went into immediate production [16]. The Zeppelin Company had made Würzburg dishes from the beginning, but it was with the Riese that the unique structures noted in airships became evident. Kammhuber organized a belt of Freya–Würzburg–searchlight stations 50 to 75 km back from the coast where the early warning stations were located. British intelligence determined who the commander was and dubbed it the Kammhuber Line. The line had no anti-aircraft (AA) guns, the cities no fighter defense, which allowed each weapon to function without concern about the other. As stations began using Dunkle Nachtjagd their searchlights were moved farther back to take over with Helle Nachtjagd once the bomber left the dark zone [17]. The combination of a Freya with lobe switching, two Würzburgs and a fighter formed a defense cell 45 km long and 22 km wide and code named Himmelbett [18], which translates as four-poster bed. Bombers at night did not fly in tight formations because of the danger of collision, and they passed the

Kammhuber line almost as individuals, just right for Himmelbett. They also had to re-cross the line on their return home.

Crucial to the effectiveness of Himmelbett was the eternal problem of IFF. Without it the controllers could—and at times did—send one night fighter to shoot down another. The IFF Erstling was still in trial and was not making converts in the Luftwaffe. They preferred the Y-Gerät used during the Blitz for beam navigation. The strict—and not too well fulfilled—requirements for blind bombing were not required for following the fighter. The Würzburg would hold its position accurately and the Y-Gerät, in its new form called Y-Verfahren, would be close enough to identify the fighter to the controller. It had a singular advantage over IFF of yielding range and effectively solved the problem of getting the fighter back to a landing field in the dark. Two changes in procedure were required: the ground antennas that formed the guidance beam tracked the return signal as direction finders; the returned audio signal that determined range by comparing the outgoing with the incoming phase was transmitted 20 seconds out of every minute on the radio telephone on which the pilot obtained instructions, in open imitation of the British Pipsqueak [19]. The system was much preferred by the fliers because it was reliable IFF and could get them back home. It was also used for day-fighter control. Attempts were made to dispense with what seemed a redundant Würzburg, but Y-Verfahren was too inaccurate [20].

German air defense had a component called Y-Dienst (Y-Service) that must not be confused with Y-Gerät and Y-Verfahren. It was a system of directional antennas that picked up the bomber's radar, radio-telephone, IFF and jamming emissions and located the approaching formations by triangulations. With the cooperation of Bomber Command in providing a generous supply of assorted radiation it proved to be an important part of German early warning. It had for the defenders the singular virtue of being un-jammable. The RAF also had a Y-Service, although probably not named with the mischievous intent of confusing future readers of history, whose duties encompassed monitoring with skilled use of special receivers all German radio emissions [21].

An un-jammable early-warning radar that covered the English coast very well was Klein-Heidelberg. This was a series of passive stations in the lowlands that observed the arrival times of the direct pulses from a CH station and those reflected from a bomber formation. The time delay between the two established the position of the aircraft on an ellipse. There were two receiver antennas: a small synchronizer for the direct CH pulse and a very large steerable dipole array mounted on a Wassermann-S tower. The array gave a directional accuracy of 3° for the echo signal; range resulted from the intersection of this line and the ellipse [22]. This was the price paid in CH using floodlight rather than searchlight transmission.

The Germans believed that a spy penetrated a Klein-Heidelberg station in the form of a mysterious Leutnant Kunkel. No one knew him, whence he came or where he went—hardly a glowing commentary on security—but after that the CH stations applied some kind of time jitter to their pulses. This proved to be no problem but convinced everyone of Kunkel's villainous nature. R V Jones says that Kunkel was not one of his men and that knowledge of Klein-Heidelberg's function came from the analysis of communications traffic [23].

German air defense required a means for the control to follow the bombing formations as they proceeded across the Reich. The Himmelbett stations were too short ranged and much too involved with their own problems to furnish reliable information to higher command levels. In 1940 and 1941 GEMA designed a prototype set built by Siemens und Halske, called Panorama and placed on a large tower near Berlin to sweep the region. It had 18 dipoles arranged on a 20 m supporting beam to produce a narrow vertical fan-shaped beam at the Freya wavelength band. In addition to overcoming design problems this set taught the lessons of how best to site a meter-wave set on land without being swamped with local echoes. From these experiments Siemens built Jagdschloss (hunting lodge) on the same basic pattern but with two alterations: (1) Jagdschloss incorporated a PPI display, which was transmitted from the station to command headquarters by high-frequency cable or the widely used 50 cm directed-beam communication links; (2) a broad-band antenna allowed modest changes of frequency and specifically allowed a change to excite the IFF set Erstling, which then served as a secondary radar that enhanced the positions of the German aircraft at the closing of a switch. Jagdschloss began entering service toward the end of 1943 [24].

In September 1940 the Luftwaffe initiated a successful technique of attacking the bombers as they took off or landed at their bases in England or in the early or late stages of their journey. One of many decisions made by Hitler that caused some to question his stature as a great, universal military leader was to transfer the Fernnachtjagdgruppe (long-range night-fighter wing), which had built up these careful and destructive techniques, to the Mediterranean in October 1941 and to forbid such methods in the future [25].

Flak also began preparing for the coming onslaught. The original Würzburg had not been intended for gun laying, but when this important capability was recognized Telefunken's engineers began to work on a version that improved accuracy for direction and range by factors of three. Other virtues were a smoother flow of data to the gun director, assisted-manual tracking and automatic tracking in a few sets. This new design was named the Mannheim and used the same Würzburg wavelength [26]. The tongue had so accustomed itself to the name Würzburg that it was generally applied to Mannheim, which looked similar.

The commando raid on Bruneval had been recognized as the preliminary to some form of countermeasure, and both the Würzburg and the Mannheim systems were adapted to the expected active jamming. This was done by providing both with a front-end adapter called Wismar that allowed a 15% change in wavelength that could be accomplished in about half a minute. The active jamming came in the form of Carpet, which was dealt with reasonably well with Wismar, but the excessive secrecy placed on Düppel prevented any advance action for the coming passive jamming by Window, and that was another story [27].

Perched at the top of a narrow and steep-sided peak in Swabia about 10 km south of Reutlingen sits the tiny Castle of Lichtenstein. Its small size and high location may have inspired Telefunken's engineers to name their radar series for airborne interception after it, although this clue does not seem to have been picked up by British intelligence as the names for Freya and Wotan had been. Kammhuber had recognized the need for such equipment shortly after assuming command of night air defense and by early 1941 had presented his requirements to Runge at Telefunken. Runge proposed a 20 cm design but was told to utilize the company's experience with the 50 cm techniques to which they had now devoted nearly a decade, as there was no time for experimentation [28]. By summer 1941 he had a prototype flying [29] and by February 1942 the first night fighters were so equipped [30]. It made its entrance as a contract development by Telefunken's engineers, moving from drawing board to prototype to production in a normal manner.

The Lichtenstein had a rectangular array of four dipoles with reflectors mounted at the front of a twin-engine plane, the Me-110 and Ju-88 being initial favorites. This minimum array gave a wide beam, which provided a wide angular field for searching but also a short range, 3 to 4 km. Lobe switching in two directions would have provided enough directional accuracy for the radio operator to guide the pilot to the target; however, Runge and Hans Muth found a more elegant solution, one that imitated the Würzburg's rotating dipole by using a rotating phase-shifter in the transmission lines to the dipoles that produced the same twirling beam [31]. The vital minimum range achieved was 200 m, which fit perfectly with the Freya–Würzburg system of fighter direction and greatly improved Dunkle Nachtjagd.

A number of different Lichtenstein models were put into the field as criticism of the first ones returned to Telefunken. The Lichtenstein B/C went into active service in September 1942, but production was slow, primarily caused by a bottleneck in manufacturing the special vacuum tubes, and production was not generally of high quality, causing many sets to be returned. The biggest complaints were in the range, the angular region searched and indicators that were too complicated for combat. Several modifications corrected some of these faults, but to increase the range it was necessary to go to a 3.75 m band, the Lichtenstein SN-2, which allowed more power than possible at 50 cm [32]. The long waves were to prove an adroit choice for reasons yet to come.

The intitial response of the pilots was remarkably cold. The dipole array detracted not only from the graceful appearance of the aircraft but also reduced its speed quite a bit, and initially fighters so equipped were not flown. A technically minded captain, Ludwig Becker, who had worked with Diehl and Falck in their Zuider Zee operations [33], changed this attitude by using the new device effectively [34].

Lichtenstein was Runge's last important contribution to radar. On graduating from the Darmstadt Technische Hochschule in 1923 he had selected Telefunken in part to avoid working with Dr Karl Rottgardt at another company at which he had been interviewed, as his dislike for Rottgardt was immediate and enduring. When the object of his loathing became Director of Telefunken years later, mutual animosity found fertile ground, and by April 1942 Runge no longer had real responsibility. He left Telefunken on 1 November 1944 [35].

An episode in German radar worth noting took place in October 1940 when Manfred von Ardenne and Hans Hollmann, each the owner of neighboring private laboratories, proposed the plan of a panoramic radar based on microwaves and the plan position indicator. They first approached Göring, then a representative of Admiral Dönitz and finally Hans Plendl, the Plenipotentiary for High Frequency Research. All rejected the proposal [36]. There is reason to believe that the project as presented was not well founded technically. They apparently had no suitable microwave generator, and a drawing that accompanied the proposal could easily have left Plendl unimpressed, as it suggested that the two had lost contact with radar research since the mid-1930s—the obvious result of the tight secrecy imposed—but these were defects that two such men would have quickly made good. The incident is revealing about both the Nazi state and the two engineers. Hollmann had withdrawn from work on radar with GEMA because of misgivings about the direction of the regime, and it is unlikely that this had been ignored at top levels of the government. Von Ardenne, a shrewd business man, had come near to joining the NSDAP in 1933 but rejected the offer, as had his father. He had also come under the anti-Nazi influence of Max Planck, Max von Laue and Graf Georg Arco, founder and long director of Telefunken [37]. Neither of the two Lichterfelde laboratories was engaged in war work; Hollmann was doing medical research for which von Ardenne was developing the use of tracer isotopes. Hollmann had, nevertheless, continued research in microwaves with emphasis on magnetrons, and it is not certain what Germany's—and until recently Europe's—leading microwave specialist had in mind. The regime thus rejected the help of two of the nation's outstanding electronic experts for political reasons. What impelled the two to propose something of military use was their concern about the air attacks they feared were coming [38]. There is an adage: 'When war comes, you go with your people'.

While Germany was preparing the weapons for defending the Reich from night attack TRE set out to furnish Bomber Command the means for navigation; they had by mid-1941 finally come to realize this was an absolute necessity. Beam systems had two faults. For the pilot to know his position he had to be on the line that the beam defined, which was a restriction of no small consequence, as it is difficult to conduct a raid, even a peaceful flight, that requires the aircraft keep to an aerial railway. The flier needs to know his position wherever he may stray. Second, a beam is simple for the enemy to interdict, either electronically or by having the defending fighters use it to locate the attackers.

R J Dippy approached A P Rowe, Director of TRE, shortly after that organization had reached Swanage in the summer of 1940 with an idea for a radio navigation system, which came to be called Gee, a curious non-abbreviation of G, which in turn stood for grid. Dippy had discussed this idea while at Bawdsey, but there had been no push to follow it then [39].

Gee was fundamentally simple but had complicating details. A chain of stations consisted of a master and two or three slaves, all located at distances of about 150 km from one another. The master broadcast radar-like pulses on wavelengths chosen from 3.5 to 15 m. The slaves had receivers for the master pulses, which they then re-broadcast after predetermined and accurately timed delays. Master and slaves used the same wavelength, so the navigators received all relevant signals on the same receiver, limited, of course, by line-of-sight requirements.

On receiving signals from any master–slave pair the operator measured the delay between them. A given time delay did not determine position but did locate the receiver on a hyperbola. Repeating the process for another pair located the receiver on a second curve. The navigator was provided with a map on which these curious grid systems were superimposed and located his position from the intersection of the two curves he had determined. For identification, the master and slave pulse rates differed. The master had a repetition rate of 500, the first two slaves had rates of 250 transmitted alternately and the third transmitted double pulses either at 500 or 500/3. Complicated as it would appear, it presented an easily recognizable display on the receiver cathode ray tube. A fix could generally be accomplished by any two station pairs; the third pair resolved ambiguities and extended the range when its slave was the closest to the aircraft. Accuracy decreased with distance, primarily because the intersecting curves became more nearly parallel; the uncertainty varied from one to a few kilometers [40]. There was much enemy territory not covered by the new system because of the line-of-sight restriction, but even for those raids it furnished a secure starting point. It also reached out to guide the tired, often wounded fliers home, a function of value comparable to sending them toward their targets.

Small-scale tests using low-power stations in fall 1940 showed the idea to be practical. By August 1941 three full-power stations tried operation with 12 aircraft that had prototype receivers, and the results were an outstanding success [41]. Unfortunately, something disturbed the euphoria. The tests were extended to a raid on the Ruhr, and one Gee-equipped bomber failed to return, leaving the distinct possibility that a receiver was being examined by Martini's engineers, who would be onto the secret in short order with jamming techniques running through their minds. This, of course, was to be expected in normal operations, but these sets were laboratory-built prototypes and production was months off, despite all pressures to expedite it—months during which countermeasures could be devised.

The matter was turned over to R V Jones as an inversion of his normal duties. Now he must do what he could to obfuscate any intelligence the enemy might have gained. Here Jones, a notorious practical joker, was in his element. He had the Gee transmitters continue broadcasting but with unsynchronized pulses; he had them disguised to look like radar stations; he had some navigation beams set up, which the Air Force found quite useful, they having nothing else [42], and which he called J in the hopes that the nearness in pronunciation to German ears would confuse interrogators of Air Force prisoners [43]. From whatever cause, secrecy was retained.

On 22 February 1942 Air Marshal Arthur Harris became the Commander-in-Chief of Bomber Command. He brought to this position a vigor and ability that his long service had amply demonstrated. He also brought to it the conviction that strategic bombing was to be decisive and the determination to make it so. He knew the destructive ability of his command but preferred to ignore its manifest weaknesses by pushing ahead, surmounting difficulties as they arose rather than worrying overly about them in advance.

On the other side of the Channel leadership was virtually absent. Kammhuber had devised what was to develop into a good initial mode of defense, but he was allowed the power of a technician rather than of a commander and was swept aside when new tactics were demanded. Göring bore complete responsibility, but his long series of wrong decisions crippled the arm that had been so terribly feared in 1939 and led to its ultimate destruction. Unwilling to turn authority over to Erhard Milch, alone among Germany's ruling clique with the competence and drive necessary to retrieve her fortunes in the air, and retaining the dictator's support, Göring blundered to the end. The air defense of Germany settled onto subordinate commanders, radar engineers and brave, resolute air crews who fought each new attack with whatever resources were given them but without a sure, guiding hand directing a strategic response.

The air war against Germany could now begin in earnest. Both sides had prepared for the event and completed their preliminary actions. The Luftwaffe had perfected its air warning, was improving its night-fighter direction and Flak rapidly and Bomber Command finally had a method of navigating a few hundred kilometers into enemy territory and enough bombers to begin. The first attack with Gee on Essen on 8/9 March 1942 opened Harris's campaign of strategic bombing.

And so it was that March 1942 bore a strong resemblance to July 1940. Britain, later joined by America, was to attempt what the Luftwaffe had failed to do. In this they were successful because the invasion succeeded two years later, whereas the German invasion had not been attempted. But they failed to bring victory without invasion—the central doctrine on which the proponents of air power had based demand for resources—and the methods employed in the attempt generated more contention than any other part of the Allied war effort.

6.2. COUNTERMEASURES

Watson Watt's gigantic CH towers offered mute testimony to early concern about radar countermeasures. The steel transmitter towers had many antenna configurations during the course of the war, with horizontal dipoles mounted along the sides in the beginning. The radiation pattern from the side-mounted dipoles was sufficiently poor that curtain arrays became the standard transmitter antenna for the later, West Coast, stations. Many of the initial, East Coast stations were re-fitted with one or at most two curtain arrays hung between the fortuitously placed towers that had resulted from the original CH design of four operational frequencies. Owing to the narrow bandwidth of these antennas, extra arrays had to be available for the rapid changes of frequency needed to evade jammings. (Jamming is an enemy transmission on the radar frequency of a signal modulated to fill the operator's indicator with obscuring traces.) This fitted well with the need to have spare capability in the event of units not functioning, either from malfunction or enemy action. Multiple wooden receiver towers mounted dipoles of wavelengths corresponding to those of the transmitters. A design anticipating countermeasures was but one characteristic of British radar that set it off from early American and German equipment: it had been intended from the first to be part of an air-defense system. Colonel Martini attempted to jam CH during the Battle of Britain from ground stations across the Channel but with little success [1].

Fighting the radar war required a knowledge of what devices had to be countered, which became the province for a new specialty, electronic intelligence. The information so gained then went to the laboratories where equipment was designed to thwart what the enemy had brought forth. Destruction of the enemy radar was the natural and traditional way of dealing with any kind of new weapon, but countering it electronically was often easier—and somehow seemed a more appropriate, less brutish way for the engineers to fight—and if one was clever the enemy might be fed choice bits of misinformation and thereby led astray.

Use of radio countermeasures slacked during 1941 after the lessening of night attacks on England. Bomber Command had begun its attacks on Germany but was still unaware of any serious radar threat and had no interest in electronic warfare. As R V Jones' scientific intelligence began piecing together the dangers of German radar through the actions of 80 Signals Wing and the RAF Y-Service, the need for countermeasures became more apparent to the people at TRE. It was the knowledge that the Channel Dash of the *Scharnhorst* and *Gneisenau* on 12 February 1942 had been aided by Martini's jamming of the CD and CHL equipment expected to monitor navigation that grasped the attention of the technical people on both sides and showed them what was coming. The radar-intelligence reason for the Bruneval raid that followed on 27/28 February served to emphasize matters. An official organization, the Radio Countermeasures Board, met for the first time on 24 March with all manner of service organizations represented, including one from the United States. It was only an organizing group; TRE had to do the work.

It was Fighter Command, trained and skilled in the use of radar, that made first use of a radar-spoofing device. Moonshine was TRE's first offensive interference. It responded to a Freya interrogation by returning a much stronger echo, thereby giving the radar operators the appearance of a large formation. It was first used mounted in an otherwise useless two-man fighter, the Boulton-Paul Defiant; it was first used operationally in a feint to cover the US 8th Air Force's debut in a daylight raid on Rouen on 17 August 1942 and remained in service for another two months, after which the ruse no longer deceived with assurance and was set aside to be brought out again for the invasion [2].

Bomber Command's first active countermeasures were undertaken unexpectedly and wrongly by the flight crews. This also initiated Bomber Command's greatest electronic weakness, one in which it persisted for two more years—filling space with unnecessary radiation. If their commander was not alert to the need for countermeasures, the flight crews were. They knew by then that Flak was radar controlled, at least in part; they also knew that it could be deadly accurate. In their informal discussions the flight crews came to the conclusion that the IFF mark II, which should have been switched off when away from possible interrogation by Allied radar, worked as an anti-Würzburg jammer, if left switched on. In this the desperate crews followed the paths of others having terminal diseases and who seek delusive cures, basing their faith on the statistics of small numbers highly skewed by the selection effects in acquiring them.

Astoundingly, the Command found no harm in this, which was true inasmuch as none of the German sets of the time were interrogating on wavelengths to which IFF mark II responded and hence the transponder emitted no radiation or did anything else, but they compounded this foolishness by permitting the installation of a switch, called the J-switch, that caused the set to transmit even when not interrogated [3]! The object of this was to boost morale by showing understanding for the crews' anxieties; fortunately for them it came too early to become the delight of the newly organizing Y-Dienst, but it destroyed any remaining air crew respect for the value of radio silence, although it also removed the last resistance of

Bomber Command to the use of active countermeasures.

Active airborne jamming of the German 2.4 m air warning equipment, at the time restricted to Freya but soon to include Wassermann and Mammut on the same wavelength, took the form of the TRE-designed Mandrel, first used during the night of 5/6 December 1942. By June 1943 the activities of Y-Dienst were known, and Mandrel's utility as a jammer had to be weighed against its disservice as an alarming beacon. Various modifications in its use never resulted in a suitable balance between the two. The hope that Mandrel would prevent the Himmelbett Freya from directing the Würzburg GCI onto the bombers proved illusory when there were so many bombers that the Würzburg could locate a target unassisted [4]. Another device, called Tinsel, went along with Mandrel. It was simply a slight modification of the communication transmitter tuned to the German night-fighter-control radio frequencies and modulated by a microphone that picked up engine noise, the first of many attempts to disturb the German pilots' communication with their controllers.

The radar designers wanted to do more than just interfere with enemy electronics, and they did this by furnishing two unfortunate pieces of equipment: Boozer and Monica. The former was a passive receiver for German radar signals that set off a warning alarm for the crew, the latter a small radar in the tail to warn them of the approach of a night fighter. Neither remained welcome for long aboard the aircraft. There were soon so many radar emissions over and on the way to Germany that Boozer continually gave alarms to crews perplexed as to what to do with the information. When Bomber Command began to enter Germany in streams rather than the earlier extended formations, Monica began to alert the crews about the presence of nearby bombers. The Germans extracted Monica from a grounded bomber, and night fighters began to home on it with the simple receiver Flensburg [5]. Not all inventions help.

American countermeasures began with Luis Alvarez taking the initiative of having one of his staff at Rad Lab, the Canadian Dr Don Sinclair, work with the General Radio Company in modifying their field-strength measuring set P-540 into radar intercept receiver SCR-587 (Army) and ARC-1 (Navy). Its extraordinarily wide band, from 10 to 300 cm, was to make it useful throughout the war, but it suffered at the time from not having single-knob tuning, so it was not yet ready for the field. It was the second American receiver to prove of value in the new field of electronic intelligence; the Hallicrafter S-27, available in England for sale to amateurs, had picked up the emissions of Seetakt and Freya in early 1941. In an early lesson Rad Lab had been made receptive to the idea of countermeasure work rather early when the transmission by a nearby amateur had overridden a poorly shielded intermediate-frequency amplifier of one of the experimental radars.

Four days after the attack on Pearl Harbor American countermeasures began at a meeting among representatives of the National Defense

Research Committee, the Rad Lab and the Navy. They decided without dissent to establish a special laboratory for this purpose and proposed Alvarez as its head because of his intercept-receiver initiative and general interest in the subject. Commitments to work in progress at Rad Lab caused him to decline the offer but with the strong suggestion to offer the job to Professor Frederick Terman, head of the Department of Electrical Engineering at Stanford University. This proved a fortuitous choice owing to Terman's knowledge and stature in the field of radio and his ability to approach his many former students in building a staff. In August 1942 he secured an especially valuable addition in the team that was working for Columbia Broadcasting developing color television. Better yet, they brought with them their own laboratory equipment, exactly the type needed and already familiar to them [6]. Terman's general arrangement for obtaining industry people was to have them remain employees of their parent corporation, thereby retaining seniority and pension rights, and to pay the corporation for their services. In addition to the satisfaction of contributing to the war effort, the company thereby gained electronic skills through their engineers that they might easily have missed, had they remained on the sidelines. By January 1944 Terman's lab had 744 employees, 214 of whom were research personnel [7].

Terman went to Britain in April 1942 for six weeks and established excellent rapport with Robert Cockburn, who was chief of RAF countermeasures, while he learned the characteristics of German radar and British response. On return he organized work into three sections: building a jammer for Freya under John Byrne, a jammer for Würzburg under Bob Sorrell and continued work on the intercept receiver by Sinclair. By then his organization had been given the obligatory deceptive name, the Radio Research Laboratory, and had taken up quarters in a wing of the Biological Laboratory at Harvard [8]. Whether for security or to confuse future historians, the jammers for Freya and Würzburg were given the same names used by TRE, Mandrel and Carpet, but they were different designs [9].

Early jammers on both sides emitted the radar frequency modulated with a sine wave, but experience showed that it was better to modulate with noise, as this gave the appearance of a deterioration in the performance of the targeted radar receiver rather than some disturbing agency. This was the technique Martini had employed so successfully in the Channel Dash. Gas regulator tubes were common circuit components that formed a moderately constant reference voltage when current within some range of values passed through them. Without filter capacitors they showed significant noise levels and took on a new function as noise generators. Another noise generator was the photomultiplier tube. This device was capable of producing an extremely short output pulse on absorbing an optical photon and was the heart of all television cameras. By keeping the multiplier in the dark and increasing the voltage beyond that required for normal operation, thermal electrons from the photocathode and first electrodes produced a shower of pulses at the output and a wonderful, high-amplitude source of high-frequency noise was obtained very simply. Both the Mandrel and Carpet jammers of the Radio Research Laboratory used the RCA 931 photomultiplier for this. Specially designed gas tubes later proved equally good and more convenient, as they needed much lower voltages than photomultipliers [10].

The Radio Research Laboratory produced one of the most curious jamming devices of the war, Tuba. The Lab's American contingent at TRE noticed a prototype ground jammer called Ground Grocer that worked on 50 cm, designed to get as much power as possible out of the Micropup triodes and intended to interfere with the new Lichtenstein radar of the night fighters. Beams were directed from the coast of England at regions where they would encounter the night fighters pursuing returning bombers [11]. The Americans recalled the very-high-frequency beam tetrode called the resnatron, invented by Sloan and Marshall, that had almost entered the open scientific literature[1] It could generate decimeter waves of very high power with a modest range of adjustable wavelength. Radio Research Lab proposed a jammer based on it, and the British ordered one immediately [12]. The tubes that formed the heart of Tuba were demountable and vacuum pumped; on-site repairs and maintenance were possible from a truck-mounted machine shop. The power was radiated through waveguide-fed horns of the large size that 50 cm required, and its 80 kW of continuous power could light a fluorescent tube a mile away. A canvas cover once burst into flame when placed over the exit horn. Initial deployment proved troublesome to German fighters near the Channel that headed into Tuba's radiation but not to those further away, regardless of their direction. All effort was soon pointless when the airborne radar wavelength changed to 3.5 m just as Tuba became operational, a wavelength completely outside its capabilities [13].

At times nature provided radar operators with disturbances far more intense than did the enemy. This was particularly true of long-wave equipment. Engineers and Air Force officers often saw alarming signals on the oscilloscopes of CH that the seasoned women operators summarily dismissed as 'ionospherics' or some such, yet saw them fasten onto an imperceptible blip for which a squadron scrambled to intercept. On the other side, the versatile Telefunken engineer Wilhelm Stepp commented that 'Initiating new equipment, and the long-wave set "Heidelberg" comes to mind, was like an expedition into a wonder world of new and at the time unknown phenomena and difficulties' [14]. Out of these anomalies emerged some new science. J S Hey located the ions in the upper atmosphere that marked the path of an incoming meteor [15] and discovered the greatly enhanced radio energy of solar flares using GL equipment; he at first thought the solar radiation was very clever jamming by the Ger-

---

[1] See Chapter 4.1 (p 150).

mans [16]. Radio propagation peculiarities near the ocean surface often produced unaccountably and unexpectedly long ranges to targets that confused the operators and sometimes generated phantom targets—fired on by batteries from both sides of the Channel. Clouds and humidity gradients added to the confusion, and sidereal noise limited the signal-to-noise ratio obtainable with CH [17].

When Watt and Wilkins made their calculations in 1935 of the intensity of the re-radiation by a dipole, intended to approximate the resonance behavior of the wing structure of an airplane, they inadvertently investigated a very effective means for interfering with radar. A modern radar operator is by no means exempt from the confusion that this early countermeasure produces when a shy target fills the space surrounding itself with dipoles cut to lengths that resonate at the wavelength of the incident radiation. There are ways for him to deal with this nuisance but a nuisance it remains, one that can saturate his equipment. When one considers how obvious the idea is and how many persons thought of it independently, how puzzling seems the extreme secrecy placed on it before its first use in 1943.

Lindemann, ever ready to find fault with early radar, had seized on the effect of releasing dipoles in the CH beam that his student, R V Jones, who was studying infrared detection, had pointed out to him [18]. One of the virtues of CH, however, was its resistance to this form of interference. The long wavelength and horizontal polarization required that the dipoles had to be some meters long and suspended horizontally, which never proved feasible. Knowledge of the efficacy of dipoles spread when the Bawdsey researchers used dipoles suspended by balloons in testing the 5 m GL sets [19]. American operators of SCR-268 learned early to have someone climb a tower with an attached dipole in order to test their equipment.

Thinking about these things was not restricted to the learned and was especially encouraged among those most closely trapped by events. Thus when 148 Squadron was making electronic intelligence flights in North Africa in September 1941 and found themselves subjected to intense AA fire, they suspected that their antennas were enhancing their radar echo. In a subsequent raid on Benghazi they had the bomber crews throw out 46 cm long aluminum strips, the dimension being that of the antenna suspected of having drawn fire, but found no effect on the accuracy of the Flak [20]. Obvious speculation about what might have resulted had the crews been informed about the Würzburg wavelengths and cut the strips to 27 cm must be tempered by the knowledge that there were no Würzburgs in North Africa until April 1942, indeed no German radar at all until January of that year [21].

In March 1942 Joan Curran, the only woman scientist at TRE, reported her own investigation of the idea, which she found practical and compelling. By 4 April its use had Air Ministry approval and the code name of Window. Fighter Command requested that its use be held back until its effect against British radar had been established. These tests did, in fact, show that the AI marks VII and VIII, the first 10 cm sets, were affected. Their beams spiraled around the direction of the fighter, and Window confused the radial indicator display; the tests also showed the GCI sets then being introduced, which used 50 cm radiation, could also be seriously confused, greatly complicating night interception [22]. Bomber Command was beginning to find the losses attributed to Würzburg-directed Flak unacceptable, which led to a dispute between the two commands, and Window introduction was postponed when it was learned that AI mark X or SCR-720, which had a different kind of indicator, was much less susceptible. Neither command was aware at the time that the German airborne interception radar was being designed for the Würzburg wavelength. Cockburn was permitted to experiment but under the strictest security.

Window was the perfect countermeasure for the Würzburg wavelength. To be effective Window must fall through the radar beam to insure that the operator's indicator is overwhelmed. If the radar beam is narrow, as is the case for microwaves, the chance of a sufficient number of dipoles lying in it will be correspondingly small. The Würzburg 50 cm beam was large enough to pick up thousands of the little dipoles as they drifted through the sky, whereas a 10 cm beam would intercept only about 1/25 as many. The Würzburg used a rotating dipole at the feed, which provided a continuous sweep of all polarizations, so the random orientations of the falling strips were well scanned [23].

Telefunken tested the idea at the Luftwaffe research station at Rechlin in early 1940 with results so shocking to the top levels of the Luftwaffe that the tightest secrecy was imposed on the knowledge; a ban was even imposed on further experimentation, effectively cutting off the development of protective measures. In accordance with Telefunken's use of geographical locations as code names, Leo Brandt called the metal foil technique Düppel after a Flak battery stationed on an estate by that name near Berlin-Zehlendorf [24], an exceptional way of being secretive—the German word for dipole is similar to the English [25].

When Terman returned from his stay with Cockburn he brought knowledge of Window with him. He instituted research in what the Americans chose to call Chaff and assigned Dr L J Chu, a noted antenna expert, to study the matter theoretically. Theory was important because the ratio of the width of the foil to its length determined the bandwidth of its resonance. A very sharp resonance could miss the wavelength of the incident radar and generate only a small echo. To extract the practical aspects Terman turned the theory over to Fred Whipple, an astronomer. Practical manufacture set constraints on a design for the huge quantities that would be needed, and Whipple soon worked out formulae that gave radar cross section for a given bandwidth per kilogram. His results were confirmed in experiments at TRE, and large quantities were stockpiled against the day of use [26]. By the end of the war three-quarters of all American aluminum foil production went for Window [27].

The introduction of Window caught the Würzburg-equipped gunners and Lichtenstein-equipped night fighters completely by surprise. Ever since the Bruneval raid, jamming had been expected for Würzburg, and a modification, Wismar, that allowed relatively rapid changes of wavelength was ready, but an air filled with dipoles had been officially banished from thought and plan. What might have been accomplished in calm now had to be done in extreme haste.

The night fighters were already in the process of changing their 50 cm equipment for 3.5 m sets (for unrelated reasons) and thereby became immune to the new clouds obscuring their prey, but a substantial change of wavelength was not an immediate option open to the AA radar. The basis for working through Window came from the very first reflection experiments of the early 1930s—the Doppler effect. The wavelength of the reflected signal was altered by motion of the target, and bombers moved fast whereas the drifting dipoles moved with the wind. The Würzburg used special high-frequency triodes in its transmitter, which were stable enough in frequency to allow filters in the receiver to distinguish shifted from unshifted echoes [28]. This was the first use of the technique that has come to be called 'pulsed Doppler'. Unfortunately, echoes from aircraft were not just shifted, they were shifted to either side of the transmitter wavelength depending on whether the targets were approaching or departing and by an amount dependent on the velocity.

A suitable device, called Würzlaus, invented at the Max-Wien-Institut within two weeks after the alarm [29], suppressed the unshifted signal that came from the cloud of dipoles, which made it easier for the operator to see the unsuppressed echoes from the target. It allowed some degree of success in distinguishing bombers from aluminum foil, and in skilled hands under the right conditions this restored much of Flak's accuracy. It worked on the approach, where there was a Doppler shift to shorter wavelengths, and on the departure, where the shift was to longer wavelengths, but failed, of course, in the important mid-course region when the bombers flew at right angles to the radar line-of-sight and where the Doppler shift was too small for discrimination. But Würzlaus had the unfortunate characteristic of requiring a fixed wavelength, which made it incompatible with Wismar's frequency agility. This meant that formations throwing out packages of Window and operating their Carpet jammers, something that became common in fall 1944, made radar-directed gunfire sometimes impossible. When overcome in this manner the Würzburg made use of Stendal A and became a passive device that at least determined the direction to the Carpet transmitter [30].

Another expedient made use of an audible signal produced on the reflected pulses by the propellers and general vibration of airframes, an effect noted by Lorenz investigators when observing a windmill in exper-

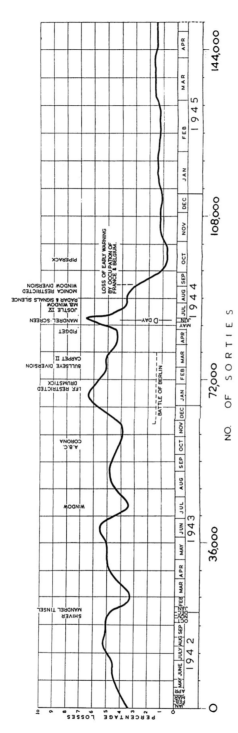

The effect of countermeasures on the loss rate of RAF Bomber Command. The percentage loss averaged over each 3000 sorties is plotted against the number of sorties for attacks on Germany. Note that the number of sorties was not uniform in time, the consequence of the Command having been given missions other than over Germany and from restrictions forced by circumstances. The introduction of the various radio countermeasures is noted by their code names (not all of which are described in the text) as are additional factors that affected losses. The introduction of each new countermeasure generally resulted in an immediate reduction of losses followed by the defenders countering the countermeasure. The loss rate dropped significantly after the liberation of France, after which the Luftwaffe was greatly weakened and their air-warning capability significantly reduced. Adapted from RAF Signals vol 7, p 190. Crown copyright.

iments just before the war. The radar receiver output was passed through an audio-frequency filter that suppressed the pulse-repetition frequency before transmitting it to a pair of headphones; the operator then attempted to 'hear' the airplane. The device, called Nürnberg, was manufactured between September and December 1943 [31].

In early 1944 Dr H Pöhlmann of the Reichs-Luftfahrt-Ministerium removed the frequency rigidity and other deficiencies of Würzlaus with an improvement called Tastlaus, which worked with Wismar and thereby greatly improved the ability of radar control to work through Window and Carpet [32]. Despite all Allied precautions, Flak took ever greater numbers of the attacking formations. During the last seven months of the war the US Army Strategic Air Forces in Europe lost 1566 aircraft to AA fire [33]. What would the total have been without Carpet and Window?

The unsatisfactory nature of the methods to counter Window was reflected in a competition offered by Göring for the best technical solution with tax-free prizes up to 300 000 Rm depending on the value of the invention. Entries were to have been delivered to the Air Ministry Technical Office GL/C-F4 by 1 April 1944. Staatsrat Professor Abraham Esau, who would shortly replace Plendl as Plenipotentiary of High-frequency Research, was to head the judges [34]. There is no record of an award.

The Pacific War was almost a separate war, and this held particularly for radar; thus the introduction of Window by the Japanese during a night raid on Guadalcanal in May 1943 went unreported to Europe and did not enter into the discussions that led to its introduction in the July attack on Hamburg. The Japanese dropped 75 cm strips of 'deceiving paper' (Giman-shi) to disrupt the SCR-268 that was proving to be a problem. They considered the use successful in reducing losses [35].

Observing enemy radar and radio navigation equipment in order to discern its characteristics was the earliest function of countermeasures people. The Royal Navy entered quickly into these transactions, setting up an electronic countermeasures station at Dover during summer 1941 [36]. Such services quickly became airborne, and their missions, though inherently passive, were by no means safe. The direct observation of a Lichtenstein during the night of 3 December 1942 proved to be a very near thing for an air crew that had deliberately positioned themselves as bait [37]. Much more tangible information about the Lichtenstein came the following May in the form of a Ju-88 equipped with 50 cm B/C model, thanks to the carefully planned defection of a night-fighter pilot and radioman who secured the acquiescence of the mechanic with the aid of a pistol. This gave Britain not only the details of the radar and radio altimeter but allowed the machine to be used in simulated fights with British aircraft, which helped them locate weaknesses to exploit [38].

The vast expanse of the Pacific resulted in a number of intercept-receiver-equipped long-range aircraft, called Ferrets, probing the Japanese-held islands, but the first solid data came by way of a submarine, not an airplane. USS *Drum* went on patrol in September 1942 with an ARC-1 receiver on board. The radio operator was told how to use it when the boat was on the surface near Japan, but only if its employment did not interfere with other operations. The strip chart records brought back were the first interceptions of Japanese radar. A newly equipped Ferret in the South Pacific at about the same time failed to be first, presumably because of a poor receiver. The Ferrets soon improved their technique and successfully mapped the radiation from the Japanese mark I model 1 set on Kiska, Alaska in March 1943 [39]. These kinds of flight became a normal part of fleet reconnaissance.

When radar sets pointed their beams according to the will of the operator, intercept receivers could be tuned by hand, as the interrogating beam could be expected to remain long enough to be recorded. When the PPI technique became widespread, the receiver could expect to be in the sweeping beam only a small fraction of the time, and this required automatic scanning receivers. The ultimate of such receivers was the TRE frequency-indicating receiver that presented received signals on a cathode-ray tube with their amplitudes shown as straight lines radiating from the origin and the frequencies given by the angles of trace orientation. Signals that might be missed during most receiver scans stood out and were easily identified by the operator [40].

Such were the principal means by which the adversaries came to grips over Germany during the night and, after the arrival of the Americans, during the day. There was variation that will occupy us yet, and deception was not restricted to radio waves. German radar engineers found themselves devoting ever more time and resources to devising the means of countering Allied countermeasures, but they understood the peril of the time and gave a full measure of ingenuity to protect their homes while the regime that controlled their country seemed to the end incapable of comprehending the need for an extensive development of defensive weapons.

6.3. AN AIR WAR OF ATTRITION

Lübeck is an ancient Hanseatic city closely tied through commerce to the other old ports of northern Europe. Until the Nazis incorporated it into adjoining Prussia in 1937 it was a separate political entity with a republican government, expanded after 1848 to include democratic representation. The trading families of Lübeck felt more akin to people in the sister cities of Hamburg, Amsterdam, Antwerp and London than to those in Berlin. But the cultural stamp that attracted Arthur Harris in March 1942 was Lübeck's old buildings, not their splendid medieval architecture but their half-timber construction, for the city had been selected to be destroyed. In addition to its combustibility the distinctions that made its selection obvious as a target were a relatively light air defense and a location near the sea easily recognized on a moonlit night, necessary because it was beyond the range of Gee. It was small enough for Harris to be able to demonstrate the destruction of a city with the force available to him, yet large enough to count in the headlines.

So mankind increased the nightmare of total war by yet another degree, but by then who would have urged the planners at Bomber Command to desist? It is doubtful that such restraint would have come from many inhabitants of Coventry or burghers of Rotterdam or Londoners who had spent the winter nights in the underground stations, wondering whether they would find their homes or work places standing when they made their way out the following morning.

There were skeptics to Bomber Command's capability, and Harris saw the need to have results that would impress top levels of the Air Force as well as political leaders. This was the reason for destroying Lübeck and, shortly thereafter for the same reason and under similar circumstances, Rostock. He was also clearly playing to the press in making the attack with 1,000 bombers on Cologne at the end of May 1942, for which he had had to strip the training squadrons in order to achieve the magic number. The city was severely damaged, and it was a success for Gee.

Gee was a great navigational advance but had three weaknesses: (1) its range was limited by approximate line-of-sight transmission, (2) its accuracy was of the order of kilometers and (3) jamming, expected from the start, began on 4 August 1942 by jammers called Heinrich [1]. Foreknowledge of these faults had set TRE planning other navigational aids, two of which become important: Oboe and H2S, both introduced in December 1942. Oboe would prove to be a remarkable blind-bombing device that provided skilled users with the ability to drop bombs from very high altitude to within a hundred meters of the target. H2S[2] was the most inaccurate of all—indeed accuracy is probably not the right word—but was not restricted in range by the curvature of the Earth as were Oboe and Gee. Gee remained in service, as it was not always jammed and was excellent for navigation, a use for which the Luftwaffe soon found affection.

Using one another's radio navigation systems became relatively common. When the Germans introduced Elektra Sonne and set up a station in Spain to work with a second in Brest, Britain quietly accepted this neutrality violation because it was useful to Coastal Command flying over the Bay of Biscay, where Gee became inaccurate or did not reach. Elektra Sonne's intersecting beams required no special kind of receiver and were extremely simple to use; it remained in civilian service after the war as Consol, its British wartime code name [2].

Oboe made use of two widely spaced radar stations in England, called Cat and Mouse, that could control one aircraft, generally one of the remarkable Mosquitoes that flew very high and very fast. The aircraft was equipped with a transponder that allowed the Cat to determine its range with high accuracy. On approaching the target the plane was constrained

---

[2] See Chapter 4.5 (pp 188-90).

to fly at a constant radius from Cat, whose transmitted signals were coded to give the pilot a pure tone if on course, dots if he were on the near side and dashes on the far side. The Germans noticed the curved approach and called the new system Bumerang (boomerang). Mouse measured the range between it and the aircraft, determined the ground speed and calculated the release time for the bomb. It transmitted coded signals that allowed the crew to prepare and indicated exactly when to release [3]. The reader need not be told that skill, practice and cool nerves were required for success, but all became available.

Oboe originated with A H Reeves, whose enthusiasm was picked up by F E Jones, who helped him work out details in spring 1941. Given knowledge of the usefulness of Oboe, especially in the destruction of the Ruhr industries, it is difficult to comprehend the vehemence with which its development was opposed at TRE and in the RAF. One objection was that it would be suicidal to fly an aircraft on the required preparatory course. This would have been a valid objection except for the remarkable characteristics of the Mosquito that flew at altitudes and speeds that made it almost impossible to be intercepted by fighters or shot down with AA fire, and the Mosquitos that flew Oboe had astoundingly light losses. The objection that only one aircraft could be controlled at a time was overcome by having the Mosquito drop marker bombs, a method which came to be accepted in summer 1942 [4]. Another objection was the general impression that Oboe would be extremely easy to jam, which was true for mark I that operated on a meter-wave band but not for marks II and III that used 10 cm radiation. The other problem Reeves and Jones faced at TRE was the general acceptance that H2S would do the job and at ranges far beyond 400 km from the English coast.

Despite opposition, the director of TRE, A P Rowe, allowed the work to go forward, and in September 1941 optical tests made at a distance of 130 km from Cat gave bombing accuracies of 50 m. These were followed by blind bombing that showed Oboe superior to release by optical sights in good visibility. A demonstration on 2 July 1942 before a number of senior officers yielded a bombing accuracy of 65 m [5]. It was an accuracy that was transferred to the Krupp works.

In December 1942 Mosquitos began to try Oboe on the enemy. There had been reason to suspect problems of accuracy because of the poor control linking British and continental grid systems, but when Clifford Cornford, who was responsible for the mathematics of Gee and Oboe, looked into the matter, he found that the Germans had supplied him with the data he needed. During World War I they had triangulated points across the Channel in order to bring the two grid systems into agreement with long-range artillery fire the objective, and the results had come into British hands. A Mosquito raid on the easily recognizable Wehrmacht Headquarters in Florennes, Belgium had demonstrated the validity of the correction [6].

In practice things were a little more complicated than when displayed to the senior officers. In addition to the hostility of the recipients there was the need to operate at extremes of altitude and range, 10 000 m and 400 km. There was the much more difficult matter of having the main force find and see the marker bombs, which cloud cover might completely obscure although industrial haze would not. The first part of the practice used isolated Mosquitos dropping sticks of bombs on targets clearly identified on maps to confirm that Oboe was properly calibrated.

The defense was puzzled over these isolated flights and at first assumed they were just harassment to stop production through the air alarm. On 7 January 1943 Hauptmann Alexander Dahl, artillery officer from World War I, free balloonist and radio amateur when he had dared, noted that the explosion of a stick of three bombs had been quickly followed by an immense flash bomb high in the sky. A quarter-century earlier Dahl had observed artillery fire from a kite balloon and thought the RAF was 'adjusting fire' [7]. The photograph exposed at the time of the flash would show the exploding bombs in the target area, and he assumed this was being used to determine and remove systematic errors. It was a reasonable conclusion but wrong. The accuracy already sufficed; it was a training flight.

After training his crews for two months Harris began his Battle of the Ruhr on the night of 5 March. He had created a Pathfinder Force that contained a small number of heavy bombers to lay many marker bombs visually on those dropped by the Oboe-equipped Mosquitos leading the way. This insured a sufficient number of markers for the main force to aim at. The Battle of the Ruhr continued through July, although other parts of Germany received attacks during that period [8]. The Battle of the Ruhr ended as an unquestioned victory for Bomber Command, a victory for which the necessary conditions were the accuracy of Oboe and the inability of the air defense to destroy or even harass the high-flying Mosquitos. The validity of both conditions had been completely rejected by Air Force command two years earlier and viewed with considerable suspicion even one year before.

H2S was introduced at the same time as Oboe and was used in some of the Ruhr attacks as a means of gaining experience with it. Initially, at least, H2S used Pathfinder techniques, not because it was restricted to a single plane but because there were not enough sets for the many bombers. Its function was strongly dependent on ground characteristics, especially the presence of bodies of water, which reflected the 10 cm waves like a mirror and presented the PPI operator with a welcome dark image of a river or estuary. Thus the easiest targets had bodies of water as part of their make-up; the next best were small cities surrounded by farmland; the worst were large, distended industrial regions, and Berlin was the worst target of all. H2S was not in the same class as Oboe for accuracy regardless of the target conditions. If conditions were right, one could hit the town, occasionally doing better than that.

The growing success of the defenders' Himmelbett system and the need of the main bomber force to release on marker bombs combined to introduce the bomber stream. This sent so many aircraft through a Himmelbett cell that its two Würzburgs and one night fighter were overwhelmed. In the 1000-bomber attack on Cologne in May 1942 only eight Himmelbett cells were touched and only 25 night fighters guided to targets. Fighters and Flak destroyed 44 aircraft, a large number for a night's work but small relative to the number of attackers. It also left a large number of fighters tied to inactive Himmelbett cells.

The immediate response was Kammhuber's call for more radar and more Lichtenstein-equipped night fighters. The call was answered at well below the requested amount. Neither Göring nor Hitler, having taught themselves that war must be won by attack and ignoring air defense for which their understanding was slender at best, had yet come to perceive the seriousness of the threat. Kammhuber received much less than requested; nevertheless by August there were 96 night-fighter stations and 30 000 signals personnel. Each night-fighter sortie required 140 persons operating radar sets, working communications links and evaluating the vast amount of data. With a second summer in Russia demanding extremes of manpower, ground radar and control centers began receiving large numbers of women; by January 1943 there were 14 000 Luftnachrichten-Helferinnen in service [9]. The searchlights that had allowed Helle Nachtjagd to take over after the planes had left Dunkle Nachtjagd were removed and placed at the Flak batteries near target centers.

By early 1943 fighter pilots were increasingly critical of the Himmelbett system. Part of their complaints had roots in the inadequacy of the Würzburg-Riese. It had gained the additional range over the small set simply by increasing the size of the dish, but transmitter and receiver had not been improved, and its range was inadequate for a proper GCI radar, so the resultant small size of the Himmelbett cells constrained the maneuvers of the night fighters.

New tactics began to take a freer form over the objections of Kammhuber, who favored tight control. He was initially bypassed and then replaced. Rather than guiding individual fighters to individual targets the control officers located the bomber stream and guided the fighters to it. There were enough targets that the flight crews could pick them up with their Lichtenstein sets, something at which they became skillful. Over burning cities the single-engine day fighters would fly above the bombers to find them illuminated from below by the fires they had created. This was particularly successful, if there was a thin layer of clouds below bombers, whose outlines were easily seen on this illuminated screen. It was wild for these fighters, who had to dodge their own Flak and called the tactics Wilde Sau (wild boar). The tactic of night fighters joining the bomber stream took by contrast the name of Zahme Sau (tame boar). The Himmelbett system remained. It was effective on the retiring bombers that were scattered from their original streams.

One of the curious outcomes of these new tactics was a call by the fighter pilots for radar beams that had a wider angular pattern and greater range. The 50 cm Lichtensteins had functioned well enough when the Würzburgs had guided them to within a few kilometer of the target, but now they had to seek their targets with less-accurate guidance. Telefunken's engineers were quick to come up with a satisfactory new design, Lichtenstein SN 2, which worked in the 3.5 m band; a wider angular spread and increased range came naturally from increased wavelength. The first of this type was introduced in July 1943 [10] and became a very effective piece of equipment. The reader will recall that the main British inducement for pressing the development of microwaves was the idea, seldom questioned at the time, that they were essential for airborne interception (AI). The earlier argument that the range of meter-wave AI was limited by the ground-return of meter waves and thus by the altitude of the fighter became irrelevant when combat was so high. The air war in 1943 was quite different from the air war in 1941.

The air war against Germany was consuming a large portion of British and American resources by January 1943 and took its form from the constraints that radar imposed. Radar now guided the bombers at night and increasingly by day, as the Americans began to realize that not only can one not see at night but frequently not during the day either. As an offensive weapon radar was proving weak, but as a defensive weapon it was still strong. That it was strong by day had been accepted by both combatants by 1941; that it was strong by night was coming as an uncomfortable surprise to Bomber Command. The losses of the attacking bombers would grow until some new countermeasure was introduced; the defense would then master it and losses would grow again. It became a repetition in the sky of the trench warfare that air power advocates had hoped to avoid. The RAF decided losses up to 5% in an attack were acceptable; crewmen could calculate their chances of completing the required 30 missions. For the defending pilots and their radar operators the arithmetic was less straightforward but just as deadly. Just as German and Briton climbed out of their trenches with steadfast loyalty in 1916, so their sons just as coldly flew towards their deadly, confusing encounters in 1943.

When Prime Minister and President met at Casablanca in January 1943 the air war had been shaped by events. The conference stated the purpose of the air offensive to be 'the progressive destruction and dislocation of the German military, industrial and economic system, and the undermining of the morale of the German people to a point where their capacity for armed resistance is fatally weakened' [11]. It was policy following events. It was policy that formalized a war of attrition. The air power talk of a 'knock-out blow' was forgotten. Another doctrine came from the Casablanca meeting, almost, it would seem, as an afterthought:

unconditional surrender as Germany's only way out of the war. This doctrine, combined with the unremitting destruction from the sky, the threat from the east and the memories of 1919 stiffened the backbone of the German people markedly.

As spring changed into summer of 1943 the Lichtenstein-equipped night fighters became ever more proficient. Attempts to counter them with Beaufighters equipped with a homing device for Lichtenstein, called Serrate, was a failure because of the deficiencies of the Beaufighter relative to the German night fighters, and losses from AA fire continued to mount. This situation finally forced the introduction of Window. It had been repeatedly delayed as a consequence of the fear of how German bombers might use it in attacking England, a fear that tells one that the superiority in the air achieved by the Allies had not yet been fully grasped.

The German bomber force of 1943 could not be compared to the British or American even taken separately, but memories were fresh as time raced, and British night fighters were busy, nevertheless. German night raids had never ceased, and the RAF fought each one. When GCI/AI began to take a heavy toll of the He-111s that came over, the faster Ju-88 and fighter–bomber FW-190 were substituted with almost daily changes in tactics, sometimes flying high, sometimes low. A particularly difficult tactic to counter was the intermixing of Me-410s equipped with the radar Neptun R-2, a 1.8 m AI set of limited production that allowed the detection of a very closely pursuing fighter [12]. These fast, agile planes, the equal of the Mosquitoes in speedy acrobatics, drew the night fighters on merry chases often allowing the bombers to pass unmolested [13]. It all mimicked in small format the kind of radar war being fought over the Reich with each measure generating a countermeasure.

During the Battle of the Ruhr estimates of causes of Bomber Command losses were 35% to radar-equipped night fighters, 35% to other night fighters, 20% to radar-controlled Flak and 10% to other Flak [14]. (This demonstrated the growing efficiency of the night fighters; 87% of the losses during the 1000-bomber attack on Cologne were attributed to Flak [15].) It was reasonable to assume that 55% of the losses would be countered by Window. The debate in council was divided, and Churchill took personal responsibility for the introduction of the new countermeasure, which was used against Hamburg for the first time 24/25 July 1943. Damage to the city was extreme, resulting in the first fire storm, and losses of the bombers were remarkably light. It was a major defeat for the Luftwaffe. It not only justified the use of Window but showed the value of H2S when ground conditions were favorable. Despite the severe damage done, half of the bombs missed the city. Attacks by night and day followed to complete the destruction, which deeply affected the German high command.

When the American 8th Air Force began assembling in England under Major General Ira C Eaker to join the RAF in the attacks on Germany the two forces shared the common belief that strategic bombing could win the war alone but disagreed completely on the next item of dogma, how to do it. Bomber Command had been forced from its early belief in the ability of self-defending bombers to penetrate German air defense, which had left them nothing else but saturation bombing at night using navigation techniques inadequate for hitting a target much smaller than a city. The Americans were convinced that the B-17s, which flew higher and were more heavily armed than the British bombers, could beat off the attacking fighters during the day by the combined guns of tight formations [16]. By early 1944 both delusions had been dispelled, and both air forces had been defeated by the air defense made possible by radar. They were to defeat the Luftwaffe yet, but the means for doing it, though available from the beginning, had been rejected earlier and was not ready.

The Americans had no alternative to daylight bombing. Their crews had no training in night flying and the much lower bomb load of the B-17s could not compete with the Lancasters, a disadvantage they compensated by an ability to hit the target instead of scattering explosives and incendiaries almost indiscriminately—compensated, that is, if they could see the target. Disagreements naturally arose between Harris and the Americans and intensified when Lieutenant General Carl Spaatz became commander of the US Strategic Air Forces in Europe in December 1943. Spaatz would concentrate, or such was the plan, on specific parts of the German economy that were expected to have great effect: petroleum, transport, ball bearings. Harris rejected these 'panaceas' out of hand. He would break the German will, not their economy.

The Americans encountered an uncomfortable fact: the atmosphere of northern Europe was nothing like that at the Bombardier School in west Texas where 'the skies are not cloudy all day'. The skies over Germany were not only cloudy, they were richly filled with smoky haze, especially around industrial targets, and could sometimes be made impenetrable by smoke generators. They found themselves dependent on radar navigation and bombing much as Bomber Command.

By October 1942 the overcast of European skies had convinced Eaker that visual bombing might be the exception rather than the rule and the only alternative was the blind-bombing techniques of the RAF. Oboe and H2S were in advanced prototype stages, and the Americans asked their hosts for some of each. This was not a welcome request. Production of both was rather low and would be for a time; besides that, there was reluctance to put Oboe into anything but the Mosquitos in order to delay its inevitable capture, but in March Air Chief Marshal Portal agreed to Eaker's request for eight H2S units for American pathfinder units. The Americans quickly accepted Gee as the standard for navigation and ordered 2000 sets for 1943 for which British production was adequate [17].

At about this time Assistant Secretary of War Robert A Lovett and radar advisor David T Griggs were studying British blind-bombing techniques. Their discussions with the upper command levels of the 8th Air

Force produced a near commitment based on Grigg's knowledge of things at Rad Lab for American H2S equipment, available tentatively in September 1943 [18].

Rad Lab elected to produce a 3 cm set rather than an improved duplicate of H2S, as experimental work at that wavelength was advanced and an improved resolution would accrue. Twelve B-17s with this equipment, called H2X [19] but officially carrying one of the new Army–Navy nomenclatures, AN/APS-15, arrived for the 482nd Pathfinder Group at Alconbury, located on a Roman Road 25 km northwest of Cambridge. The new radar was not long out of the experimental stages and was accompanied by a significant number of design and maintenance personnel.

The 8th Air Force had made its first use of H2S on 27 September in an attack on Emden, a principal port since the severe damage inflicted on Hamburg two months before. They made their first use of H2X on 3 November at Wilhelmshaven. By the standards of the method the attacks were considered successful [20].

Both Americans and British strove to improve the performance of H2S and H2X. Much depended on the rapid identification by the navigator of landmarks from the 'map' that the PPI presented. Comparison with a topographic map seldom proved particularly helpful and more often was confusing. Operators used photographs of PPI displays of English towns and cities in their training to develop the near-intuitive skills required to trace their course deep into Germany, and photographs of PPI displays from raids were pieced together to make radar maps. In mid-1944, when the fabulous SCR-584 became available, they could practice on English towns and cities. The 584 would track the test aircraft accurately in three dimensions and record when the bombardier said he would have made his release, allowing the error to be determined [21].

An analysis in September 1944 of relative accuracies for various bombing methods showed Oboe and visual aiming to be about the same; H2X was slightly better than H2S, hardly significant. Oboe–visual had a typical error of 350 m, H2S–H2X 1800 m. Transformed into areas this made Oboe–visual approximately 25 times more effective [22].

As Rad Lab's equipment began to come into wide use by American forces there was a clear need for a British Branch of the Radiation Laboratory, and on 9 September 1943 Lee DuBridge submitted a memorandum to Vannevar Bush requesting its establishment. In less than two weeks L C Marshall had arrived in England to establish the outpost. The choice of location was between the TRE labs at Malvern and a site near the US Air Forces Headquarters. Technical concerns won over tactical, so Malvern was selected. Douglas H Ewing became the director, administratively heading one of Rad Lab's Divisions. This group formed extremely productive and cordial ties with the TRE personnel. They helped adapt American equipment for the RAF and British equipment for the Army Air Forces. They contributed greatly to the development of the second, microwave generation of Oboe [23].

While matters progressed smoothly in outfitting the two air powers with the latest in radar, things were going badly in the skies over Germany—day and night. The first American attacks were a form of advanced training and were on targets in the occupied countries, which were not nearly so heavily defended as those in the Reich and which were often within range of escorting fighters, something that obviously made a lot of difference. As the 8th Air Force bombed targets within Germany beyond the range of escorts, losses assumed rates impossible to sustain. There should have been no questioning the fallacy of the doctrine of penetration by tight formations of self-defending bombers after the attacks on Schweinfurt and Regensburg of 17 August 1943 in which 60 out of 315 planes attacking were lost, but Eaker was not yet ready to accept defeat. After two months devoted to peripheral targets and re-thinking tactics a second attack on Schweinfurt was attempted on 14 October in which 60 out of 229 were lost [24].

Bomber Command watched Eaker's dilemma with the sympathetic understanding of a teacher who had correctly predicted the consequence of the disregard by his pupil of proffered guidance. But things were not going well in the night either. The introduction of Window at Hamburg had thrown the defense into serious confusion, and another success in the 17/18 August attack on Peenemünde, the flying bomb and rocket experimental station, introduced an optimism that brought from Harris's pen a statement that was to be quoted in nearly every book written on the European air war: 'We can wreck Berlin from end to end if the USAAF will come in on it. It will cost us between 400–500 aircraft. It will cost Germany the war'. Whatever possessed Harris to think the Americans would join him after their recent experiences is difficult to fathom, but Bomber Command had reached a peak in its strength, and he set out alone on the Battle of Berlin, a continuing action that began in November, went through March 1944 and was extended to many other cities. It did not cost Germany the war and almost wrecked Bomber Command, but it gave a decided boost to the morale of the British public.

The night fighters had been equipped with 3.5 m Lichtensteins that were unaffected by Window and were well suited for individual hunting on the long road between England and the deep targets. They acquired upward-firing cannon that allowed them to fire on the bomber from an unexpected quarter. They preserved communication between controllers and pilots in a continual radio battle of voices, noise and music. Flak gunners found that the bombers were not always enveloped in aluminum foil, another consequence of the long passages. Ground radar improved continually and had centralized control by February 1944 [25]. There seemed to be some kind of counter for every new countermeasure, even if not perfect. Losses grew beyond the 'acceptable' 5%, finally becoming 12% during a March attack on Nuremberg. In April 1944 Harris terminated the battle because of 'casualty rates which could not in the long run be sustained'. The results may have been far short of what had been expected but they were disastrous enough for the Berliners [26] for whom the worst was yet to come.

Crew morale in the bomber forces of both nations had been dealt hard blows. Pathfinders reported far too few bombs hitting their markers; planes landing in neutral countries had not always lost their way or been seriously damaged. It was just as well that the bombers had to prepare for the invasion, which required bombing targets in France that had much lighter defense. The end of the Battle of Berlin came at the beginning of a new phase of the air war—the introduction of the long-range fighter that led to the absolute defeat of the Luftwaffe.

6.4. ARBEITSGEMEINSCHAFT ROTTERDAM

The extraction by technical personnel of the remains of an H2S from a Stirling bomber downed near Rotterdam on 3 February 1943 sent a shock wave through the German radar community, one that triggered a number of people to say 'I told you so'. Kühnhold together with Dr Anton Röhrl had continued microwave research into the first two years of the war, but as the heavy requirements of an unexpected air defense for the Reich became apparent the sparse engineering resources went to designs having better prognoses of success. After the 1939 failure to detect CH with airship espionage, Martini had expanded the listening stations along the coast both in number and in bandwidth, but none had microwave capability and consequently had missed observing the 10 cm radiation emitted by the coast defense stations across the Channel that had the NT 271 sets (also called CD mark IV). Thus H2S realized Martini's dire fears, painted all the blacker because of his inability to secure the research he had favored.

Telefunken had continued projects for the development of 26 and 5 cm tubes, although without noticeable success. There remained those who continued to dismiss the value of microwaves for radar because of predictions that far too little of the incident radiation would be reflected back to the radar set, being mirrored off to the side instead. These were old, widely held arguments that rested on simplistic theory and idealized experiment: reflections of such wavelengths from a flat plane did show specular reflection—but airframes had far more complicated surfaces for which the computational capabilities of the time were inadequate.

On 22 November 1942 Karl Rottgardt, Director of Telefunken, presided over a demonstration of the latest Würzburg equipment and asserted that 50 cm radar was, given the shortage of research personnel, the best answer for gun-laying radar. Shortly thereafter he terminated the company's 26 cm research, and less than a month later the Technical Bureau of the Air Ministry, over the objections of Martini, ended all microwave investigations because of the shortage of technical specialists, this despite

Martini's and Hans Plendl's success in having a few thousand [1] electronic specialists returned to industry from Air Signals [2].

At just the moment Germany had decided that microwave research was no longer worth the effort, Bomber Command introduced H2S in flights over the Reich. At just the moment the Wehrmacht realized a crucial defeat at Stalingrad, the radar men realized a crucial defeat in the laboratory. The extracted H2S equipment was severely damaged, but the function of the magnetron was recognized, indeed all essential design elements were quickly understood. The first question was: what was its purpose?

Not that uses failed to spring to mind now that the British had summarily disposed of specular reflection. Gen Martini, who had acquired in December 1941 the post of Special Commissioner for Radar in addition to his duties as Chief of Air Signals, understood the gravity of the situation even without knowing the exact purpose of the new weapon. He responded quickly to the request of Leo Brandt, who had replaced Runge by then as Chief of Development at Telefunken, to form a special committee to exploit the Rotterdam-Gerät, as it came to be called. Plendl, by this time Staatsrat and Plenipotentiary of High-Frequency Research, concurred in this. The Nazi state produced organizations that tended to overlap and that prove difficult for an historian to untangle, and electronics was not immune to this. Plendl and Martini shared responsibilities with Generalmajor Erich Fellgiebel, Supervisor of Technical Communications [3].

The committee, Arbeitsgemeinschaft-Rotterdam [4], first met on 23 February 1943 at Telefunken in Berlin with Brandt as chairman. Brandt had already demonstrated qualities of organization and leadership that were hoped would turn the committee into a useful mechanism. Twelve persons attended the first meeting, two of whom, Plendl and Runge, had figured importantly in radar but were to be removed from active participation within a year. Two companies, Telefunken and Lorenz, were represented as was the Physikalisch-technische Reichsanstalt, the German standards laboratory, which like its British and American counterparts had excellent general scientific competence. Wehrmacht and Air Ministry representatives completed the group. Representation expanded in subsequent meetings, 43 attending by the end of the year.

GEMA was not represented at the first meeting and took essentially no part. At that time GEMA was heavily involved in providing early warning radar for which microwaves had little to offer. That Rotterdam was obviously going to be dominated by Telefunken provided no incentives for GEMA to participate. Less easy to understand is the complete absence from the meetings of Kühnhold and Röhol, who had conducted microwave research at NVK longer and more tenaciously than any of the others. Also absent was Hans Hollmann, who had written the authoritative text on microwaves, used to advantage by the British [5]. He had withdrawn from military work after observing the first years of Hitler's Third Reich.

The operation of H2S was thought well enough understood at the first meeting for them to order six similar devices to be built at Telefunken for experiment. In the meantime Lorenz, Pintsch, Blaupunkt, Telefunken and the Reichsanstalt pulled out all of their old experimental microwave equipment, generally split-anode magnetrons, to start laboratory work.

Passive receivers to detect the approach of H2S received the highest priority, and it was here that engineers met what was to be the main obstacle for the whole project—crystal diodes! Two receiver designs were proposed: a simple, low-gain device called Naxos and a heterodyne, high-gain set called Korfu, both of which required diodes. Naxos, like nearly all German radar equipment, was to appear in various forms and sub-variations. Naxos Z was primarily a homing device for night fighters that would use H2S as a beacon for locating the formations, although not individual aircraft; Naxos U was to warn U-boats of the approach of patrol planes [6]. The Reichsanstalt succeeded in making crystal diodes, but high production proved elusive and vibration caused many to fail. Naxos in all its forms was for several months a very fragile and disappointing thing and would soon be at the center of serious technical confusion in the Bay of Biscay. It owed much of its fragility to a ceramic band-pass filter at the front end, which often broke from shock, a defect not often recognized at the time that gave Naxos an evil name. The cure was simple: remove the filter. This also made Naxos capable of detecting 3 cm radiation. Korfu used a magnetron as local oscillator, as it was initially easier to produce than klystrons.

Next to the magnetron the most interesting item pulled out of that wrecked Stirling was some co-axial cable. German high-frequency cables had used Opanol (polyisobutylene) or for the highest frequencies cup-shaped ceramics linked together in a daisy chain that was flexible but fragile [7], so finding a plastic insulator used for these highest frequencies was startling. Analysis showed it to be I G Farben's Lupolen H, known to the Allies as polyethylene, whose splendid electrical properties had been overlooked[3]. I G Farben promised an initial production of 100 kg per month [8].

Interrogation of prisoners established within a few weeks that H2S was a navigation and blind-bombing device [9], which added methods for jamming and camouflage to the fund of problems for which solutions had to be found. That the PPI gave a maplike representation of the ground came as a surprise. The first thoughts had been that it would pick up a few readily identifiable targets. Its use by attacking bombers was clear. Much less clear was why Brandt continued to put so much effort into duplicating a device of little value for Germany, as the Luftwaffe was in no position to carry out night strategic bombing, while Flak had serious need of microwave radar for gun laying because of the effects of Window.

---

[3] See Chapter 2.1 (pp 38–40).

The principles of H2S design may have been appreciated early, but the construction of a functioning device did not come quickly, despite strong support. The magnetron and the duplexing tube were the easiest components to fabricate, and by May the company Sanitas produced the first German production equivalents: the magnetron LMS10 and the TR switch (Sperrohr) LG76. By the middle of June a complete Rotterdam made of British and German components was completed after a delay imposed by air-attack damage to the Telefunken shops where the work was being carried out. It was mounted in an He-111 at the Rechlin Laboratories and Development Establishment and gave a reasonable representation of the ground when flown at altitudes below 6000 m, not an outstanding performance but at least a start. Telefunken's production of six Rotterdam sets continually fell behind the expected schedule with the last delivered in December 1943 [10]. Plans were made to extend the new techniques with improved designs of their own in a series of equipment called Berlin. Magnetrons for other wavelengths were obviously to be tried, and LMS11 (5.8 cm, 15 kW) became available in December [11]. During these months technicians searched wrecked bombers not just for equipment disclosing new techniques but for components to use. They especially valued the nearly indestructible magnetron magnets because the Deutsche Edelstahlwerke, the normal source of permanent magnets, had been seriously damaged by air attack. Electromagnets with the inconvenience of another power supply became the norm.

The wide use of countermeasures undermined a human element of German radar. The high engineering quality of German equipment had allowed station personnel to be sub-standard, as operation and maintenance were simple. Now there was a need for a much higher level of ability to work through foils and noise with the rapidly designed countermeasure equipment. The question was where to obtain them. In addition to combing people from the military, Hans Plendl saw three sources: teenage boys who were radio hobbyists and who were being inducted into AA defense as Flakhelfer, prisoners with technical knowledge who were now residing in concentration camps and technically trained men in the occupied countries. His actions hint that he not only saw these people as a resource for wartime radar but as a resource for the future that needed preservation.

Britain and America had seen very early the great value of radio amateurs, whose knowledge and enthusiasm permeated whatever quarter of the electronic war they fought. Amateurs were not forbidden in Germany, but unrestricted communication was outside the narrow limits of a society in which everything not explicitly allowed was forbidden [12]. But there were still hobbyists who built and repaired radios, and Plendl as Plenipotentiary of High-Frequency Research intended to draw on this group. He had first-hand knowledge—his son was in this group. In the summer of 1943 Flakhelfer were allowed to compete in examinations for positions in a course of extensive radar training to be held at a camp on the

Stegskopf in the Westerwald. On 23 October 1943 the first of four classes of 'Stegskopfer' began their training as radar mechanics and operators [13].

That scientists and electrical engineers were being removed from society by the Gestapo could hardly be kept a secret from men such as Plendl and Martini, despite the terrible individual isolation imposed by a police state. One of those known to Plendl was Dr Hans Mayer, Director of the Central Laboratory of Siemens und Halske, who had been denounced by the maid of a neighbor for unguarded talk [14]. Following his arrest in August 1943 Plendl was able to have a high-frequency research institute formed at the Dachau concentration camp with Mayer in charge of about 25 technically trained prisoners [15]. There is no evidence of it having produced significant high-frequency research, but it probably helped Mayer to survive [16]. Had it been known that he had written the Oslo Report, he would have never even reached Dachau.

By securing contract work for engineers in the occupied territories Plendl obtained tangible research results, and those so engaged were saved from forced labor in Germany [17].

Whether the top levels of the Nazi state were favorably impressed by these measures is impossible to say, but Plendl was dismissed from his post in March 1944 for disregard of an order of the Führer: 'No one, no office or officer, may learn of a secret matter unless this is absolutely necessary in the line of duty' [18]. The reason behind his dismissal has been given by Plendl and is instructive about the attitudes toward research and development in the Reich. Plendl had naturally become very concerned about the effect of the air attacks, about which he was, of course, extremely well informed. He conceived the idea of an improved shell for AA guns and assembled a small group of ballistic experts to work up the idea. Initial tests were favorable, and he took the design with the results of the tests to the Chief of Flak, Generaloberst Weise, whose reaction was that Flak ammunition was his affair and that Plendl should concern himself with radio. The resulting discussion was not marked by restraint, and Hitler dismissed Plendl from his post.

Plendl was not alone in trying to protect colleagues from a regime that combined malevolence toward individuals and institutions with indifference to the future well-being of the people. Hans Hollmann was able to prevent deportation of people at the Kammerling-Onnes-Institute in Leyden by having them awarded a contract for investigating photographic film development at low temperature. The SS saw no distinction between temperatures suitable for photography and those near absolute zero, the Institute's specialty [19].

The terrible pounding Germany was taking from the air was creating uncomfortable times for the radar men, aggravated by serious mistrust of the whole scientist–engineer caste by the Nazis. There was also a growing group of young, Nazi-indoctrinated engineers at the Air Ministry Technical Bureau, party members, some members of the SS, who had been expressing strong criticism against Martini and Telefunken, which they took to be under the influence of its earlier, Jewish leadership. The result was Himmler's intent in early 1944 to indict for treason Martini, some of Telefunken's management and Vizeadmiral Erhard Maertens, the former Chief of Naval Communications, who had been sacked in May 1943 as a consequence of the radar debacle in the Bay of Biscay[4]. Martini did not lack defenders, and Göring, not that he did not share in the enmity addressed to these parties, stopped the process, partly because he realized the weak basis of the charges and partly to keep Himmler out of his territory [20].

These accusations against intermediate management levels in Telefunken—they were certainly not directed at Rottgardt—arose from the atmosphere that the company's founder Graf Georg Arco had instilled and that were expressed in the acknowledged friendship of many on the staff for Emil Mayer, the Jewish director who had been forced out in 1933 without even a ceremonial farewell, as well as for others who had had to leave. The accusations must have included Runge, who by 1944 no longer had specific responsibilities and who left the company and war work toward the end of the year. An incident had clarified his attitude on this key Nazi point. After Otto Böhm left Telefunken, Runge kept a photograph of him in his office. One day he found it on the floor with 'Böhm—Jew!—Bloodsucker!' written on the back. Runge then hung the reversed picture showing the inscription. Soon he was visited by a colleague he respected who told him things were getting out of hand and that he should take the picture down. Runge later wrote 'and that is what happened' [21].

The action intended against Telefunken was in fact out of character, for the Nazi regime's evaluation of political reliability was the inverse of that of the United States, which screened industrial research workers with care but took no interest in university faculties. Political purity was highly valued for Germany's universities, party membership becoming a requirement for new academic appointments, but scientists and engineers who had had during their student years problems owing to membership in organizations hateful to the government generally found a desired seclusion in industry [22]. Göring's quashing the whole thing fits with his use of jewish scientists at his Rechlin research institute and protecting their families: 'Wer Jude ist, das bestimme ich!' (I determine who is a Jew!) [23].

While Brandt's group attempted to understand and design 10 cm radar equipment and passive detectors, those engaged in air defense had immediate and pressing wants. Alexander Dahl, who had grasped the significance of the isolated Mosquito flights over the Ruhr in January 1943, had received a Naxos receiver for observing the approach of the attacking formations by summer but found its directional accuracy and range so poor that it gave no really useful information. He discussed the matter with Feldwebel (Master Sergeant) Robert Kaufmann, another former

---

[4] See Chapter 7.1 (pp 334–48).

radio amateur, and they quickly came up with an excellent but illegal solution: combine the Naxos and a Würzburg. Such field modifications were definitely not allowed, nor were violations of these regulations taken lightly, especially when they involved radar, which was entangled in near paranoic secrecy, but these men were defending their homes in the literal meaning of the word. Dahl's apartment in Wuppertal had been destroyed in September, although his wife had escaped injury.

They saw that what was needed was a high-gain antenna, such as a Würzburg dish, but also saw that the Naxos, although small, could not be accommodated at the focus of the dish, so they decided to extract the crystal diode from the receiver, incorporate it into a dipole for 10 cm radiation and place this assembly at the focus of the dish, which they had been able to 'acquire' because of a local Flak battery's recently developed Window blindness. The resultant signal, if it proved to be strong enough, would be audible at the pulse repetition frequency of the H2S set, coming in bursts corresponding to the rotation speed of the sweeping antenna. Nothing but a high-gain audio amplifier following the diode was required. For testing, Dahl made a 10 cm sparking dipole that he found he could pick up easily at 1000 m. On 23 September they set up on a high point and observed Mosquitos over the Zuider Zee with directional accuracy of 1° and a range that soon proved to be limited only by the curvature of the Earth.

The next step was obvious. Similar equipment, soon to be called Naxburgs, had to be established at widely spaced locations to allow triangulation. This required cooperation from air defense and this required confession of the destruction of a Würzburg and a Naxos. The disclosure made the next layer of authority uncomfortable but common sense triumphed over bureaucracy. Kaufmann and four other amateurs set up a small factory in a barracks. By 16 October a second station allowed the first triangulations, and the network soon joined the honest ranks of air defense [24]. Air Signals installed a chain of Naxburg observation posts stretching from the northern tip of Jutland to the Swiss border. It provided Y-Dienst an extremely reliable early warning system against the Mosquitos, which gave weak radar echoes and whose approach was the first indication of an attack [25]. Naxburg was without question the simplest and most elegant piece of equipment used in the radar war, and probably the cheapest. It continued to the end of the war to give reliable positions of the pathfinders and may have been the most effective use to which the Germans put their knowledge of microwaves.

Having made magnetrons of shorter wavelength than Rotterdam's 10 cm, the German engineers were hardly surprised when a British H2X and an American AN/APS-15 came into their hands at the end of 1943. They called the British set Rotterdam-X; there is no evidence that this name came from any knowledge that 3 cm equipment was classified as X-band by the Allies, but for whatever reasons it was aptly named. The American set acquired the name of the Dutch village of Meddo where it was found. Knowledge that the enemy was using X-band radar was immediately important for the need of suitable passive receivers, which was not an additional problem once the production of crystal diodes was in hand.

By the end of 1943 the Germans understood the basic elements of microwave radar. The production of a few copies of H2S diverted engineering skills from other development, possibly through uncertainties as to how the magnetron might best be employed in satisfying Germany's needs. After a failed attempt Telefunken turned the design of a gun-laying set to A-E Hoffmann-Heyden. They succeeded in picking up a He-111 at 8 km with an 80 cm paraboloid, then changed to the 300 cm dish of a Mannheim set and achieved a range of 30 km. Their experiments against Window were so successful that they recommended microwaves for extensive use. The gun-laying set took the name of Rotterheim and finally Marbach, which eventually saw a successful use in the defense of Hanover and Hamburg before the war ended [26].

An advanced AA radar system, Egerland, combined the Marbach with a 10 cm panoramic search radar, Kulmbach. This set used a cylindrical parabolic reflector fed from a slotted waveguide that, owing to phase considerations, produced a beam oriented 30° relative to the cylinder axis. Peak power was 10 to 15 kW with ranges of 20 to 30 km [27]. Introduction of this system in early 1944 in sufficient numbers might have changed the course of the air war, but only two were deployed before the end. By January 1945 there were prototypes of airborne sets of advanced designs, but production, if it could have been fulfilled, was scheduled for spring 1945. A few Berlin N1a sets were successfully used by night fighters in March [28].

In May 1944 the decision-making functions of Arbeitsgemeinschaft Rotterdam were taken over by the Sonderkommision für Funkmesstechnik (Special Commissioner for Radar), a new sub-branch of Albert Speer's War Production Ministry and headed by Karl Rottgardt, Director of Telefunken [29]. The Rotterdam Group held meetings until 1 September, but the function became one of providing instruction to high-level persons through lectures on various radar topics, and the number of participants grew markedly.

6.5. THE DESTRUCTION OF GERMAN CITIES COMPLETED

'London can take it!' Churchill used this exultant phrase as the name of a chapter in *Their Finest Hour* wherein he described the Blitz, the unrestricted bombing of Britain during 1940–1941. Those words must have expressed extreme relief at the disproof of the air power doctrine that had crippled western politicians for two decades, that the bombardment of cities would produce panic and uncontrollable pressures to sue for peace. It had not happened. Just the opposite. The people wanted the Air Force to 'give it back to 'em!'. But this elemental fact had made remarkably little impression on British strategy and certainly none at all on Arthur Harris, leader of

Bomber Command, for London was not alone in being able to take it, so could Berlin and Hamburg, which would provide no grounds for the Prime Minister's jubilation. On the other hand, it was irrelevant whether Britain's top levels believed the German civilians could take it or not, because the imperatives of war required that the RAF bomb the enemy's cities. If not by day, then by night; if not precisely, then by saturation; if not effectively, then as a gesture of defiance.

The air attacks on German cities grew in intensity while German production grew in mocking proportion. This was primarily because German production did not go on a war footing until after the defeat of the Wehrmacht before Moscow in December 1941 and Hitler had declared war on the United States, but it also grew because the bombings had remarkably little effect on industrial output. Those bombed out returned to their ruined homes and built some kind of shelter in them. City services returned and work went on. Heavy machinery in many smashed factories generally functioned after minor repairs even though the buildings were in ruins. Provisional shelter was provided, and work began again. None of this showed on the aerial photographs [1].

The bombing hurt the civilian population well enough, but it stiffened their resistance, just as the same treatment had affected their English cousins. As the attacks became ever more terrible and the defeats at Stalingrad and Kursk indicated the war was lost, a fatalistic determination to fight to the end arose. Many gave thought to the meaning of the terror from the skies, the unconditional surrender ultimatum from the Casablanca conference, the menace from the Soviet Union and the gradually growing knowledge of Nazi crimes for which the whole country would be called to account. When these thoughts were compounded with memories of the terrible post-war year of 1919 they yielded the grim evaluation: 'Enjoy the war; the peace is going to be tough'. Gestapo terror stiffened more than a few backbones.

In early 1944 the Luftwaffe could mark a victory over the attackers. They had shown the Americans the folly of their belief in the ability of formations of Flying Fortresses to defend themselves against fighters and to bomb accurately, and they had stripped the cover of darkness from Bomber Command through the ingenuity and industriousness of their radar engineers and fliers. It was a victory that did not, could not hold against the thousands of aircraft coming against them, and the turn came with the introduction of long-range fighters to protect the bombers in mid-1944. When the Americans returned with fighter protection to deep daytime flights into the Reich, the defenders rose to meet them and in so doing suffered the attrition that wore them down. Replacement aircraft came along, but replacement pilots were more difficult to produce; there was ever less time, less fuel, fewer instructors. The daytime attacks became so serious that the specialized night fighters were thrown into the struggle with consequent loss of these skilled men.

The introduction of the long-range day fighters had to be matched by long-range night fighters as well. The Luftwaffe tactics of Zahme Sau became particularly effective after the introduction of Lichtenstein SN-2, the 3.5 m set with wide beams and a relatively long range. For Zahme Sau the fighter directors would use ground radar and Y-Dienst triangulations to guide the fighters to the bomber stream, where they followed the bombers and located targets by airborne radar. Extreme efforts were made to block or confuse the directions the controllers gave over the radio in an electronic war of its own that changed from week to week, but the fighters were equipped with Naxos-Z that allowed them to find the bomber streams from the H2S emissions and without controller direction. They also had the passive receiver Flensburg that homed on a bomber's tail-warning radar Monica. Adding these two devices to the Lichtenstein resulted in a deadly combination.

Beaufighters proved too short ranged for countering the night fighters, so Mosquitoes were used instead. They carried the 1.5 m AI mark IV, which was replaced by the 10 cm AI mark X, the SCR-720, as use of it over Germany was allowed [2]. They also carried a version of the passive receiver Serrate tuned to the new 3.5 m band that homed on the Lichtenstein. Knowledge of this had come about through a bit of ill luck for a German pilot. He had flown in July 1944 his Ju-88 on a reciprocal course and landed his machine with a Lichtenstein SN-2 and a Flensburg on an RAF airfield, thereby ending the mystery of the new airborne radar in a most direct manner and further alerting the RAF to the dangers of using Monica [3]. Another device, Perfectos, actuated the German IFF thereby obtaining range as well as direction. Knowledge or suspicion soon resulted in the IFFs being turned off, putting an end to that advantage but complicating the task of the German controllers.

When Bomber Command began intensive attacks on the Reich after the invasion, they imposed radio silence on aircraft communication, radars and jammers until close enough to the targets to make them useful. Each attack was conducted with an intricate pattern of feints to steer the night fighters away from the main force, sometimes by nuisance raids preceding the main attack, sometimes by Mosquitoes pretending to mark the path to a city that was not bombed, sometimes by using the large force of training bombers that turned back at the last moment [4]. That Britain had sufficient bombers to do this tells much about the course that the air war was taking.

Bomber Command also had an increasing number of B-17s flown by 100 Group whose only function was jamming with all of the power and finesse that could be brought to bear. Their equipment allowed the frequencies of German radar and communications channels to be determined and disrupted immediately [5].

When deep daylight bombing became effective in late 1944 and 1945, the destruction of Germany's synthetic gasoline plants—the only sources of fuel after the loss of Romania—became a reality, and the plight of the

Luftwaffe became nearly hopeless. Fighters rose to defend their homeland to the very end, but most were flown by inexperienced pilots whose spirit and determination were no match for the well trained and well equipped Allies. The introduction of a modicum of precision bombing not only brought down the synthetic oil industry, it also brought the railways to collapse. The result of these two 'panaceas' was the end of significant industrial production.

It came to pass during the last months of the war that radar began to be less important for the defenders. Not that there were not plenty of ground stations. In the territory that one might describe as the core of Germany's air defense in 1944–1945, including the Benelux countries, Denmark, Austria, Czechoslovakia, western Hungary and western Poland, were approximately 65 first- and 126 second-class stations. First-class stations had one or two Wassermänner with some coastal stations substituting a Mammut and some inland stations a Jagdschloss in place of a Wassermann, two Freyas and one or two Würzburg-Giants; a second-class station had one or two Freyas and one or two Würzburg-Giants [6]. Germany's air-warning and fighter-control systems functioned with ever growing proficiency, but with fewer and fewer fighters its function came to be more and more to alert Flak and the cities. Flak became increasingly strong as guns were pulled back from the occupied territories retaken by the Allies and accounted for the major proportion of Allied losses. Gunners learned to shoot through jamming and Window and obtained electronic aids for doing so.

Flak suffered from the combination of Window and Carpet, which became extremely strong from the US forces toward the end of 1944, reducing the effectiveness of radar-controlled gunfire to about a quarter of what it was without these countermeasures. Frequently batteries were ordered to resort to predicted barrage fire when controlled fire was no longer possible. This reason for filling the sky with undirected bursts was justified much as had similar expedient on the other side during the Battle of Britain and the Blitz: (1) make US air crews think their countermeasures were not working, (2) maintain civilian and military morale through sound effects and fireworks and (3) to achieve an occasional lucky hit. When caught in good visibility Flak was up to its best accuracy, which was somewhat better than their purely radar-controlled fire [7]. Jamming the air-warning and fighter-control equipment was much less effective than jamming Flak because of the redundancy of the information available to the defenders.

Keeping radar going became more difficult. Stations were manned by what was left after numerous 'combing' operations to procure men for the front. Old men and young women operated the sets, often developing commendable skills, but maintenance personnel for faults that could not be removed by replacing modules were in short supply. The Stegskopfer, the teenaged radio enthusiasts who had received crash courses in radar, helped bring ailing equipment back to service with the most difficult repairs done by senior mechanics who raced from station to station by train, bicycle and hitch-hiking with lodging secured as they might. Feldwebel Kaufmann, the co-inventor of the Naxburg, found himself during the last months setting up a jamming station and keeping it and others operating. On occasion he obtained meals by repairing the radios of the farmers with whom he stayed. Travel in those days entailed a high probability of being strafed by long-range fighters, who shot at anything on the ground that moved, if there were no Luftwaffe planes to fight. Kaufmann had his tool box shot to pieces from this source [8]. Not the least problem such men had to face was the danger of the Feldgendarmerie, the vicious military police who pursued deserters, real and imagined, whom they left hanging with a sardonic sign to deter future malefaction.

By late 1944 the Flak units defending the Reich had taken on an irregular character. Personnel were 45% foreigners, teenaged boys and young women (Flakhelfer and Flakhelferinnen) with assistance from the Labor Service (Reichs Arbeitsdienst). There were, however, enough experienced men to keep the Allied fliers unaware of the loss of elite status [9].

German emphasis went toward the radio-navigational technique Gee. The most effective station for this was located on the Feldberg in the Taunus Mountains about 25 km north of Frankfurt am Main, which utilized a pre-war television station. It also jammed meter-wave Oboe, but with time Oboe worked on microwaves and the meter-wave transmissions were only a cover. The jammers used Gee-like pulses to confuse the receivers, but these transmissions disclosed the position of the station [10], and photo reconnaissance planes confirmed it. In a deviation from the normal, soft procedures of radio countermeasures, the Feldberg station was attacked by a force of P-47 fighters and put out of action on 2 March 1945 [11].

Jamming radio navigation equipment that operated on meter waves was fairly straightforward, but microwave Oboe proved essentially unjammable. Indeed jamming microwaves in any application proved difficult and was never satisfactorily accomplished by the defenders. It was the fundamental problem of placing the jammer in the narrow beam of a microwave set. For H2S an alternative was to illuminate the target area being observed by the bomber with radiation from a ground jammer station, thereby scattering extraneous radiation into the H2S receiver. For a while large spark transmitters with their great power looked hopeful, and Professor E Marx at Braunschweig High Voltage Institute, whose tremendous spark demonstrations have fascinated—and deafened—visitors to Deutsches Museum in Munich for years, worked on this method. It failed because the broad spectrum of Hertz's oscillator injected too little energy into the relatively narrow wavelength band of the H2S set.

Professor P. Gorcke worked with Telefunken to use the most powerful magnetron available, the LMS100 that gave 100 kW peak power, and might thereby circumvent the spark transmitter problems by concentrating the radiation in the H2S bandwidth. An He-111 with a German-made H2S unit allowed testing of their equipment, which was able to cover the PPI with numerous little spots, but these dubious results came too late to have any effect on the bombers [12].

It was not until the first copy of H2S, Rotterdam No 1, had been tested in flight that the possibilities of radar camouflage could be examined in any serious way. The first flight in June 1943 showed the contrasts between water, towns and flat land on the PPI scope, which automatically presented the question of altering the reflecting conditions sufficiently to confuse the H2S operator. By September a number of experiments had been tried, both to determine what might weaken the return signal and what might enhance it. They learned that flat land appeared darker than plowed fields and that meter-sized corner reflectors fashioned from sheet metal were the best means for enhancing a reflected signal; it was further determined that mounted on floats these might prevent lakes and streams from appearing dark. Spacing them about 150 m apart did remove the darkness and hide the distinguishing features of bodies of water, but they showed up as points of light, if the radar receiver sensitivity were reduced [13]. It is difficult to say what radar camouflage contributed or even in whose favor.

The extent to which German electronics was devoted to countermeasures toward the end of the war was estimated by Professor Abraham Esau, who replaced Plendl as Plenipotentiary for High-Frequency Research, as 90%, a total of 4000. This prevented any serious development of 10 cm capabilities [14].

If radar became less important to the Germans, it became more important to the Americans and in a way that completely turned their air policy upside down. As they initiated deep daylight raids again the US 8th and 15th Air Forces found that visual bombing was the rarity not the rule. Only four clear days in a month was not unusual during fall and winter. This produced a remarkable change in American bombing doctrine: the vaunted precision bombing allowed by the Norden bombsight, which had held fast the minds of the Army Air Corps during its entire preparation for war, was used only as occasional opportunity allowed. Radar bombing was to be the rule, as it had been for Bomber Command, and the inherent inaccuracy of H2X meant that this would be carpet bombing. In practice the 8th Air Force found the error of radar bombing was even worse than expected from the controlled practice studies made in Britain, being 3 km on the average, a rather large pickle barrel [15]! So it was that the two bomber forces approached one another in tactics as the war groaned to a close. Bomber Command began attacking in daylight along with the Americans, and both bombed by radar as a rule [16]. Increased daylight operation made it easier for Flak.

Some improvement in the accuracy of radar bombing came from the extended range achieved for new Oboe stations placed in France and the lowlands. Oboe was enhanced by a new technique from TRE called Gee-H. This made use of the existing networks of Gee chains. The transmitting stations were modified so they could function as beacons, and an additional unit was added to the aircraft's Gee receiver to allow it to operate as interrogator of the ground stations. Ranges could be determined as accurately as with Oboe and used to the same effect. It had the advantage of permitting several aircraft to use the system simultaneously by interrogating with coded pulses yet leave Gee remaining functional for its normal hyperbolic navigation [17].

The H2X systems, AN/APQ-13 and APS-15, were designed to work with ground beacons and had, in addition to the PPI display, an accurate range unit used in a manner called Micro-H [18]. It required, of course, special 3 cm beacons that responded to coded interrogation. Both Gee-H and Micro-H allowed blind bombing as accurate as Oboe, but were limited by the distance to the radio horizon. However, as the war progressed and the Germans fell back this allowed ever deeper penetrations.

Both air forces sent electronic intelligence flights frequently over Germany. The British chose the Mosquito, the Americans a two-man version of the P-38. One of the P-38s was equipped with the new AN/APR-7 microwave receiver developed by Terman's Radio Research Laboratory, but German interceptors using microwaves were very rare [19].

The news of the bombing of Dresden and its resulting destruction by firestorm attracted little attention in the United States, where audiences at the newsreels would soon be cheering the fire bombing of Japan, but the reaction in Britain was surprisingly strong. In happier times Dresden had been a favorite city of the English; some had maintained residences there and many had visited to savor the music, architecture, art, history and general charm that had gained the capital of Saxony the reputation as the German Florence. The contending forces had worked out an informal agreement to spare Italy's treasures. Could not the same have been done for Dresden? And for the Allies to have willfully destroyed a marvel of European culture shocked people into examining the horror that was being conducted in their name. By February 1945 area bombing had been altered in the minds of those conducting the war from an act of barbarism to an accepted practice; it had become the ultimate expression of outrage at Germany's refusal to give up—Ludendorff would have accepted the logic of the situation after the invasion had succeeded. As an example of radar bombing with H2S, H2X and countermeasures, however, Dresden was a notable success.

So ended the Great Radar War. It was a struggle lofty in its technical sophistication yet vile in its indiscriminate destruction of population and civil organization. It was a struggle in which an air battle was won or lost according to the appearance of an otherwise insignificant trace on an oscilloscope located far from the scene of combat. It was a struggle that at times seemed to remove control from the military commanders and give it to an academic caste untrained in war who forced alterations of tactics with bewildering rapidity. All of these remarkable changes in the nature of war came about from radar.

In the Battle of Britain radar had been exclusively a defensive weapon, one scarcely appreciated by Germans in command positions. As a consequence of this ignorance they employed only trivial or half-hearted countermeasures and did not press home attacks on the stations. When the weight of bombs began to flow east and south rather than west and north the new defenders put their own radar to immediate use—for not everyone on the German side had been oblivious to its potential—and the new attackers set about just as quickly to counter it. The four-year struggle had begun.

At squadron level the Luftwaffe had begun to experiment with ground-controlled fighter direction during the first weeks of the war, well before their superiors showed interest or understanding. These individuals soon caught the attention of Generals Martini and Kammhuber, who began to construct the Reich's air defense around radar. The engineers at GEMA built the necessary long-range equipment, Wassermann and Mammut, and those at Telefunken improved the Würzburg as a gun-laying set and adapted their 50 cm skills to the airborne equipment, Lichtenstein, that Kammhuber requested. During 1940 Bomber Command had been able to fly over Germany at night with light losses but little effect, owing to their severe navigational problems. As Bomber Command began to have the concealment of darkness taken from it, TRE proceeded to invent countermeasures to restore it.

In all this, radar was a defensive weapon being countered, but the attackers needed more than the ability to protect themselves while over Germany; they needed some way of hitting the enemy's vitals. To this end they strove to make radar an offensive weapon. The most effective outcome of these efforts were the radio-navigational equipments Gee, Gee-H and Oboe. The latter two proved to be blind bombing equipment of startling accuracy—so long as the bombers were above the radio horizon of their control stations, a criterion that left most of Germany's industrial potential safely out of range. Had these techniques been amenable to long-range use, World War II would have had a different, possibly happier conclusion, because the selective destruction of key industrial elements would have been possible, as adequately demonstrated in the Ruhr.

The hopes that H2S would provide such a possibility did not long survive the tests of reality. It was not 'the turning-point of the war' as A P Rowe had pronounced [20]. Under ideal bombing conditions H2S could assure that a particular quarter of a large city might be hit, but under combat conditions it could give only moderate assurance that the city as a whole could be hit—a decided improvement over 'celestial' navigation but hardly what was needed.

If H2S was unable to deliver much more than saturation bombing, was it worth it? If German morale did not collapse and industrial production was not seriously affected until long-range fighters allowed some degree of accuracy through visual aiming, was radar bombing anything more than a way of punishing Germans? The answer to this is yes. Whether militarily or industrially effective, the attacks hurt the population and had to be fought. The air offensive against Germany amounted rather early to a second front, against which the defense required Flak batteries for the cities, elaborate air warning and fighter control, squadrons of highly skilled night fighters, bunkers for the population, greatly enhanced fire-fighting organization and equipment, services to assist those who had lost their homes and continual readjustment of transport and industry. These things required a huge military force and large numbers of administrative and support personnel.

If the war could not be won in the air, then it had to be won on the land, and for this the invasion was necessary. The success of this unparalleled undertaking turned on control of the air over the ships and beaches, and this happened because the Luftwaffe had been severely weakened by defending the Reich. That the air offensive was a descent into barbarity cannot be denied. That the Allies had an alternative is a matter that only subsequent generations have had the serenity to consider.

PHOTOGRAPHS: RADAR FOR AA ARTILLERY

An American SCR-268 1.5 m gun-laying radar on Kwajalein Island. This was the Signal Corps's first radar design. Its original function was to give AA batteries air-warning capability and to point searchlights, but experience showed it capable of directing blind fire. The center array of dipoles was the transmitter antenna, the left array determined the azimuth (horizontal direction) to the target and the right array the elevation (vertical direction). It remained the US Army's AA radar until replaced by the 10 cm SCR-584 in 1944. National Archives photograph 80-G-400984.

An example of the technical imperative, Flakleit g (FuMO 201). This radar was built for the Kriegsmarine by GEMA. It used the 80 cm Seetakt wavelength and was capable of directing fire on surface or air targets. It had an antenna configuration very similar to SCR-268 but was developed independently. It was mounted on a rotatable, underground, armored optical range finder for shore batteries. Photograph courtesy of Fritz Trenkle and Bernd Röde.

Receiver for British GL mark II. This early AA set worked on the 5 m band. The wooden structures at the sides of the cabin held two dipoles that allowed the determination of horizontal direction through rotation of the cabin and lobe switching. The transmitter was a separate unit located about 100 m distant. Elevation could be determined through the elevator-mounted dipole. This set was noted for its robust construction and reliable electronics but not for blind-fire capability for which it had little. It proved a great favorite among the Russians, who built copies known as SON-2, but British imports overwhelmed native production. National Archives photograph 111-SC 242266.

An American SCR-584 at Montalbano, Italy on 4 December 1944 and used by Battery B, 403rd AAA Gun Battalion. This auto-tracking 10 cm set is easily recognized by the paraboloid antenna sitting on top of the van that housed the electronics and secured the dish during transport. The separate antenna in the foreground is for IFF. The 584 was the most versatile radar developed during the war. The best tribute to it is that when this was written (1998) hundreds were still in operation and that it could be purchased from stock as could supplies for maintenance. National Archives photograph 111-SC 229897.

British Army GL mark III. This 10 cm gun-laying radar saw deployment in 1944 when it became the standard for British AA batteries. It employed manual tracking and, lacking panoramic capability, generally had to be 'put on' by other data. It was linked to gun directors by selsyn transmission. Manual tracking greatly diminished its accuracy when following the rapidly moving, low-flying V-1s. That its qualities were inferior to the SCR-584 and that it came into service so late after the British invention of the cavity magnetron were the direct result of the low-priority given to AA artillery in Britain. Historical Radar Archives. Crown Copyright.

Royal Navy type 284 50 cm radar mounted on a mark 4(GB) HA director for both main armament and AA fire-control. The navy began fitting ships with sets of this basic design in 1940. High power came from the Micropup triodes, and beam formation came from combinations of Yagis, called fishbones. It was with this kind of radar that the Suffolk tracked the Bismarck during the initial phase of a memorable surface action. Churchill Archives Centre, Royal Navy photograph. Crown Copyright.

German Würzburg C (FuMG 39T-C) gun-laying radar. This 50 cm set was originally designed for following aircraft by having a large number scattered about the countryside. When its directional accuracy, even without the rotating off-axis dipole shown uncovered here at the feed, indicated its value for directing AA guns and searchlights, its function changed; it went through various modifications. It proved the best radar until the Allies introduced 10 cm equipment for this in 1944. Bundesarchiv photograph 594/266/31A.

329

German Würzburg at a typical 88 mm AA battery position. AA batteries generally had four to six guns firing with identical or near identical settings from the same director, whether optical or radar. Bundesarchiv photograph 356/1845/8.

Railway radar. Poor road conditions in the Soviet Union led the Germans to use radar trains. A small Würzburg D, rather than a giant probably meant that it was used for a rail AA battery. Such batteries were also useful in rushing AA defense to important locations, especially harbors, until permanent defense could be established. Bundesarchiv photograph 621/2943/24.

Japanese Army 1.5 m Tachi-2 searchlight and AA radar. This set had four dipole receiver antennas placed before a metal screen and connected to the receiver through a rotating capacitor that generated a conical scan. A fifth dipole was located at the center for the transmitter. This set was successful enough in directing AA fire at the B-29s bombing Japan that serious countermeasures were undertaken. It had similarities with the British SLC but was designed before the capture of one of these at Singapore. Tachi-4 grew out of this set and incorporated aspects of SLC, replacing dipoles with Yagis. National Archives photograph 111-SC 290064.

Japanese Army Tachi-3. The transmitter of a 4 m searchlight and gun-laying radar having design elements suggested by the British GL mark II. The dipole array was mounted above an underground shelter that rotated in azimuth. Altering the phase allowed the elevation of the radiation pattern to be adjusted. National Archives photograph 111-SC 290063.

Japanese Army Tachi-3. The receiver of a 4 m searchlight and gun-laying radar having a design suggested by the British GL mark II. The dipole array was mounted above an underground shelter that rotated in azimuth. The signals from four dipoles were connected through a rotating capacitor to the receiver forming a conical scan. National Archives photograph 111-SC 290062.

Japanese Army Tachi-4 searchlight and AA radar. This set had four Yagi antennas, removed here to prevent capture, placed before the metal screens for vertical and horizontal adjustment. The horn is a speaking tube for oral communication. National Archives photograph 111-SC 231308.

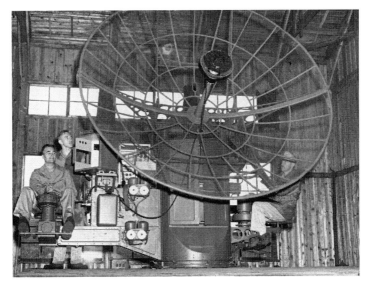

Americans operate the Japanese Army Würzburg, Tachi-24. The construction of this prototype was not completed until just before the end of the war and saw essentially no useful service. Its requirements for components, especially the Telefunken LS180 decimeter triodes, strained the capabilities of Japanese electronics industry. Manufacture was by Nihon Musen (JRC) under the supervision of Heinrich Foders. Photograph courtesy of Marvin Hobbs.

American and Japanese radar men discuss equipment and their recent belligerency. The location of the meeting was the Aircraft Control and Warning Center inside the grounds of the Imperial Palace, which had been off limits to US attacks. Photograph courtesy of Marvin Hobbs.

CHAPTER 7

ALLIED VICTORY IN SIGHT

7.1. THE BATTLE OF THE ATLANTIC, 1939–1945

On entering the Second World War the four great naval powers were surprisingly unmindful of the menace or value of submarines, despite the instruction provided in the waters around Great Britain during 1917–1918. For their part the Royal Navy accepted their unqualified victory over the U-boats with escorted convoys as reason to discount them as a major threat to the nation's vital supply lines in the future, a view shared by Grossadmiral Raeder and most of the upper levels of the Kreigsmarine. The Imperial Japanese Navy kept 'the decisive battle' uppermost in their planning and gave little or no thought to the use of submarines as commerce raiders whether by Japan—hardly the proper calling for a warrior—or by the United States. The Americans thought they had little to fear from submarines attacking their shipping and rejected a form of warfare that had been the cause of their joining the Allies in 1917.

In the high levels of world-wide naval command one man, Karl Dönitz, later Grossadmiral, alone held a contrary opinion. Dönitz was cut from the same kind of cloth as Air Chief Marshal Sir Arthur Harris. Each was confident he was capable of winning the war alone with the weapon he had so assiduously refined; each sought to destroy the enemy's capability to fight, the one by stopping the flow of supplies, the other by destroying industry and the people's spirit; each was a superb officer able to inspire his command and organize its complex activities into a coherent whole; each had in his nature a hardness founded on his concept of duty that gave no pause in using the terrors at his disposal, if necessary pushing the hideous logic of war to extremes that most soldiers reject in their hearts even though accepting in their minds; each led bands of young men possessed of a courage fashioned from the idealism of youth, a courage that allowed them to reckon coldly their poor chances of survival yet to continue unflinching; each vigorously opposed the diversion of his forces from what he insisted was their true mission. Both finished the war in disrepute. Dönitz was imprisoned for ten years as a war criminal, although for the use of slave labor in shipyards, not for the activities of the Kriegsmarine.

Harris found himself omitted from the honors generously distributed at the end of the war, but probably as much from personal animosity within the Labour Government as from reaction to his policy of saturation bombing. One of them saw in radar the cause of his failure; the other had placed in it his hope for the success that eluded him.

Dönitz based his confidence in the submarine on tactics, not on technical improvements in the weapon, as the characteristics of his boats in 1939 were but little advanced over the ones in which he served in 1918. His approach had two elements: (1) attack by a group of U-boats, the wolf pack, ordered to the scene by radio from a central headquarters and (2) attack on the surface at night rather than under water by day. He had tested the latter during the final weeks of 1918, noting that at night a surfaced submarine was almost as hard to find as one submerged during the day and that its surface speed enabled rapid positioning for launching torpedoes. If discovered, a quick dive could follow. He also noted that asdic and sonar, the acoustical location techniques developed by the British and Americans after World War I, were useless against a surfaced vessel. He had tried out the wolf-pack tactic in pre-war exercises. He could be sure that Ireland would remain neutral, removing valuable bases for anti-submarine patrols, although he probably expected this inconvenience to be dealt with brusquely by Britain.

By mid-1940 Germany's naval position had changed in a manner that shook the British command: the coasts of Norway and France had become bases for U-boat operation. By mid-1941 commerce raiding by surface vessels was clearly a failure, and Dönitz was being given greatly increased support and doing quite well.

The Battle of the Atlantic, as Churchill named the struggle of the Allies against the U-boats, was without question the most complicated and technical form of warfare the world had ever seen. From relatively quiet beginnings it grew to a raging crescendo both in violence and cunning until Dönitz's defeat in 1943, yet continued for two more years. It was a technical battle, but no less cruel for that.

Intelligence was the key for both sides. Dönitz directed his boats from a shore station that collected information about Allied sailings from two principal sources: decryption of enemy signals by the B-Dienst and reports from the U-boats themselves. Agents listening to seamen's conversations in pubs, the subject of countless admonishing posters, figured not at all. The British had succeded in working the German encrypting machine, the Enigma, through a major effort at Bletchley Park. This was Britain's most closely guarded secret and is generally referred to as Ultra, a name taken from the special security classification introduced to protect this source— Top Secret Ultra [1]. The Germans balanced things by generally being able to read the British Merchant Navy codes. There were periods when signals were secure, but for much of the time encryption was pointless.

Radar entered the battle early but had little effect until comparatively late. The 1.5 m type 286M was a combination air–surface search set based on ASV mark I that began to be mounted on British vessels toward the end of 1940, and by September 1941 177 destroyers and 40 sloops and corvettes were so equipped [2], but it presented little danger to the U-boats, owing to its lobe structure, fixed-direction antenna and simple indicator. A submarine running with little more than the conning tower above water was observable no farther than 1 km [3]. In March 1941 the destroyer *Vanoc* sighted U-100 with a 286M and rammed it. Thus the first sinking marked to radar's account was effected by the method of Roman galleys [4]. The type 286P had a directable antenna, which was a slight improvement, but only twelve destroyers had it in September 1941.

The Navy had rushed to exploit the advantages of 10 cm radiation as quickly as possible, succeeding in placing about 30 sets of this type 271 into use by September 1941 [5]. Microwaves extended the radar range, but the lack of PPI indicators until mid-1943 [6] severely reduced its use in the confusion of a convoy action. Possibly a more important contribution of PPI for convoys was its enormous help in keeping such large collections of ships on station at night [7]. The 271 gained endearment from its ability to locate lifeboats at night or in fog.

An ASV mark I had succeeded in picking up a submarine from an airplane on 2 December 1939 [8], but even the introduction of ASV mark II radar, which was delayed because of the priority given the production of radar for night fighters, had not had a dramatic effect. In general it was more useful during days of poor visibility because it was difficult, nearly impossible to attack a surfaced U-boat at night from meter-wave radar data alone. This came about because the target reflection became confused by the ocean-surface reflection as the aircraft approached the submarine; this generally gave a minimum range of 1.5 km. It was the same effect that limited the maximum range of meter-wave AI equipment. A notable exception was the success of a radar-equipped Swordfish squadron at Gibraltar that combined the slow speed and the better night vision offered by open cockpits with the predictable courses of U-boats attempting to enter the Mediterranean on the surface at night; in November and December 1941 they sank one and damaged five others sufficiently to force them to return to their French bases for repair, deterring further passage of the straits [9]. But Swordfish crews always seemed to do impossible things.

Coastal Command Squadron Leader Sydney Lugg, remembered for his invention of the radar beacon, came into discussion with Squadron Leader Humphrey de Verde Leigh, a pilot with anti-submarine experience in the previous war, and told him of the excitement and failure of ASV. In doing this Lugg violated security regulations, as Leigh's duties were administrative without any need to know things about the highly secret equipment, but out of this violation came a memorandum that Leigh submitted on 23 October 1940 that was to alter the duel between aircraft and submarine.

Leigh proposed mounting a searchlight on the airplane and trying to illuminate the target when the radar image became obscured by surface reflection. The following month was filled with meetings and exchanges of letters and memoranda; at the end of it he found himself charged with developing the idea, and on 4 May 1941 he made several successful runs on a submarine. He then experienced an infuriating delay of months when Coastal Command considered another searchlight technique, the consequence of an unfortunate and later regretted decision by Air Chief Marshal Sir Philip Joubert when he took charge of Coastal Command, but saw his design proved the better and in production by the end of summer [10]. By June 1942 five Wellingtons had been equipped with Leigh Lights and crews trained for their use. During the night of 4/5 June one attacked and seriously damaged an Italian submarine at the southwest corner of the Bay of Biscay, initiating a series of adventures for her and her stalwart crew, none of which led to her sinking. It was the first of many such engagements, but the first sinking did not take place until a month later [11].

To follow the Battle of the Atlantic one must have an understanding of U-boat tactics. The World War II submarine was a surface raider that could hide under water. On the surface it was faster than many merchantmen, submerged slower than the poorest freighter. Thus a submerged attack required prior positioning, which barring luck meant running on the surface. This was not particularly dangerous, if there were no airplanes on patrol, because the U-boat could not be seen by the convoy at distances that allowed its lookout to see the top hampers and smoke of ships. Spotting a convoy could be relatively easy when B-Dienst had extracted its route.

The tactics of defense swung between two opposites of thought: win by sinking submarines or win by saving merchantmen. In practice, combinations of the two eventually did win, but it was a long time in the doing. The first U-boat successes were countered by organizing ships into convoys, the age-old method of protecting ships from raiders of any sort. Procuring sufficient escorts and trained crews held back the start, but they were in hand by late 1941. A properly protected convoy combined the two approaches by forcing the submarines to fight—and thereby be sunk—while protecting the freighters and tankers. It was not perfect for the defense, but the convoy fights, often conducted in raging winter gales, ended the first of the submariners 'happy times'. Ship-mounted radar was one of many weapons in this phase, all of which were desperately needed.

President Roosevelt's bold and hostile extension on 18 April 1941 of the boundary of the Pan-American Neutrality Zone, placing it closer to Europe than to America, was followed by his ordering the US Navy to escort convoys within this region. Hitler did not want a repetition of 1917 and ordered his forces not to attack ships so accompanied. It is difficult to ascertain what the Navy learned during this on-the-job training, but whatever it was, it was not apparent to the American merchant seamen who died in record numbers during 1942. The unescorted, indeed completely unprotected coastal shipping of the United States suffered astounding losses after Germany's declaration of war, making it a second, extremely happy time for U-boat skippers, with occasional relaxation in tropic climes.

(The extreme ineffectiveness of American antisubmarine activity in 1942 led to the most contentious disagreement of the war between Britain and America. All British and most American historians have laid the blame on Admiral Ernest King, who held the post of Commander in Chief, US Fleet. An evaluation of the controversy clearly lies beyond the bounds of this book, but the reader must be informed that King has found an able defender in Clay Blair [12].)

During 1941 Dönitz had greatly extended the range of his raiders through the use of supply ships, and when the Royal Navy eliminated these, as it had the surface raiders, he replaced them with large supply submarines. During 1942 this mode of supply allowed the attacks to be changed quickly from convoyed regions to those of individual sailings. For a time, arrivals and departures from New Orleans became risky matters.

For Coastal Command the main battlefield in the war against the U-boat was the Bay of Biscay, where the boats entered and left their patrols from the French bases. By the end of summer 1942 enough aircraft had been equipped with ASV mark II and Leigh Lights to have impact. At the beginning there were never enough planes for the job, and very few of the four-engine bombers that Joubert eagerly sought. A German radio-listening station at Boulogne had detected 1.5 m ASV radiation from patrol aircraft in the fall of 1941 [13], and Telefunken is reported to have examined an ASV set in May of that year [14], so it is not surprising that the radar secret of the Leigh Light was quickly recognized. Submarines began to mount the radar receiver R600A, made by the French companies Metox and Grandin and generally called the 'Metox', that swept a wavelength band from 0.9 to 3.8 m. The pulse repetition of the radar yielded a characteristic tone at the receiver output that was monitored by headphones. The first operational boat to carry a Metox left Brest on 9 August 1942, only a month after the first sinking generated by the light [15]. Metox was a great success, as it could pick up patrolling planes at ranges far beyond the capabilities of the radar that was looking for them. Fliers soon began to notice radar contacts disappearing from their screens as they approached, and in October only one U-boat was sunk using the radar-light technique, bringing the total for the method to only four, although many more had been attacked [16], an event that did not contribute to feelings of well-being of the U-boat crews, on whom the light was particularly unnerving.

The advantages of microwaves over meter waves for ASV were as compelling as for AI. The potential was so strong that the Radiation Laboratory had begun the design of the low-altitude bombing radar that was to allow bombers to attack Japanese ships in the Pacific at night using radar unassisted by light. Use of microwaves against submarines became in late 1942 a matter of cabinet level policy in the British government because the system proposed for anti-submarine warfare differed but slightly from H2S, the navigation radar that was the hope for bombing German cities beyond the range of the radio navigation systems Gee and Oboe.

Those favoring initial use of microwaves by Coastal Command alone argued on the basis of a rapidly advancing science that used physics and statistics to analyze problems of engagement and that became a permanent military discipline called operational analysis. P M S Blackett, physicist and naval officer, demonstrated that the merchant ships saved by a few long-range patrol planes would contribute far more to victory than their use in bombing Germany [17]. Based on that analysis he, Tizard and the Admiralty proposed diverting a large number of long-range bombers to the Bay of Biscay action, but the plan was rejected as unnecessary by Air Marshal John Slessor, who had just taken over Coastal Command, and by Lord Cherwell and the Air Ministry [18]. In considering this dispute one must be aware that Blackett and Tizard were temperamentally opposed to the bomber offensive and that Slessor favored it. Tizard, Blackett and their associates further argued that using H2S over Germany insured that enough of a set would be recovered from a wrecked bomber to disclose its essence to the engineers of Telefunken within days of its recovery, and although copies might be slow in coming, receivers could be at sea in short order. The proponents of H2S replied that it was by no means certain that the Germans had not already learned of the microwave work and might have receivers ready as quickly as they had countered the success of the Leigh Light. (Neither party would have guessed that this receiver would take so long to produce or be so unsatisfactory.) The resolution of this conflict between Tizard on one side and Lord Cherwell on the other was resolved by the War Cabinet Chiefs of Staff favoring immediate use of H2S by Bomber Command.

Additional long-range aircraft for the Biscay fight came to the aid of Coastal Command from another, unexpected source, one that was quite agreeable to Churchill—the United States Army. Admiral King's inaction at the massacre of American coastal shipping had so affected his Army counterpart, George Marshall, that Army bombers had been organized to protect off-shore shipping [19]. By the end of 1942 American antisubmarine efforts had become sufficiently menacing to force the U-boats to seek ships in less hostile waters than the American coasts. By that time the American Army was operating in North Africa and had the desire to give their ocean supply lines some protection of their own, so one of the groups of Liberator bombers was ordered there, the other to follow in time. The movement of these aircraft, which were equipped with the new SCR-517 microwave surface search radar, caught the attention of Churchill who proposed through Harry Hopkins, Roosevelt's chief confidant, that they be based in England instead to help Coastal Command patrol the Bay. Roosevelt agreed, so the units substituted a harsh winter in Cornwall for a mild one in Morocco but with clothing intended for the latter [20].

Leo Brandt heard reports on 17 March 1943 at the second meeting of the Rotterdam Committee of design efforts for a series of microwave receivers intended for 10 cm air warning [21]. They were needed at ground stations to help work through countermeasures, to guide night fighters to the bombers, and to alert submarines; all were named Naxos, Naxos-U being for U-boats. All sets were delayed because of the difficulties in fabricating the crystal diodes for the detectors, and the submarine design had additional problems, ones that had not troubled Metox. The 1.5 m receiver had a wooden frame to hold the antenna, the Biscay Cross, which could be taken below in the few seconds allowed for a crash dive. A similar arrangement for Naxos-U, issued to the U-boat men in October 1943, required coaxial cables capable of operating at higher frequencies, and the cables available had fragile insulation easily damaged by the rough treatment of moving the antenna below in a crash dive [22]. The mistreated cables generated standing waves with lowered transmission the consequence. This added measurably to the difficulty of poor sensitivity in the experimental sets tried in the summer of 1943 and helped compound the confusion [23].

Coastal Command began using the 10 cm ASV mark III in the Bay of Biscay in March 1943, the month after Air Marshal Sir John Slessor replaced Joubert. Joubert had built Coastal Command into a splendid weapon and led it through its worst times but could not evade the onus attached to its performance in the escape of the capital ships from Brest, and he disappears from the commanding pages chronicling the history of World War II.

The first aircraft equipped with microwave radar for the Bay of Biscay were Liberators of the 480th Group of the Army Air Forces, which began operating in cooperation with Coastal Command from late December 1942 to early March 1943. They had the 10-cm SCR-517 that had emerged from a collaboration of Rad Lab and Bell Labs, but their planes had no Leigh Lights, so radar only augmented eyes. Crews had difficulty in maintaining the sets at first and found the range-bearing indicator [24] confusing and of little value in navigation. Of their ten sightings only one was by radar first, although two were simultaneous. An attack followed each sighting; one submarine was thought to have been sunk [25].

Owing to the priority given H2S, Coastal Command did not have an aircraft fitted with mark III until the end of January 1943 with no more until March, when 20 were installed [26], and the first attack by the new radar-light combination was on 18 March [27]. Mark III performed much as had the mark II before being countered by Metox; it brought more sightings than attacks and far more attacks than kills, but confidence in the Metox receivers, which now gave no warning for microwaves, was shattered. A fierce secrecy restricted knowledge of Allied microwave capability within the Kriegsmarine [28]. Nothing was withheld from the top command levels, but they lacked the electronic knowledge and experience that would have allowed them to evaluate the rapidly changing stream of information from which they had to make decisions. Those who did have the knowledge and experience lacked an easy channel of communication with those at the top and were unable to help them sort out fact from fiction. If those at the top were confused, it is understandable that the boat commanders were mistrustful. The records of the Befehlshaber der U-Boote show no mention of microwaves until late 1943. The earliest evidence of Dönitz's comprehension of the microwave problem is in his report at meeting with Hitler on 13 January 1944 [29].

This crisis led to the dismissal of Vizeadmiral Erhard Maertens, Chief of Naval Signals and caused the Kreigsmarine to organize the high-frequency research of naval and industrial groups into the Organisation der deutschen Schwingungsforschung für den U-Boot-Krieg having nine co-working sections [30]. Conspicuous successes of this effort have escaped record.

The 480th Group was sent to Morocco before the Battle of the Bay began, which was generally fought during the day, for which they were well equipped. They came under the control of the US Navy in North Africa where acrimonious dispute replaced cordial RAF cooperation [31]. By then their radar skills had greatly improved, and sightings were generally by radar first. An analysis of the realistic ranges for sightings by ASV marks II and III and by vision on a clear day shows them to be equal at about 8 km [32].

Two possibilities came to the minds of the German radar engineers who had to deal with the problem in ignorance of Rotterdam: the Allied ASV set was switched on only very briefly so that the signal on the Metox was so brief as to pass unnoticed or the pulse repetition rate had been increased so that it generated a tone in the observer's earphones above audible limits, conceivable because of the short usable range of ASV, which allowed short periods between pulses. The answer to both of these possibilities was determined by replacing the earphones with a 'magic eye', a small tube with cathode-ray lighting that responded to high frequencies and that was used for tuning radios visually. It quickly eliminated these reasons as the cause of the submariners' distress [33].

A third possibility was that the aircraft were using some means of homing on the surfaced boats, and emanations from the Metox receivers immediately came to mind. All heterodyne receivers radiate to some degree at the local oscillator frequency, detectable if one seeks it. At about this time an unknown Coastal Command prisoner told his interrogators that the patrol planes were indeed homing on the receivers. Coming at the time that it did, this removed all doubt about the origin of the mysterious attacks, and on 13 August Dönitz ordered all Metox sets removed from service. A replacement receiver that radiated substantially less and that was also specially designed to detect short bursts of radar signal, the Wanz G-1, was quickly put into service and just as quickly removed, yet most patrol aircraft were still equipped with the 1.5 m radar. By the end of 1943 U-boat skippers mistrusted all radio equipment [34]. The identity of the prisoner who 'confirmed' the Metox radiation theory has never come to light. Evidence points to individual imagination and action, not to an intelligence plant [35]. Patrol aircraft crashed in the ocean, so examination of the equipment they employed could not be made.

There is another prisoner interrogation that contrasts with that of the crewman from Coastal Command. An American, almost certainly from the Army Air Forces antisubmarine service, with a radar specialty and considerable experience, did not just answer questions, he gave a short course on microwave radar and its use against U-boats that fills 11 pages of report [36]. He explained the operation of both 10 and 3 cm equipment; he described the under-water microphones (sonobuoys) that transmitted the sound picked up to aircraft in the vicinity through small radios; he gave the sensitivities for various targets at different ranges; he explained the tactics used in attacking a submarine; he even chided them for having stopped an earlier 10 cm jamming to the west of Spain in October 1943 when it had been very effective, a statement difficult to reconcile with German capabilities[1].

The most puzzling aspect of the entire local-oscillator affair was that it had been proposed by none other than Wilhelm Runge, sent in July from his technical exile at Telefunken to examine the failure of the Metox receivers. He suggested the local oscillator as a homing signal for the patrol aircraft and detected it in an airplane at a distance of 80 km. Runge had attended the first four meetings of the Rotterdam Committee and was well informed about H2S and of the work being done on the Naxos microwave receiver and yet appears to have approached the problem with the remarkable idea that the Allies would not employ this new weapon at sea [37].

A grueling battle in the Bay began in late winter 1943 with tactics changing from week to week. The terror of the lights greatly exceeded their danger but led Dönitz to order surface passage during the day with enhanced antiaircraft (AA) armament to fight off the attackers. When this failed for the passage of single boats he ordered them to move in formations so as to offer maximum AA fire. These changes were serious mistakes because attacks during daylight were much more effective than those at night with Leigh Lights, and Coastal Command was not deterred by AA fire; the grim arithmetic of war favored the sacrifice of a few patrol planes to sink one submarine.

Luftwaffe Ju-88s from France joined to attack the bombers, which were soon protected by Beaufighters. It became rough over, on and under the Bay's surface. During June and July 14 boats were sunk in the Bay by aircraft (out of the total of 54 sunk everywhere) [38] but mostly from visual, not radar sightings. Since March 1942 there had been five Freya stations on the west coast of France [39], which assured that at least the last hundred or so kilometers would be free of patrol planes. A map [40]

---

[1] See Chapter 6.5 (pp 321–2).

showing the locations of submarines sunk for June 1943 through May 1945 indicates how the Bay was a dangerous part of the sea but less so than the mid-ocean convoy routes.

High losses soon put an end to surface passage defended by AA fire, and Dönitz adopted the method of maximum avoidance, having his boats creep along submerged except for the four hours in the 24 needed to charge batteries, and finally had them hug the coast of Spain until in the open ocean. Maximum avoidance and the path by the coast of Spain proved moderately successful. In addition to allowing occasional incursions into neutral territorial waters, the Spanish route had inherent radar protection: the coast itself caused confusing reflections, especially for the still plentiful mark II sets, and the region had far more fishing boats than the open Bay. These looked like U-boats on any kind of radar screen and had to be examined with the Leigh Light, which could be seen at great distances, warning watch officers that a patrol plane was in the vicinity [41].

Such was the famous Battle of the Bay. Its main characteristic is a panic induced in Dönitz and his brave captains by the Leigh Light and the Metox local oscillator business. It is impossible for anyone now to determine why Runge did not immediately grasp that the shift to microwave radar was the cause of the Metox failure. Perhaps he was so intrigued with a clever technical solution and its experimental verification that he accepted the prisoner's 'confirmation' without confronting a simpler reality. Perhaps the startlingly long range of 80 km at which he had observed the radiation—was it a fluke?—caused him to think that homing on the local oscillator was a better method of attacking the U-boats than radar, a method so clever that the Allies were using H2S to conceal it. It is impossible for anyone now to know the extent to which secrecy prevented a wider discussion. It is one of those times in history when one just does not know what to think.

May 1943 marked the turning point in the Battle of the Atlantic. This can be seen through graph or table in every book published on the matter. Sinkings of submarines rose. Almost as many were sunk during the last eight months of that year as during the entire war before May [42]. Sinkings of merchant shipping decreased accordingly. These undisputed facts are used by radar enthusiasts in asserting that the introduction of microwave radar was decisive. Given the coincidence in the turn of the battle and the introduction of ASV mark III, it is a conclusion worth entertaining but not a true one, the testimony of Dönitz and Hitler notwithstanding. For evidence siding with ASV the reader is referred to a paper by Russell Burns, who relies primarily on the correlation of U-boats sunk and merchant shipping not sunk with the introduction of microwave radar [43].

By the spring of 1943 the U-boats had acquired just too many adversaries, who were equipped with a remarkable array of new weapons. Of these adversaries the airplane and the convoy were foremost [44]. When nearly all shipping could move in convoys with air cover, the submarine's day was over. The reason lies in the surface nature of U-boat warfare. The issue of the Bay of Biscay aside, the raiders required undisturbed daytime movement on the surface in order to locate their prey and maneuver into position. This could not be done under water because of low speed and limited vision, and by spring the number of long-range patrol aircraft and escort aircraft carriers kept the water around convoys under nearly continual visual surveillance. A sighting led to an attack, more often than not frustrated by a crash dive, but often followed by surface vessels and more aircraft.

Sighting was a game for two, however, and a submarine's watch usually saw an aircraft in time to submerge discretely or prepare to fight it out; the mere presence of a patrol plane in the sky—with or without radar—had accomplished the objective of forcing the raider down. Some aircraft flying outside the Bay of Biscay began carrying Leigh Lights in 1944 [45], but daylight sightings predominated. The U-boats found themselves having increasing difficulty getting into firing position and found their role changing from hunter to hunted. These were also the same months when ship building overtook ship sinking, an ultimate reason for U-boat defeat.

Worse yet, the U-boats encountered an enemy who was finally organized, equipped and trained for the task. The Americans had taken their whipping, eventually listened to their elders and become rather professional as a consequence. The Royal Canadian Navy had shed illusions about a cruiser fleet and put a formidable escort force together. There was finally the necessary minimum of escort vessels and trained crews. Much of Allied success came from using Dönitz's own methods. Dönitz applied strict control from map-filled rooms in Paris, and Commander Roger Winn, a reserve officer crippled by polio and as such disqualified from more active service, directed Britain's war against the U-boat from similar rooms in London.

Unified command, the outstanding mark of the British–American alliance and the single most important element of the struggle, was slow in coming, however. There were in 1942 six 'nations' fighting the U-boats: the Royal Navy, the RAF Coastal Command, the US Navy, the Army Air Forces, the Royal Canadian Navy and RAF Bomber Command, the last trying to destroy submarine production. Having Admiral Ernest King as US Chief of Naval Operations was not conducive to the correction of this awkward business, but cooperation did improve as one descended the chain of command and with time at higher levels, although true unified command never came about.

Allied technical advances were there too [46]. Aircraft had torpedoes—so secret they masqueraded as the mark 24 mine—that sought the sounds of U-boat propellers, duplicated by a similar German device for seeking surface vessels. Aircraft could drop buoys having underwater microphones and pick up the sounds by radio, so that a U-boat sometimes found itself in an ocean filled with ears. Aircraft flew with the ability to locate a submarine by its magnetic field, although it was a marginal technique. Escorts could fire depth bombs from mortars to the front, removing the problem of sonar losing the target just when the attacker was getting close. Rockets fired from aircraft became as effective, perhaps more so, than aircraft depth bombs. Not surprisingly they were fired first from a Swordfish, which had proved itself a useful machine flying from British escort carriers.

High-frequency direction finders, HF/DF, for which only a minimum of inspiration was required to dub Huff-Duff, began triangulating submarines rather early from shore stations and from aboard ship. The method adapted Watson Watt's old cathode-ray tube method of obtaining the direction to a lightning strike. Having been designed for recording transient events with a wide band of frequencies, it was altered into a device capable of responding to messages on a single frequency, if ever so short in duration. Direction finding is accurate for low frequencies or for distances only slightly beyond the horizon for all frequencies, but suffers for high frequencies from erratic changes in the wave reflected from the ionosphere, corrections for which are imperfect. Its use on ship by skilled operators for close contacts was as valuable as ship-board radar, as it gave an accurate bearing on which to dispatch an escort vessel [47], something that proved dependable because of the incessant radio chatter that U-boat tactics required.

Shore-based HF/DF was useful, if not particularly accurate, but had an important secondary function in hiding the source of information learned with Ultra [48]. HF/DF was secret, of course, but had to be known by the many people operating the equipment and using the data. It made, therefore, the ideal cover for Ultra and gave shore-based Huff-Duff an enhanced reputation. Naval officers with knowledge of Huff-Duff's directional accuracy of about 5° began to become suspicious of the locations attributed to it [49]; nevertheless it continued to cover for Ultra in the decades after the war. Mutual concealment was inverted when Dönitz's slowness in realizing the effectiveness of radar changed into an exaggeration of its capability and so caused radar in turn to help hide the value of shipboard HF/DF from him until June 1944 [50].

The technical and tactical complexity of the Battle of the Atlantic is made all the more bewildering by its extent over space and time. Statements concerning radar's effectiveness necessarily rest on imperfect records wanting statistical surety. Y'Blood's study of the American escort carrier hunter–killer groups comes as close to satisfying the desired statistical criteria as any. Such groups were comprised of a carrier and a few escort vessels and were used offensively on the open ocean. The aircraft were a mixture of fighters and bombers equipped with the 3 cm AN/APS-3 designed by Rad Lab and manufactured by Philco. A hunting tactic had been favored early in the war by those unfamiliar with operational analysis and skeptical of the defensive doctrine of convoying. It was tried without carriers and was a complete failure because the ocean is big and a submarine small. Later success had its origin in the escort carrier that greatly extended the area of observation and Ultra that localized the region to be searched [51].

Y'Blood describes in detail the 55 sinkings by American carrier groups, and from them one learns that the initial contact was made visually for 29, with sonar for 11, with HF/DF for seven and the remaining eight with radar. In the subsequent, often extended battles every kind of technique was used, but eyes were invariably the most common means of observation [52]. In one of these fights the carrier USS *Card* and four 1919-vintage destroyers engaged U-boats sent to protect the blockade runner *Osorno* carrying an invaluable cargo of rubber; in the fight the destroyer *Leary* was torpedoed and sunk by U-275 during the night of 23/24 December 1943 [53]. The reader will remember that she was the vessel on which Page had tested his rudimentary radar for the first time afloat in 1937.

It was not in the nature of the Germans, and certainly not of Dönitz, to hold back in the unequal struggle that marked the last two years of the war. Microwaves did not block passage through the Bay, although they certainly enhanced the danger there, and U-boats fought on the open sea until the end. Losses were high but accepted. Desperate improvisations went forward, and radar, Dönitz's special technical devil, received what ingenuity with limited resources could provide. The simplest countermeasures were decoys, buoys set at sea with reflectors to lure patrol planes onto false tracks. One in which dipoles for 1.5 m radiation formed the tails of a kite balloon carried the attractive name Aphrodite. Although such objects were pursued from time to time, they were at most trivial perturbations on great events.

The advantages that might accrue from equipping submarines with radar, used with such excellent effect by the Americans in the Pacific War, had not escaped German thought, and in March 1942 five boats had been equipped with a modification of the 80 cm Seetakt. It had a six-dipole array each for transmitter and receiver mounted in arcs on the conning tower. The fixed antenna required swinging ship for direction, although phase-shift circuits to the dipoles allowed 10° swings to the left and right [54]. In operation its range proved to be significantly less than the 9–13 km specified by the manufacturer. It also suffered from the unreliability that generally attended the introduction of new naval radar, the combined result of design that did not properly account for operation aboard a submarine and of not adequately training members of the crew in its operation. The bulk it added to an already crowded interior and its failure to achieve notable success led to its rejection by boat commanders, as was a modification with a steerable antenna tried out a few months later [55]. A 50 cm airborne sea-search radar, the Hohentwiel, was adapted to submarines in December 1943 and had good use in a few cases, but in general the skippers refused to turn it on, so great was the fear of detection through emanations [56].

In efforts to determine what detection methods the Allies were using, U-boat command routinely ordered submarines equipped for monitoring radio and infrared emissions to sea under the code name Feldwache. As the 1943 crisis developed, one was sent in the summer followed by two more in February and March. The February cruise ended badly with the vessel sunk and the specialist taken prisoner [57].

Research in camouflaging the conning tower of a submarine by covering it with a radar absorbent material, a 'net shirt', to reduce the reflected signal gave way to something more distinct, the schnorkel submarine. This kind of submarine had a tall breathing pipe that extended to periscope height, allowing air for the diesel engines to be drawn while the boat was under water. It was invented by the Dutch before the war and known to both the German and British navies, but only one of them became so needful of extended under-water navigation as to build it. The underwater speed was much slower than on the surface; the diesel fumes could be observed from aircraft; vision was much poorer through the periscope than from the bridge; the all-important abilities to listen for ship screws and move silently were eliminated when the diesel engines were running; radio communication was greatly restricted and they were hated by their crews because of sudden and unexpected reductions in the inside air pressure caused by automatic valves on the air intake that shut to prevent water from a too-robust wave drowning the engines. Nevertheless they were used, as anything else had become suicidal by the last months of the war [58]. The Allies countered with 3 cm airborne radar, which was moderately effective against the schnorkel but equally so on small waterspouts, blowing whales and the many items of detritus that covered the war-stricken sea.

Radar camouflage techniques succeeded toward the end of the war in hiding schnorkels—the only success of the Kriegsmarine in the electronic war—but it was a success of no consequence; the war was lost. Professor J Jaumann developed with I G Farben Co. a combination of semiconducting layers arranged for destructive interference into a rubber-like substance that not only proved capable of greatly reducing an already small radar echo but was also practical in application. The first of 150 U-boats so equipped went on patrol 5 October 1944 [59].

Radar's greatest value may well have been its psychological impact. When ASV mark III began pointing the Leigh Light, the submariners fell to the defensive, and the confusion caused by false information, ignorance, and overwrought imaginations created an atmosphere that made regaining the initiative difficult. Most of their sinkings took place on the open ocean, but the psychology of defeat was reinforced with every tedious crossing of the Bay of Biscay. One of many idle speculations one can make about the war questions the difference that a rapid and successful design of the Naxos-U might have made—ten months elapsed between knowledge of Allied microwave capability and the outfitting of submarines with even substandard receivers. All that was needed were good semiconductor diodes! A less idle speculation questions what effect a clear explanation to the U-boat men about what was known of microwaves might have had. One cannot but wonder about the purpose of this rigid secrecy.

Would Germany have won the Battle of the Atlantic, if the Allies had had no radar? The answer is no. The turn would have been delayed a few months, more ships would have been lost, but as soon as shipping traveled in convoys with continuous air cover, the submarine had lost as surely as had the surface raider. It was the airplane that was decisive. During the years when air cover was sparse, Allied defeat had been held off by Ultra and HF/DF in determining the location of the wolf packs and routing the convoys away from them. When this failed, as it often did, the fate of the convoys was determined by the efficiency and pugnaciousness of the escort, characteristics that grew with experience—but complete air cover was best. A reasonable estimate of the relative value of the observational methods can be best estimated from Y'Blood's report of escort carrier groups [60]. Radar, HF/DF and sonar proved about equally effective, but alert eyes were equal to all of them together.

7.2. RADAR IN ARCTIC WATERS

When Stalin's gigantic military force showed every sign of defeat, possibly of an early collapse, Churchill pledged help to his ally. Stalin wished this in the form of supplies and equipment but specifically not in the form of non-Soviet troops within Russia. Indeed the aversion to foreigners made life so difficult for the liaison and technical personnel required for providing aid that they often felt more like prisoners of war than comrades in arms. His appetite for supplies, however, was great and calls for them loud. That he had offered nothing to Britain in her hour of need was now conveniently forgotten—by both parties. There were only three ways to send goods to Russia: (1) an extremely long sea route to the Persian Gulf with trans-shipping onto a long and hopelessly inadequate railroad through Iran to Armenia, (2) a shorter sea route through the Norwegian and Barents Seas to Murmansk and (3) the extraordinarily long route by sea to Vladivostok and the trans-Siberian railway, placing undue strains on British shipping and Russian rails.

Convoys passing the northern waters in summer had to contend with almost continual daylight with little but fog and otherwise unappreciated foul weather in which to hide. The winter had brief, weak daylight and storms that concealed them from attack but which were nearly as bad as the enemy. Arctic mariners needed radar, and it was cherished even in its primitive forms by those who had it. It proved crucial in one arctic engagement.

A strategic element having a strong effect on these northern convoys was the disposition of German capital ships. A battleship or cruiser let loose among a convoy without comparable protection was a serious matter for the defenders, and the mere presence of such vessels in nearby ports required the deployment of substantial naval force. Hitler had convinced himself that the Allies were planning to invade Norway and had ordered most of his big ships to be stationed there. This had been the reason behind the Channel Dash in February 1942, which had relieved the requirements for defending Atlantic shipping but augmented those of the north.

Early in 1942 Luftflotte V began organizing anti-convoy operations, training the first German squadrons skilled in torpedoing and dive bombing ships [1]. None of their aircraft had radar at the time [2], an awkward disadvantage above seas where poor visibility was frequent, but during the long summer days these fliers proved extraordinarily dangerous. In July they sank well over half of convoy PQ-17. The losses were made much worse through an order for the ships to scatter as the best response to an attack expected from a battleship surface force—one that never sailed. Scattering gave the best opportunities for air attack, as individual freighters had little with which to fight off the raiders [3]. The next convoy was well provided with AA defense, including the carrier *Avenger* and two AA ships, all equipped with air warning radar, but nevertheless the bombers and torpedo planes sank ten of 39 merchant ships at a cost of only five of the 40 attacking aircraft [4]. Radar provided the defenders early warning but little else, as none of their AA guns were radar controlled. The onset of winter with its few daylight hours and dreadful flying weather curtailed the efforts of Luftflotte V, which was sent to the Mediterranean to make life difficult for the inhabitants of Malta and helpful for Rommel.

Convoy JW-51B, composed of 16 merchant ships escorted by seven destroyers and five lesser vessels, left Iceland on Christmas Eve 1942 bound for Murmansk [5]. The destroyers were equipped with type 286 (1.5 m fixed-antenna search) and type 271 (10 cm surface search) radars; of the smaller vessels a minesweeper had a type 271. None of the merchant ships was radar equipped. Two British cruisers, *Sheffield* and *Jamaica*, that came to the convoy's rescue had type 286s, 50 cm gun-laying type 285s and the new 10 cm search type 273s [6]. Attacking the convoy were the pocket battleship *Lützow* and the heavy cruiser *Admiral Hipper*, both with 80 cm Seetakt sets on forward and aft gun directors.

On the 28th and 29th the convoy endured gales that made station keeping without PPI sets extraordinarily difficult, and five vessels became separated. The ships became so coated with ice that several might have capsized had the storm not abated. The attack by the two large ships accompanied by six destroyers had been planned by the Kriegsmarine in direct consultation with Hitler, who shared at least one characteristic of his predecessor, Wilhelm II: the loss of a capital ship was considered to be a serious loss in prestige, to be avoided if at all possible. The orders for the attacking force were thus coached in cautionary terms that took initiative away from the commander. Their opponents followed an old rule of the

Royal Navy: 'When in doubt, steer for the sound of guns'.

The attackers had overwhelming strength against the escort but had suffered from long periods of inactivity during which the members of the escort had been at sea almost continually. The escort was led by Captain R St V Sherbrooke, who had thought out a plan for just the eventuality that took place and had prepared his command for it.

Naval battles are complicated by continual changes of position and confusion about who is an enemy; they are consequently complicated to describe. The details of the attack and defense of JW-51B need not concern us here. The escort prevented the heavy ships from getting among the merchant ships, employing destroyer tactics much admired, even by Sherbrooke's opponent Vizeadmiral Oskar Kummetz. British radar range extended only to 10 km, and the opening of the fight followed the visual sighting of three destroyers from the *Hipper* group. As gunfire erupted from both sides during the next few hours radar was used effectively for ranging, but none of the equipment present allowed blind fire. This was particularly hurtful to the Germans, as the convoy was quickly hidden by smoke. It is unlikely that any of the officers in the German squadron knew that the Kriegsmarine had rejected lobe-switching in 1936 as too complicated for combat use. The confusion of identity was not helped by IFF, as the few sets installed generally did not respond. Radar ranging made fire by both sides remarkably accurate, and initial fire was invariably close. Deficient as it was, ship commanders kept radar reports in mind continually. Both sides were fortunate in having equipment that experience had made reliable. Only one set, a type 273 of the *Jamaica*, failed as a result of the ship's own gunfire [7]. Both sides had well trained operators and mechanics.

Radar did not save JW-51B: Sherbrooke's bold and well executed plan did. Radar was relied on for accurate ranges and confirmations of sightings in poor visibility, but human eyes made the critical first sightings. Radar's main contribution was to make the battle bloodier. The Germans lost a destroyer with all hands, and *Hipper* suffered damage with casualties. The British lost a minesweeper with all hands, a destroyer with severe loss of life and a destroyer damaged with moderate loss of life. All this came from brief but accurate cruiser gunfire. There is no evidence that the outcome would have been different had none of these stricken vessels been hit [8]. The damaged British destroyer, Sherbrooke's flagship *Onslow*, might have been sunk but for the alertness of her radar men. They observed the salvoes leaving the *Hipper* and determined whether they were coming towards, going right or going left, allowing the ship to steer away and possibly avoiding more serious injuries—at least they thought so at the time [9].

This small action had annoying consequences for the Kriegsmarine because the failure to have sunk a single merchant ship had put Hitler into a fury. He summoned Raeder and announced that the big ships would be paid off and their guns used in coast defense. Raeder resigned and was replaced by Dönitz, who with time was able to have the imbecilic order modified.

The Allies were not the only ones having convoy difficulties in the far north. German Radar had been posted on the north cape of Norway—desolate, isolated stations—to detect enemy aircraft or direct those of the few Luftwaffe bases [10]. These German bases received most of their supplies by costal convoys, which encountered trouble primarily from Soviet submarines. After April 1943 these attacks increased significantly, bringing the suspicion that there was a secret submarine base nearby. Remedy came about by means of a Würzburg located at Vardø on the northeast tip of Norway. It observed morning and evening that a Russian plane disappeared into the ground clutter at the same place. Aerial photography disclosed the base in an isolated Norwegian fjord, and a surprise army–navy operation not only eliminated the base but uncovered a Norwegian underground group that had been providing reports of German activity. They also captured a Russian code book with which misleading information was propagated for a few weeks [11].

When Lorenz lost to Telefunken in the competition for the best gun-laying and searchlight radar they did not choose to leave the field entirely. In particular, they had improved the decimeter triode, DS310, into the much more powerful RD12Tf for the otherwise unsuccessful Kurmarkt to give 50 kW pulses at 55 cm. Around this tube they built, at the urging of their principal radar engineer, Gotthard Müller, the excellent airborne sea-search radar Hohentwiel. It used a variety of dipole antenna configurations, generally arranged to give either forward or lateral coverage. Display used the standard dual-beam oscilloscope with the right and left traces showing alternately right–left coverage or lobe switching for the forward direction. The time base was logarithmic. Hohentwiel could observe a 5000 ton ship at 80 km and a submarine periscope at 6 km [12].

As the end of 1943 approached and the days became short, convoys again began to go to Murmansk, and both sides prepared for attempts by surface ships to break them up. When reconnaissance planes, now equipped with Hohentwiel [13], spotted JW-55B on 22 December Dönitz decided to have *Scharnhorst* and six destroyers attack. The Royal Navy expected the appearance of surface raiders under such conditions and came well armed and forewarned by Ultra; their available strength consisted of the battleship *Duke of York* with cruiser consort and an independently maneuvering squadron of three cruisers. The *Scharnhorst* lost touch with her destroyers early; they had no radar and were unable to join the fight or find the convoy, so the big ship fought alone [14].

The British battleship and cruisers had types 281 (3.5 m air warning), 284 (50 cm main armament), 285 (50 cm heavy AA) and 273Q (10 cm surface search), and the *Duke of York's* 273Q had a PPI indicator [15]. The *Scharnhorst* had the long-standard arrangement of an 80 cm Seetakt on the forward and aft gun directors [16]. Radar had had little influence on the outcome of the action the year before, but this time it dominated and spelled doom for the German ship more surely than the disparity in artillery. The British radar was much improved over the previous year. *Scharnhorst* was sighted by *Belfast*'s radar 20 km distant at 0840 hours from which an opening, star-shell-directed volley carried away the *Scharnhorst*'s forward Seetakt, leaving her half blind in a battle that took place almost entirely in the dark. *Scharnhorst* eluded the pursuers until 1221 hours when she again encountered the squadron of three cruisers and headed south to lose them with her 5 knot speed advantage. She could watch the pursuers with her aft Seetakt but was blind forward. The *Duke of York* lay directly in her path and had seen her on the oscilloscope screens well in advance. The position of the cruisers to the north and the battleship to the south left a free path to the east, but her electronic blindness allowed no warning of the danger into which she fell, after which her speed advantage was unable to save her. At 1650 hours *Scharnhorst* was completely surprised by illumination followed by accurately placed 14 inch shells. It was one of the dramatic moments in naval history, etched in the memories of all who saw it and remembered in a painting by Charles Pears at the National Maritime Museum, Greenwich.

No one knows how many hits the *Scharnhorst* took, but they were many during the two-hour struggle. Like the *Bismarck* and her World War I namesake, she went down fighting courageously to the end. Only 36 survivors were taken from the dark, icy waters.

The invasion of North Africa put an end to the possibility that the Allies might invade Norway, which together with the failure of German surface ships to break up the convoys, removed much of the function of the big ships in the north. *Lützow* was sent to deal with more pressing needs in the Baltic and the Gulf of Finland, and Dönitz dismantled the badly damaged *Hipper* as he maneuvered to retain big ships without direct disobedience to Hitler's orders. After the sinking of the *Scharnhorst* only the *Tirpitz*, sister ship of the *Bismarck*, was left to threaten the northern convoys.

Bringing about the end of the *Tirpitz* strained British ingenuity. It began with carrier planes of the *Victorious* in March 1942 to little effect, followed in October with some curious devices called sea chariots that had been used with great success by the Italians, but British imitation failed. Two of six midget submarines succeeded in September 1943 in placing large bombs beneath the vessel and causing serious damage, although the degree of hurt was unknown to the British. The Fleet Air Arm tried again in April 1944 and succeeded in inflicting casualties but not in removing the ship as a threat. The approaching air strike had naturally been observed with a Freya, but the warning only allowed the crew to man guns and start smoke generators, as Göring had no fighters to spare for *Tirpitz* [17].

Sinking the *Tirpitz* then became a special project for Bomber Command that was to occupy them for three months. The task was assigned to Squadrons 617 and 9, the former famous as the Dam-Busters. Lancasters were modified to take special 6 ton bombs that were certain to penetrate the armored deck. Three attacks, each in excess of 30 bombers, were needed to complete the job with two fatal hits finally secured on 11 November 1944. In order to obtain the maximum of surprise for the third and last raid, electronic intelligence flights mapped the radar of the Norwegian coast and found a location through which aircraft at 450 m could pass across Norway into neutral Sweden, whence they attacked from the east [18].

This effort wrote a final chapter to the controversy between naval and air power men that had raged during the two decades before the war. An immobile battleship, defended only by her own AA guns and smoke, required three attacks with 100 superbombs to be sunk. Although marked up as a success, it in fact underscored the fallacy of believing surface ships had been made obsolete by airplanes. Of the demonstrated failures of air power to deliver on pre-war promises, the failure of land-based heavy bombers to destroy naval power is the most grotesque.

7.3. THE MEDITERRANEAN, 1943–1945

'For all its awesome history as a battleground between civilizations, the Middle East did not strike American strategists as an area in which the European war could be expeditiously won. On the other hand, they recognized it as an area in which the global war could be very speedily lost' [1]. Given the British stake in the region these sentiments were certainly not the kind to cause Anglo-American friction and were to lead to useful lessons in cooperation. The shores of the Mediterranean also proved excellent places for the much needed apprenticeship of the green American forces. There was enough resistance to the North African landings, greatly helped by a defense lacking conviction or spirit, to provide useful intermediate-level training. When the French forces surrendered the Germans occupied Tunisia and quickly gave the Allies—the British ground troops there had had no battle experience either—a taste of Luftwaffe ground support with demonstrations of Rommel's desert skills not far behind.

An understanding of radar did not accompany most US commanders ashore. The 560th and 561st Signal Corps Air Warning Battalions encountered officers who had no conception of how these units were to protect air bases, and they did not obtain their equipment from the holds of transports until eight days after the first landing, when they were able to set up an SCR-602, the new lightweight air-warning set [2].

RAF radar landed proficient and organized. A light-weight air-warning (LW) set came ashore with the first wave followed by four COL/GCI sets (COL designated the overseas version of CHL) three days later. Additional equipment of the same kind soon covered the coast of Algeria and two MRUs arrived after four weeks. All stations reported to a filter center established in Oran [3].

An understanding of radar did accompany the commanders of the US Army's AA units. Among the first units ashore was the 62nd Coast

Artillery, which had long furnished the air defense of New York and was the regiment that had provided the first troops for training in radar use in late 1937. They had received a production SCR-268 for each gun battery and had had time to hone their skills. With them came other units that provided gun and automatic-weapon defense of North African ports and air bases [4]. Battery commanders soon found shooting under radar control much more effective than with searchlights [5], and their accuracy grew with practice. The SCR-268 provided batteries with local air-warning as well as gun-laying capability, justifying the Army's decision not to adopt the Bell Laboratory 60 cm CXAS.

The pattern of air attack on sites defended by American 90 mm guns became fixed throughout the Mediterranean: a few trial raids at night on a new target tested the defenses and found losses too high to be sustained, after which raids were rare. An attack on Palermo on 4 August 1943, for which the enemy numbers and losses were confirmed by prisoner inter-rogation, showed what the SCR-268 could do. Shooting on radar control at 29 night bombers, defending guns brought down five aircraft over the port and damaged two sufficiently that they crashed on the return flight. The prisoner asserted that the AA fire was the most accurate these fliers had ever encountered. Despite the importance of Palermo as a port, there was only one more raid; in it two out of 20 planes were seen to come down and bombs were jettisoned before the target [6]. Microwave engineers, steeped in the heady mysteries of the magnificent SCR-584, often speak with disdain of the 268, but from November 1942 until March 1944, years when the Luftwaffe was a potent force in the Mediterranean, it was the 268 that gained respect for American gunfire [7].

A large air-warning and fighter-direction system was soon in opera-tion with a record amount of equipment. Night attackers had to deal with GCI stations that guided AI-equipped fighters. Radar was secret and air-borne radar was really secret, so the Air Ministry required this equipment be sent by sea, and the attempts to bring down night bombers with GCI alone failed. The first raid on Algiers on 20 November 1942 was met by gunfire and did little damage, but General Eisenhower emphasized the need and an AI-equipped Beaufighter squadron arrived and shot down five unsuspecting bombers a week later [8]. A British GCI unit near Bône was credited with aiding Beaufighters in the destruction of 23 aircraft, but Freyas helped make December 1942 a wretched month for the RAF. The 1.5 m AI sets displayed the ground-return limitation that had originally indicated the need for microwaves. Night raiders, Allied or Axis, learned to fly very low to lose themselves in ground clutter. It was a different war than the one over Germany where heavy bombers were forced to the high-est altitudes to escape gunfire and in so doing made themselves perfect targets for the 50 cm and later 3.5 m Lichtenstein sets.

The star of North African radar was unquestionably SCR-582, a 10 cm harbor-surveillance radar that arrived at Oran on 27 January 1943 [9]. This set, the second Rad Lab set to see combat (SG had entered a couple of months earlier off Guadalcanal), startled all the American and British radar men who saw it. Although intended for harbor defense, its 120 cm diameter dish allowed it to pick up low-flying aircraft at 40 km. Its PPI indicator allowed the operators to guide ships entering harbor through the protecting minefields, and it proved perfect for detecting German motor torpedo boats. All of these functions were nearly impossible for meter-wave sets. SCR-582 was soon modified in the field for air defense functions [10].

US Secretary of War Henry Stimson was an early and enthusiastic apostle for radar and had had Dr Edward L Bowles appointed as his special radar advisor. Bowles requested that a senior engineer from Rad Lab examine the use of radar on the North African front to keep design in close touch with the realities of combat, and DuBridge sent Dr Louis N Ridenour. He confirmed the correct emphasis of Rad Lab on microwaves; he noted the deficient understanding of proper radar use by Army Air Forces personnel; he learned the extent and quality of German radar and countermeasures [11].

Rather than extract the Axis forces from Africa, Hitler chose to build up his forces in Tunisia. The painful experience of supplying the much smaller Africa Corps was thought to be offset by a much shorter supply route, and the troops that came by the hundreds of thousands by air and sea brought radar with them. Of particular note were three stations set up for fighter control around the Gulf of Tunis, each using the Himmelbett system of one Freya and two Würzburgs and directed by an 'ace' controller. Like their Allied counterparts, they found using meter-wave radar in the mountains to the west more exasperating than rewarding. The error of the decision to defend Tunisia became all too apparent in May 1943 when a larger force surrendered than at Stalingrad. A few pieces of radar equipment were evacuated, as were the aces, but most sets had to be destroyed and their personnel became prisoners [12].

Infantrymen on both sides experienced little comfort and less knowledge of the wonders of radar. They were always convinced that their own air power was gone. 'Where is this bloody Air Force of ours? Why do we see nothing but Heinies?' Except for the defense of ports and air bases, ground troops had little reason that they could see to thank radar—in the unlikely event that they knew about it [13]. Radar was the weapon of an air fight; its introduction to ground warfare had to wait two years.

The French surrender of North Africa caused not only the German occupation of Tunisia but also of the remaining part of Metropolitan France controlled by the Vichy Government. This brought French radar development to the end. With defeat in spring 1940 radar research had moved from Paris to Toulon, where it continued in cooperation with the Constructions Navales, and to Lyon at the laboratory of the Société LMT. Two designs reached prototype stage: a 16 cm sea-search set capable of detecting capital ships at 25 km and torpedo boats at 10, and a 50 cm set for sea and air search that attained ranges on aircraft of 17 km. The ships at Toulon were scuttled before German forces arrived and the radar equipment was destroyed. Some clandestine research was continued in Lyon [14].

With Tunisia secured by the Allies the next step was Sicily. Here was a more difficult operation, one that required a higher degree of ability, almost as if it were the next step of a training schedule. North Africa had been defended entirely by French units, many of whom—the Foreign Legion no doubt excepted—had no wish to fight. Sicily was defended by a mixture of Italians and Germans. Except for the units that had benefitted from Rommel's command, Italian troops had made an extraordinarily poor showing, and those defending Sicily were green, ill trained, ill equipped and ill disposed toward the Axis. The very opposites of these descriptions described the Germans there.

Commanders planning the invasion took radar as a serious threat to their air and naval operations, so electronic intelligence set about to map out the Axis radar screen. The Americans dispatched B-17 Ferret aircraft loaded with equipment provided by Frederick Terman's Radio Research Laboratory on 22 April 1943. Their equipment proved of value before it had even reached the coast of Africa, as the flight's navigation was in serious error, and they located the mid-destination, Ascension Island, by homing on its radar. They set up operations at Blida, south of Algiers, where their strangely painted aircraft and super-secret nature filled the beer conversation of neighboring fliers. Secrecy played them false later when planes filled with the jamming sets intended for the invasion arrived only to have maintenance crews, who were not instructed about their special nature, begin making them 'normal' by removing non-standard electronics.

Flights in June established the locations of Freyas and a Wassermann or two on Sicily; another flight circumnavigated Sardinia. After the invasion the locations of the stations mapped by the Ferrets were confirmed, either by a wrecked set or evidence of an earlier presence. The effects of American and British jammers, Mandrel (for Freya) and Carpet (for Würzburg), were never determined. Any effect escaped mention in available German records [15].

The Italian fortress island of Pantelleria had German radar that had to be eliminated before the invasion could proceed. The Italian garrison failed to fulfill Mussolini's boasts of the impregnable strength of the base—indeed they did not fight at all—and it was quickly converted into a British radar post. A floating filter center allowed fighter control during the landings [16]. IFF was its usual sorry self with fighters wasting much time following up unidentified plots [17]. Ground radar units went ashore with the landing forces.

Sicily saw the introduction of a new technique contributed by TRE that made possible the landing of airborne troops behind enemy lines at night. It consisted of a very light interrogation radar, called Rebecca (calling her children) and a responding beacon, Eureka ('I have found it!'), thereby providing the troop-carrier pilot with range and bearing. Eureka differed from beacons generally in responding only to Rebecca's call [18]. The interrogator could be mounted in any army support plane and the responder was light enough to be carried by a parachutist. Advance parties set up Eureka beacons in Sicily to mark the dropping zones for parachute and glider troops. The mere idea of attempting such a navigational feat without Rebecca/Eureka would have brought the most accomplished flier to despair. Success in Sicily ensured it an important place in the airborne component of the Normandy invasion and before that in the infiltration of secret agents and their supplies into occupied Europe [19].

Completing the occupation of Sicily was a ground operation with extensive air activity. Allied air was by then dominant and remained so thereafter, but the Luftwaffe fought back tenaciously and skillfully; the Mediterranean remained a major front for them and a major drain on their resources. Radar was now a routine component of the air operations for both sides, although secrecy often kept fliers from knowing the source of their guidance.

The occupation of Sicily precipitated the fall of Mussolini and his Fascist government followed by Italy's surrender on 3 September 1943 followed in turn by the invasion of the Italian peninsula nine days later. Whether this was the best application of Allied resources has remained a matter of dispute, but the question of whether the Germans should have resisted this new invasion was not considered by Marshal Kesselring. He rightly saw Italy as a dangerous base for air attack on the Reich and decided such an operating area must be kept as far away as possible. That the Alps formed an impenetrable radar barrier for concealing attacks from the south featured in his reasoning. This decision turned Italy into a major theater of ground fighting for the next 21 months. German ground radar covered the entire sea coast of continental Europe from Spain to Turkey and including Sardinia, Crete and the Greek islands with 86 Freya, 18 Wassermann and 54 Seetakt stations [20], all carefully located by Allied electronic intelligence. The sources of this information do not disclose the number of Freyas and Würzburgs, great and small, that belonged to Flak and fighter direction, but they must have been correspondingly great.

British forces landed on the sole of the Italian boot, and the Americans went ashore at Salerno, 45 km south-east of Naples. Both landing places were within fighter range of Sicily; German air power was not adequately situated for a strong response and was unable to hinder the landings seriously [21].

Much of British and American radar was concentrated in the 1.5 m band: airborne AI mark IV and ASV mark II; ground-based radar COL/GCI, LW and SCR-602; naval type 286, CXAM and SC; AA SCR-268. Inasmuch as Window had been used in the July Hamburg attacks it should have hardly come as a surprise that the German bombers made a distribution of Düppel cut to 75 cm at the Salerno landings, but it proved nearly as effective in blinding radar as Window had been to the Würzburg and Lichtenstein equipment months earlier. The call went out for microwaves, and Allied pilots and gunners began to learn how to distinguish clouds of foil from aircraft, as had their enemy counterparts a few weeks earlier [22]. The 7.5 m MRUs and naval type 279s were immune to 75 cm Düppel, which redeemed the otherwise increasingly obsolescent sets. Düppel cut to 75 cm became part of the Luftwaffe technique throughout the theater [23].

After four weeks the advance came to a halt along the Velturno River and, air superiority or no, with or without armor, progress became very slow, finally coming to a complete stop at the German fortified line called Gustav that stretched from the Tyrrhenian to the Adriatic Seas. This was just the kind of war that the tank and the airplane were supposed to have relegated in 1918 to that location of favorite cliche, 'the ash can of history', but the line held.

To break this stalemate another invasion was planned to get behind the Gustav Line. It was a technique now well exercised, and the newly acquired territory allowed fighters to range up the coast to cover it. The place selected was near a small port 50 km south of Rome called Anzio. The landing proceeded very much as had the one at Salerno but failed to accomplish its objective of turning the enemy flank, in part through hesitant American command and by a timely and effective German response. It was marked by two radar matters of note: a radar feint and the introduction of the SCR-584.

The Luftwaffe was in much better position to react to this landing than the one at Salerno, and Kesselring had suspected such a move. The landing was preceded by strong air and naval bombardment and established a bridgehead in the early hours of 22 January 1944. The Luftwaffe immediately began shifting units from other theaters—one had bombed London on the 21st—to attack the fragile concentrations on the beach. The US Twelfth Air Force removed much of the power of this blow before it had a chance to strike. The most remarkable of these preventive raids took place on 23 January when five American bomber groups, well escorted, flew on courses obviously intended for the landing field of some of these squadrons. They were intended to be picked up by radar and they were, but a group of P-47s went to the scene of combat by another route flying close to the ground to avoid radar and so timed as to arrive as the defending fighters were beginning their take off to meet the radar-heralded force. The defenders were thrown into complete confusion, and the bombers carried out their attack free of harassment. A total of 140 German aircraft were destroyed in the air and on the ground at the cost of six bombers and three fighters [24].

Allied forces failed to break out of their constrictive location on the beach and expected a strong counterattack. Hitler ordered it for 1 February, but Allied bombers had been working on the Italian railways, making tedious work of the assembly. The counterattack, planned with personal but not helpful intervention by the Führer himself, began on the 16th. Air strikes were naturally a part of this action and were matched by radar countermeasures. Martini had set up powerful land jammers for the 1.5 m wavelength of the SCR-268, the airborne sets of the night fighters and any naval radar that got in the way. Ground-based jammers that are near the action can be particularly effective, owing to the high power and the ability to remain in place, characteristics not possible in airborne jammers. The Anzio AA guns and night fighters were blinded. To complete things the bombers carried airborne jammers and let out large quantities of 75 cm Düppel.

This situation brought response from Signal Corps personnel in Algiers where new SCR-584s and 545s (the Bell Labs alternate 10 cm gun-laying set) were at hand, equipment immune to the 1.5-m jamming and to Düppel cut to that wavelength. Under intense pressure crews were trained in the use of these two, and by the 24th a night attack by 12 Ju-88s lost five of their number to the newly arrived microwave radar [25]. The counterattack had failed by then because American and British infantry held their ground, but relief from bombers was appreciated for all that. A British GCI station got use of a 584 and found it excellent for their function. Countermeasures became increasingly effective against SCR-268 both as a gun-laying and a ground-control-intercept radar, but months were to pass before there were enough SCR-584s for even vital positions.

The war continued in the Mediterranean Theater without further novelty in radar. Allied forces pounded their way up the Italian peninsula until the bitter end, securing more advanced bases for bombers and tying down a significant German force while American and British commands argued about a 'soft underbelly' approach. The invasion of southern France successfully followed Normandy and used all of the paraphernalia of radar and its countermeasures [26].

7.4. JAPANESE SHIPPING DESTROYED

Japan discarded an ancient feudalistic past in preference for the modern world in a truly breathtaking change called the Meiji Restoration and settled its government in 1889 on a constitutional monarchy based on that of 1871 Imperial Germany. As the century progressed power fell into the hands of the Army, negating any traces of representative government; ministers who entertained delusions about having real control, and who did not respond to instruction were assassinated. By the 1930s the Prime Minister was a dictator of the Empire chosen by the Army, an office that Hideki Tojo combined with that of Chief of Staff from 16 October 1941 until the loss of the Mariannas led to his resignation. Aside from the manifest evils of this regime, found in the brutal subjugation of Asian people unfortunate enough to come under its heel, it maintained the credulous belief that the warrior spirit would carry the Empire through war and that little thought need be given such servile matters as the economic and commercial basis of industry with its extreme dependence on shipping. This blindness is all the stranger given an insular need of imports, and is stranger yet when the seriousness of the growing blockade went well past alarming indications of ruin. But Japan's adversary was almost as slow in recognizing the fatal weakness that lay in shipping. When the American Navy finally understood what was to be done, it proceeded with efficiency to destroy Japan's merchant shipping.

7.4.1. Destruction with submarines

At the end of the Pacific War the US Navy's submarine service could be favorably compared to the U-boat fleet that Admiral Dönitz had commanded three years earlier. Boat commanders were young and daring and had accomplished against Japan what Dönitz's men had failed to do against Britain. But things had not begun that way. The US began with a fair sized fleet of submarines, many of construction growing out of World War I but serviceable nevertheless, and to which a large number of modern vessels, called 'fleet submarines', were joined. Compared with the older American S-boats and the German U-boats these were almost luxurious. Two defects of the service were serious and slow of correction. (1) Submarine command had followed traditional peacetime promotion patterns, and by 1941 most boat commanders were into their middle or late thirties. The baleful effects of age on audacity were compounded by the Navy belief that sonar and aircraft had made the submarine a highly vulnerable weapon and that caution was an important virtue. The near-complete failure of the Navy to meet the challenge of the U-boats in 1942 had much of its origin in this doctrine. (2) The new mark XIV torpedo was poor. It failed in three ways, all of which had to be independently discovered and corrected in the field when the Bureau of Ordnance refused either to admit there were problems or even test torpedoes for alleged faults, asserting until counter-proof was forced on them that the deficiencies lay with crew performance. First, the mark XIV ran significantly deeper than set; next, the vaunted magnetic exploder frequently either failed to explode at all or detonated prematurely; last, the contact exploder failed for a broadside collision—the condition for a perfect shot! Some of the new mark XVIII electric torpedoes proved capable of exasperating the users with exciting new ways to fail, such as running straight down or following terrifyingly erratic paths. Its defects were corrected in less time and with a higher degree of Bureau cooperation than for the mark XIV. Boat commanders were relieved after an unsatisfactory patrol or two but were all too often replaced according to old rules of seniority. It required two years before a first-class group of boat commanders was assembled and about the same time for the torpedoes to be made reliable.

The American submarine war was initially hindered by wrong goals perversely made worse by the availability of decoded Japanese messages. The Navy had immediately pronounced that the international agreement forbidding submarine commerce raiders from attacking without warning—the basis for America's declaration of war in 1917—had been invalidated by the Japanese in their attack on Pearl Harbor, yet the destruction of enemy shipping was not given urgency despite Japan's obvious vulnerability to a submarine blockade. Instead boats were dispatched on the basis of decoded messages to sink important naval units plus any freighters or tankers that might get in the way. This policy produced many wild chases but few sinkings of the targeted ships. About two years were needed to have a submarine blockade take top priority.

The Navy was greatly helped by the ineptness with which the enemy fought back. Individual destroyer commanders were aggressive and competent in the use of sonar, but they seldom pressed their attacks until there was clear evidence of a sinking and seemed to think that laying a full barrage of depth bombs around the sonar contact destroyed the boat. Thus it frequently sufficed for the submarine to lie still and wait. The Japanese never organized anti-submarine warfare in a way that evaluated the many bits of information that led to a boat being tracked and sunk. Tactics drawn from operations research were never employed.

The Naval Research Laboratory began thinking of providing submarines with radar before the CXAM was in production. Their first concern was something to warn the sub of the approach of aircraft, allowing sufficient time to dive. The first set, the SD, was with minor modifications the standard air-warning set through most of the war; it began production in late 1941 and was the ultimate in simplicity. Having warning as its function the antenna gave no bearing information and range accurate to only 1000 m. In order to provide the 100 kW power needed for its broadcast mode it had a pulse repetition frequency of only 60 Hz. The wavelength was 2.65 m, the longest used by the Navy [1].

The SD was useful for guarding the boat from surprise while on the surface but of little value in the quest for maritime victims at night or in fog or as an aid to navigation. Given the constraints on antenna size for a submarine this function could hardly be met with meter waves, but the unexpected appearance of microwaves in 1940 changed that immediately. Bell Labs, which designed most of the Navy's microwave radar, began working on a submarine radar almost immediately after learning of the magnetron. It was designated SJ.

The achieved design goals of the SJ were (1) sweeping the horizon with a narrow beam from an antenna at periscope position with the boat submerged or on the surface, (2) observing a destroyer at tactically useful distances, (3) determining the range with an accuracy useful for the torpedo data computer and (4) presenting the observations on a PPI display, although this last followed the first deployment by about a year. The wavelength selected was naturally the 10 cm band being so actively exploited by Bell and Rad Labs. The antenna took a radiative form similar to the Rad Lab Raytheon SG but had the more troublesome requirement of having to keep water out of the waveguide that connected it with the electronics in the hull and doing that for the pressures experienced. The Naval Research Laboratory contributed the necessary stock of knowledge to make SJ seaworthy. The prototype, which was completed in December 1941, had a beam 9° wide and 29° in elevation, could detect a destroyer at 10 km with an accuracy of 25 m and a low-flying bomber at worthwhile ranges. The cut-parabolic reflector could be subjected to strong forces from depth bombs and from collision with the detritus that covered the wartime ocean, so it was made to pivot harmlessly out of the way, if a replaceable shear pin snapped [2].

Boat commanders had mixed reactions about SD, which left on its first war patrol on 13 December 1941. They were restricted to radio silence except under special circumstances as a precaution against direction finders [3], so orders that prescribed having SD emit a steady stream of very-high-frequency pulses yet forbade even the briefest of radio messages impressed many submariners as another example of the weakness of intellect they had long suspected inhabited certain elements of naval command. It was thought by many that Japanese signals intelligence did learn of the SD emission and provided homing equipment for patrol planes, but the evidence is apocryphal.

Boat commanders had quite a different reaction to SJ, which went on its first patrol in August 1942 aboard USS *Haddock*. The commander used it to follow a ship at night, concluded with a successful attack, and sank another in the same way two weeks later. The new radar had not performed flawlessly, being out of calibration at times and out of operation for extended periods, but these shake-down ills were recognized as such and known to be correctable with adequately trained personnel. There was apprehension of the 10 cm radiation being used as a homing signal, but its narrow beam diminished this concern, and the set received an enthusiastic endorsement from its first users [4]. The SJ radar was probably as important as the periscope, as it turned night into day for the raiders in addition to providing accurate range and a more secure air warning.

The Japanese fleet and its merchant shipping were destroyed in 1944. Two epic battles saw the end of the principal fleet units: Philippine Sea in June, Leyte Gulf in October. By the beginning of 1944 the American submarine service had removed defective torpedoes, defective skippers and defective tactical doctrine from its baggage. These changes brought about a remarkably fast strangulation of Japan's oil supply with immediate strategic consequences and sank or disabled most of the remaining fleet units.

In the Battle of the Atlantic the technical struggle changed from month to month, measure followed by countermeasure. The Pacific war was not so marked. The American boats began to be equipped in large numbers during 1942 with SD and SJ sets and retained them with slight modifications until the end; the SD was often ignored, the SJ treasured [5]. American boats were warned in December 1943 that Japanese escorts had 'excellent radar' [6], the 10 cm mark 2 model 2, but its introduction simply added another form of alertness that commanders had to have. An evaluation of the Japanese radar's effectiveness requires analysis of data that do not exist. It certainly had no dramatic effect. In this way it was very much like the introduction of the 10 cm type 271 by the Royal Navy in the Atlantic without PPI.

A submarine–escort fight, although certainly not a typical one, illustrates the way in which the SJ affected operations. Near noon of 18 October 1944 USS *Raton* under Commander Michael Shea encountered a nine-ship convoy with three escorts southwest of Manila. *Raton* was forced down almost immediately by an escort and on resurfacing found herself in a completely obscuring torrential downpour, the onset of a typhoon. Shea regained contact through the SJ and had maneuvered on the surface to the middle of the cluster of ships by about 2030 hours, concealed by darkness, driving rain and a first rate electrical storm. On obtaining a favorable position he began to turn the boat, sending torpedoes at targets at different points of the compass until his six bow tubes were empty. At ranges of less than 1000 m it required his radar skill not only to aim torpedoes but to avoid ramming or being rammed in the confusion caused by his attack and the increasingly heavy sea. He then emptied his stern tubes and cleared the dangerous traffic to take stock and reload. He pressed home two more attacks in seas with waves approaching 10 m. When the intensity of the typhoon forced him to break off, he had watched five ships disappear from the PPI [7]. One might suspect that one or more of the escorts had 10 cm radar, but the Japanese set did not have a PPI display and, as similar experience had shown in the Atlantic, anything less is not much help in the confusion of several moving vessels.

Japanese escort vessels were equipped first with radar, but the Bureau of Naval Construction opposed putting it on the submarines even though it was the same 10 cm set as used for surface vessels and required little modification. Commanders, now familiar with its use by the Americans, prevailed and the boats sent to the Battle of the Philippine Sea were so equipped. They also received a modification of the mark 1 model 1 for air warning in which the array was replaced by a single vertical dipole for the transmitter and a steerable Yagi for the receiver. Japanese boat commanders found their radar more useful in surviving than in locating the enemy, as an example illustrates. Commander Hashimoto took I-58 with human torpedoes (kaiten) for suicide attacks on Allied fleet units during the Battle of Okinawa, but his progress was slowed to such an extent that he could not keep his batteries charged as a result of the large number of crash dives necessary to avoid the aircraft that he picked up on his radar. He finally gave up and headed out to sea to find other targets for his spirited kaiten pilots [8].

The attitude of Japanese boat commanders contrasts with that of the German, who wanted nothing to do with the Seetakt sets that were initially adapted to U-boats in small numbers. It is true they were not as suitable as either American or Japanese sets, but with a little encouragement GEMA could have probably made them useful. Absence of surface-search radar robbed them of the use of the night for maneuvering into firing positions, which was fatal when they lost the day. American and Japanese boats found night and day equally useful.

7.4.2. Destruction with bombers

It was obsolescent at the beginning of the war but still in production, although a new design was in the works; it was slow, an advantage in launching torpedoes but not in making bombing runs; it had weak defensive armament and preferred to work at night; it could carry a heavy load; it had the long range required for reconnaissance; it could absorb a lot of punishment and still fly; although not ugly, as some claimed, it certainly did not have the graceful lines that distinguished such wartime stars as the Spitfire, the Me-109 or the B-17; equipped with ASV mark II it proved to be one of the most valuable aircraft to fly for the Allies. An alert reader will associate these words with the Royal Navy's Swordfish torpedo bomber, but they apply just as well to the US Navy's PBY Catalina flying boat.

When E G Bowen wanted to demonstrate ASV mark II as part of the Tizard Mission, it was only natural to select the PBY for the demonstration. The Navy was so impressed that they ordered 7000 from Philco, designated ASE [9], so PBYs were the first Navy planes to carry radar, and they made it felt in the Battle of Midway and the Guadalcanal campaign. In the South Pacific PBY crews had to learn how best to fight their great birds and survive. To be caught by almost any Japanese plane in the daytime was a serious affair, so they restricted such flying mostly to rescue work for which radar could not substitute for vision and in so doing filled war diaries with examples of heroism. But to attack an armed enemy during the day was foolish, barring unusual circumstances, so they acquired night skills. In doing this they had something nearly as important as their radar, the radio altimeter that allowed them to approach the surface of the sea close enough for a night torpedo attack.

Their skills in attacking ships at night evolved during the early part of the Guadalcanal campaign, when cutting the Japanese line of supply to that island was vital and any method was worth trying. Radar could find the target but because of the reflection from the surface of the sea lost it as the aircraft approached. This was the same problem that Coastal Command had had in the Atlantic and that had led to the introduction of the Leigh light. Such a light was not within the improvisation capabilities of the crews in the South Pacific, but unlike a submarine a surface ship cannot escape combat by diving, so the attacking bomber has time to make repeat attacks. The difficulties of seeing an air target at night are much greater than seeing a ship on the ocean surface. Obviously, this kind of fight would result in AA fire from the targeted ship, which without radar-controlled guns would be shooting at shadows unless the plane could be fixed in the beam of a searchlight, yet bloody duels were not rare.

One of the first adaptations to this new role was the painting of the PBYs black, with the appellation Black Cats following almost immediately. This made them much more difficult to see and gave them a sinister appearance that reinforced the normal aversion to surprise nocturnal visits by an enemy. The torpedo was the most accurately aimed missile for this work, but the mark XIII aerial torpedo had as many faults as the mark XIV submarine torpedo, so good aim even at the favored close ranges frequently failed to cause an explosion; bombs could be expected to explode but were hard to aim for masthead attacks. The withering fire of four 0.50 inch machine guns mounted in the bow proved more than enough for some small craft, giving the Cat some properties of a fighter but none of the agility. These homemade techniques were extended throughout the South Pacific and accounted for the loss of hundreds of thousands of naval and merchant tonnage [10].

Most of the credit for the destruction of Japanese shipping deservedly goes to the Navy, but a large fraction must be credited to Army Air Forces bombers. If anything had been proved concerning air power during 1942, it was that high and medium level bombing of any nautical target was useless. Army fliers and their Australian comrades had demonstrated in the Battle of the *Bismarck* Sea that low-level daylight attacks by large numbers of aircraft using skip bombing and strafing with heavy machine guns and cannon were very effective against merchant ships and escorts, but this method was ill advised for single bombers seeking to sink the many ships in the southwest Pacific that did not move in convoys. In order to have assurance of a hit, the bomber had to approach at altitudes that made the AA fire of even a freighter deadly. The Black Cats had shown the way, but the Army fliers thought they could do better and had cause to do so, for reasons other than as part of the overall Allied effort to sink enemy shipping. Single vessels, generally traveling at night and hiding during the day were supplying the island garrisons that the ground troops had to fight, and the Army fliers wanted to isolate the battlefield [11].

As the 10 cm search radar SCR-517 came into use in that theater, crews of patrol planes noted the ease with which these vessels could be observed at night yet could not use it to place a bomb on one. This was not the fault of the radar, which did not suffer the surface reflection limits of ASV mark II; there was just no way to aim a bomb with the 517. To remove this deficiency Bell Telephone Labs began in July 1942 the design of auxiliary equipment to be used with various microwave sets to provide automatic bomb release at altitudes above 20 m, designated AN/APQ-5 but generally called LAB for low-altitude bombing. This was somewhat similar to the problem that Alvarez was facing at that time in designing Eagle for high levels, but general aerial bombing required a mechanical analog computer of considerable complexity and weight, and Alvarez settled on the Norden. At low altitudes the slant range as measured by the radar is a good representation of the ground range, which allows an approximation to be made that is much simpler to calculate than the general case and sufficiently accurate. After this approach was accepted in July 1942 the project moved rapidly through the use of an electric computer designed along lines that Bell had employed in the M-9 AA gun director. The bombardier observed the target on a scope display showing range against bearing for a beam swept from side to side. Range on the scope could be 30 km for search and reduced to 1.5 km for the final bombing run; a marker on the scope indicated the moment of release. Steering commands to the pilot ensued the same way as for visual bombing [12]. It proved to be remarkably accurate. The approach of the attacker at night made defense without radar-controlled guns, which the Japanese lacked, impossible. Generally the only warning obtained by the ship was the bursting of bombs.

The reason LAB was such an accurate blind-bombing technique and H2S and H2X so poor resulted from a ship being a single object on a relatively smooth surface, much like an airplane in the sky. Radar bombing of land targets suffered from an over-abundance of reflections from the objects that make up a civilized landscape and that confuse the person looking at the PPI scope.

A wing of B-24s was organized under Colonel Stuart P Wright, who had vigorously encouraged the project, and equipped with 22 pre-production models of these devices adapted to the new SCR-717 radar. After intensive training at Langley Field they went, together with a Western Electric engineer, to Guadalcanal and dispatched during the night of 27/28 August 1943 their first operational 'snooper', which located and destroyed a small freighter a few hours after takeoff [13]. During the period from December 1943 through April 1944 the 'Wright Project' destroyed 47% of the enemy shipping credited to aircraft in that region with only 20% of the machines and personnel having that assignment [14].

The LAB aircraft on Guadalcanal became part of the 13th Air Force, and a similar unit was established in the 5th Air Force operating out of Buna over the *Bismarck* Sea, but the champions of maritime destruction were those flying with the 14th Air Force from Liuchow, China over the Formosa Straits and the South China Sea. In the five months after their first flight on 24/25 May 1944 this unit sank 250 000 tons of merchant shipping, three cruisers and three destroyers in addition to damaging or possibly sinking another 100 000 tons of shipping and seven naval vessels. Shipping was so dense that targets frequently had to be selected from several, and they had to ignore 60% of the ships sighted during their first two months.

Destruction of these ships cut off supplies sorely needed for the defense of the Philippines. And this from a base where operations had to be 'measured in pints of gasoline and ounces of bombs', owing to the extreme difficulty of flying supplies from Burma over the Himalayas [15]. The destruction caused by this handful of bombers became so painful that the Japanese Army forced the Chinese back and captured Liuchow on 11 November 1944 thereby ending the 14th's commerce raiding [16]. That such a measure was necessary to stop the sinkings demonstrates the complete absence of Japanese night-fighter capability, which could have worked easily from Formosa and would have found no simpler adversary than a snooper. For want of airborne radar an army had to move.

In early 1945 the United States had brought Japan to the state that Dönitz had planned for Britain. It was a condition for which there was no correction, as Japan had no allies to intervene and no conceivable means of breaking the blockade. The hopeless condition of the nation had become obvious to the Imperial Navy and the civilian ministers of the government after the Battle of the Philippine Sea, but rule was not by reason but by the bushido code of honor that sent hundreds of fine young men to their deaths in knowingly suicidal gestures against an all-powerful enemy.

7.5. THE WIDE PACIFIC

By the end of 1942 the Allies could begin to think about victory rather than about staving off defeat, but their resources in the Pacific hardly allowed extravagant plans for the coming year. It was during 1943 that the US Navy grew to great strength and presented the Japanese with the ugly circumstance of which Admiral Isoroku Yamamoto had forewarned them. The United States had not been defeated by the wave of victories that had followed Japan's forces during the first half of 1942, and now the dreaded industrial capacity was making itself felt. The 'decisive' victory at sea might still come, or the war in Europe might favor the Axis and draw resources from the Pacific, but worry about the future was now predominantly an Axis, not an Allied occupation.

For Admiral Nimitz the arrival in Hawaii of the new fast carrier *Essex* at the end of May 1943 was a turning point. Every movement of his command during the preceding 18 months had been adjusted by fears for his carriers. Not that they had been held back; there had been plenty of action, as evidenced by five having been sunk. Replacements were following, but *Essex* had been laid down as part of the pre-Pearl Harbor building program that had given battleships and cruisers priority, and not many big carriers were on the way. On 10 January 1942 a change in plans sought to relieve what was obviously going to be a serious weakness by converting a number of cruisers under construction into light, fast carriers, a design that came into being quickly because cruiser hulls were available.

Much more ominous for Japan was the blunder in having not reckoned the high attrition of their elite corps of naval fliers. Intended, as the strategy demanded, to carry the Empire to early and dramatic victory, they had been superbly trained but now few remained. Their skill had taken advantage of the long range and agility of light-weight aircraft having neither bullet-proof fuel tanks nor armor, but their unskilled replacements found these advantages elusive when sent against well trained and rapidly seasoned adversaries flying heavy, tough machines. The grand mistake of not having provided adequate flying schools, while America turned out thousands of pilots, spelled doom for the island Empire.

7.5.1. *American radar, 1943–1944*

As Nimitz's fleet expanded in ships, planes and men, it also grew in the capabilities for electronic warfare. The 1.5 m XAF that had so impressed line officers during the 1939 Caribbean exercises had gone into production as CXAM and made a simple evolution into the SK with peak power increased to 200 kW and a PPI display for its new panoramic searching. It could pick up a single bomber 3000 m high at 160 km; installation began in January 1943 [1]. The 'flying bedsprings' were recognized by every seaman as sure guardians.

Valuable as the CXAM had been in the carrier battles of 1942, it was incapable of helping the fighter direction officers when things moved close to the ship. Missing were resolution of even moderately spaced aircraft or any height information. It was a task for which only microwaves sufficed. When the problem reached the Radiation Laboratory the solution was obvious. The 10 cm SCR-584 was moving toward production, and its ability to track a plane in three dimensions was exactly what was needed. Transferring the 584 to sea duty confronted the designers with two requirements not encountered for the Army application: (1) a carrier is not stable, and stability is essential for obtaining meaningful tracking data, and (2) an antenna aboard ship must be strong enough to function after having endured a storm, the soldier's option of retracting the dish into the trailer not being practical. The first problem was solved with a platform that maintained stable directions for three axes so as to eliminate roll, pitch and yaw, achieved through servo-mechanisms guided by gyroscopes; the second was solved by making a strong paraboloid. The resulting prototype, the CXBL, was mounted on the new *Lexington* in March 1943, and the first two General Electric production sets, designated SM, went on *Bunker Hill* and *Enterprise* in October [2]—several months before SCR-584 reached a similar stage of production. Unfortunately, the solution of the stability and strength problems made SM rather heavy, specifically 9 tons. It was just one radar too many for the light carriers, making them top heavy, a property not admired by seamen, and the light carriers rolled enough as it was, so only the heavy carriers received this fighter-direction capability. Rad Lab reduced the weight of SM by half in the SP, which was installed on lighter vessels in late 1944 [3].

Radar use progressed so rapidly during these early years that many things were left to the users. Admiral Nimitz had established schools for operators, mechanics, fighter-direction officers and commanders at Pearl Harbor, but doctrine was not settled. Radar was so new that even the most recent graduates of the Naval Academy had learned nothing about it, and this made it increasingly the province of reserve officers. This led from time to time to the delicate task for the radar men of educating officers who were both regulars and their superiors in rank [4].

On carriers the Combat Information Center had grown out of the Radar Plot room that had naturally formed. The radar set was standard issue, but everything else had to be worked out by the new users. All required a method of plotting; some chose horizontal tables that mimicked in much reduced space the Battle of Britain filter rooms, some vertical sheets of clear plastic on which a seaman wrote mirror-image text. Radar personnel arranged their accessories for most efficient operation according to their own concept, much as one arranges the furniture of an apartment for ease of living, and practice varied from ship to ship. Alert teams soon had their maintenance shops outfitted with lots of non-issue tools and test equipment, and those who knew anything about doing tedious work in the tropics managed to get air conditioning [5].

By the end of the war Combat Information Centers had become big business. They were provided large compartments with space for as many as 50 inhabitants. It became the location where knowledge of an action was concentrated and replaced the bridge as nerve center [6].

Determination of target height with air-warning sets was an art studied assiduously, as it was the most difficult coordinate to determine. Height had to be extracted from the manner in which signal amplitude changed with range as the target passed the different vertical lobes of the SK. There was a general theory, but it had to be checked against calibration flights in which an aircraft flew at a given altitude, then again at another altitude and so on. This was a time-consuming affair that was not popular with the flight officers, and success of a ship's radar group turned on the eloquence of their expression of the need and importance of this flying [7]. This kind of height determination needed a not-furnished stable vertical axis for the antenna or the motion of the lobes would confuse the lobe patterns, but the surface conditions of the Pacific were on the average the best encountered for meter-wave equipment at sea.

Secure line-of-sight communication was a minor side benefit of radar employed by some vessels in the Pacific. The CXAM had a switch for changing the pulse repetition frequency as an aid to determining whether a pip had been returned from beyond the maximum display range. Replacing this switch with a telegraph key and listening to the receiver with ear phones allowed two CXAMs to pass signals between one another in code. The very high frequency and the narrow beam made interception by the enemy unlikely [8]. When submarines began to fight as wolf packs they used the SJ in this way to communicate with one another when submerged [9].

The SG had become the loved one of deck officers for new and unexpected reasons. The charts of the Pacific islands were distinguished for their gross and dangerous inaccuracy. Close approaches to shore were the rule not the exception, which made the maplike PPI display of SG a beautiful sight for a skipper proceeding, perhaps a bit faster than he would have preferred, toward an unknown, poorly mapped coast. It could be a tricky even with a perfect radar because low-lying land in the foreground could be missed and higher ground farther back mistaken as the shore [10]. The adoration of SG did not withhold criticism, if anything it amplified it; waveguides sometimes became lossy when misaligned by the shock of gunfire, and more power was needed to increase range. Raytheon began production of the SG-1, having tightened the joints and raised peak power to 50 kW, in May 1943 [11].

Main battery fire control received the 10 cm mark 8 (a Bureau of Ordnance designation that replaced the seldom used FH), which had gone to sea in prototype from Bell Labs as CXBA. It was the first phased-array microwave radar, allowing electrical rather than mechanical scanning. Its 15 to 20 kW peak power allowed any vessel within gun range to be taken under blind fire. Electrical scanning gave the operator a continuous presentation of the 30° sector toward which it was pointed and simplified lobe switching. It allowed fall-of-shot corrections to be made while keeping other targets under observation, preserving thereby in more convenient form an advantage of the FC and FD. The beam was only 2° wide, and the pulse width only 0.4 μs, giving excellent resolution and removing the confusion that often accompanied multiple targets with the 40 cm sets. Data were presented on two scopes: one gave target amplitude as a function of range, the other range plotted against bearing. The mark 8 drew enthusiastic responses from all gunnery officers who had the opportunity to compare it with the FC and FD. Production began at Western Electric in October 1942 [12].

The antenna had a curious appearance that resulted from an ingenious design. Bell had used a phased-array antenna for steerable shortwave transatlantic telephony before the war, and the engineers saw possibilities for using the basic idea with microwaves. They also introduced a new kind of radiating unit, the polyrod, a properly dimensioned rod of polystyrene inserted into the open end of a waveguide, which radiated continuously along its length. George Southworth had observed the propagation properties of dielectrics in his earliest microwave work. A few experiments showed that rods could be shaped to give the directivity needed for the mark 8 [13]. The fixed array of the mark 8 consisted of 14 identical horizontally arranged elements, each a vertical array of three polyrods; the phase to each element was controlled by mechanically rotating phase changers. The design accomplished nicely the important goal of suppressing unwanted side lobes [14]. It also seems to have been the only use by the Allies of the polyrod radiator. It is worth noting that the limited German microwave work employed it [15].

The Navy's version of ASV mark II, the ASE, had become standard equipment for the PBY flying boats but was ill suited for the torpedo bomber TBF Avengers. NRL had had their own 60-cm air–surface radar in development when the ASE was adopted and completed it for these new carrier planes. Like the ASE this set, the ASB, used two Yagis directed 7.5° right and left of the forward direction [16].

The introduction of radar-equipped night fighters onto carriers was a problematical thing. The first sets available were the 3 cm ASH (AN/APS-4), search equipment that needed a separate crew member as radar operator because of the complexity of operation. Attempts were undertaken with the Avenger, which satisfied the requirements as the radar platform but was too slow to serve as the gun platform. Fliers from the *Enterprise* attempted to correct this by having two F6F Hellcats accompany the radar plane in breaking up a night torpedo attack on the task group in November 1943. The result was successful in shooting down two bombers and thwarting the attempts of the third, but the confusion presented by the two fighters without radar may have led to one of the American fighters, flown by a distinguished Ace, E H O'Hara, being shot down in error [17].

Later a new set, AIA-1 (AN/APS-6), was mounted in single-seat fighters and designed as an airborne interceptor. The set was simplified so that the pilot could operate it alone and a pair of red-filtered goggles preserved his night vision when viewing the 50 mm diameter scope on the instrument panel [18]. Carrier night fighters were used sparingly because keeping them aloft put additional requirements on already overworked deck crews. Other ways were found to be effective against the dangers of night torpedo-bomber attack.

A weakness of American electronic arms that was closely associated with radar was the continued use of high-frequency radio for combat aircraft—more than four years after the Royal Air Force had replaced it in Fighter Command with crucial very-high-frequency equipment! These old radios had been the curse of the carrier operations off Guadalcanal because of their inherent noise, erratic transmissions and the limited number of channels they could carry. Change was in progress, but fighter directors still had to deal with mixed systems in June 1944. Fortunately, communication discipline had been greatly tightened by then.

7.5.2. *Japanese radar, 1943–1944*

Japanese surface units were being equipped during these years with 1.5 m mark 2 model 1 air warning, a dipole array following the design of the land-based set, mark 1 model 1, and as such was similar to the general principles of Freya and SCR-270, although of shorter range. These installations were not notably successful, owing to their not being sturdy enough for use aboard warships. The set mounted on the 18 inch gun battleship *Musashi* suffered broken antenna insulators and the failure of transmitter tubes when the vessel underwent firing practice off Katsuru Island in September 1942, which naturally enough sent criticism to the Naval Technical Research Department (NTRD) [19].

The 10 cm mark 2 model 2 had characteristics that distinguished it from Allied microwave equipment. It used a low-power magnetron instead of a klystron as local oscillator, the anode of which had nearly the same arrangement of holes as the main oscillator but was made of thin metal that could be deformed; this allowed the frequency to be altered just enough to hold the intermediate frequency of the heterodyne receiver constant. The waveguides used in all Japanese sets were circular; transmission and reception were in separate horns; parabolic reflectors were not used.

This unit had made a good name for radar in the Imperial Navy. As a result of the insistence of Commander Yoshio Sakura, the earnest Communications Officer of the Chief of Staff, First Fleet, an experimental model of mark 2 model 2 had been installed in the cruiser *Atago* while she was undergoing repairs for injuries received in action off Guadalcanal in November 1942. This radar had been of great use during an attempt by seven cruisers to disrupt the November 1943 Allied landings at Empress Augusta Bay on Bougainville. The cruisers had been driven away by aircraft, but the *Atago*'s well trained radar crew had allowed them to thread their way unscathed out of the extremely hostile zone [20].

The strains of production and development caused reorganization at the NTRD. The Electric Wave Research Department became its largest element and the one with the largest share of troubles. Having only begun the first serious radar work in 1941 this group now had to compete with an enemy that had begun in 1935. The Imperial Navy's concern is reflected in the placement of Vice Admiral Takeshi Nawa as NTRD chief in mid–1943. His appointment resulted in part from dissatisfaction in the way Ito conducted the more practical aspects of radar, and Nawa made an extensive reorganization. He faced the problem of keeping design up to the latest requirements and production adequate for a huge demand as well as training operation and maintenance personnel [21]. It was a substantial order.

The new organization did not produce the results desired, and the Navy Minister created in March 1944 the Electronics Development Department to assume responsibility for all research and testing of radar, communications and sonar equipment. This move was strongly opposed by the units affected, their argument being that slow progress had come from the lack of qualified personnel and material, not from organizational defect. The new department still controlled the same operating laboratories as before; the absence of noticeable progress would bring another, equally futile change the following year [22].

Matters were not helped by the attitudes of the authorities. Shigeru Nakajima was responsible for microwave development at Japan Radio where the cavity magnetrons had been invented in cooperation with NTRD. At the beginning of the war his staff numbered 800 engineers but was progressively reduced by inductions into the services to half that by the end [23].

Knowledge of the Metox receiver arrived in Japan aboard the German U-511 in August 1942, and more than 2000 of them were soon distributed within the fleet. Inasmuch as they did not respond to microwaves, the same deficiency that soon proved so troublesome in the Bay of Biscay, they were of no more use against the ever increasing numbers of SGs than the U-boat sets were against ASV mark III and quickly lost favor [24].

In February 1944 the Americans learned that the Japanese twin-engine Betties had radar, the 2 m mark 6 model 2 (known in prototype as the H-6), with a Yagi antenna on the nose and horizontal dipoles on each flank of the fuselage [25]. This general search set was developed at the Naval Air Technical Depot at Oppama to a great degree independently of NTRD. It had a peak power of 2 kW and weighed 60 kg. The Depot also produced the 60 cm FD-2 for night fighters having a peak power of 2.5 kW and weighing 70 kg. There is little evidence of the employment of FD-2 [26].

It is much more difficult for countermeasures people to identify airborne radar. A ground radar station remains fixed while its wavelength, pulse width and repetition rate are studied, but an airborne set is quickly gone. The usual method is for a Ferret to fly as bait and, if attacked, use the advance warning to evade destruction. Hunting for Japanese airborne radar occupied numerous Ferret flights in late 1943, precipitated by reports of night attacks of mysterious origin, but no Japanese planes were so detected. The first information about the mark 6 model 2 mounted on Betties, which was ill suited for use as an interceptor, came from the same source that disclosed the existence of the 10 cm mark 2 model 2, a large number of documents captured in February 1944 on Kwajalein Atoll in the Marshalls [27].

Japanese radar suffered seriously from a shortage of most important electronic components and of the outstandingly poor quality of much of what could be obtained. To meet the increased demands vacuum-tube manufacture had to be subcontracted to small companies with little experience and with unsatisfactory results. Nickel was so short that coins of that composition were bought pretty much to extinction and recast. In August 1943 the two services were so alarmed at the shortage of materials that they set aside their grudges and formed the Army–Navy Radio Technology Committee (Riku-Kaigun Denpa Gijutsu Iin Kai) made up of Professor Hidetsugu Yagi of Osaka University, Professor Yasushi Watanabe of Tokyo Imperial University and Professor Hantaro Nagaoka of the Institute of Physical and Chemical Research [28].

7.5.3. *Carrier raids*

The carrier raids that had preceded the Battles of the Coral Sea and Midway had left a strong and very positive impression on nearly all members of the Pacific command. They had kept the enemy guessing, had kept the precious carriers away from prying eyes and had provided an unparalleled form of training for well schooled but green fliers—all possible because of the CXAM at the foremast. So carrier raids became the norm when an invasion or fleet action was not impending. They also served to reduce the aircraft population of the island bases. The process picked up a name in South Sea pidgin of 'makee learn'.

The element of complete surprise seldom came with these strikes because the islands soon had air-warning radar, but the defenders were still rattled when weeks of quiet duty were shaken by a 20 minute radar warning. One of these—and it was on what Americans incorrectly thought to be the Pacific Gibraltar of Truk—caught the defenders with most planes on the ground. After the war it was learned that a lonely freighter loaded with radar for Truk had gone down from a torpedo of USS *Trout* a couple of months earlier [29]. The radar defenses of these bases did not include fighter control, not even for the fortress that Rabaul had become [30]. The reason for this probably lay with Japanese doctrine that was similar to the German of 1940. Furthermore, no radars capable of returning the three-dimensional coordinates needed by controllers were available until 1944. No operational Japanese set had PPI.

One of these raids afforded an unexpected and valuable technical evaluation. A strike on Saipan and Tinian was ordered for Mitscher's group on 22 February 1944. On the afternoon of the preceding day a snooper had spotted them and got away, so it was obvious that surprise had been lost, and experience had taught them to expect a night torpedo-bomber attack for which the Japanese had demonstrated competence. One of the early 1942 carrier raids had been called off when it was realized that they had been seen, but two years had altered matters. There were night fighters aboard, but Mitscher did not wish to miss the advantages of a dawn attack by reversing course to launch and recover by sailing into the prevailing easterly winds. He decided that night would force the attackers to fly relatively straightforward paths, perfect for his FD radars, directors and guns firing proximity fuzes. He ordered the carriers surrounded by an inner circle of cruisers and an outer circle of destroyers—all having the most modern AA weapons. As the enemy aircraft appeared as expected, the fleet maneuvered by order from the knowledge gained by radar to present the most difficult torpedo targets and the guns opened fire. None of the ships suffered damage, and only five of the 20 attackers returned to the home base. The raid proceeded on schedule and accomplished the desired results [31]. This way of dealing with night torpedo bombers became standard [32]. The improvement of AA artillery over that of only three years earlier was startling to an extreme degree.

7.5.4. *The invasions of islands*

A consequence of the Doolittle Raid on Tokyo was a residue of doubt in the Japanese Command as to the origin of the attack. Aircraft carriers were certainly considered as the most likely base, but because the planes were the normally land-based B-25s the thought remained that they might have come from the Aleutians. The desire to remove such a threat affected the planning of the June 1942 offensive against Midway by supporting the idea of sending a diverting force against these islands. Two carriers were sent to attack the known garrison and airfield at Dutch Harbor while troops were to be landed on the western-most islands of Attu and Kiska. The two carriers accomplished little and were sorely missed at Midway.

The attack on Dutch Harbor did not catch the defenders by surprise, as they had been alerted by intelligence from Nimitz and a timely warning by the SC radar of a seaplane tender. The attacking planes from the *Ryujo* met lively AA fire, which distracted them enough from their work to keep the bomb damage minor, if smokey. Planes from the *Junyo* were unable to find Dutch Harbor, hardly surprising given the weather conditions. It was, in fact, remarkable and showed excellent skill that all but one of the aircraft launched were able to return to their ships, the one having been shot down by the waist gunner of a PBY [33].

Kiska became the first enemy radar to be accurately mapped by a fledgling electronic countermeasures unit. A B-24 equipped with an SCR-587 intercept receiver flew three Ferret missions around the two western-most islands of the Aleutians in early March 1943 and returned with the knowledge that Kiska had two radar sets but Attu none. They produced an excellent contour map of signal strength that showed the most favorable directions of approach. Their efforts were followed by attempts to destroy the stations, but the Japanese were able to repair the damage relatively quickly [34]. Determining the radar potentials of island bases became a standard component of military intelligence for the remainder of the war, generally using cast-off aircraft but with growing importance and support.

So began a three-way war between the United States, Japan and the weather, with the weather the most vicious and dangerous. It was a campaign greatly assisted for the Americans through the rapid equipping of PBY flying boats with ASV mark II, and it was from them that Admiral Charles McMorris learned on 26 March 1943 that a large, well escorted convoy led by Admiral Boshiro Hosogaya was making a major attempt to break the blockade imposed on Attu and Kiska. Action, called the Battle of Komadorskis, opened at 20 km with accurate early salvos of the Americans as a result of radar fire control. In the long-range gun battle that ran for nearly four hours the out-numbered American squadron ended badly damaged with the heaviest unit dead in the water but concealed by smoke. Hosogaya, who had hurts of his own, was unaware of his superior position, broke off the action and ordered the convoy back [35].

The Pacific War was to have many landings by Allied troops on the beaches of defended islands. The first of these was the invasion of Attu [36], begun on 11 May 1943 by the 7th Infantry Division, fresh from tropical training that was intended to throw spies off the trail. Whether freedom from spying served these men better than preparation for the conditions they were to encounter does not seem to have been evaluated. The Japanese and the weather provided the invaders ample resistance during an ugly, 18 day fight that showed what things would be like on such places as Iwo Jima.

After the Battle of Komandorskis and the loss of Attu the Japanese decided to evacuate the 1200-man garrison on Kiska. This was accomplished on 28 July by 19 vessels under Admiral Masatomi Kimura that evaded the American blockade, piloted themselves into the harbor, loaded all personnel and departed just as quietly. It was a brilliant performance followed by an elaborately planned and executed American invasion on 11 August that found nothing but a couple of dogs. Despite PBYs with ASV and patrol ships, the evacuation had evaded interception. The key technical element for the Japanese, used both for navigating through fog and evading the blockaders, was the 10 cm mark 2 model 2 radar [37].

Associated with the evacuation of the Japanese garrison was a strange radar naval engagement—the Battle of the Pips. On 22 July 1943 a PBY reported radar contact with seven vessels, and an American group including two battleships and five cruisers headed for the reported position on a wide front. At 0043 hours on 26 July the *Mississippi* picked up three or four large contacts on the SG at 30 km. This was followed shortly by the FC fire-direction radar reporting five or six. The contacts looked similar to those returned by cruisers or destroyers but were intermittent and unsteady, which was reported to the bridge. At about the same time other vessels reported similar radar sightings. Fire was opened shortly after 0100 at a range of about 23 km by those vessels capable of firing under radar control. It became increasingly clear to the radar crew of the *Mississippi* that something was seriously amiss. As they had narrowed range the SG contacts retained the same amplitude instead of strengthening, but, much more serious, the echoes of the projectiles and the splashes were stronger than those of the presumed targets. Fire ceased after 25 minutes when it became clear there was no enemy there [38]. When visual observation cleared, moonlight and star shells disclosed no targets. Examination of the ocean surface after daylight found no wreckage, no oil slicks, no survivors—nothing but clean ocean surface [39].

This embarrassing action was caused by anomalous propagation to ranges such that the blips on the scope had returned a few sweeps after the original. They had encountered this phenomenon frequently with the 40 cm of the FC but never with the 10 cm of the SG. The Radar Officer of the *Mississippi* later reconciled the contact bearings as echoes from Kiska Volcano and the islands of Segula, Little Sitkin and Semisopochnoi [40]. Had the SG and FC sets had a switch to alter the pulse repetition rate as did the CXAM and the SK, the Battle of the Pips would not have taken place. Pulses from targets within the first sweep are unaffected by such a change, but those returning from targets beyond jump on the display as soon as the switch is pressed.

Observing critically the Attu and Kiska landings was Marine Lieutenant General Holland Smith, who was to become the master of amphibious warfare. With army, marine, naval and air forces he would develop this technique to its ultimate. The Gilbert Islands were to be his first test.

Two strategic directions had evolved by late 1943. MacArthur was to clear the Solomons and parts of the Bismarck Archipelago in order to move across the northern part of New Guinea through the Halmaheras to the Philippines. Nimitz was to drive through the rings of island bases to join with MacArthur for the final drive at a location to be decided. In the original expansion Japan had garrisoned many locations throughout the Pacific and was confident that each would demand a dear price for each insignificant atoll. But once these bases lost their aircraft and could receive no resupply, they became prison camps administered by Japan. It then became clear that only those that afforded harbor and air fields for the next step across the Pacific's continental-sized distances need be taken.

The invasion of the Gilberts took place in late November 1943 as the first step in Nimitz's plan. The two islands of Tarawa and Makin received a heavy pummeling by bombers and ships' artillery, but the effect seemed to have hardened rather than softened the resistance of the garrisons. The attackers made no small number of blunders, and Tarawa became inscribed as a Marine Corps memory of courage and blood. The Marshall Islands were taken in early 1944 in a more workmanlike manner. The Marianas were to be next, and it was assumed and hoped that attack on them would precipitate a major fleet action.

While this was going on MacArthur secured the northern coast of New Guinea with troops of the US, Australia and New Zealand. His path was eased by much of the enemy's naval and air forces that might have faced him having been diverted to help check Nimitz's drive from the east.

The FC and FD fire-control radars were of little use in identifying land targets for the increasingly important artillery function of bombardment preparatory to landing. Saturation bombing and rocket attacks were spectacular and could be demoralizing, but the need was for accurate fire on targets known to present danger. Such targets could often be identified from aerial photographs but not recognized from the ship either visually or electronically. A solution presented itself as forces began moving up the Solomons in 1943. Prominent land features could be identified on the aerial photographs and seen by radar. This allowed offsets to be calculated so that fire could be directed on dangerous emplacements. The method worked optically too, but accurate ranging made radar preferable even on clear days [41]. It worked even better when the 10 cm mark 8 fire-control radar became available.

Wherever a landing was made the Australian LW/AW was on the beach soon, augmented in early 1944 by the light-weight SCR-602. They were particularly valued in the New Guinea theater where control of the air was more hotly disputed than in the Central Pacific. Once the islands were secured, local air defense relied on a combination of the SCR-270 and SCR-268; the SCR-602 was considered expendable, but the LW/AW often served when the heavy 270s and 268s could not be got into position. The defenders liked the long range of the 3 m 270 for warning and found the resolution and height data of the 1.5 m 268 good for fighter direction. Australian Mobile Fighter Control Units and American Signals Warning Battalions provided fighter direction whenever it was needed. They naturally assumed a function that endeared them to all Allied airmen—guiding lost fliers home and locating downed aircraft. The number saved is not on record, but it is very large.

The SCR-268 may have had its most successful day in directing the 90 mm guns of the 9th Marine Defense Battalion in defending the Rendova (New Georgia) landings in July 1943. Of 16 attacking bombers gun-fire downed 12 with an expenditure of only 88 rounds, leaving the four that escaped to be destroyed by fighters [42]. The island locations were ideal for meter-wave equipment, and crews became so proficient in their use that, unlike their comrades in Europe, who seldom had an excellent sea surface to form consistent vertical lobe patterns, they greeted the arrival of microwaves with indifference and relied on meter waves till the end [43].

7.5.5. The Battle of the Philippine Sea

Vice Admiral Jisaburo Ozawa began assembling the Mobile Fleet in May 1944 at anchorages at Tawi Tawi, an island off the easternmost tip of Borneo. It was by any previous measure of naval might a powerful body and clearly intended to fight the 'decisive' battle that would dispose of the American fleet with one blow. It was obvious the United States was going to make an important move, most likely the invasion of strategically important islands. The Mobile Fleet would proceed to attack the landing, forcing the American into the desired major action. The American Admirals planned things the same way. Both sides sought a big fight, each confident of victory.

The Japanese were inferior in numbers of carriers and aircraft, disadvantages that could have been made good by adroit tactics, but they suffered other disadvantages that were to make the outcome for them an overwhelming defeat. The splendid carrier fliers with which Japan began the war were largely gone, and their replacements were ill trained, and training faced the same serious fuel problem as the Navy as a whole. Although Japan possessed huge petroleum resources in Borneo, American submarines were keeping it out of the homeland and the refineries with the result that both ships and planes were desperately short by mid 1944. Naval aviation suffered doubly because the skills needed by carrier pilots had to be acquired by taking off and landing on carriers, something that could only be done with the ships under way. Training carrier pilots consumed enormous amounts of fuel.

The Japanese were also behind in the techniques of submarine warfare. Their anti-submarine forces never approached the skill of American and British, which had proved themselves slow but assiduous pupils in the hard school maintained by Karl Dönitz. Furthermore, the actions in the Solomons had cost Japan a disproportionate number of destroyers. Thus when an American submarine's contact report of the assembly of the Mobile Fleet off Borneo brought enough submarines to cause the fleet losses and keep Nimitz well informed, the Imperial Navy could not keep them off. A line of Japanese submarines posted to the east for the same purpose was discovered and decimated by US hunter–killer groups taught by Atlantic veterans.

Ozawa knew he was outnumbered in carriers and carrier planes but was confident he held an important advantage in the large number of land-based aircraft located at island bases capable of cooperating with the fleet. These unsinkable aircraft carriers had long figured in Japanese and American war plans, initially being highly regarded by both sides, but the success of the carrier raids had reduced American fears. Unsinkable they were to remain, but immobile as well and thus waiting for attack. As the Battle of the Philippine Sea was to emphasize, a carrier without aircraft is worthless, and unless it can be resupplied, as was unsinkable Malta, it remains worthless. Ozawa's unsinkable carriers were placed in this category in a series of attacks over much of the southwest by bombers of the Army Air Forces and the carriers, combining to destroy much of Osawa's air power as well as confuse him about the goal of the obviously forthcoming invasion of someplace, revealed later to be Saipan.

There were five major carrier actions in the Pacific War, two of them American victories of overwhelming degree, both under the command of Raymond Spruance, a battleship admiral possessed of an austere personality remarkably similar to that of George Marshall. The situations for him at Midway and the Philippine Sea were inverted. At Midway he knew the position of the Japanese and was able to make the crucial first strike; in the Philippine Sea he was until the very end in the dark about the location of Osawa's ships, the consequence of the significantly shorter range of American carrier planes and the prevailing easterlies that allowed the Japanese to launch and recover aircraft without reversing course—the modern seaman's equivalent of the weather-gauge advantage of the age of sail. By 1944 Spruance no longer looked on this with the concern of 1942 because he understood and was confident in the power of defense: radar would give him adequate warning to intercept attacking formations well away from the ships; planes that passed this line of defense would encounter radar-directed 5 inch AA fire using the proven and respected proximity fuzes; survivors would be received next by 40 mm guns having the excellent Draper sights and finished with the 20 mms. His confidence proved well founded.

On the morning of 19 June Osawa dispatched four major attacks, all of which were spotted by the SKs in time for the fighter directors to place ambushes of Hellcats, which removed a significant number of the attackers. The play followed the script in each successive phase with little damage or loss of life to the defenders resulting. Of particular importance, all carriers remained operative. Much credit goes to Lieutenant Joseph Eggert who functioned as task-force fighter director aboard the *Lexington*. He was aided by Lieutenant Charles Sims, who was fluent in Japanese and who listened to the commands of the airborne Japanese air coordinator and informed Eggert of his every word.

Despite the serious losses Osawa was known to have suffered, the Americans wanted desperately to sink enemy carriers. We can now say this was a waste of effort for the battle was over; Japanese carrier power was irretrievably ruined whether empty carriers existed or not, but such talk is hindsight and would have been violently rejected on 20 June, when the enemy fleet was finally located near the end of the day. A dramatic attack was launched that succeeded in destroying a carrier. The retrieval after dark of the returning planes marks a romantic peak in naval arms, but the strike's effects were not worth the losses. Three Japanese carriers were sunk in this great battle, but two went down from submarines.

The 19th and 20th of June 1944 encompassed the most intense air battle ever fought. It is impossible to give the number of aircraft involved both from the uncertainty in deciding who was and who was not engaged and from the uncertainty of the number of Japanese planes retained by the island bases at the start, but one can give the numbers carried by the two fleets with some assurance. Osawa had a total of 473 aircraft at the beginning, Spruance had 956; of these Osawa lost 426 to which might be added about 50 more from the island bases, Spruance lost 130, three quarters in the attack of the second day with its high loss of returning planes [44]. It is more difficult to establish what portion of Japanese losses fell to AA fire, although significantly less than those shot down by fighters, and much more difficult to separate the results of radar-directed proximity-fuzed 5 inch guns from the 40 and 20 mm automatic guns. The radar-directed fire was effective at ranges beyond those for the automatic guns, which were also the ranges at which the courses were slow enough for the manual-tracking FDs and especially for the mechanical-analog mark 1 director to follow. These directors had become by then the recognized weak link in the radar-director–gun-fuze chain. Dramatic improvement was on the way in the form of the mark 57 being designed at the Johns Hopkins Applied Physics Laboratory by the team that had provided the proximity fuze, but it would not be seen in the war zone until January 1945 [45].

Radar had transformed carrier warfare yet again. Its great strength throughout the war was defense, and the SK, the improvement over the XAF bedsprings demonstrated in the Caribbean five years before, had defeated the might of Imperial Japan off the Marianas as surely as the long bow had defeated the French at Agincourt [46].

This great carrier battle had settled the fate of the Marianas. Without control of the sea the garrisons on the various islands had no hope of relief, but this in no way affected the determination with which they fought off the invaders. In the early morning of 7 July the few thousand Japanese troops that remained made a convulsive and suicidal attack on the American lines that caught some Army units off guard but which was cut to pieces within a few hours at the usual ten-to-one losses. After that their commander killed himself. Separately the naval commander of Saipan did the same. His command was sufficiently small, a few patrol boats and shore personnel, that we would normally not record his passing in this mere outline of the military struggle, but he was Vice Admiral Chuichi Nagumo, 26 months earlier hailed as one of the greatest admirals the world had ever known. What must have been his emotions in seeing the battleships he had sunk at Pearl Harbor firing at Saipan from off shore?

Saipan had a large civilian population, a large fraction of which— men, women and children, young and old—killed themselves in mass suicides, although in nothing like the proportion of the military personnel. Such was the Pacific War to be.

The word 'radar' entered newspaper articles in June 1943, having been recently made public by Britain and America in a joint release. At that time the Navy allowed an interview of NRL engineers that disclosed many theretofore secret details [47]. In November 1944 a US Navy recruiting poster stated: 'Young men wanted by the Navy for training in RADAR, one of the newest and most exciting developments of the war—with a great postwar future [48]'.

PHOTOGRAPHS: AIRBORNE RADAR

An AN/APS-4 mounted under the wing of a Curtiss Seahawk. This 3 cm radar, which functioned for surface search or aircraft interception, was enclosed in an air-tight container that could be fastened to a bomb rack, indicated by an arrow. It was designed by the Bell Telephone Laboratories and manufactured by Western Electric. The photograph shows the aircraft hoisted aboard USS Chicago on 14 August 1945. National Archives photograph 80-G 700315.

1.5 m ASV mark II radar mounted on a Swordfish. This is a combination of two of the most valuable weapons the Allies had in the war at sea. It was a combination that cut off Axis supplies to North Africa in 1941 and 1942 and stopped the Bismarck for destruction. When flown from escort carriers it became an important weapon against the submarine. The 'towel rail' transmitter dipole is seen at the center of the leading edge of the top wing. One of the two receiver antennas can be seen attached to the outer starboard wing strut; the receiver antennas were Yagis directed off axis to allow lobe switching. The photograph was taken aboard USS Wasp in April 1942. National Archives photograph 80-G 7093.

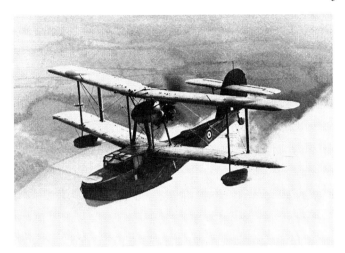

1.5 m ASV mark II radar mounted on a Supermarine Walrus reconnaissance plane. The 'towel rail' transmitter dipole is seen at the center of the leading edge of the top wing. The two receiver Yagis are seen attached to the outer wing struts and are directed outward to allow lobe switching for determining direction. The photograph is dated 3 November 1941. National Archives photograph 80-G 25014.

1.5 m ASV mark II radar mounted on a PBY. This is a combination of two very valuable weapons the Allies had in the war at sea. This British-designed radar gave the US Navy its night vision when mounted in these long-range patrol bombers. The transmitter antennas were mounted on both sides of the hull, below and behind the pilot's window. The receiver antennas were mounted beneath the wing on both sides; they were Yagis that lay in a plane parallel to the one defined by the wing. The date of the picture is 8 March 1942. National Archives photograph 80-G 403256.

The first 1.5 m ASV mark II radar mounted on a PBY. This is a combination of two very valuable weapons the Allies had in the war at sea. This British-designed radar gave the US Navy its night vision when mounted in these long-range patrol bombers. The transmitter antennas were mounted on both sides of the hull, below and behind the pilot's window. The receiver antennas were mounted in various configurations, often Yagis that lay in a plane parallel to the one defined by the wing and beneath it on both sides, although it appears that in this case it was mounted on top of the aft section of the hull. National Archives photograph 80-G 700269.

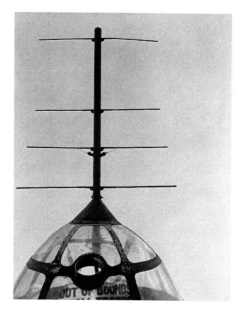

The antenna of a Japanese 2.0 m mark 6 airborne ship- search radar mounted on a Mitsubishi-01 two-engine medium bomber cap- tured by American forces in February 1945. This very suc- cessful ship-search radar used a common antenna for transmitter and receiver. Shown here is the Yagi used for forward search. A pair of dipoles mounted on the sides of the fuselage permitted scanning to the sides. Although the Yagi was a Japanese invention, they did not incorporate it into their designs until discovering it on British Army searchlight radars captured at Singapore. National Archives photograph 111-SC 289080.

A German Me-110 equipped with Lichtenstein airborne radar. The nose of the two-engine machine shows four dipoles with reflectors for each of two radar sets, the 50 cm Lichtenstein C-1 and the Lichtenstein SN-2, normally described as 3.3 m but shown here nearer 2 m. The 50 cm version proved to have too short a range for picking up targets independently of ground radar but was better for the final stages of the tracking. Imperial War Museum photograph CL 3299. Crown Copyright.

A Bristol Beaufighter. This became the standard RAF night fighter and was equipped with radars for airborne interception using 1.5 m and 10 cm. It is not possible to ascertain the radar type from the photograph, except to say that it was microwave, evidenced by the plastic nose that shielded the paraboloid reflector. From the date, 27 March 1949, one would assume the set to be an AI mark X, identical to the SCR-720. National Archives photograph 80-G 403264.

CHAPTER 8

THE END IN EUROPE

8.1. INVASION

One could, without great scholarly exertion, assemble quotations from many leaders of World War II emphasizing the difficulty of landing and maintaining an invasion force on defended shores. For the Americans in 1942 such a stratagem for the European continent seemed a straight-forward way of preventing Soviet collapse. For the British it recalled a memory of the Dardanelles and Gallipoli in 1915, a memory burned particularly into the mind of Winston Churchill and reinforced with the later horrors of the Somme and Passchendaele. Thus General Marshall's call for a cross-Channel invasion in 1942 by green American and British troops was received with incredulity in London. The disastrous outcome of the division-sized raid on Dieppe made a quick end to such plans but did not remove the goal from American and, though retained more reluctantly, British minds.

The cross-Channel invasion was Marshall's unalterable intention from the moment the United States entered the war, and it assumed the same importance for his deputy in the War Plans Division, Brigadier General Dwight D Eisenhower. That they maintained for a period what in retrospect would have been a highly flawed course can be understood from the dire situation in which the Allies found themselves in 1942. There was every reason to fear that the Soviet Union would be defeated by the seemingly unstoppable drive of the Wehrmacht into southern Russia, and a landing in France was a way to help prevent this collapse. The risks were well understood by the two American generals; they estimated the chances of a successful 1942 landing to be 'only one in two and of maintaining the beachhead at one in five'. These were desperate times [1].

When Allied troops finally went ashore in Normandy in June 1944, they had behind them the experience of the North African and Mediterranean landings as well as the early ones of the Pacific. The knowledge so gained allowed the invasion to be a success, but of equal, perhaps greater importance was the absolute command of the skies by spring 1944. For whatever its strategic value, the bombing of the Reich by night and day had forced the Luftwaffe to fight, and although they had inflicted serious damage on the attackers—one can say they had defeated the bombers by early 1944—it had been accomplished at a terrible loss in trained air crews and machines. Allied losses were also great but could be made up; German losses could not.

By spring 1944 Allied superiority in radar was equally well established, but its affect on the battle is less easy to evaluate. Its most important contributions were in navigation and countermeasures.

The German radar chain occupied a special part of the invasion plans, a part that produced some curiously contradictory actions. The task of cataloguing the characteristics and locations of German radar and jamming stations, much helped by the easily recognized fortifications with which they had been provided after the Bruneval raid, was shared by R V Jones' Scientific Intelligence and Claude Wavell's Central Interpretations Units [2]. By locating the invasion fleet and its air cover these sets could provide crucial warning of where and by inference when the landing would take place. Ideally, the stations should have been destroyed, but starting with Göring's half-hearted attempt to destroy Chain Home at the beginning of the Battle of Britain, it had been learned that radar stations were not so easy to eliminate as it might seem. The electronics were well entrenched and required a direct hit by a heavy projectile, and the spindly antennas of Seetakt and Freya were hard to bring down and often capable of repair. Rocket- and cannon-firing fighter bombers attacked the sites and proved much better suited to the task than had Göring's dive bombers. It was a nasty business for the fliers because the radar operator clearly read their intention, and the stations were well defended by 37 and 20 mm automatic guns [3]. Destruction was attempted during the preparation along the whole coast where landings might be possible, so as not to tip off the selected beaches, and of 92 naval and air sets operating before the assault only 18 were either left in operating condition or restored during the critical time between 0100 and 0400 hours of 6 June. On top of this the general failure of the electric power grid forced the use of engine-driven generators for which fuel was scarce to non- existent [4].

Knowing that complete elimination was not possible, the attackers followed with jamming and deception. Jamming the many sets from ground and airborne transmitters could only hope to reduce the range at which detection would occur, and that with uncertain reliability. The amount of warning radar gave the defenders is not certain, reports vary from three hours [5] to almost nothing [6].

Three squadrons of the airborne Mandrel screen were assigned the protection of the advancing fleet by flying racecourse patterns at ten locations along a line from the Bill of Portland to Littlehampton [7]. A reinforced squadron set up a similar screen along the Somme estuary to protect the very vulnerable 1000 aircraft transporting airborne troops for the landings east of Caen and on the Cherbourg peninsula. These jammers dropped for the first time Window cut for Freya, often called Rope because of its 1.7 m length, in addition to using Mandrel. They also jammed the radio-telephone channels of the German night fighters dispatched toward what they took to be a bomber stream. The total effect was so strong that the night fighters returned to their control points. One complication for the counter-measure aircraft was that their Mandrel jammers seriously interfered with their own Gee receivers, so navigation had to be by dead reckoning [8].

The vessels of the fleet did not have to depend on airborne jammers; they were outfitted so as to produce cacophony on the Seetakt, Freya and Würzburg wavelengths. The Channel radar war had seen the use of the Royal Navy's type 91, which by 1944 was installed on major warships as well as on the English shore. Lighter vessels were equipped with light-weight jammers designed by the Harvard Radio Research Laboratory, origi-nally for airborne use. This was the first time that the Royal Navy per-mitted jammers modulated with noise to be used, rather than sinusoidal and repetitive-pulse modulation, and the type 91 sets were modified to conform to this American practice. Noise jamming is harder to filter out and requires a transmitter with a wider pass band, which makes it corre-spondingly difficult for the jammed radar operator to evade the noise by making a small change in frequency. Noise modulation had been avoided for fear the Germans would take it up too (a curious fear because Martini had used it during the Channel Dash), but by June 1944 their jamming was no longer so greatly feared. Some 800 jammers were distributed among the assault vessels [9].

The limitations on destruction and jamming caused much effort to go into deception. The closest point between England and France was across from Dover. It was a natural place to cross, and was close to Germany. Normandy was selected for various reasons, primarily for the ease with which it was hoped to take ports with sufficient capacity to sustain a conti-nental army, but from the beginning an elaborate ruse had been exercised to make Calais seem the selection. This had involved the creation of a ficti-tious army group in Kent, manifested by discreet radio transmissions, the deceptive meanings of which were left for astute German intelligence of-ficers to penetrate in a manner conforming to Allied wishes. At some time the fleet approaching Normandy would be observed, more than likely by a few functioning Seetakt sets, and the enemy had to believe it a feint. Here Robert Cockburn's skills came into full play.

To reinforce the impression that the real invasion would be at Calais required that a fleet be detected by the radar stations that remained after the fighter bombers had done with them. The requisite number of vessels could hardly be released for such a purpose, so a radar spoof, the greatest of all time and Cockburn's masterpiece, took its place. To ensure that no unsuspected radar wavelengths would be encountered, Martin Ryle, who did the detailed planning, made an electronic espionage flight along the coast [10]. Finding no surprises, he loaded the aircraft with Window cut to the half-wavelengths of the Seetakts and Würzburgs. Following a series of parallel racetrack courses, eight Lancasters directed an 'invasion' toward Le Havre; they dropped the foil at locations that continually advanced at the speed of an approaching fleet with each turn of their course. Six Lancasters provided the same spoof directed toward Calais [11]. Enough Mandrel jamming was added to give the operation credibility without completely obscuring the 'ghost fleet'. The distribution of the drops in time and space were matched to Seetakt's beam width, pulse width and lack of height data. It was another instance where Seetakt's lack of lobe switching hurt as the cloud of foil could not be distinguished from a number of small individual targets. It was all interpreted by the radar operators as designed, although one can well imagine that their analytical skills may have been somewhat diminished by the recent visit of fighter bombers.

Insuring that the advance of the foil fleet proceeded at the stately speed of water craft when dropped from aircraft flying forty times faster required extraordinarily accurate dumping, which in turn required extraordinarily accurate navigation. This had three components. One navigator computed dead-reckoning positions; a second located the aircraft with Gee; the third fixed the exact position for the drop with Gee-H, using the approximate positions from the other two in order to make his fix at just the right moment, as Gee-H required interrogation by the operator [12].

A few launches equipped with Moonshine repeaters replied to the radar interrogation by an enhanced return signal, thereby creating the illusion of the reflection from a large vessel and covering the possibility that some of the equipment had enough resolution to distinguish individual targets from the mass. Moonshine had been introduced earlier as a feint for the first daylight flights by the US 8th Air Force. The method, a ruse fairly easily penetrated, had then been withdrawn after a few successes for use on D-Day[1]. The launches also towed balloon-borne reflectors. As the 'fleet' came closer to shore powerful loudspeakers provided the sound of ships' winches and anchors, enveloped in copious amounts of smoke. A fake airborne drop, complete with fireworks for sound effects, completed the Calais spoof [13]. The naval bombardment force off Le Havre had a diversion of its own made up of radar reflectors on balloons that were both anchored and towed by minesweepers [14].

It was learned later that the feint toward Le Havre had not been detected because the attacks on the radar stations had been too effective, but the stations at Calais reported the menacing fleet [15]. A few hours of confusion resulted from these games with the result that forces in the vicinity of Calais were given an invasion alarm order at 2300 hours on 5 June whereas those in Normandy received no such order until the landing was in progress [16].

By June 1944 navigation for surface vessels around Britain had come

---

[1] See Chapter 6.2 (p 291).

to rely on Gee routinely. It functioned well in all kinds of weather and gave positions accurate to less than a kilometer; jamming was less a problem at locations well removed from the air over Germany. One of the few successes of the Dieppe Raid was the accuracy with which the assigned beaches were reached, the result of Gee. There would be confusion enough on the beaches without units landing well away from their goals, and Gee kept things sorted out. Indeed, the channels of approach followed the hyperbolic Gee grid lines. The Channel had huge belts of mines that had to be swept accurately just before the arrival of transports and landing craft, and the minesweepers used Gee. These were navigational requirements that would have been quite impossible three years earlier and were a fundamental requirement for 6 June 1944. Gee made it seem easy.

The importance to the landings of this navigational system, which had been known to the Germans long enough for Allied bombers to encounter jamming routinely, does not seem to have been given major consideration in the plans for defending the beaches. It figured highly when planning the attacks on radar and jammers, and the five known jammers were eliminated in the same fighter–bomber raids that hit the radar chain and no jamming for the Channel was encountered [17]. One suspects that the oversight of not preparing reserve jammers, kept quiet until the invasion was looming, was the result of defensive plans having been made primarily by men versed in war on land who had little appreciation of the navigational difficulties of the landing craft. The attackers had expected jamming despite all, so an alternative system was mounted on principal ships, minesweepers and lead landing craft. This was another system using hyperbolic coordinates that was manufactured by the Decca Company and called Decca Navigator. Although not without fault, it was more accurate and easier to use than Gee. When no jamming was encountered, it was removed from service because of its secret nature [18].

In the initial movement of the attacking fleet, radio emissions of any kind were severely restricted, but once it was clear the enemy knew what was under way, even if imperfectly informed of certain key elements, these restrictions were removed. Transports moved to about 6 km from the shore to transfer men and equipment to the landing craft. These were guided by control landing craft that depended primarily on a gyro compass and a 10 cm surface-search radar for guidance in the final approach but also had fathometer, accurate clock and underwater sound equipment. They had special radar maps of the shore that showed the effects of prominent coast features and allowed the operators to compare their PPI screens with a chart showing a reasonable expectation of how the shore line would appear. Landings were expected to be accurate to within 60 m in position and 1 minute in time [19]. Minesweepers dropped radar-reflecting buoys to mark the boundaries of swept zones and shore parties erected corner reflectors to serve as fixed navigation points [20].

The flanks of the landing were covered by parachute and glider troops, who dropped under cover of darkness. The aircraft used Gee for initial navigation but followed radar beacons set up by undercover agents to locate the drop sites. This was the great triumph of Rebecca (US manufacture AN/APN-2), a modification of ASV mark II that sought the portable beacon, Eureka (AN/PPN-1). Eureka had not only the happy characteristic of speaking only when spoken to, it also allowed the operator to transmit signals at a slow rate to Rebecca, thereby providing crucial last-minute information about local wind or enemy actions [21]. This airborne action was fraught with the possibilities of disaster and was opposed by many in high command. The success in getting the airborne forces moderately close to their assigned target areas at night was possible only through navigational accuracies considered impossible by many not privy to the secrets. The success of these forces was limited and the casualties high. Landing division-sized units from the air, which had seemed the most modern form of warfare, was already obsolete, but unfortunately the lesson was not yet learned and the great airborne disaster at Arnhem was not avoided [22].

Defense of the fleet and the beachhead against air attack had been prominent in the thoughts of the planners. To this end the British LW and its American copy, SCR-602 went ashore in early landings [23], although air superiority left them little work. Fighter control radar units went ashore on D-Day plus 1 [24], but during the interim their duties were assumed by three fighter-direction tenders accompanying the fleet with RAF teams manning them. Each tender had two 1.5 m GCI sets (AMES type 15) and one 50 cm early warning set (AMES type 11). These ships proved of little value. The confusion caused by so many aircraft with the usual IFF mix-ups, by the large number of ships and by the reflections from land made it difficult to identify German planes, and those that did inflict damage on the fleet—the low-flying mine layers—were seldom detected with 1.5 m and 50 cm equipment [25].

It was during the invasion that MEW (microwave early warning, AN/CPS-1), the ultimate surveillance radar, made its first contribution[2]. With its great range, 300 km under good conditions, it provided a target resolution in horizontal coordinates superior to any air-warning equipment. Its 66 tons, four 12 kW engine-driven generators and crew of 30 to 50 provided unparalleled information at five indicator positions, each with a 30 cm PPI scope and two smaller scopes for range plotted against signal and range plotted against direction. The PPI scopes were not only large but were so devised that the indication zone could be moved off center, thereby allowing the active sector to be seen in more detail. MEW came with a fighter-direction center equipped with a large vertical ground-glass plotting panel on which coast, azimuth and range lines were marked along with the air-controllers' grids; data were recorded with colored chalk by plotters standing behind and writing in mirror image [26].

---

[2] See Chapter 4.5 (pp 195–6).

Shortly after MEW had been set up and before it had operational communications installed, it gave a dramatic demonstration of its capability. On 20 March the operators in training noted a large formation at 270 km headed over the Atlantic from France. A telephone call to a nearby GCI station, which could not see the aircraft, disclosed that 14 B-17s with 140 men had just reported themselves hopelessly lost and were planning to ditch. They were advised of their true position and vectored back to England [27].

Luftwaffe radar troops gave a demonstration of what the subjugation of Germany was going to entail. A radar station near Douvres, about 12 km north of Caen and 4.5 km from the beach, was encircled as the invaders moved inland. The garrison consisted of only five officers and about 200 men, but they were well dug in and had acquired a nice assortment of antitank and machine guns. The station finally surrendered to a major armored attack after 11 days' siege [28].

In addition to bombardment by major warships, the British and American bomber forces contributed their devastating power to softening the beach defenses. RAF Bomber Command dropped 5300 tons of bombs on beaches and coast defenses during the night of 5/6 June using H2S to aim. The US 8th Air Force had the task of repeating the exercise five minutes before the first troops were to land. Although intended to be a daylight attack, the deteriorating weather forced the use of H2X. Bombardiers received radar maps and extensive instruction; the sharp water–land boundary was helpful to give hope of accuracy. Given that their bombs were intended to fall only 1000 m ahead of the landing craft and that a one second error in bomb release translated into about a 100 m error on the ground, it is not surprising that caution was in the air [29]. The bombardment killed no Allied personnel, but then few bombs fell on the beach defenses; they did not, however, contribute to German good humor in positions immediately behind the beach.

In looking back on the course of World War II one tends to see the Allies on a steady path to victory after the beginning of 1943, which is indeed how it came to pass, but it was not foreordained. Failure in any one of the campaigns preceding the cross-Channel invasion would have meant costly delays, but failure of the invasion would have had consequences that would have given us a completely different world. The success of the landings makes it appear as if there could have been no other outcome, but this was certainly not the view on either side in early 1944. Many in high places of Allied command were openly skeptical of an invasion holding, and just as many on the other side were confident of repelling it [30].

8.2. FLYING BOMBS

During the early years of the 20th century ordnance experts gave consideration to what was generally referred to as the '100-mile gun'. The ballistic knowledge of the time indicated that it should be possible to construct a cannon with such a range but also indicated that the size of the projectile would be so small and the accuracy so poor that it would have no tactical value commensurate with the great expense and the short useful life of the tube [1]. So matters stood until the stalemate of the trenches of World War I gave desperate ideas special appeal, and Germany built and employed the famous Paris gun. Its purpose was to cause panic in the city, thus replacing tactical justification with strategic. If the Paris gun had any effect on the outcome of the war, it was to add to the French bill of particulars against Germany.

It is therefore strange that the German Army kept the idea alive after its proved worthlessness, but the belief remained that such projectiles, greatly increased in size and in the number delivered, would produce panic that would force the desired capitulation. The Army had lifted a page from Douhet's air power doctrine, although it is doubtful whether they were affected by his writings or even knew of them. The ballistic arguments against a gun had been accepted by then, and plans for a rocket took its place. In 1936, when Rm100 000 ($24 000) looked big to the radar people, about 100 times more was spent on the development of what was to be the A-4 rocket [2], subsequently given the propaganda designation V-2 by Josef Goebbels, the V standing for Vergeltungswaffe (retaliation or vengeance weapon). The Luftwaffe became interested in the Army project because of their interest in rocket propulsion for aircraft, and the two services collaborated to build an experimental station at Peenemünde on the island of Usedom off the coast of Pomerania.

The A-4 project quickly became gigantic. The Army built a special town for the thousands of workers employed. When experiments on rocket-propelled aircraft soured, the Luftwaffe turned their attention to the militarily much more effective pilotless flying bomb, the Fieseler Fi-103, propaganda number V-1. Its development was at Fieseler with production scattered, but Peenemünde served well as a test station. The Oslo Report had described, not too clearly, activities at Peenemünde, but the report's general reliability had attracted R V Jones' attention, so he kept it ever in mind as a guide for the future. Thus when reports of flying bombs and rockets began to arrive from the wide range of sources on which he had established his intelligence network, his studies took a serious look at Peenemünde.

By December 1943 tests of the V-1, also known as FZG-76 began. (The V-1 acquired a number of names. In addition to FZG-76, which stood for Flakzielgeraet and rose out of earlier intentions, it also had the German code name Cherry Stone; the British code name was Diver, and the recipients of the device called it the buzz bomb and the doodle bug.) Flights were launched from a site on the eastern side of the island and were directed along the coast where Würzburg-Riesen could follow their courses and determine their three-dimensional coordinates. As months went by, modifications in design improved accuracy to the point where the device could be used for targets the size of a city. These data were of extreme value to the designers and just as valuable to Jones—for he received the data immediately. The tracking stations transmitted by radio the results of each flight in a very low-grade code that did not even require the services of Bletchley Park for decryption. The data told all one needed to know about the flight characteristics of FZG-76 [3].

Given what was learned it required little effort to recognize the many launching sites being constructed across the Channel for what they were. Members of the underground provided details that aerial photographs did not show, risking and often suffering the fate such activities brought with them. The radar plots of May 1944 showed the accuracy had reached a stage satisfactory for attacking London [4].

Preparing a defense against the two V-weapons was a challenge of the first order. Bombing attacks on the production and launch sites were obvious and came first. Production of the V-1 was dispersed, so the attacks concentrated in the launch sites, which proved difficult to destroy. Production of the V-2 was concentrated at Peenemünde and was hit by a major and costly attack by Bomber Command during the night of 17/18 August 1943, which seriously affected production. Except for the chief designer of rocket engines, Dr Walter Thiel, few of the highly skilled designers and workers were killed, but many of the foreign and concentration-camp workers were [5].

These rockets, capable of placing only a few tons of explosives daily on London, had become in the minds of Hitler and his accomplices weapons of immense power. The measures taken to insure production show the grip it had taken on the high levels of command. Most astonishing was the rapid creation of an underground factory near Nordhausen from an unprofitable gypsum mine. This plant retains a standing in industrial history that may be unique—it had its own crematorium to dispose of the corpses of thousands of forced laborers.

The flight characteristics of the V-1 allowed it, unlike the V-2, to be brought down by both fighters and anti-aircraft (AA) guns. Its speed was known to be sufficiently high to require the fastest machines. It was a nasty target because the explosive used in the warhead was easily exploded by a fighter's projectiles with the resulting blast often destroying the destroyer.

The flight characteristics had come about from the aerodynamic requirements that the designers faced, but for whatever reason, these characteristics were unquestionably the best for avoiding the AA gun defenses then in place. The altitude was high enough to be out of the range of the automatic cannon but low enough to provide large ground returns for meter-wave radar. For the heavy guns the situation was very bad. The combination of high speed and low altitude meant that tracking by whatever mechanism would have to be fast, too fast for most crews doing manual tracking. The tracking problem was compounded: radar or optical aiming equipment was manual, the directors were not only manual but slow mechanical analogue computers, and the mobile guns also manual. These units had greatly improved their efficiency since the Blitz but were hard pressed to engage the V-1s [6].

Guns placed close to London were worse than useless because in downing a bomb—unless they caused it to explode in the air—they caused it to fall on metropolitan London, whereas it might have continued beyond, had it not been molested. Since October 1943 some of the heavy guns had begun receiving the GL mark IIIC, the Canadian 10 cm set together with an electronic predictor. It was a tremendous improvement over the 5 m GL mark II but was a manual tracking set [7].

In preparation for the coming onslaught, the Air Ministry ordered Air Marshal Roderic Hill to prepare the defenses of RAF Fighter Command and Army AA Command. Night- and day-fighter squadrons were detailed for this service with machines stripped of armor and other non-essentials to give them greater speed. The bulk of the heavy guns moved to the edge of Greater London along an arc at the southeast perimeter. Moving them to the coast was rejected because of the expectation of jamming from across the Channel. This gun belt soon became well organized and entrenched [8].

The battle entered the first phase on 13 June and continued for a month. During those days the fighters were more successful than the guns, both together bringing down about 40%, but it was apparent that the gunners were improving and that the position of the gun belt was unfortunate. The CHL stations would guide the fighters onto the bombs while over the Channel, leading to pursuit. If a bomb had not been destroyed by the time the two had reached the gun belt, the fighter had either to relinquish the chase or risk the fire of excited gunners. On the recommendation of General Pile, the long-suffering AA Commander, Hill ordered the gun belt moved to the coast and instituted new rules of engagement: fighters were to operate over the Channel and between the gun belt and London and not to enter the gun belt no matter how hot the pursuit. Thus fighters would get first crack over the Channel; gunners would form a tough gauntlet after which any escapees could be hunted down before reaching London; finally there was a line of barrage balloons. The transfer of the gun belt on what amounted to a moment's notice was achieved during three days in mid July; it required the movement of 4.5 million vehicle km, re-deployment of 23 000 men and women, and re-laying telephone cable sufficient to stretch from London to New York. Dawn of 17 July saw all guns in action in their new positions, and the second phase began [9].

During the weeks before the move to the coast General Pile had pressed for American SCR-584 automatic-tracking radar with the Bell Labs M-9 electronic director. In this he was helped by Churchill, and when things began to get tough a special emissary of Pile's to Washington secured 165 of the new gun-laying radars. General Eisenhower took interest in the situation and lent 20 American four-gun 90 mm batteries [10]. Unfortunately, most of this equipment came into the hands of troops, whether British or American, who had to begin shooting at the Divers while studying the operating manuals. Into this stepped a few Rad Lab men who went from battery to battery adjusting equipment and instructing personnel in its use. Among them were Ivan Getting's co-designers, Hurach (later Henry) Abajian and Lee Davenport [11].

The scores of the guns went up markedly as robot began shooting robot. The SCR-584 tracked automatically and its data were converted automatically to gun orders by the M-9 director, and the American 90 mm and the fixed-mount British 3.7 inch guns followed these order automatically. At the end of the chain came the final element in this completely new air defense weapon—the proximity fuze. The fuzes had been adjusted to the small size of the V-1 from firings in New Mexico at a model built from information gained through Jones' intelligence network [12]. The results of all these technical advances reached a climax on 28 August when 97 V-1s were launched: fighters shot down 23, guns 65, balloons removed two more, and only four of the remaining seven reached London [13].

Advancing British forces soon occupied the launch sites. This did not free London from attacks because V-1s were carried beneath He-111s and launched over the North Sea, which required moving batteries again, but the number reaching London was very small. Antwerp, however, was less easy to defend and received a heavy bombardment, far more than England. Except for relatively minor fighter action the defense of Antwerp relied on the guns but without proximity fuzes, which had not been cleared for use over the continent. The 584 with time fuzes destroyed 40those engaged, but the gunners wanted proximity fuzes. Opposition to their use came from Admiral King, who looked on them as private property of the Navy, and it took a confrontation between him and Vannevar Bush to free them for use on the continent—just in time for the Battle of the Bulge [14].

The MEW saw action against the V-1s; the first one sent to Europe had served to monitor the sky over the invasion and was then moved to track the flying bombs. The single complaint about MEW was its lack of height information, which the SCR-270 provided, but height information was unimportant for the V-1s because of the limited range of altitude in which they flew. This made the MEW a perfect GCI set for the fighters as well as giving accurate warning to the batteries [15].

That the capabilities of Britain's AA artillery lagged so far behind both Germany and the United States has its origins, as discussed in Chapter 1.5, in attitudes found at various levels of British command, but Lord Cherwell, Churchill's scientific advisor carries no small amount of blame. The talent found at the War Ministry's Air Defence and Research Establishment at Christchurch was adequate to have equalled the American achievement, but it was blocked at every turn by Cherwell's bias against that kind of weapon. Less hurting to the war effort but adding to a record of poor counsel was his attitude about the rocket bombs. As intelligence continued to pour in from Jones's network about the V-2, Cherwell continued to dismiss it as a ruse to divert resources, a position he retained for a remarkably long time—a repetition of his earlier refusal to believe the Germans had radar.

Once the danger was too clear to be ignored the matter of defense became acute. The V-2 was immune from fighters and guns, although there was a plan for use of the latter that was never implemented. Attacking the locations for production and launching was the only immediate recourse, but after the Peenemünde raid of August 1943 production became dispersed and underground, and the launch sites were simpler than those of the V-1s and harder to destroy. Some results came from attacking train loads of the rockets, but the overall effect was not great. London and Antwerp simply had to take it.

Radar played an inconsequential role in this part of the play. Some CH stations were trained to watch for ascent of the rockets. The long wavelengths gave satisfactory reflections, but the accuracy was too poor for locating the insignificant launch sites. Coupled with tracking by specially altered gun-laying sets that gave last minute information about the descending phase, this brief warning had an important function. A hit on the Charing Cross river tunnel of the London Underground would have flooded the entire tube portion, almost certainly with great loss of life and serious disruption of economic and administrative functions. The tunnel was equipped with floodgates that were shut when the radar warning was received [16]. Oddly enough a German copy of CH, Elefant-Rüssel, a few sets of which appeared late in the war, was used to help the engineers determine where the bombs fell. It was able to observe descent at a range of 800 km [17] but with an accuracy of no use to either designers or users [18].

Elefant-Rüssel was an updated imitation of CH, built by the Reichspostzentralamt with Telefunken help. The reason for its introduction is to be found in the increasing chaos that marked the high levels of German radar direction late in the war and that led to many useless radar developments while ignoring the vital. There is no evidence that the Peenemünde group asked for it. The transmitter, Elefant, broadcast over 120° on the 10 to 15 m band from an array mounted on a high tower. The receiver, Rüssel (elephant's trunk), was located about a kilometer away and differed from CH in using a high-gain steerable array for direction finding rather than pairs of crossed dipoles. Its range was less than Wassermann despite much greater power and direction was significantly worse, much as its CH counterpart. Very few were deployed, all of them along the coast.

The means of guidance for the V-2 during its ascent—once the motor stopped it was a ballistic projectile—was initially not known in England, and there was reason to suspect a radio method. To this end a tremendous jamming effort was undertaken [19] but to no effect of course, owing to the German use of inertial guidance.

The development of the V-2 must rank as one of the most curious military events of modern time. The total weight of explosives it delivered to London, Antwerp and Liege amounted to little more than a single attack by Bomber Command and was even less accurate. To achieve it Germany expended talent and resources equivalent to the atomic bomb project of the Allies [20]. Peenemünde was in a way a mirror image of Los Alamos. Both projects achieved scientific and technical results that surpassed dreams of 1935. Both projects were driven in part by the fear that the other side was pursuing a similar goal. Both projects were managed by generals of uncommon ability and had technical leaders of brilliant intellect. Both projects produced weapons that were, in so far as the outcome of World War II was concerned, of minor importance. Both projects combined during the postwar decades to provide two states with ultimate weapons and their peoples with ultimate nightmares. The German project proved a serious drain on manpower and resources despite the use of slave labor. In the end it was merely the Paris gun writ large.

8.3. THE BATTLEFIELD TRANSFORMED

For all its overwhelming effect in the air and at sea, radar had altered the war of infantry, field artillery and armor only secondarily through its effects on the air war. The dive bomber and later the rocket- and cannon-firing fighter–bombers had shown themselves to be excellent antitank weapons, but radar seldom had a part to play in their encounters. Ground radar had been employed in the deserts of North Africa where the RAF had relearned its army support role, but its function had been constrained by an unyielding secrecy that prevented soldiers and even other airmen from learning its capabilities and discussing among themselves how best to use it, so its impact on the ground fighting had been peripheral.

All this began to change as armies moved from the Normandy beachheads toward Germany, although not as a part of the enormous amount of planning that had preceded that campaign. It grew out of experience and resulted primarily from the characteristics of two pieces of equipment: MEW and SCR-584. These two 10 cm radars had already made a strong impression on those who had encountered them. Both made strong contributions to the destruction of the V-1s, MEW by accurately locating the incoming low-flying bombs and the 584 by directing AA gunfire. Both belonged to the new, wartime radar generation.

Ground troops received much closer air cooperation during their movement through France than they had in previous campaigns, and MEW was one way the American forces found to accomplish it. The British Branch of Rad Lab, the Massachusetts extension at TRE, converted one of the precious, laboratory-made MEWs into a portable rig during 11 days in April, and it went ashore in Normandy six days after the first landing parties. By August it was placed at Cherbourg at an elevation of 225 m from which its view of the terrain reached to the Eiffel Tower. Installation at the highest elevation available was a complete break in the technique used by meter-wave ground radar, which sought emplacement in a shallow bowl-like terrain in order to restrict ground returns to nearby objects, thereby clearing the oscilloscopes of fixed reflections at more distant ranges where targets were expected. To eliminate the oft-remarked deficiency of no height information, a British 10 cm height finder (AMES type 13) was joined to the station. The extremely thin horizontal fan-shaped beam of this set allowed the height of selected targets to be determined with a resolution equal to that of MEW's horizontal components. Absence of the Luftwaffe frequently left MEW serving only as an air traffic control, but the positions of even very distant enemy planes caused the few that did venture over the battlefield to be attacked promptly [1].

8.3.1. *Tactical Air Control*

Infantry and armored forces received fire support from field artillery by calling on forward observers who went with them and shared their fate. It was a well proved technique, so Tactical Air Control copied the method by forming teams of ground observers, who could discuss requests for air strikes with the MEW crews that had the big picture of the air situation and talk to the pilots of the fighter bombers once they had reached the spot. At this point the other star performer entered the play, SCR-584 with a new, unforeseen role [2].

Early tactical air support had generated a fragment of dough-boy wisdom: 'When you want them, they can't come. When they can come, they can't find you. When they can find you, they can't identify the target. When they can identify the target, they miss it and drop the bombs on you'. The employment of MEW and SCR-584 could not remove the first part of this incantation, but they could remove the remainder—if the right equipment was near the spot. The main problem for the pilots was seeing the often well camouflaged target on a confusing and unfamiliar ground during the brief period available. To be effective, indeed to survive the attack the pilot could not invest much time studying the terrain. If the ground observers had access to a 584 from which range and direction to the target was known, then the radar set could track the incoming fighters and radio corrections to their course that put them onto the target with near infallibility. This not only insured that the enemy received the fire but also gave the attack near complete surprise.

The organization of this new form of warfare was worked out by the US 555th Signal Aircraft Warning Battalion. At the heart of a fighter control center sat an MEW with AMES type 13 Height Finder that was located 20 to 40 km behind the front lines. The center was connected to three 584s located closer to the front and connected to the ground observers through the center or through one of two intermediate forward director posts. The 584s had been equipped with special plotting boards having reverse illumination so the plots could be seen on a military scale map [3]. Calls for support could be examined in terms of the complete air picture, and attacks dispatched with control originating with the MEW and ending with the ground controllers and the 584s. The battalion deployed other radars positioned as necessary for coverage in MEW gaps resulting from hilly terrain. These were AMES types 11 (50 cm), 15 (1.5 m) and 22 (50 cm) with some light-weight air-warning sets near the front, a few of which were the new AN/TPS-3 [4].

These auxiliary sets were, in fact, the result of the history of the organization, as they were the original equipment with which the battalion had been organized and trained before the invasion and used in the early weeks. A remarkable improvement in the performance of the battalion came about when the key radars changed to the MEW and the SCR-584, transforming the technique from desultory to vigorous [5].

Similar British units, made up of more experienced personnel, had long used what by then had become auxiliary sets for the Americans. This left habits acquired in North Africa that caused them to reject the capabilities of their own AMES type 26, which had capabilities similar to MEW. Specifically, the fighter control center continued to rely on the reports of the intermediate forward director posts rather than looking at the screens of the type 26. These procedures even denied the information to the fliers that they were, or more correctly could be under radar control. There were successful but surreptitious demonstrations by restive young officers of the new techniques, but they were unable to add education of their superiors to their wartime accomplishments [6]. How curious that the open relationship between fliers and controllers in Fighter Command was not picked up by Army Cooperation Command.

Officers skilled in GCI attained reputations among their comrades much as fighter pilots became Aces. Such a one was Squadron Leader John Lawrence Brown, who guided the first GCI interception on 26 February 1941. Brownie, as he was known to fighter pilots, had 'that genial, well fed look that one usually associates with gentlemen farmers' [7]. His exploits earned him the appellation among ground radar officers of 'The Great Brown'. He was killed in the fighting at Arnhem, that disastrous result of Montgomery's bad judgement, where he was directing Light Warning units [8].

8.3.2. *Radar observation of the battlefield*

As alert operators of the forward-positioned SCR-584s began to accustom themselves to service in the front lines they noticed that ground returns sometimes showed movement at night or in fog of vehicles, even individual soldiers. Thus when there was little activity in the control of fighter–bombers, they aimed their beams on various bits of terrain to act as electronic sentinels, and occasionally put an explosive end to movements thought to have been concealed by providing accurate coordinates to the artillery [9]. Even MEW had a few successes of this kind, which helped cement a sound relationship with the ground troops [10].

In February 1945 during action to secure a bridgehead over the Saar a 584 was placed within 1500 m of the front. Its easily identifiable antenna was exposed only at night, at which time it assembled information about enemy and friendly troops: echoes from shells disclosed gun emplacements, infantry and tank movements showed up, and in one case vehicles were located at 26 km [11].

8.3.3. Counter-mortar capability

The prototype XT-1 had disclosed a capability that would alter the techniques of field artillery. The early radars had observed the flight of shells toward the target and the fall of shot, but the 584 had shown very early that it could track the path of projectiles. This had led to the discovery that the firing tables for the American 90 mm AA gun were in error, which necessarily meant that the directors were calculating incorrect firing orders. The error was verified by other means, and the tables were replaced [12].

It was noted on the Italian front that the 584 could pick up and track mortar shells in flight, effectively plotting their trajectories in three-dimensional coordinates. This allowed the projectile's path to be extrapolated back to its point of origin with good accuracy, so a judiciously placed 584 could locate enemy weapons otherwise well concealed [13].

The infantry mortar is a light-weight device that is easily concealed in a hole or trench. Its accuracy and high rate of fire had made it one of the great killers of ground warfare since its introduction in World War I. In tracking the shell's parabolic trajectory the 584 could determine the ground position of the weapon onto which artillery fire could then be placed, but the 584 was ill suited by its large size for this kind of work. Experimentation showed that the new generation of light-weight air-warning sets, which were much more mobile and inconspicuous, could accomplish the task. The 50 cm AN/TPQ-3 had a wire-mesh paraboloid antenna, and although it could not track the shell as did the 584, it could sweep the horizon through a sector and obtain range-azimuth plots of it ascending and descending. A special slide rule allowed the point of origin to be determined. This powerful new method saw only experimental use during the war but clearly pointed the way for the future.

8.3.4. Radar for AA guns

While these new uses for SCR-584 began to disclose themselves, uses that soon had Rad Lab sending special modification kits for better adaptation to the tasks, the set's original purpose proved itself in countless actions. The combination of 584, M-9 director, automatic tracking 90 mm guns and proximity fuzes (during and after the Battle of the Bulge) had made AA fire deadly accurate. This was the weapon viewed in some quarters just a few years earlier as a hopeless waste of resources, valuable only for its psychological effect on the enemy and the defended.

As an example, in August a bridgehead 40 km southeast of Paris was attacked at night by 35 German bombers preceded by three pathfinders and defended by the 109th and 413th AAA Gun Battalions. The attacking planes reached the bridgehead without drawing fire because they responded to IFF interrogation as Allied aircraft, disclosing their true identity only after dropping flares. The two battalions brought down all three pathfinders and 13 of the main forces with eight probables, and this without proximity fuzes.

On New Year's Day 1945 the Luftwaffe launched a strong attack with single-engine fighters against Allied airfields in response to requests from ground commanders whose troops were being dealt with severely by tactical air. AA batteries, by then armed with proximity fuzes, were credited with bringing down 394 with 112 probables during action later referred to as the 'AA Battle of 1 January' [14]. It proved to be the Luftwaffe's last major attempt.

When the US 9th Armored Division captured intact a bridge across the Rhine at Remagen, it was protected by an extraordinary air defense with SCR-584 a key element, but this equipment served another function as well. One was placed upstream to watch for swimmers, boats, mines or whatever else might be floated downstream to attack the bridge [15].

8.3.5. Saturation bombing

The German attack on Allied lines on 21 March 1918 was preceded by a brief but paralyzing bombardment by 6000 guns that resulted in the near extermination of the British lines. The initial penetration succeeded, but the Allies' rear lines of communication were intact and the attackers found the transport of reinforcements and supplies extremely difficult over the moon landscape that the bombardment had created. The Allied line gave ground but ultimately held. The same technique was revived by the Allies 26 years later, but with bombing aircraft furnishing the explosives.

Artillery still provided the soldier his most dependable support fire, both in its accuracy and availability, but shells from light and medium field artillery are smaller than all but the lightest bombs. A 200 kg artillery shell is considered large, whereas a 1000 kg bomb is not. Starting in Italy the Allies began concentrating the bombardment of German entrenchments by using hundreds of bombers.

Such massive aerial bombardments preceded the American breakout from Normandy at St Lo, and the inability of the Germans to recover, owing to the complete Allied dominance of the air, and the use of tracked vehicles to move through the shattered land spelled the difference from

March 1918. Similar bombardments were used in front of the British lines, although with less success. Accuracy was of extreme importance and it was with Oboe and Gee-H to which success could be ascribed [16].

8.3.6. *The proximity fuze over land*

From the time of its introduction into combat in the Solomon Islands in January 1943, the proximity fuze remained an exclusive weapon of the US Navy. This insured that any duds would fall in the ocean, virtually eliminating the possibility that one might be captured. Their use against the V-1s during summer 1944 was the first deviation from this strict rule. Fuzes had been designed for use in field artillery pieces against ground targets with release to follow as the strategic and tactical situation demanded. The German December offensive in the Ardennes, the Battle of the Bulge, was the crisis that called for employment of the new fuze in ground warfare.

Artillery shells bursting in the air over the heads of infantry in the open was long recognized as particularly dangerous, but such fire had to be observed and adjusted to have any effect, because the effectiveness of a shell burst is very sensitive to the height at which it explodes. Blind fire was possible, but only in a static front where the artillery had had time to determine the coordinates of battery positions and targets, conditions that definitely did not apply to the retreating American forces and that had allowed the Germans to advance openly under cover of darkness. On release of the new fuze for use over land these attacks were broken up, leading General George Patton to remark: 'The funny fuze won the Battle of the Bulge for us' [17]. Henceforth the proximity became standard with Allied field and AA artillery. It allowed air bursts to be placed on targets beyond the effective range of time fuzes and under all kinds of condition. It enhanced the already high reputation of US artillery held by the Germans.

8.4. POST MORTEM

The radar war between Germany and Great Britain began before the official outbreak of hostilities and continued after the surrender. The British sides of the before-and-after events were controlled, curiously enough, by the same man: W P G Pretty. In August 1939 Pretty, then a Flight Lieutenant, was on duty at Fighter Command Operations Room when LZ-130 approached on the airship electronic espionage flight. Just two months short of six years later, Pretty, by then an Air Commodore, took command of what must rank as one of the strangest military operations in history—Exercise Post Mortem.

As German forces had retreated they had destroyed nearly all of the radar equipment that had to be left behind, but the surrender on 7 May 1945 left the stations in Schleswig-Holstein and Denmark intact, and their retention in operating condition with the crews held together was stipulated in the terms for surrender. It had long been planned, of course, to examine the equipment and interrogate personnel involved with air warning, raid reporting, night fighting and its direction, but the Air Ministry decided to attempt something significantly beyond such routine measures: have the Germans put this system back into operation with RAF officers observing how well their recent enemy gauged the nature and weight of attacks, which would be made by formations of heavy bombers using all the cunning of electronic countermeasures.

German cooperation was essential and needed to be obtained voluntarily because the tests would be meaningful only if the Germans went at it with a will. The response of the prisoners was remarkably positive, allowing plans for the exercise to proceed apace. Two reasons can be advanced for this degree of cooperation: first, there was the pride of professionals in demonstrating what they could do; second, there was the conviction among the Germans that the Soviets were bound to clash with the western powers, and that they would make use of captured German radar. A choice between Britain and Russia was no choice at all for the prisoners [1], who now wanted Britain to know all their tricks.

The performance of radar against the night attacks was the matter of primary interest, but it was decided to make the flights during daylight because the bomber streams crowded aircraft so closely that collisions were a significant danger at night, and it was all the same to radar. This made tests with night fighters and their controllers meaningless, but the idea of having fighters rise to meet the bombers seemed to be pushing things a bit far anyway, so only ground radar was tested. Flak was also excluded from the exercise.

The radar deployed amounted to 16 Freya, ten Würzburg-Riese, two Dreh-Freya, six Wassermann, three Mammut, two Jagdschloss and one Elefant-Rüssel. Also to be evaluated were two stations of the direction-finding Y-Dienst and 21 visual and aural observation posts. The total system could not be expected to work with top efficiency because the Denmark chain had not had to deal with particularly heavy attacks during the last few months and so were not as experienced as their comrades to the south; furthermore, the confusion at the end of hostilities had made it impossible to reassemble all crews as before with a corresponding reduction in efficiency. Nevertheless, the personnel were qualified and experienced. They were commanded by Generalmajor Alfred Boner, Chief Signals Officer, who strongly favored the project [2]. Boner and Pretty had feared that the personnel, after having been subjected to years of vicious propaganda with hatred of the Allies as its object, would not apply the skills necessary to make the operation a success. It was not a situation where forced obedience would suffice. In this they were pleasantly surprised because for whatever reasons cooperation was very good [3].

Fourteen exercises were planned of which 11 were flown during nine days beginning 25 June and ending 5 July. Three were canceled owing to bad weather [4]. Seven of the exercises flew more than 200 heavy bombers, and during four of these about 24 aircraft from 100 Group accompanied them to test the effectiveness of concurrent airborne jamming and interference. Five exercises were made up only of jammers from 100 Group, and two were flown entirely by 31 fast, high-flying Mosquitoes [5].

Impressions came thick and fast to the observers of the system that had been the principal weapon against them through so many hard air battles. Two contradictory observations came first. The quality of German electronic engineering was found to be remarkably good, but it had been in part so designed to compensate for the generally poor quality of the operating personnel. This was the inverse of the British approach, which was to get new models of equipment into the field rapidly and rely on personnel of the highest intelligence and training to overcome problems with them. The German policy, initiated at the start and held to the end, was to engineer the sets for ultimate simplicity of operation and maintenance.

The observers were surprised to learn that some of the long-range air warning sets such as Wassermann and Mammut had the weak transmitters of Freya and depended on high-gain antennas for range. It also came as a surprise to learn how little had been done with microwaves. The Würzburg equipment that was relied on for the Himmelbett close-in fighter control data was still found to be overwhelmed by Window. It had been expected that 10 cm sets would have been deployed by then to replace the old 50 cm equipment, yet none were found, and this more than two years after the Rotterdam H2S had been recovered and nearly two years since the introduction of Window. The Himmelbett system was overwhelmed by the bomber streams but worked well against returning planes that had become separated from the stream and frequently had no Window cover. Window and jamming of the night-fighter-communication radio were seen to be the most effective of interference methods. Here the poor quality of operating personnel made itself felt, as the anti-Window modules added to the sets proved complicated to operate and were frequently ignored by the crews.

The observers noted the relatively late application of PPI and only for Jagdschloss and Dreh-Freya, the latter a Freya modified to rotate and present panoramic display. That PPI had not been applied to the Würzburgs for improved GCI, the first British application, was a definite surprise. It was with a tinge of satisfaction that the observers noted the miserable cabins in which the German operators had to work [6].

Compensating for these faults, however, was the sobering realization of how easy it had been to pick up the attacking force at extreme range. When Bomber Command terminated the continuous attacks loosely called the Battle of Berlin in early 1944, which had debilitated them as well as hammered the enemy, they had come to realize that many of their problems had resulted from the bombers emitting far too much radiation: radio communications, jammers, tail-warning radars, IFF and H2S. On resuming heavy attacks on the Reich after the invasion was secure, they reduced these emissions to what was thought to be the absolute minimum, but there were always signals for Y-Dienst to use for triangulation, and they showed themselves masters of the technique. Mandrel jammers often made it difficult for the sets working on the Freya band to track the formations, but Mandrel proved to be an excellent signal for Y-Dienst as did H2S! The first two simulated attacks by main-force heavy bombers were conducted without the use of any interference methods and the IFFs were ordered switched off. It came as a shock to see one of the Freya sets, which was equipped with the Flamme attachment, pick up the one single bomber that had its IFF inadvertently operating as if to proclaim 'I'm English, I'm not a bloody foreigner!'. The IFF had functioned as a secondary radar for the defender at ranges much greater than Wassermann, in this case at 310 km. The main value of Mandrel for the attackers was found to be in confusing the defenders as to the composition of an attack but certainly not in hiding it. Mandrel also helped in deceiving the defenders as to which was the real attack and which was a spoof, but the only certain way of avoiding early detection was to approach very low [7].

There was an Elefant-Rüssel, the imitation of CH, at the Robbe Station. It was found to have about the same range capability as Wassermann and Mammut for bombers but was much worse for Mosquitoes; its receiver array had no lobe switching and was, of course much worse in directional accuracy than the two older sets.

A non-concentrated force of Mosquitoes flying without countermeasures at 7000 m was detected by Mammut and Wassermann at ranges in excess of 200 km only minutes after Y-Dienst reported their H2S emissions. Initial pick-up was for an individual aircraft; the estimate of the number in the formation became confused as more followed. The long-wave Elefant-Rüssel observed them first at only 50 km, a consequence of their wooden construction and Elefant's long wavelength [8].

The operating personnel and their officers cooperated fully and were eager to demonstrate their system. The women operators at one station were an exception [9]. This was generally attributed to their being more thoroughly indoctrinated in Nazi ideology than their male comrades, but a Danish source reports another reason: 'During the exercise the female personnel were not very cooperative. This has later been interpreted as their being Nazi-minded, but the truth is that an English soldier had assaulted one of them immediately before the start of the exercise, and he was in the bunker during the exercise' [10].

With this exception the prisoners were fortunate, primarily for having come into British hands. Of the four allies, Britain took the most relaxed attitude toward prisoners, allowing many to proceed home quickly. Those taken by the Soviets suffered the worst, although erratically, some released soon, others after many years of hard servitude, if they survived. Those taken by the Americans had either a fair time of it or suffered the hell of the infamous—and oft denied—100 000-man pens in which huge numbers of men were enclosed for months without shelter, medical treatment or adequate food.

The teenage radar technicians from the Stegskopf, who were scattered all over the Reich, experienced the dreadful variety of the chaotic end of German resistance. The detachment assigned to the Kriegsmarine was pulled from their radar sets and sent in their blue uniforms with Mauser rifles to defend the last ditch in Berlin, nearly all dying in it. One died remaining behind to destroy a Freya against the calls of his comrades to be sensible and get away while there was time. One simply got off the British truck that was transporting him when it passed through his home town. Some died in the pens. One secured early release from Russians by repairing their radios. Hans Plendl's son went from an American pen to a French prisoner of war camp and was released in May 1946 [11].

When the exercise was complete the equipment was dismantled and shipped to England or the United States for further examination with Freya serial No 1 going to the Royal Aircraft Establishment at Farnborough [12]. Station Robbe on the island of Rømø was operated by the RAF for several months in order to make detailed studies of Freya, Wassermann, Mammut, Elefant-Rüssel and Würzburg-Riese [13]. For unknown reasons the RAF gave none of this equipment to the Danes or the Norwegians [14], and equipment not shipped out was destroyed, although a substantial amount of German radar and electronics equipment was removed surreptitiously for the Danish Army by members of the underground, who had been given the responsibility for guarding it by the British [15].

As the war moved toward a conclusion that was obvious to all Germans with insight and whose minds were not enslaved by Party fanaticism, the radar engineers feared the consequences of defeat for Germany's technical history and possibly even its future. This gave rise to the curious 'Aktion Kindersarg' (Operation Child's Coffin), which was organized towards the end of 1944 by General Martini with the participation of Leo Brandt and a very few top radar men. It was clandestine and dangerous, initially from the Nazi government, which had draconian punishments for defeatism in any form, and potentially from the Allies, whose benevolence was a most uncertain quantity and who could easily be believed capable of punishing any who withheld information they wanted.

The conspirators made a selection of key documents describing radar and packed them into a tight metal casket. The original idea of it being the container for a child gave way to a larger one for a soldier worthy of burial with honors. A bit of scurrying during times of administrative confusion that often held toward the end of the war allowed the material to be given a military funeral with proper burial papers [16]. The name 'Kindersarg' was also the name of a minor piece of radar equipment, which allowed discussion by telephone without arousing suspicion. Removing the coffin from the Russian Zone in the early 1950s proved to have difficulties and risks of its own, described, perhaps a little dramatically, by Cajus Bekker

[17]. Some of the material so preserved was later published by the Deutsche Gesellschaft für Ortung und Navigation in a very limited edition [18].

The postwar fates of the leading German radar engineers was as varied as those of the Stegskopfer. Some attained high positions in the new West German military services, once they and their country got on their feet again. Wolfgang Martini and Josef Kammhuber attained high rank in the new air force and in NATO; Rudolf Kühnhold returned to the navy [19]. Gotthard Müller returned to the successor-company of Lorenz, SEL, after ten years as a Soviet prisoner [20]. Wilhelm Runge quit Telefunken in late 1944 but returned to it after the departure into a Russian prison of Karl Rottgardt, the company's director whom Runge had despised so long. The determination of Telefunken's employees to save their company during such difficult times created a working atmosphere that he described as 'the happiest of my life at Telefunken' [21]. GEMA's factories fell within the Soviet zone of occupation and ceased to be a part of western economy. Freiherr von Willisen attempted to form a company with a number of former GEMA workers, but it failed in 1948 [22]. Hans Plendl emigrated to the United States where he was employed at the Air Force Cambridge Research Laboratories [23]. Leo Brandt directed to reconstruction and electrification of railways in the Ruhr and lower Rhein districts [24].

PHOTOGRAPHS: TUBES

Japanese cavity magnetron M-312. The split-anode magnetron, which was capable of generating microwaves, was the invention of Kinjiro Okabe in 1927, and research with magnetrons was continued in Japan. In May 1939 Shigeru Nakajima operated a 10 cm water-cooled magnetron that delivered 500 W of continuous power and that quickly evolved into this tube. A smaller, tunable version with the same electrode configuration, the M 60, served as a local oscillator. The pair formed the heart of the mark 2 model 2 sea-search radar that the Imperial Navy deployed in 1942. Photograph courtesy of Dr Shigeru Nakajima.

Anode of the Japanese M-312 cavity magnetron. This anode was water cooled, hence the groove around the outside for the tubing, and was capable of delivering 500 W continuous with 10 cm wavelength. Diameter was 52 mm. The alternation of circular with flask-shaped cavities, called the Mandarin configuration, provided frequency stability much as did strapping on British designs. Inaccurate production machining of the large cavity pulled the frequency sufficiently to make the matching with a local oscillator tube, the M-60, difficult. Anode courtesy of S Nakajima and N Koizumi; photograph courtesy of John Bryant.

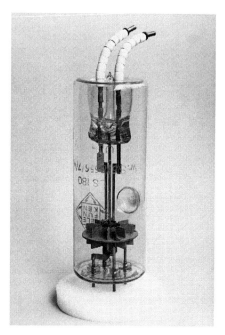

The Telefunken decimeter-wave triode, LS180. Transmitter triodes for the 50 cm band were developed independently by five different tube manufacturers around 1939. Others originated at General Electric Company (Britain), Western Electric, Lorenz and GEMA. In tubes of this kind the leads were positioned and dimensioned so that they formed part of a resonating structure. The LS180 was not the most powerful even among the German tubes but, as the heart of the Würzburg series, probably shares with GEC's Micropup the honor of being most widely used. Photograph courtesy of Bernd Röde.

CHAPTER 9

THE END IN ASIA

9.1. THE PHILIPPINES AND OKINAWA

In leading his squadron into Manila Bay in 1898 Commodore George Dewey led his country into the affairs of the western Pacific, a region concerning which few of his countrymen had ever given thought, and in so doing altered America's position in the world. Dewey's defeat of the Spanish warships was followed by little more than a *pro forma* defense by the Spanish army, and the Philippines came under American control. These events had interrupted a rebellion of the Philippine people against Spanish rule, and they merely transferred the fight on learning that there had only been a change of masters.

The resulting Philippine Insurrection was fought according to the martial traditions of the archipelago, and although the Americans scorned them as the ways of an inferior people, they adapted with alacrity to local custom. The result was an ugly war, but not an ugly 'little' war, for it was the largest the United States had fought in more than a hundred years, the Civil War being the obvious exception. It compelled quadrupling the army and dragged on for seven years. It had abundant examples of courage and endured hardship but has left no trace in the mythology of the Great Republic, seldom even catalogued in the lists of its wars.

Surprisingly, the outcome was positive. The Insurrecto leaders were taken into the new colonial administration, and the people were promised independence in 1946. The Japanese invasion came at a time when things American were highly valued in the Philippines, and their newly created army fought bravely beside the Americans against the intruders; their highly revered commander, a Field Marshal of the Philippine Army, was General Douglas MacArthur, whose father had been instrumental in securing amicable relations between the two peoples. These historical matters were now to determine the strategy of the last year of the war against Japan.

MacArthur's paternalistic feelings towards the Philippinos had been reinforced by the steadfast loyalty that they had maintained. Americans who had avoided or escaped Japanese prison were hidden and incorporated into the growing number of guerilla bands that fought the forces of

411

occupation with a violence and unity of purpose far greater than anything observed in the underground movements of western Europe. MacArthur insisted that the liberation of the Philippines had to be the next step on the road to Tokyo, initially opposed by the Joint Chiefs of Staff and by Admiral King in favor of an attack on Formosa. A decision for the Philippines came about through Nimitz's analysis, which convinced King, of the need for more service troops for Formosa than would be available.

9.1.1. Leyte and Luzon

The campaign began with an attack on airfields on Formosa, Luzon and Okinawa by Admiral Halsey's task force of nine fleet and eight light carriers. Their purpose was to destroy as much land-based Japanese airpower as possible and to cover the target of the invasion, the large island of Leyte located between Samar and Mindanao. Large numbers of defending fighters met the raiders but suffered severe losses and could not prevent the destruction of many aircraft on the ground. Japan's aviation manufacture was still delivering aircraft in large numbers, but the quality of the replacement pilots was by then very poor. These altered circumstances brought about a change in Japanese tactics.

The small fraction of these green pilots to return from a raid and the meager results they achieved combined to cause Vice Admiral Takijiro Ohnishi, the Navy's senior air officer, to employ suicide fliers, the kamikazes [1]. He reasoned that inasmuch as most of the fliers did not return anyway, suicide missions would improve the probability of their destroying the enemy by having them crash their bomb loads onto the ships, a maneuver requiring much less flying skill than dropping a bomb or launching a torpedo and causing a gasoline fire as well. He also reasoned that this would prevent them from dying in vain, logic that perplexed western minds. This relaxation of training requirements for kamikazes also inverted the logistics of Japanese airpower by providing more pilots than aircraft. This decision, triggered by the Formosa raids, would soon give the Pacific War a new and terrifying phase, one greeted by the Emperor with 'Was it necessary to go to this extreme?' [2].

Japan planned to meet whatever landing took place with the full remaining power of their fleet, despite it being by then woefully inferior in numbers and efficiency and completely lacking in squadrons for the carriers. The Americans expected this response, and the result was a battle that saw almost every conceivable combination of naval units in conflict with one another. The Battle for Leyte Gulf was the largest naval battle of history. It was also decisive, although the decisiveness was not accepted in Tokyo.

The US 7th Fleet under Vice Admiral Thomas C Kinkaid was to place MacArthur's soldiers ashore and provide bombardment from six old battleships and air support from ten escort carriers. Halsey's 3rd Fleet with its ten fast carriers and six fast battleships would stand ready to counter the Imperial Navy when it attempted to interfere. There was, however, no local unity of command. The two fleets were subordinate to Admiral Nimitz in Hawaii, and Nimitz and MacArthur were subordinate to the President through the Joint Chiefs of Staff in Washington.

Japanese interference began as soon as it was clear that Leyte was the point of invasion. Two forces of battleships and cruisers converged on Leyte Gulf, where the transports and landing craft were to be found. One, under Vice Admiral Shoji Nishimura, approached from the south, to enter the Gulf through Surigao Strait; the other, under Vice Admiral Takeo Kurita, which included the 18 inch gun super battleships *Yamato* and *Musashi*, was to pass through the San Bernadino Strait from the west and attack the landing sites from the north. The latter expected to meet Halsey's ships, hence the presence of the strongest units. A third force came from Japan under Vice Admiral Jisaburo Ozawa with a fleet made up of useless carriers intended as bait to draw Halsey from his duties of protecting the landing.

The Battle of Surigao Strait took place during the night of 24/25 October 1944, more than two years after the US Navy's defeat at Savo Island. The two engagements had the similarity of action in a broad channel formed by islands, but the outcomes were vastly different. The use of radar illustrates how the two navies had changed by 1944. In 1942 two American radar-equipped destroyers had been positioned as pickets, but for uncertain reasons the 1.5 m SC on the crucial guard ship failed to observe the large fleet. No such incident marked the 1944 action. Nishimura was first reported and harassed by motor torpedo boats equipped with the Raytheon 10 cm SO [3], who then turned them over to attack by destroyers [4]. The American radar had PPI, which removed the confusion of the reflections that proved so troublesome for the A-scopes. The Japanese ships had 10 cm radar but no PPI, a serious disadvantage in the narrow waters filled with rapidly moving ships, islands and a jagged coastline. Radar-equipped PBYs, the 'Black Cats', had searched for the fleet but had failed to find it; one was shot down by friendly fire [5].

Forewarned, the Americans had lines of Kinkaid's battleships and cruisers at the top of the Strait, barring entrance to the Gulf of Leyte. The force was stronger and the superiority of radar made it overwhelming, and it rapidly dispatched Nishimura's ships. Three of the battleships had the mark 8 fire-control radar, the Bell Labs 10 cm phased-array set that swept a 30° sector and allowed the battle to be kept under observation when firing at a single target. It worked perfectly. The other three had the 40 cm FC (new designation mark 3) and had difficulty in sorting out the confusion on their screens, which retained the A-scopes. USS *Pennsylvania* was unable to identify a target and fired none of her 14 inch projectiles. To be without effective radar was to fight without effect [6].

When Kurita exited the San Bernadino Strait he expected to meet the full force of Halsey's ships and aircraft. The previous hours had certainly been difficult for him. In the approach to the strait he had encountered aggressive submarines and had to swim from his sinking flagship. Then he received the full force of the Third Fleet's vast air power that cost him the super battleship *Musashi*, sunk from the impact of bombs and air-launched torpedoes. But when he entered the open ocean the overwhelming force that he bravely expected to meet was not there. Indeed, there was nothing there! Not so much as a motor torpedo boat or a PBY patrol plane circling overhead. Halsey, who thought Kurita had turned back, had swallowed the bait of useless carriers to a degree the Japanese could scarcely believe and had raced north with everything he had. With Kinkaid's heavy units to the south of the Gulf of Leyte nothing protected the landing but the slow escort carriers, being used for air support of the troops ashore, and their protective screen of three destroyers and three destroyer escorts, all under the command of Rear Admiral Thomas L Sprague, who had not been told of Halsey's departure from station.

The collision of Kurita's overwhelming force with Sprague's thin line was a confusing fight in which the planes of the escort carriers went aggressively after the attacking ships, smoke was applied to the scene in liberal quantities and the destroyers and destroyer escorts pretended they were capital ships. The vehemence of the defense by aircraft that attacked with machine guns once their ground support bombs were gone and destroyers that used 5 inch [7] guns once their torpedoes were gone caused Kurita to think he had finally encountered the Third Fleet and to repeat the Japanese mistake of Savo Island. He did not steam resolutely into the Gulf and make quick work of the fragile ships he would have encountered; he turned back and left the scene through the strait whence he had come. Radar functioned as an accustomed weapon during these exciting actions, although the heavy punishment inflicted on the American destroyers eliminated much of their electronic equipment before they were sunk.

The conquest of Leyte followed the long-established methods of Pacific ground fighting but with a somewhat larger radar contribution. Leyte put an end to the peculiar reluctance of Army anti-aircraft (AA) units to make use of SCR-584. On previous island locations they had been satisfied with the SCR-268, but Leyte was mountainous, and the ground returns, which had been no problem when fighting off an attack from over the sea, were now serious when planes came erratically and suicidally over the hilltops. The combined 90 and 40 mm units of Leyte brought down over 300 planes during the first few weeks, mostly from 90 mm fire, still without proximity fuzes, which had not been released for use over land. It was the most amazing example of AA fire seen in the Pacific and brought admiration from General MacArthur [8].

The AN/TPS-2 was one of the most advanced radars to be placed in the hands of troops in 1944: a portable 75 cm set designed and produced by General Electric with a total weight of 300 kg, capable of observing a bomber at 100 km and being placed into operation in 20 min [9]. This modern set saw deployment under conditions in austere contrast with its sophisticated origins, for it accompanied a detachment from the 597th Signal Air Warning battalion which was set ashore in Mindanao from a submarine to monitor Japanese shipping and air traffic before the invasion of Leyte, guided and protected by Filipino guerrillas. A similar action went with the invasion of Leyte. The 583rd Battalion placed similar detachments starting by submarine and continuing by jungle trails and native sail boats to advantageous positions on the islands of Panay, Negros and Cebu, effectively putting an end to the radar blindness imposed on the east coast of Leyte by the mountains to the west [10].

With practice the AA units expanded their new skills and found a local use for the 584 in the ground fighting. An artillery observation plane would fly directly over a target well behind the front while being tracked by radar, thereby furnishing coordinates of the target for a future concentration of unexpected artillery fire. This technique became more accurate as it was refined by having the spotting plane make multiple passes from different directions.

American field artillery had become a surprise weapon during the war by the technique of placing all batteries for a given front on a common coordinate system, thereby allowing fire from any gun within range of a target to fire on it. This created the terrifying 'time-on-target', wherein every gun within range placed a shell on the target within a second or two of the same time. The basis for this was a rapid and accurate survey of the gun positions, not always an easy task, and the forests and hills of Leyte increased the difficulty until the AA artillery units did the job with the SCR-584 [11].

While the Army was beginning to demonstrate excellent AA fire, the Navy's performance with FD radar with the mechanical-analog predictor, although greatly improved with the introduction of the proximity fuze, was failing all too often. The high-flying, level formations that characterized air attacks in the first years of the war were being replaced in the Philippines, where there were many dispersed landing fields, by many small formations [12] coming from who knew where and being hidden by mountains from electronic eyes of the ships during part of their approach. Attacks on ships often came in very low, evading air-warning radar until late and making it impossible for the 40 cm FD to determine height accurately. When air attacks began to come from kamikazes the deficiencies of air defense became serious. The proximity fuze was of no avail, if the shell was not placed within 20 m of the target.

Bell Labs had recognized the weakness of the FD in tracking low-flying aircraft and designed a 3 cm adjunct height finder for mounting beside it, a vertical 'orange peel' reflector to form the radiation pattern into a thin horizontal fan. This fan-shaped beam had an advantage over a narrow conical beam, such as found in SCR-584, because it was less susceptible to locking onto the image of the target mirrored by the ocean, the problem Luis Alvarez and Alfred Loomis had had to solve for Ground-Controlled Approach. Bell also undertook to design a replacement for the FD, the mark 12, using the same basic structure but on 33 cm and with automatic tracking in range; automatic tracking in direction was to follow in a later modification [13]. This still left the mark 12 with manual tracking, not suited to the agile attackers and still followed by a slow, mechanical predictor, and most ships retained the mark 4 (FD). No mark 57 directors[1] appeared until early 1945. Once the kamikazes were within their range, the non-radar 40 and 20 mm guns had to save the ship and often did.

This situation seemed nothing short of scandalous to Ivan Getting, who had been working on a Navy 3 cm fire-direction radar, the mark 35, and a compatible and highly advanced director, the mark 56. When there was no evident move by April 1945 to place this system into service, he wrote a sharp letter to the Coordinator of Research and Development for the Navy Department, pointing out the superiority of the Army's radar, thinking that tweaking inter-service rivalry would suffice to bring about the desired conversion from 40 cm manual sets to 3-cm automatic. Given the size of the US fleet by the end of 1944 and the time required for conversion compared to the expected length of the war, it is hardly surprising that this suggestion was rejected, which with afterthought Getting conceded to have been the right decision [14].

The Navy's radar equipment was not their only air-defense problem. The five great carrier battles had been conducted at an almost stately pace. The SK radar and its predecessors could assure a carrier's Combat Information Center (CIC) of 30 to 40 minutes warning before the attacking formations arrived—at which time 'stately' hardly described things. This warning was the carrier's armor plate, because it allowed the assorted collection of explosives and gasoline on and below the thin flight deck to be removed. Bombs might hit, but the American carriers did not turn into floating infernos as did the unwarned Japanese. Off the Philippines this comforting situation vanished.

This use of radar did fairly well for the fights between carriers and Luzon-based planes in the first phase of the Battle for Leyte Gulf, but when air attacks came at almost any hour of the day with only minutes warning, it was not possible for the ships off Luzon to function and keep their decks free of gasoline and explosives [15], and on 24 October the light carrier *Princeton* suffered the dreaded fate and became an exploding, blazing hulk. The calm assurance of American air dominance that had been built up earlier in the year and crowned at the battle of the Philippine Sea was becoming distinctly less assured.

Things became worse with the invasion of Luzon, the principal island of the Philippines. Preparatory air attacks on 3–6 January 1945 by carriers

---

[1] See Chapter 7.5 (p 380).

for the landings at Lingayen Gulf were met by a ferocious defense made up of some experienced fighters and many kamikazes, which left one ship sunk and 11 damaged. Further, the efficiency of the carriers for working their assigned land targets was seriously disrupted by the confusion in the Combat Information Centers that these many attacks caused.

The need to decentralize the radar defense was evident to all. Destroyers and other light craft equipped with SC and SG (useful against the low fliers) were spread out to pick up the attackers as soon as possible and vector fighters onto them independently of the saturated CICs of the carriers [16]. Before, the CIC would marshal its forces for an ambush; now the destroyer controllers would call upon available fighters to go after intruders much as a radio dispatcher sends taxis to passengers. In the past destroyers had screened capital ships against submarines; now screening against aircraft was added to their duties. Unfortunately, the new assignment did not bring with it the increased space for the plotting boards essential to well run CICs. Destroyers are crowded vessels. The new system was just going into use when the Japanese air strikes stopped—they had run out of planes on Luzon and could not get new ones. The fleet began to breathe more easily—but not for long.

While preparing in December 1944 to support the Luzon landings, the 3rd Fleet was struck by the full fury of a well remembered typhoon, which made the SG radar all the more beloved by seamen. When the storm hit, the fleet was attempting to refuel, and the relative proximity of the ships to one another became a serious hazard, one compounded when at times visibility shrank to less than the distance from the bridge to the bow and when some small vessels lost the ability to steer. During these hours station keeping and avoidance of collision resulted through the efforts of the SG operators [17]. It was the first such storm to leave behind a record of PPI photographs that clearly showed its eye [18]. Three destroyers foundered, 146 planes were lost and 790 men died.

9.1.2. *Okinawa*

After the Battle of the Philippine Sea in June 1944 perceptive authorities on both sides knew that Japan had lost the war. But Japan was ruled by a dictatorship controlled by the Army, and the Army could appease their war gods only if defeat ended in death. Were there doubts among the Allies about this determination, they were dispelled by battles on islands whose names, theretofore unknown, oppress the memory: Iwo Jima and Okinawa.

Iwo Jima is a bit of unstable volcanic rock and proto-soil midway between Saipan, the base of the B-29s attacking Japan, and Honshu. It sustained airfields that were used to attack the bombers and the new Saipan base, and bombardment from sea and air had not been able to neutralize them. Seizing the island, an operation not expected to be more costly than for other Pacific islands, would eliminate this and allow fighters to accompany the bombers in their attacks and provide a refuge for damaged B-29s. Seizing Iwo Jima had no parallel. There had been vicious fights over Pacific islands, but nothing like the month-long struggle in February and March for Iwo Jima. It was not a modern fight, and radar was of little consequence, but the technical advancement of the times and the importance of the island for the bombing of Japan insured that it would have a strong radar complement. An AN/TPS-10, a light-weight, 3.3 cm height finder named Li'l Abner, was quickly placed on the top of Mount Suribachi [19].

A base in the Ryukyus was necessary for the invasion of Japan, and Okinawa was selected. Experience with the kamikazes at Luzon had prepared the minds of the invaders for a difficult time off shore; Iwo Jima taught them to expect the worst once they were on shore.

The Battle of Okinawa engaged a larger part of the British Commonwealth than the Australian units that had fought with the Americans since 1942. Naval requirements of the European war had slacked, and in December 1944 Admiral Sir Bruce Fraser conferred with Admiral Nimitz on deploying the British Pacific Fleet, which included New Zealand vessels and consisted of two battleships, four fleet carriers, five cruisers and 15 destroyers. (One of the carriers was a veteran the reader will remember from the tough Mediterranean of 1940–1941, HMS *Illustrious*.) They came with different but effective radar and were greeted warmly. The overwhelming naval superiority of the Allies would soon be put to the test.

Preceding the 1 April landing on Okinawa was a carrier strike against all possible Japanese landing fields, helped by B-29 raids on bases in Kyushu. During these preparatory raids radar failed the Essex-class carrier *Franklin*, which was heavily damaged by two 250 kg bombs dropped on a busy flight deck by a single Aichi dive bomber that approached low and was undetected until too late [20]. Japanese aircraft losses were staggering and provided the beachhead with four days relatively free of kamikazes, the time the Japanese needed to restock their fields. This was expected and predicted. The absence of resistance to the landing parties was neither expected nor predicted. Also not predicted was a garrison almost double the reported size. That it was deeply entrenched in the southern hills was the final surprise. The defending garrison had its own surprise. They had planned to watch the ships struck by an overwhelming kamikaze attack, but none came [21]. What followed was morosely predictable, following plans laid out by the two contending generals.

The key to defense against kamikazes was a ring of radar picket boats, destroyers or other light vessels that remained at station 60 to 80 km from the location of the main fleet. Each had a fighter-control officer aboard in communication with fighters patrolling in the vicinity, whom he vectored onto the approaching fliers. The method worked moderately well in protecting the beach and main elements of the fleet, but the pickets had to look out for themselves. The Japanese occasionally made use of Window [22], probably limited by the availability of cut foils.

Göring had given little thought to Chain Home (CH) on initiating the Battle of Britain, dropping the idea of destroying the stations after a few attempts seemed to have failed, and instinctively mistrusting the whole idea anyway. Five years later no commander held such beliefs, and the radar pickets—sentries terribly alone—bore the brunt of the Okinawa kamikaze and bomber attacks. This led to a tactic of hitting a picket with multiple attacks, which succeeded all too often. Many sank from the hard blows received but none can match the ordeal of USS *Laffey*, a 2200 ton Sumner-class destroyer launched in 1944. She, her AA consort LCS-51 (a landing support craft) and the combat air patrol fought 22 attackers, in a wild mixture of aircraft and gunfire. *Laffey* was hit by six kamikazes and four bombs in addition to receiving intermittent showers of machine-gun fire. Radar may have been the object of the struggle but soon became irrelevant to these desperate warriors. Only one of the attackers escaped. The *Laffey* was salvaged [23].

To relieve the pickets the island of Ie Sima and Hedo Misaki, the north tip of Okinawa, were taken and by mid-April functioned as air warning stations, which allowed the number of pickets to be reduced to five [24]. Two other islands, Iheya and Aguni, were taken in June to relieve the pickets further [25]. Great ships also suffered. The Essex-class carrier *Bunker Hill* was the recipient of skillful use of rain clouds and Window that planted a single kamikaze on a deck full of refueling and rearming aircraft with the inevitable result [26].

The great battleship *Yamato* with a few remaining elements of the fleet sailed from the Inland Sea that April and perished in an attempt to hit the Allied ships off Okinawa. Neither fuel for a return voyage nor air cover were provided. The pride of the Imperial Navy served primarily as a decoy to lure carrier planes away from the kamikaze attacks.

On 19 June the last organized resistance of the defenders was overcome. This last great battle of World War II, fought in the air, on and under the sea, and on land, was the clash of two hardened, professional forces. When the kamikazes failed to destroy the invasion fleet during the first few days, Lieutenant General Mitsuru Ushijima, commander of the island garrison, and his staff knew they had no hope of defeating the invaders or of relief, circumstances that in no way altered their dogged resistance, which was countered by an equally determined opponent, commanded by General Simon Bolivar Buckner. Ushijima and Buckner died within less than a week of one another, one by suicide, the other killed in action. Few Japanese survived. An American who survived the 90 days of combat 'could consider himself a fugitive from the law of averages'. Both sides supplied history with countless examples of courage and devotion to comrades and country, seldom recorded, most now forgotten. The victors pondered what the future held for them in the next battle—known by all to be the invasion of the Japanese homeland.

9.2. THE DESTRUCTION OF JAPANESE CITIES

America's relations with China stand at the heart of the Pacific War. Drawn by the mysteries of its distant civilization, recounted by returning missionaries and traders, Americans found sympathy for China's pain during the disintegration of the ancient empire at the beginning of the 20th century and enmity for Japan's brutal attempts at domination in the 1937 invasion that filled American magazines and newsreels with graphic accounts of savagery, specifically the unrelenting bombing of defenseless cities and the delivering of Nanking to weeks of indiscriminate slaughter, rape and plunder. The sinking of an American gunboat at the same time by Japanese bombers hardened American anger. These events, tangled by diplomatic matters touching every imaginable issue, eventually led in December 1940 to the American embargo on metal and petroleum products that forced the Japanese decision for a war directed to the south against the United States and Britain rather than to the north against Russia.

If China was at the center of America's entrance into the Pacific War, it was at the periphery of the conduct of that war, the consequence of communications strangled by the mountain passes to Burma that required maintenance by air and an inadequate road. China's Generalissimo Chiang Kai-shek, a political leader of limited military skill who found internal enemies more important than the Japanese, had the assistance of two American generals whose opinions conflicted. Major General Claire Chennault had organized an effective group of American-financed volunteer fighter pilots for Chiang before Pearl Harbor, which had been expanded into the 14th Air Force in March 1944. One of its objectives was to provide refueling fields for the 20th Bomber Command, based in Kharagpur, India, to bomb the Japanese homeland with the new long-range B-29 bombers. Lieutenant General Joseph W Stilwell, fluent in the language and with years of experience in China, was the obvious but unwilling choice for commander of US forces in the China–Burma–India theater. He organized Chinese divisions in Burma and demonstrated that, given proper leaders and training, they were a match for the Japanese. He considered the leadership of Chiang's armies facing the Japanese on the mainland to be inept and corrupt.

Chennault was confident that American air power from Chinese bases would perform the kinds of miracle the prophet Mitchell had promised, and Chiang found this an agreeable way to fight. Stilwell pointed out that the Japanese had not overrun the locations for the bases simply because there was no reason to do so, and if American bombers began using them, the Imperial Army would have no trouble taking possession. This is what happened, although the logistics of operations over the eastern Himalayas, the fabled 'Hump', would have ended the B-29 raids anyway. Eliminating the 14th Air Force's B-24 radar-directed attacks on shipping in the Formosa Strait probably figured larger in the Japanese decision to take the bases[2]. No one who correctly predicts a defeat is loved, and Stilwell's inability to be tactful with the incompetent soon made him the enemy of most Allied commanders in the theater, General William Slim, his British counterpart and Britain's best field commander, being a singular exception. Stilwell was recalled [1].

The decision to build the B-29 had been a great gamble. Initial discussions with Boeing during the summer of 1939 begot a prototype in September 1942, at which time contracts for full production had been made but for which factories were not yet completed. The Superfortress became an unquestioned success, but the speed with which its production was secured resulted in countless minor design faults—minor from the engineering point of view but major to the crews of the many planes lost in the first months because of them. For a significant time these 'bugs' were a greater danger than enemy fire.

The B-29 had space reserved for radar-countermeasure equipment, and these functions were uppermost in the minds of the fliers that prepared for the 15/16 June 1944 raid on the Japanese homeland, the first since the Doolittle raid of 1942. British electronic intelligence flights from India and Ceylon had found meter-wave radar early in the year in southeast Asia, and the B-29s that had attacked Bangkok ten days earlier had logged radiation from nine sets. These were from Japanese Army Tachi-6 early warning radars that worked on the 4 m band. It was a static device with wide-angle transmission and up to four receivers, each with a steerable, directional antenna so that multiple attacking formations could be tracked. Peak powers of 10 to 50 kW and pulse widths of 25 to 35 μs gave it a range of 300 km. Special interest was attached to the absence of any evidence for the 50 cm radiation that might have indicated the presence of Würzburgs.

American forces had captured a 1.5 m mark 4 model 3 radar on Saipan in June 1944 and three months later a mark 4 model 1 on Peleliu, both of which appeared capable of functioning as searchlight or gun-laying sets, and Allied planes had been shot down under unseen conditions [2]. So it came as no surprise that the attack on the Imperial Iron and Steel Works at Yawata, Kyushu launched from Hsinching, China, the longest bombing run in history, disclosed a rich radar spectrum. The Tachi-6 came first as they passed over mainland China, followed by the familiar naval mark 1 model 1 when they reached the coast. As they prepared for their bombing run the receivers picked up a score of radars working on meterwaves, the Tachi-1, 2, 3 and 4 searchlight and gun-laying radars. (A 1.5 m set made by Toshiba, the Tachi-31 that incorporated the Würzburg indicator, would make its appearance by the end of the year, although not distinguished from earlier models on the intercept receiver.) The inaccuracy of the fire and the few times when bombers were illuminated by searchlights did not demand the conclusion that the equipment was capable of locating the

---

[2] See Chapter 7.4 (pp 366–7).

bombers in three coordinates. Again no sign of the Würzburg [3]. The effect of their bombs on the Imperial Iron and Steel Works was trivial [4].

The weakness of the defense led to a reconnaissance by a modified B-29 during the daylight of 21 August flying very high over Japan and Korea. It returned with excellent photographs, which had been sorely lacking to intelligence, and was the start of routine coverage. The plane had the usual AN/APR-4 receiver with analysis equipment and directional antennas, and it found plenty of signals to analyze, but no fighters rose to intercept, presumably because the speed and altitude were too great [5].

Other attacks on Japan from China followed the Yawata raid, none yielding results worth mentioning except to raise Chinese and lower Japanese morale. In August 1944 Major General Curtis LeMay took command of the 20th.

The Yawata raid had preceded the Battle of the Philippine Sea by four days, so the Saipan base would soon replace the need for the circuitous route to Japan. The 21st Bomber Command began to occupy Saipan in October 1944 with Brigadier General Haywood Hansell commanding. Attacks from the Mariannas could be carried out more frequently than those from India by way of China, but the results were in no way commensurate with the cost of operating such a long bombing run with a supply line a third of the circumference of the globe. Although carried out during daylight hours, the problems of accurate bombing that had been ignored from the planning of the late 1930s made themselves felt in the Asian theater as in the European. The cloudy skies not only forced radar bombing, but the unexpected winds of the jet stream at times made accurate bombing impossible from the high altitudes needed for safety. Hansell, a strict believer in precision bombing, did not deliver the expected results and was replaced by LeMay on 20 January 1945, who faced the same problems as Hansell and saw them as a crisis.

On the night of 9/10 March 1945 the bombing of Japan changed from an expensive nuisance to the most destructive force the world had ever seen. LeMay decided that the night air defenses were so weak that he could risk low-altitude bombardment. There was no evidence of radar-equipped night fighters, so he stripped his planes of machine guns to lighten and streamline them. Combined with the elimination of the fuel-expensive climb to high altitudes, the tactic allowed much heavier loads of incendiary bombs. Its first application was the most destructive air attack of all time, which made a holocaust of Tokyo. Other cities followed, and Tokyo received more fire from the sky. LeMay's estimate of the defense was correct. There were no night fighters and the guns were no more effective than the British guns had been during the Blitz, but these were not the puny bomb loads of a hundred or so He-111s, these were the immense loads of hundreds of B-29s.

Japan's air defense was weak in material and trained personnel, but the rivalry between the Army and Navy confounded the hope of using efficiently what little there was. In the homeland the Navy provided the defense of their own, extensive bases while the Army provided for the defense of the cities. Each maintained separate air control centers, separate fighter squadrons and separate radars. They had separate IFF equipment and interrogation signals, so that in addition to the normal problems of this device they had the severe one of not knowing whether an aircraft that did not respond was the enemy or the other service. Each service maintained its own gun engagement zones [6].

The weakness of the defense caused no letup in American countermeasures activities. There were false alarms about Japanese fighters homing on bombers by activating American IFF that had to be investigated. Some of the Japanese early-warning sets did turn the IFF into range-enhancing secondary radar, but early warning was not the defenders' problem [7]. Gun-laying and airborne radar was. These elements might become effective, so the attackers deployed countermeasures, using techniques honed in Europe. At first reliance was placed on Rope, the meter-wave version of Window, which was dispensed liberally and which interfered demonstrably with searchlight operation. Nevertheless, the radar-directed flak improved and learned to work through the drifting foil, bringing down 4.4% of the bombers during two May night attacks. This was answered by special B-29s carrying high-power jammers, called Porcupines because of their numerous blade antennas or Guardian Angels because of their circling the target area during the attack while saturating the ether with powerful radio noise. Losses attributed to the gunners dropped [8].

The cities were destroyed one after the other with very high casualties, much higher in proportion than for the German cities that had sturdier construction and an advanced civil defense organization. The bombing of Japan was carried out by radar to a high degree and in many ways resembled the night bombing of Germany by the RAF. It was, however, much more destructive than for Germany. First, there were no navigational problems in locating the target cities. Loran navigation was in place sufficient to guide the formations to southern Japan [9][3], and the preponderance of targets located on the coast yielded PPI displays that approximated a map of the target city to a degree rarely encountered in Europe. Second, the H2X (AN/APS-15) was slightly better for rendering detail than H2S. Third, the weakness of Japan's air defense allowed bombing at altitudes half as high as had been used by RAF Bomber Command, making the PPI displays sharper and the bombing more accurate. Finally, the inflammability of the targets made area bombing inherently destructive. Daylight precision bombing was retained, in part because of a shortage of incendiary bombs, but as over Germany, formations that intended to attack visually often encountered unexpected cloud cover and had to aim by radar.

---

[3] See Chapter 10.1 (pp 430–1).

As the weakness of Japan's defense against night attack became obvious, LeMay began warning cities by radio and leaflet of impending attack. This had the advantage of demonstrating the inability of the Army to defend the people and removing some of the stigma attached to area bombing [10].

Two of Luis Alvarez's radar designs had been deployed in Europe with outstanding effect at the time and for the future: Microwave Early Warning (MEW) and Ground-Controlled Approach (GCA). The third contribution of his brief radar period was the 3 cm blind-bombing radar Eagle (AN/APQ-7)[4]. This set had electronic scanning ±30° in the forward direction that provided a much improved resolution over H2X. Its antenna was a vane mounted below the fuselage, which required specially modified B-29s. This antenna provided a beam only 0.4° wide horizontally and was the reason for Eagle's superior resolution; the beam's large vertical radiation pattern was an advantage.

Eagle was given to the 315th Wing of the 21st Command along with careful instruction by Rad Lab personnel before being committed to action, something that had been lacking for the crews using H2X. The 315th Wing conducted 15 strikes beginning 26 June and ending 14 August. They were precision attacks of the kind normally assigned for daylight raids, but being night attacks were carried out at low altitudes with accuracy equal to visual. The unit flew a total of 1200 sorties of which 1095 bombed the primary target [11].

During these trying times occurred one of the most remarkable parts of the radar story, one which in a bizarre way returned it to radar's earliest years. One aspect of Japanese research that progressed with a degree of satisfaction was the development of high-power magnetrons. Indeed it so satisfied Yoji Ito, the head of electronic research at the Naval Technical Research Department (NTRD), that he extrapolated powers to levels that convinced him a microwave death ray was possible, calling to mind Harry Wimperis's 1935 query of Watson Watt about the possibilities of a death ray. Ito's ideas found strong support from military authorities. The Army had been investigating a number of possible death rays since 1939 under the direction Major General Hideki Kusaba, and a rare example of a combined Army–Navy project resulted at the Mitaka Office of the NTRD that involved so many experts that it became known as a 'physicist super laboratory'. Tests succeeded in killing a rabbit at 30 meters in 10 minutes and stalling an engine, presumably neither protected by the sheet-metal that would surround the contents of a bomber, and plans went forward to build a 15 m reflecting dish to concentrate the beam. Needless to say, nothing affecting the war came from this project [12].

For maximum destruction atomic bombs were to be exploded about 500 m above ground. To accomplish this the device was equipped with

---

[4] See Chapter 4.5 (pp 192–3).

four radar fuzes, a barometric fuze and impact fuzes on nose and tail, in case the other five failed. The radar was the 72 cm AN/APS-13, a modified tail-warning set with Yagi directional antennas pointing forward. The squadron's electronic officer, Lieutenant Jacob Beser, who flew both missions, monitored the spectrum of Japanese radar to learn whether any matched the wavelength of the bomb's sets, including possible harmonics. If such were found, he could disable one or more of the bomb's radar fuzes, as such signals might detonate the bomb prematurely, possibly destroying the plane, but no interfering radiation was encountered and the fuzes were not altered [13].

On 6 August a B-29 carrying the uranium bomb took off from the Marianna island of Tinian accompanied by two observation planes, one of which carried Alvarez, by then a bomb designer, and preceded by others to advise them about the weather. Cloud cover over Hiroshima, the primary target, was light enough to permit the required visual bombing. Single B-29s were common in Japanese skies and seldom attacked, owing to their altitude, speed and apparent immediate harmlessness, but this one made a perfect bomb run and destroyed much of the city. Among the attacks of the past few months it would not have stood out in the statistics of destruction, except that the casualty rate was high for the moderate amount of building damage, the consequence of the population ignoring the somewhat confused alarms.

Three days later the same procedure was followed for the plutonium bomb dropped on Nagasaki; it was 67% more powerful but caused less damage, in part because less accurately laid. After Hiroshima the government had warned the population to take cover for single bombers but had not explained the terrible nature of the new explosive. As a consequence the people of Nagasaki had not taken cover and suffered high casualties, although significantly lower than Hiroshima [14]. Nagasaki was the secondary target; the primary, Kokura, was obscured by clouds. The bombardier reported that he aimed visually, as required, through an opening in the cloud cover, but the inaccuracy was typical of an H2X bombing. Given the reluctance of the crew to bring the bomb home, many have thought the hole in the clouds to have been fictitious.

The bombing of Hiroshima and Nagasaki ended the Pacific War. Of this there is no question. Whether the end would have come as soon or soon enough to prevent the dreaded invasion without the use of these two explosives is a dispute that began immediately after the end of hostilities and continues, growing in asperity as time passes. It is a curious dispute because it centers on the nature of the bombs, not the injury done by them, for there is little evidence that the arguments would rage, had Hiroshima and Nagasaki been dealt equally hard blows by a few hundred of LeMay's fire bombers. Japan surrendered through the direct and brave intervention of the Emperor, who had long wanted peace and found that the atomic bomb provided him the means to obtain it. There is little doubt that had the war continued only a few weeks longer Japan would have suffered more from the destruction of nearly every remaining seat of population than from the two atom bombs. That it was war degenerated from the chivalrous rules to which western nations had romantically thought themselves to have attained is hardly a question. That it came to this through the implacable working of military and technical imperatives is obvious. The absence of strategic bombing with nuclear explosions in the succeeding half century may have accrued from the awakening of humanity caused by the tactics of LeMay and Harris and the atomic bomb.

These questions aside, it is clear that strategic bombing was decisive in defeating Japan. Sea power was necessary to implement the bombing and had blockaded the island kingdom to the degree that it was no longer able to defend its skies or feed its people, but air power had finally delivered the promised victory. Nevertheless, the Pacific War had followed the general plan formulated by the Navy during the previous decades: (1) force the Japanese fleet to fight with the American strength for its defeat maintained through the establishment of island bases, (2) blockade the home islands and (3) if that did not force surrender, attack the cities from the air. None of these pre-war plans had foreseen an invasion of the homeland—or an atomic bomb [15].

CHAPTER 10

THE MEASURE OF RADAR

10.1. NAVIGATION TRANSFORMED

At the end of the war the British carrier HMS *Implacable*, bound for Vancouver, received the pilot for the passage through the assortment of islands south of the port. Despite the urgency of the vessel's mission, to land a large number of repatriated prisoners of war in time for the special trains that awaited them, the pilot refused to take the ship up because of fog. A consultation between the captain and his navigation officer over charts concluded with agreement that the passage could be made with radar and the skipper rang for 20 knots, whereupon the pilot announced he not only disclaimed responsibility but would not even remain on the bridge [1]. Such a decision by the captain resulted from the sure knowledge of the capabilities of his radar. The PPI provided a clear definition of the shores and established position, even if navigation markers were not visible; other ships were easily avoided.

This clash illustrates the dramatic difference in navigation that radar and associated techniques had brought about. In 1939 the position of a ship in fog or under cloud cover came from dead reckoning—an estimate based on speed and course with corrections for the effects of wind and currents. After a few days of dirty weather at sea the calculated position might be a hundred miles from reality. In 1945 the position of a ship equipped with Loran (see below) would, depending on momentary radio propagation conditions, be known to accuracy as good as a routine celestial fix.

10.1.1. Radar

Before microwave surface-search radar became a fixture on their masts, liners on the North Atlantic route ran consistently in violation of rules requiring reduced speed in fog, '...a speed which does not preclude getting all the headway off a ship in a distance less than half the range of existing visibility'. But to arrive in port late meant schedules disrupted and possibly tug boats and gangs of longshoremen idle at company expense. Captains who valued their berths took care to avoid this.

The postwar handbooks for navigators and pilots instructed their readers on the relative merits of 3 and 10 cm radar and about atmospheric effects on propagation. The shorter wavelength had the advantages of reliably sighting buoys and other small targets combined with the disadvantages of more confusing sea return and of being blinded by a rain squall. Small ships generally elected 3 cm, large ships frequently both.

With time almost anything that floated on navigable waterways carried radar, but in the decades after the war many ships did not [2]. This caused busy harbors, especially those troubled with heavy fog, to install shore-based radar. In this a 3 cm station was located so as to have the best possible view of the harbor and approaching waterways. Operators provided pilots and masters with data that allowed them to find their way rather than remain hove to, waiting for visibility. It could also advise ships equipped with radar of traffic conditions beyond their line of sight, served as a continual check on the condition of buoys and other floating seamarks and proved valuable in rescue operations or in positioning dredges or dumping. Liverpool installed the first such system in August 1948 and was quickly followed by Sunderland, the Isle of Man, Long Beach, Vancouver and Spitsbergen [3].

The use of radar for aerial navigation proved to be the inverse of maritime. At sea it was the ship that wanted radar in order to see land, other ships or the the rapidly growing number of radar beacons. Shore-based radar was useful, but not often encountered. By contrast commercial aircraft did not generally mount radar. There was certainly little in H2S or H2X to recommend it as a system of navigation, and when an airplane came within range of a radar beacon, it was probably subject to traffic control on the ground. Ground radar never allowed a competing airborne system an opportunity, it was simply too obvious a method of controlling traffic.

Pilots favored methods that allowed them to determine the bearing to a ground station of known location, often their next destination. The favorite after the war was Omnirange, officially VOR, a continuous-wave directional system similar to Sonne–Consol, except that its receiver actuated a visual indicator rather than relying on audible signals. The name is flier's parlance, as the device provided only direction information. Distance measuring equipment, officially DME, was soon paired with VOR. It drew directly on the design of Y-Gerät, with the exception that the aircraft rather than the ground station transmitted the interrogating audio-modulated radio-frequency signal, which the ground station re-transmitted on another radio frequency [4].

Under such circumstances there was a clear need for secondary radar. In addition to enhancing the range and representing the target aircraft with a strong mark on the PPI, the responding signal could be coded to give the air controller such obviously important data as the aircraft's altitude, identification and remaining fuel. It was all to happen but years after the end of the war. Secondary surveillance radar brought with it all the old problems of IFF and many new ones. Whereas primary radar depended only on the characteristics of the ground set with no problems of compatibility, secondary radar is surrounded by a great compatibility problem that includes commercial, private and military interests together with an ample portion of international disagreement [5]. The technical constraints continued to change as a result of the rapidly evolving electronics of the postwar years, continually complicating negotiations. Finally, the International Civil Aviation Organization adopted a standard based on the American IFF mark X in 1958 [6]. It is a story with its own attraction, but it is not our story [7].

Air controllers would have to wait for secondary surveillance radar but were able to receive the benefits of a technique that would remove many distracting signals from their screens, the moving target indicator. Its invention had first been forced on the Germans as a way to distinguish the echoes of moving bombers from near-stationary clouds of Window. This had been done, as described earlier, by filters that passed only return pulses that had had their frequencies shifted to some degree by the Doppler effect of the moving target. Pulses on or near the transmitter frequency were attenuated. Its value to the operator of surveillance equipment is obvious; it removes the large number of echoes from fixed objects that clutter the screen, allowing the aircraft to stand out.

TRE and Rad Lab worked on the same idea but without the air attacks that spurred the Germans. Their approach did not use filters to reject pulses that had not been Doppler shifted. Instead the received pulses were compared with the previous pulse, rejecting both if they were identical in amplitude and phase. This was accomplished by requiring the returned pulses to pass through a transmission line that delayed them by exactly the time between transmitted pulses. Such a delay line was, interestingly enough, not electrical but acoustical, generally being the transmission of sound through mercury. This proved capable of detecting planes moving at near right angles to the line of sight much better than filters [8].

The most highly valued radar for bomber crews returning to their base in fog was the ground control approach, Alvarez's GCA, which allowed a ground radar operator to guide the blind flier down. When America and Britain began the supply of blockaded Berlin by air in 1949, GCA allowed the transports to land irrespective of visibility. Indeed, the airlift would have been impossible without it. Despite these achievements, GCA was not used except for military flying, and commercial airfields were to be shut down by fog long after blind landing was an actuality. The reason was simple. An airline pilot is responsible for the safety of the plane, and a landing with GCA in effect gave the controls to an unknown radar operator on the ground. The pilots would never agree to this.

10.1.2. Radio altimeters

Barometric altimeters had a sufficient number of faults, potentially dangerous ones, that commercial aviation quickly adopted the radio altimeter in the postwar flying world. The United States left the war with two lightweight and relatively inexpensive devices: the pulsed SCR-718C and the frequency-modulated AN/APN-1, both on the 70 cm band. The pulsed set proved best for determining high altitudes, the frequency-modulated set for low altitudes [9]. Germany left the war with equally well designed equipment. The pulsed FuG 102 worked well for high altitudes, the frequency-modulated FuG 101 for low [10]. Both were adopted by eastern block nations.

10.1.3. Loran

The organization of the Radiation Laboratory had called for three high-priority projects: airborne radar, gun-laying radar and a long-range-radio navigation system to meet the needs of both aircraft and ships. One might object that the resulting Loran (long range navigation) was not radar and hence does not belong in this account, but its use of pulsed high-frequency waves, its close association with navigation methods that used radar directly and its development at Rad Lab make the inclusion reasonable. By the end of the war there were a remarkably large number of radio navigation systems in use: radar beacons, Oboe, Sonne–Consol, Rebecca–Eureka, ground radar control, Gee, Decca Navigator, Gee-H, Rebecca-H, Micro-H, Shoran and Loran. Of these Loran remained in service the longest and had the widest coverage.

Alfred Loomis proposed Loran in October 1940, and a Project 3 Committee was formed almost immediately to carry out the charge. In addition to the Rad Lab people its meetings included representatives of RCA, Sperry Gyroscope and the British Embassy. When committee management yielded what might be expected, the work was re-organized under Melville Eastman in early 1941. Loomis's original plan called for a hyperbolic coordinate system using pairs of master–slave stations in the same way as the British Gee. The Americans had decided by summer 1941 that the line-of-sight restrictions imposed by very high frequencies limited the use too severely for the kind of navigation desired. They began studying the possibility of using the reflections off the ionosphere of 40, 60 and 100 m waves that allowed a greatly extended range.

Initial tests indicated positions determined from the sky wave to be stable enough for accuracies of a kilometer or less. An operational test with 100 kW peak transmitter power was undertaken in June 1942 using a 150 m band from which amateurs had been ejected and which remained the Loran standard [11]. At that time R J Dippy, the inventor of Gee, came to Rad Lab from TRE to work out the compatibility of the aircraft mounting of Gee and Loran equipment. The Loran sets were so designed that they fitted into the slot where a Gee set had been removed, thus allowing planes of the two nations to use both kinds of equipment interchangeably.

The first priority was given to navigation through the foul weather of the North Atlantic, the route of many airplanes for transport, transit and convoy protection, so pairs of stations began operating in Newfoundland, Labrador and Greenland to form the North Atlantic Chain. By January 1943 the US Coast Guard and the Royal Canadian Navy assumed operational responsibility.

The designers originally thought the slave station had to be so situated that it received the ground wave, but evidence accumulated that the sky-wave pulse, which lags the ground wave in time, could be identified and was stable enough to synchronize the two satisfactorily. This allowed the two base stations to be situated 2000 km apart, giving wider coverage and a grid capable of greater accuracy. It became known as sky-wave synchronized or SS-Loran and had a range capable of encompassing most of continental Europe. It was used operationally by RAF Bomber Command in October 1944 [12].

The extreme distances of the Pacific placed heavy demands on any radio-navigation system and led to the decision to exploit the good long-range propagation of very long waves. During the winter of 1944–1945 an experimental system using 1700 m operated in the south-eastern United States but did not see war service [13].

By the end of the war 70 Loran stations and 75 000 receivers were providing navigational information for 30% of the Earth's surface [14]. The 150 m band Loran continued as the postwar standard. The advantages of low-frequency propagation, the frequency region favored for maritime communications because of the absence of skip zones, compelled investigators to surmount the problems of greater atmospheric noise, and a 3000 m system became operational on the American East Coast in 1957. The older system was referred to as Loran-A, the newer as Loran-C. In Loran-C the multiple sky waves can lag as much as 1000 μs, but the first arrivals were readily identified and sufficiently stable for an acceptable fix [15].

10.1.4. *Decca Navigator*

A hyperbolic-coordinate navigation system similar to Gee, officially designated QM by the Admiralty but quickly named Decca Navigator for its manufacturer, Decca Radio Ltd, was employed during the Invasion and then withdrawn from service to preserve secrecy. After the war Decca marketed the system and established chains of shore stations along the Channel. The network of chains spread over most of Northern Europe and to many coastal region in the British Commonwealth and became very popular with seamen.

It was similar to Gee in using a pair of stations to determine a hyperbolic coordinate, but differed in using low-frequency continuous waves around the 3000 m band. Positional data were carried by the relative phase shift between master and slave and attained accuracies of tens of meters. It was restricted to similar ranges as Gee but by confusion brought by the sky wave, not by line of sight propagation. The phase difference was indicated by a meter. Constant deflection indicated movement on one of the hyperbolic coordinate lines [16].

10.1.5. *Radio navigation and the velocity of light*

Shoran was a hyperbolic-coordinate navigation system that used the same principles as Gee-H but with an automatic pilot's direction indicator. It had an on-board computer allowing a continuous presentation of the data that permitted blind bombing. It proved to have the accuracy of Oboe or visual bombing but with each bomber capable of using it independently. It was introduced into combat in April 1945 and consequently had little effect on the course of the war [17].

The US Coast and Geodetic Survey learned of the accuracies suspected of the various radio-navigation systems and wished to evaluate this new method of measuring the Earth. With the Army Air Forces they set up as a test, Shoran ground stations at locations where there was surveyed control. The results showed precision of two parts in a million but systematic discrepancies from the control data of 50 parts per million, which could be effectively removed if the velocity of light was altered from the accepted value [18].

The tabulated value in 1941 was $299\,776 \pm 4$ km s^{-1}. This had resulted primarily from extensive measurements initiated by A A Michelson and carried to completion after his death by his assistants. The experiments had used long, evacuated paths and incorporated what was thought to be the ultimate in experimental accuracy [19]. It was significantly lower than Michelson's previous value, and the experimental uncertainties ascribed each measurement made them incompatible one with the other. Any warning provided by this was swept aside by a confirmation experiment at Harvard that used the Kerr-cell light switch rather than Michelson's rotating mirror and that allowed electronic timing. The result was $299\,764 \pm 15$ km s^{-1} [20]. The Coast and Geodetic Survey found that a value of $299\,792 \pm 2.4$ km s^{-1} made their Shoran data compatible with their ground-surveyed data, very nearly Michelson's older value.

William Hansen had proposed to use a resonating cavity to measure the velocity of light. By knowing the frequency at which the cavity resonated, its dimensions and the rate of its power dissipation, one could calculate the velocity of the waves. This experiment had been carried out at the National Physical Laboratories in England before the Shoran experiment, and gave $299\,793 \pm 9$ km s^{-1} [21], a value also significantly in disagreement with the then accepted value. Other confirmations of the new value were to follow with microwaves and light. The laser was eventually to take accuracy to the currently accepted value of 299 792.458 $\pm$ 0.0012 km s^{-1}, but for a time the blue ribbon belonged to microwaves.

These products of the wartime radar laboratories present us with an example of technical progress in its purest form. As a result of radar and Loran fewer ships run aground or collide. As a result of radio navigation passenger travel by aircraft has become as reliable and dependable as that by the railroads before the war, and ground control by radar has allowed traffic densities almost equal to the demand. This has transformed the world more than anything since the age of discovery. To fly across oceans for a routine business meeting has become commonplace. People of moderate incomes take a week's vacation in parts of the world that their parents would have never dreamed of visiting. Nothing has shrunk the Earth as has modern air traffic, and radar is its primary guide.

And yet this progress has come at a price. Within the lifetime of many of the readers, the navigation of air and sea required skills acquired only after exacting apprenticeships, skills for which men were justly proud and for which at times required them to draw on their reservoirs of courage. The automatic ship and airplane are now technical realities. But this is the inevitable result of technical progress. A worker's skills—skills that had given him his sense of worth, skills whose application gave him his place in the world, skills that made him a man—are being made of no importance. Whether in transport, the shop, the mill or the farm, occupations that once gave meaning to life for much of humanity are being destroyed or transformed into tractable jobs that give little satisfaction at the end of the day. It is the technical imperative. It is the consequence of mankind having taken the course of evolution into its own hands. It cannot be avoided.

10.2. SCIENCE AND THE ELECTRONIC AGE

Radar was derived from science, primarily physics, but the link was through a path of electrical engineering that had already converted the basic ideas about electromagnetic fields and the motion of electrons in vacuum and conductors into design elements. The link between physicists and electrical engineers was strong during the inter-war years. Nuclear physics was growing in the United States and Britain and demanded competence in electronic design. Cyclotrons required high-frequency oscillators of extreme power, and the radiation detectors needed pulse electronics, both generally made by physicists. These skills were directly transferable to radar and account in part for the large number of physicists employed to that end by both countries. Having forced some of her best nuclear physicists into exile and covered quantum mechanics and relativity with odium as 'Jewish physics', Germany lagged in that discipline, and only the Kaiser-Wilhelm Institut at Heidelberg had a cyclotron. Relatively few German physicists went into radar.

Physicists picked up engineering quickly but brought very little new physics with them. Scientists and engineers pushed the development of microwaves together, and an attempt to assign credit one way or the other becomes a game with words. Bell Labs was populated with engineers, Rad Lab with physicists. Both designed advanced microwave systems. One contribution science did make to radar during the war was the study of semiconductors, which allowed the Allies to produce reliable microwave detectors. In going the other way, in contributions of radar to science, the examples are many and important.

10.2.1. Astronomy

Observing celestial objects through the electromagnetic radiation they emit in wavelength bands longer than infrared had had a tentative beginning before the war. In determining experimentally the noise sources of a 14.6 m receiver at Bell Telephone Laboratories, Karl Jansky noted a source that could not be ascribed either to his equipment, atmospheric discharges or man-made interference. In order to check his suspicions that he was receiving radiation that was not of terrestrial origin, he constructed a directional antenna that could be rotated about its vertical axis and demonstrated after a long series of observations that it had a maximum in the constellation Sagittarius, which lies in the direction of the galactic center [1]. Jansky's discovery did not go unnoticed, for Bell made certain there was adequate public announcement, but astronomers did not know exactly what to do with these data, as they were neither able to apply them to any of the phenomena then under investigation nor use them to suggest new observations. Furthermore, the radio techniques employed were simply too far from their discipline to allow the matter to be properly considered. The radio amateur Grote Reber did the only continuation of Jansky's work and succeeded in having his results published in the astronomical literature [2].

Receiver noise limited the sensitivity, hence the range of a radar set, so noise studies were common, both as research and as parts of the routine maintenance of equipment. Jansky's 'cosmic noise' must have been observed by them countless times without an appreciation of its significance, but reports of it number only two: J S Hey saw it with GL mark II but had to have a colleague identify it [3]. Wilhelm Stepp, the Telefunken engineer, also encountered it [4]. Two new astronomical phenomena caught the attention of alert operators of meter-wave radar receivers: solar emission and meteors incident on the earth's atmosphere.

Radio emission from the sun, particularly strong during sunspot activity, had been observed before the war by radio amateurs, who reported a 'curious hiss' at times in their receivers [5], but there had been no scientific follow-up. This source of radiation was rediscovered in February 1942 by Hey, who had been assigned the task of evaluating the possibility of the Germans jamming the Army's GL equipment. The escape of the three capital ships in the Channel Dash a few days earlier had enhanced concern about such techniques. As it turned out he set about his task during the period of a solar flare and was rewarded by what appeared to be a very cunning form of jamming, which he quickly identified as radiation from the sun [6].

The 5 m GL mark II was the radar set with the longest wavelength that had an antenna steerable in two angular coordinates, making it ideally suited for establishing the source as the sun [7]. Hey found these characteristics just right for making another discovery. When V-2 rockets began falling on England some means of tracking them was sought as part of the desperate methods being planned for countering them, and Hey suggested equipping GL mark II with a larger antenna. None of these anti-rocket methods came to trial, but in the process of tracking them Hey made another discovery. The operators observed high-altitude echoes four to ten times an hour that had nothing to do with V-2s but sometimes triggered false alarms. As the danger of German air and rocket attacks receded, Hey requested permission from Antiaircraft Command to investigate the phenomenon systematically while an operating network still existed. General Pile's intellect was attracted to the idea, and permission came quickly. The troublesome echoes came from the trails of ions left momentarily in the upper atmosphere by meteors [8].

Bernard Lovell, who had led the TRE team in the development of H2S, intended to return to his pre-war research in cosmic rays at the University of Manchester, where, just before leaving for war service, he had observed with cloud chambers a cosmic-ray shower that had deposited an impressive amount of energy in the atmosphere. In one of his early duty assignments, he noted activity on the receiver scope at a Chain Home (CH) station, which he took to be enemy aircraft, but was told by the operator 'Oh, those are not enemy aircraft, they're ionosphere'. Finding no basis for such transient events in his knowledge of the ionosphere, Lovell thought he might be seeing cosmic-ray showers [9]. The idea so appealed to his professor at Manchester, P M S Blackett, that the two wrote a paper about the radio events, without disclosing the method by which the radio observations had been observed, speculating that they might have had their origin in cosmic-ray showers [10].

On learning of Hey's meteor studies, Lovell reasoned that the same equipment should show up the desired cosmic-ray showers, so he and Blackett, both having acquired a substantial amount of gratitude in the War Office, arranged for the transfer of Hey's modified GL equipment to the University. The electrical noise that inhabited a major industrial city caused them to shift the work to a rural site in Cheshire named Jodrell Bank. Lovell's observations showed that cosmic-ray showers were beyond the sensitivity of his equipment, and he shifted his interest to meteors, the events he had seen at the CH station in 1940. His systematic study of meteor trails after the war allowed their trajectories to be determined and showed them to have come from within the solar system, answering a question open since the time of Newton [11]. He and his associates pursued other aspects of astronomy using radio waves, and a major radio-astronomy observatory came into being. As indications of the varied possibilities of radio astronomy began to be perceived, Lovell became convinced that a large steerable paraboloid was needed as an antenna and found enthusiastic support for his wild idea of a 75 m diameter dish in Blackett.

Desires to enter the new field caused two other observatories to be opened. E G Bowen had left Rad Lab in late 1943 to join the Radiophysics Laboratory in Sydney, which greatly impressed him with its high level of competence [12]. Australians had made some solar observations with 1.5 m equipment as soon as the war had ended that indicated radiation from the sun-spot regions of the solar disc [13] and were not long in planning a 63 m diameter steerable dish for radio astronomy, whose similar appearance to the SCR-584 was not accidental. Martin Ryle at Cambridge University built a special array of dipoles that allowed him to observe solar radiation at 1.7 m in a clever way that insured that galactic radiation did not contribute [14]. These three observatories, all operated by former radar men, were for a decade the principal sources of knowledge in this new discipline.

The Moon offered a more substantial target for radar astronomy than trails of ions. It also offered the possibility for instructing the public about hitherto secret work by the Signal Corps, and the opportunity was not lost. An SCR-271 at the Corps's Evans Laboratory at Belmar, New Jersey was altered for the task so that single returns would be visible on a scope displaying range against signal. Two parameters controlled the experiment: the travel time to the Moon and back is about 2.5 s, and the half-diameter of the Moon from which signals can be returned provides a spread in return times of 0.0116 s. The former led to pulse repetition about every 3 s; the latter allowed the use of pulses of 0.05 s. The long pulses allowed the use of a receiver with a very narrow pass band and consequently very low noise. (The narrow pass band required compensation for the Doppler effect that results from the relative radial velocity of the Earth and the Moon and the Earth's rotation.) The 271 was given an extra dipole array but was unable to track the target in elevation, so the traces showed, rather dramatically, the Moon entering and passing through the radar beam at its rising or setting [15].

Two years earlier Zoltán Bay had begun to plan the use of a Hungarian-designed and built 2.4 m radar to observe reflections from the Moon. Bay was head of the laboratory of United Incandescent Lamp and Electrical Co. (Tungsram), a large manufacturer of radio tubes. He had an antenna steerable in both azimuth and elevation that allowed tracking the Moon, but he did not have the equipment that would allow the high degree of frequency stability required for a narrow-band-width receiver, so his transmitter power and antenna gain were insufficient to observe single traces on an oscilloscope. He devised instead a unique method of integrating the return signals—electrolysis of a KOH solution into hydrogen. A rotating switch formed the 0.06 s transmitter pulse, protected the receiver and routed the signals according to delay into ten 1 mm diameter coulometers that bracketed the times at which signal would be returning. This procedure left a record of receiver noise plus signal from data acquired during repeated cycles of 30 min on the Moon and 30 min off. In cell 6, where the 2.5 s pulses were expected, the gas volume from the Moon observations exceeded that of the blind cycle by 4.4%. It was a small effect, only five times the experimental uncertainty, estimated from the blind data. Wartime air attacks and the removal of his original equipment by the Soviets delayed the experiment, but he persevered and succeeded with reassembled equipment in February and May of 1946 [16].

These two experiments introduced two important techniques that were to make radar astronomy a valuable method for studying the solar system. DeWitt and Stodola showed how sensitive the observations were to Doppler shifts, and Bay showed the power of integration. The later combination of sharp time resolution, Doppler shift and signal integration would evolve into a technique that would provide remarkably accurate topographic maps of the surfaces of Venus and asteroids [17], but long before such sophisticated results could be admired, radar astronomy determined with great accuracy the astronomical constant, the distance scale of the solar system.

But there were radar observations of the Moon before either of these two experiments. In 1943 Telefunken was attempting to build a long-range early-warning radar using their 50 cm electronics, which they expected would be an improvement over GEMA's 2.4 m Wassermann and Mammut, reasoning along lines similar to those advanced at Rad Lab for making MEW. Unlike the Rad Lab project, the Würzmann, named from the combination of Würzburg and Wassermann, was a cheap trial that drew on a large stock of components used in Telefunken's widely used 50 cm directed-beam relay communication links Michael [18]. The antenna was made of 32 Michael arrays (a total of 640 dipoles), arranged side by side 16 on each of two parallel 36 m masts capable of rotation about their common vertical axis. The Würzburg electronics were used with the transmitter power boosted to 120 kW peak power and a newly designed high-frequency amplifier at the receiver input that greatly improved sensitivity. It was located overlooking the Baltic Sea at the easternmost tip of the island of Rügen [19].

After setting up this equipment Stepp left it with an assistant, Willi Thiel, for further testing. When Stepp visited the site a few weeks later, Thiel reported a strange disturbance that had troubled him from time to time, which lasted about two hours and vanished without remedial action on his part. In examining it together they noted that it disappeared when the antenna direction was turned from the east; when the antenna faced the east the signals would remain for 2.5 s after the transmitter power was switched off. Clearing visibility disclosed the Moon and settled the matter. Thiel demonstrated the phenomenon to various visitors, but Stepp made no written report at the time.

He did make a report of the matter later, but in a remarkably understated manner. Immediately after the war he entered the Technische-Hochschule Darmstadt, as his mentor, Wilhelm Runge, had done a generation before, and took his theoretical and experimental radar work as the subject for his thesis. At the bottom of a page of his dissertation is recorded: 'Als erstes von uns jemals erfasste extraterrestrische Ziel wurde von uns Anfang 1944 mit dem Würzmann auf Rügen der Mond beim Aufgang erfasst' [20]. He also mentioned it at a conference about space travel in 1951 [21] and again the following year at one on maritime navigation [22]. This curious behavior has three possible explanations: (1) during the war such a report might have been interpreted by morbid-minded officials as wasting time on non-essential work, which could have been considered sabotage; (2) Stepp was the consummate engineer, who saw the work as too rough for a formal report and easily improved, given the opportunity, or (3) he may have taken the attitude that astronomers already knew the distance to the Moon better than he could have determined it, making it nothing more than a nice stunt [23]. Years later when reports of the event gained wider circulation, Stepp wrote a short article that gave credit for the discovery to Thiel as well as providing some of the technical parameters of the equipment [24].

Radio astronomy was not restricted to meter waves, and the Sun attracted microwave experts. Bell Labs's inventor of waveguides, G C Southworth, led the way, observing the quiet Sun with microwaves in 1942 and 1943 [25], but a more important function of these short waves lay elsewhere. During the war Dutch astronomers at the Leiden Observatory had received the issue of *Astrophysical Journal* that contained Reber's article on radio astronomy and discussed its implications. The advantage of having a spectral line at radio frequencies caused H C van de Hulst to demonstrate the theoretical possibility of the hyperfine transition of atomic hydrogen, the major constituent of interstellar space, being observed at 21 cm [26]. The group set out to build detection equipment, as did E M Purcell of Harvard and Rad Lab. Both reported detection at about the same time in coordinated publications that showed mutual assistance [27]. The Leiden group used the 7.5 m diameter paraboloid antenna of a Würzburg-Riese, and a few of the thousands of these dishes continued their existence in this new peaceful service. Their data provided remarkable evidence for the rotation of the Milky Way through the Doppler shifts observed, but a hoped for interpretation in terms of the galaxy's structure proved only qualitative, owing to the absence of reliable data for distance.

Both groups used a microwave invention from Rad Lab that would become the heart of radio astronomy, the Dicke radiometer [28]. In this instrument Dicke addressed the problem of measuring thermal radiation with a superheterodyne microwave receiver and devised a simple and elegant solution in which he alternately compared 30 times a second the noise-like radiation incident on the antenna with the thermal noise of a resistor of known temperature [29], a method that eliminated drifts in gain or noise within the receiver. This device, equipped with a horn antenna, determined the temperature to a precision of 0.5 K of various objects around Rad Lab from a small microwave band of their thermal radiation, often in ways that amused his coworkers. The concept of antenna temperature was not new, having been introduced by R E Burgess a few years earlier [30], but the accuracy and simplicity of the radiometer fixed it as the unit of radio astronomy. Dicke used the radiometer in its simple form to measure the temperature of the Sun and Moon, doing with a small horn something that had required a large paraboloid for Southworth just two years earlier using a non-comparing receiver and only for the Sun [31].

10.2.2. *Laboratory physics*

Given that so many of the physicists joining radar projects in America and Britain had left research on the atom and its nucleus, it would have been surprising had they not put the skills and techniques learned in radar to use in that field.

E O Lawrence's laboratory at Berkeley, California had led the world in the design of high-energy particle accelerators before the war, of which the cyclotron had proved most efficient. As higher energies were sought, bigger—and more expensive—magnets were required, which had given rise to a parallel project by D H Sloan to use acceleration by radio-frequency fields but dispensing with the magnet by stringing out the electrodes in a straight line. The idea pre-dated the cyclotron and, in fact, had inspired it. This mode of acceleration had a series of cylindrical electrodes connected to a high-frequency source. Particles were accelerated when residing between electrodes and drifted when they resided within the shielding of the cylinder during the other half cycle, emerging again to receive another acceleration. The electrodes had to be positioned to account for the increasing speed of the particles. This idea had been the basis of the earlier Berkeley accelerator, but owing to the low-frequency and low-power oscillators then available, its use was restricted to heavy ions at experimentally uninteresting low energies [32].

Luis Alvarez began working on the idea of a linear accelerator during the last months of the war, inspired by the thought that there were thousands of SCR-268s in existence that had been made obsolete by SCR-584. The 1.5 m wavelength, pulsed high power and number of transmitters available allowed dreams to extend to fantastic particle energies [33]. A 31 MeV machine worked successfully using 30 radar transmitters, operating with 300 μs pulses, later increased to 600 μs, and a repetition rate of 15 Hz. These transmitters were replaced by specially designed units [34], probably the result of dissatisfaction with the relatively short life when driven hard of the 480 Eimac 100TS tubes in the ring oscillators and with the time required to re-tune when they were replaced [35].

William Hansen's microwave research before the war at Stanford had been directed toward constructing a linear accelerator for electrons using resonant cavities. The electron's light mass required wavelengths best measured in centimeters rather than meters, if the accelerator was to have practical dimensions. The availability of high-power microwave generators led to a number of such machines being built right after the war. Although priority can hardly be ascribed great significance [36], the Radiophysics Laboratory in Sydney won the race [37] with a design using 25 cm wavelength, quickly followed by TRE [38] using 10 cm. Hansen fulfilled his plan and completed a machine at Stanford [39] only months before his untimely death [40]. Much greater electron energies would be needed before this technique would be of use in investigating the nucleus, eventually done at Stanford and completing Hansen's pre-war goal.

I I Rabi's laboratory at Columbia University had led the world in the atomic and molecular beam experiments to which he had added radio-frequency resonance for nuclear magnetic moments in a static magnetic field, but they had been unable to perform one experiment that was high on their list as being of fundamental importance: accurately determining the hyper-fine structure of atomic hydrogen, the tiny energy difference determined by whether the atom's proton and electron magnetic moments are parallel or anti-parallel. (This is the transition that is responsible for radio astronomy's 21 cm radiation.) From a knowledge of the already measured magnetic moments of the proton and electron, quantum mechanics showed that radio frequencies were needed for which laboratory capabilities in 1939 were just approaching. This restriction fell while the members of the group were away at Rad Lab, and on return to civilian habits they concentrated on hydrogen in studies that were to carry them beyond atomic structure. By the end of 1947 they had measured the hyperfine splitting and had found a discrepancy with the theoretical values ten times their experimental error, but it was the Golden Age of physics, and virtue triumphed as the authors were able to add in the proofs of their article that others had shown the magnetic moment of the electron differed from the expected theoretical value, bringing experiment and theory into happy agreement [41].

The hyperfine splitting of hydrogen was not the only mystery this simplest of atoms had to offer. The rules of quantum mechanics that say yes or no to its changes of energy allow the atom's electron to be trapped in an excited state for periods of seconds from which it can be induced to decay by various external means. Here the electromagnetic effects of a new discipline, quantum electrodynamics, disclosed themselves through experiment in what became known as the 'Lamb shift' after its discoverer, the details of which we must forego. While determination of the hyperfine splitting went on in one part of the Columbia University Physics Department, this delicate little structure was mapped in another [42].

In both of these experiments the magnetic moments of the nucleus and the electron interact with one another and a static magnetic field imposed from without. Changes in the orientation of the little atomic magnets alter the energy of the system and are brought about through stimulation by high-frequency radiation. These experiments were performed on free atoms in space, but there are also couplings of electron and nuclear magnetic moments to external static fields when the atoms are not free in space but locked into solids or liquids. In such systems high-frequency radiation can also produce measurable effects, but here the results did not carry the observers to the dizzying heights of theory but to an analytical technique that would transform chemistry: nuclear magnetic resonance or NMR. In it the proton's resonant frequency measures the total static magnetic field which it experiences, a field that has two components: the field imposed from without by the experimenter and the local field that results from the molecular composition. This latter proved a useful quantity for unraveling molecular mysteries, and many years later in the computer age, it would become a useful medical diagnostic instrument, although cleansed of the dreaded word 'nuclear' or even its initial, becoming magnetic resonance imaging or MRI.

Two investigators invented independently complementary techniques of making these observations. E M Purcell left Rad Lab for a position at Harvard, and Felix Bloch left the Radio Research Laboratory (countermeasures) to return to Stanford. Both had conceived of a resonance formed in flipping the proton in a solid from one orientation in a static magnetic field to another. By the end of 1945 both had concluded a successful experiment. Purcell's observed the alteration of the load experienced by the oscillator when the external magnetic field passed through resonance [43]. Bloch used the vibration produced in the proton to induce an oscillatory signal in pick-up coils so arranged that they received little or nothing from the oscillator directly [44].

None of these experiments, neither the hyperfine splitting, the Lamb shift nor the nuclear magnetic resonance, used klystron or magnetron. Favorite tubes brought from radar were the RCA triodes 2C40 and 2C43, designed as microwave amplifiers but excellent for low-noise work at longer wavelengths. Klystrons did find wide laboratory use, magnetrons much less, although their magnets were encountered performing a wide variety of functions. It is unlikely that anyone expected the magnetron's future to lie in the kitchen.

Molecules are capable of rotating or vibrating as solid bodies, although restricted by the laws of quantum mechanics to discrete energy states. The quantum energies that drive them from one state to another are in the infrared or longer wavelength bands, and some of these transitions fall into the region of microwaves. A transition in water vapor presented itself uncomfortably at Rad Lab as work progressed on the 1.25 cm H2K microwave system. Tests made in spring 1944 gave significantly shorter ranges than had resulted in winter, and the cause was absorption by water vapor in the air. This was not a surprise. J H Van Vleck had predicted in 1942 possible water absorption in this region as well as by oxygen for the 0.5 cm band. The problem was, as pointed out in his report, neither the exact wavelengths nor the linewidths, determined by collisional effects, could be given with accuracy sufficient for engineering needs. The value expected from the much finer resolution of the H2K radar caused development to be pressed despite uncertainties. Data were sought from four different approaches: (1) controlled radar-range measurements, (2) infrared experiments on water lines of shorter wavelength that should have the same degree of collisional line broadening, (3) absorption in a resonant cavity with varying amounts of water vapor and (4) atmospheric measurements with Dicke's radiometer. The result of all this was that the water line was found to be right on the 1.25 cm wavelength. The limited amount of H2K equipment produced was restricted to low-altitude bombing and production was curtailed [45].

Out of Dicke's atmospheric absorption measurements came one of the first experiments of what would be a large part of postwar experimental work on molecules [46], and out of Van Vleck's studies came better understanding of collisional broadening [47]. The water resonance study was the beginning of a very large postwar field. From Oxford came the measurement of a number of resonances between 1.1 and 1.5 cm in ammonia [48], which was to lead to masers in the hands of Charles Townes [49], thence to lasers and an entirely new and rich approach to the study of atomic structure.

The microwave studies of ammonia demonstrated in a dramatic form the difference that a new experimental technique can make. Before the war the literature gave only a single broad resonance of 1.1 cm in ammonia, tediously made with home-made split-anode magnetrons, notorious for their poor frequency stability. Klystrons made the difference [50].

10.2.3. Meteorology

Mariners and aviators had noted early that fog and clouds were invisible to 10 cm radiation but that rain storms showed up clearly, the consequence of the radar reflection of raindrops and hail stones, which increased as the sixth power of their radius. Snow flakes were visible on radar too, but the description was less quantitative. With a bit of experience the plan position indicator proved to be a useful weatherman. The wartime results of US Navy experience were published [51] with a number of PPI photographs of thunderstorms, cold fronts and a striking series of the infamous typhoon of 18 December 1944 in the Philippine Sea. The Radiation Laboratory had not neglected this subject and reported similar studies with both 10 and 3 cm equipment [52].

Meteorologists found radar sets available after the war and added them to the traditional instruments of their craft. One of their first investigations mapped the vertical structure of storms with a transportable 10 cm height-finding set, AN/CPS-4, and reported a layer, called the 'bright band', straggling the 0°C isotherm of stratified clouds, thought to consist of an ice–water mixture, and a column structure, thought to consist of convective rain or hail or both [53]. Others showed the strength of the radar signal to be proportional to the amount of precipitation [54]. The field even attracted one of the radar eminences, E G Bowen, who sought to understand rain formation from ground and airborne observations [55]. Radar was to remain a highly visual instrument of this science.

Closer to the workaday routine of the meteorologist was the use of a radio direction finder, SCR-658, for tracking weather balloons [56] and whose steerable antenna often has caused it to be identified as a radar set. During the war meteorologists frequently made use of accessible radar for this.

10.2.4. Semiconductors

Of the science whose origins can be drawn in some way from radar, the beginnings of the understanding of semiconductors is easily the one that has led to the most substantial changes in human life, yet it was the wartime research that was easily forgotten at the time among the more sensational ones already reported. The crystal detector, the non-linear conducting barrier formed by a metal wire pressed against the surface of some kind of semiconducting crystal, was one of the early means of receiving radio signals. The vacuum diode, followed by scores of vacuum tube types, quickly relegated the crystal detector with its vagaries to a cabinet of discarded equipment or to a child's toy. The predictability of vacuum triumphed over surface conditions difficult to control, much less understand.

As experimenters approached microwaves they found that the electron transit times and inter-electrode capacitances of vacuum tubes made their use impossible and noted that whatever the faults the crystal detector had, they did not extend to transit times and large inter-electrode capacitance. Hans Hollmann utilized it as an element of a regenerative receiver, which he described in his 1936 textbook on high frequency [57]. At about the same time Southworth was studying waveguides at Bell Labs and had incorporated crystal diodes as rectifying detectors [58]. Rectifiers for power circuits were devised during those years using various kinds of layered surface but with indifferent understanding of the basis of their operation.

Microwave radar demanded a mixer for heterodyne reception, and crystal-detector methods were taken from vague memories or from the library. The purest grades of elemental silicon were found to be the best surface on which to place the cat's whisker, but purity was not easy to obtain. At GEC B J O'Kane and G C Edwards had used a silicon–tungsten catwhisker in a coaxial line for their 25 cm work in early 1940, and GEC furnished silicon to other researchers [59]. After experiments by H W B Skinner, British Thompson-Houston Ltd manufactured them as did a section at Rad Lab [60]. Engineers had little patience with the erratic behavior of these early diodes, and basic research was the obvious solution. Henry Torrey at Rad Lab coordinated a program that utilized outside research groups, principally the University of Pennsylvania under Frederick Seitz, who concentrated on silicon, and Purdue University, under Karl Lark-Horovitz, who concentrated on germanium. Silicon became the chosen element because of its temperature stability. The first requirement was extremely high chemical purity, obtained through fractional crystallization; the second was the discovery of the importance of controlled, microscopic amounts of an impurity, boron being found to be the best [61].

Few dreamed of the effect on the world of the subsequent invention of a three-element semiconducting device, the transistor in the 1950s, and none imagined the effect a decade later when the intrinsically small transistor was incorporated into the chip. To attribute all this to radar stretches an historical connection to the breaking point, but radar was a station on its technical progress.

Such was the rich legacy radar left the civilian world. Except for radar astronomy, all would have come into being without radar, for all had begun to grow, however haltingly, before 1939. What radar contributed was the intellectual fire in the postwar investigators. They had made radar. It had been their unique fate, and they acknowledged no limits. They sent men to walk on the Moon.

10.3. SECRECY AND THE TECHNICAL IMPERATIVE

The affairs of men have ever been entangled with secrecy. A poker player does not want the contents of his hand known to the other players nor a general his plan of maneuver known by his adversary. But whereas honorable card players do not try to penetrate the secrets of the others except through the psychological guile that is part of the game, nations consider the acquisition of such knowledge about hostile, potentially hostile or even friendly nations an honest profession. Military intelligence accumulates secrets by various means, sometimes adventurous and romantic but generally dull and repugnant; to the information obtained surreptitiously belong huge amounts of information obtained by tedious study of open sources. In modern military organizations intelligence is the occupation of staff officers at almost every level of command from battalion up; at the top it has become a substantial bureaucracy.

Radar was from the first enveloped in great secrecy by all parties. This was, of course, quite natural and an alternate course seems in retrospect hardly possible, and yet it is not far from the truth to say that there were only a few aspects that were worthy of the severe restrictions imposed. For radar itself was not a secret, amply demonstrated by its being pursued independently in eight different countries in 1939, whose differing degrees of advancement had their origins in the support provided rather than in differing stores of fundamental and clever ideas. All of the important inventions of radar were made separately as the need for them arose: extreme peak-power transmission for pulsed signals, broadband radio-frequency amplifiers, common-antenna usage, lobe switching, conical-beam scanning, triodes for decimeter transmitters, plan position indicators, IFF directly and indirectly interrogated. There were three independently conceived developments of radio-navigation systems that used hyperbolic coordinates: Gee, Decca Navigator and Loran. One might include the German Sonne as a fourth, although it placed the master–slave pair only a couple of kilometers apart and used the nearly radial asymptotic grid lines.

A similar situation applied to countermeasures. The potential of metal foils cut to dipole length for the incident radiation and thrown out by attacking aircraft so terrified British authorities that their use against Germany had to be decided at cabinet level in 1943, yet it had been suggested five years earlier and was well and early known to the three major radar powers and used first by Japan in May 1943—information that did not reach the European theater. It so terrified Göring that he forbade even research in methods to counter it. Other jamming and interference techniques were almost self-evident. On the other hand each radar group thought that it was unique in coming onto radio location, and there does not seem to have been much discussion questioning this assumption.

Keeping things from the enemy begins by restricting knowledge from all friendly personnel 'who do not need to know'. This seemingly logical principle assumes that those who set the restrictions know themselves who needs to know, and it is with this that matters become discordant. It is instructive to insert here Hitler's obsession with this.

1. No one, no office or officer, may learn of a secret matter unless this is absolutely necessary in the line of duty.

2. No office or officer may learn *more* about a secret matter than is absolutely necessary for carrying out the task in question.

3. No office or officer may learn about a secret matter or the necessary part of a secret matter *earlier* than is absolutely necessary for carrying out the task in question.

4. It is forbidden thoughtlessly to pass on orders, the secrecy of which is of decisive importance, according to some general distribution list [1].

Even a casual knowledge of Nazi Germany will have led the reader to the belief that secrecy was driven to extreme there. A radar man recalls the very strict security under which people worked.

> Even the slightest violations were punished extremely hard, so it was best not to know about things, and people got into the habit of forgetting everything they had heard when they left the office. We had a slogan: 'Secret! Burn before reading!'. And it was more than a joke. It was dangerous for unauthorized persons even to know that a secret document existed, especially those marked 'Geheime Reichssache' (secret of national importance) [2].

In Britain the radar research that was carried out for the Air Ministry at TRE, famous for Rowe's Sunday Soviets, dispensed with any 'need to know' rule for those cleared for the level of secrecy being discussed. Not surprisingly the situation in Germany was different. A prominent engineer for Lorenz writes in the company radar history of the situation created by secrecy:

> In the last years of the war there were approximately a thousand [!] different radar projects. This degenerate situation, which was completely out of control, resulted from extreme secrecy and from the orders pertaining to secrecy, which were made by persons having little or no knowledge of the technical situation. This made coordination of the needs of Army, Air Force and Navy with research and development by industry practically impossible.

> Among other things this led various services to request the development and even to purchase for the same purpose completely different pieces of equipment, secrecy covering extremes of independent administration. Immature designs, even failures were covered under this protection. Reciprocal exchange of information was practically impossible except in special cases when it had been 'filtered' through the proper offices, which generally robbed it of technical detail [3].

Especially puzzling was the slowness of the Germans to make use of their excellent universities and technical institutes, which were employed only slightly for war research until relatively late, in marked contrast to the Anglo-American case. The Kriegsmarine prevented GEMA from utilizing them until 1944 on grounds of secrecy [4].

Things were not all that great among the Allies. Production of SCR-584 did not begin until a year after the first order, and it was not until early 1944 that the first set reached combat, delays in part attributed to a misunderstanding of material priorities in the War Department that had its origin in secrecy [5]. But for secrecy London might have been able to fend off the V-1s with effective anti-aircraft (AA) fire from the beginning rather than after weeks of destruction followed by frantic equipment shipments and the hectic training of battery radar sections in its use while the bombs flew. A close-in fighter-control radar for the US Navy, the SM, designed at Rad Lab and based on techniques developed for the 584 went into service six months before the Army set [6] because the Navy Department's assignment of priorities was not so entangled.

The United States showed signs of imitating Germany in the matter of circuit diagrams and instruction manuals when distributing the 584 in the Pacific. The new equipment began to arrive in March 1944, but the manuals were marked 'secret' and for that reason did not reach the troops, so the sets remained in their crates, unused. When knowledge of this situation reached Washington, the Signal Corps dispatched in July H B Abajian, who had worked closely with Getting in designing the set, to provide instruction in the use and the great, but at the time unappreciated, advantages of the new radar [7].

It would have been to either side's advantage to relax secrecy requirements significantly, but such a decision would have to have been made at the very top level of government by someone with Solomon's wisdom and authority combined with an incredibly broad knowledge of electronics. The grip of the conventional and the weight of responsibility were simply too great. Nevertheless, a faint attempt in this direction was made in 1938. The Deputy Director of the Scientific Research and Experiment Department of the British Admiralty saw the article in the *New York Herald Tribune* of 21 March 1938 that had described an approach to radio location; he suggested that thought should be given to the desirability of announcing Britain's RDF capability and, inasmuch as radio location was known in America, of opening exchanges with the American government [8]. An absence of anything further on this tells us how far the 'thought' went.

Modern technical industry is beginning to appreciate this not obvious attitude, as demonstrated by a comparison of the two American centers of technical innovation. Silicon Valley, where there is almost no attempt to retain industrial secrets, has left Boston's Route 128, where industrial secrecy is highly valued, behind [9]. Such an attitude is rejected out of hand by Akio Morita, noted leader and co-founder of the enormously successful Sony Corporation [10], although his reasons seem to rest more on his repugnance with espionage as a way of conducting business and with employee disloyalty rather than from a careful evaluation of the effects.

For radar to be effective it had to be used by troops, and here the cold hand of such rules as The Official Secrets Act with penalties for slight infractions being 'hanged by the neck until life is extinguished' made themselves felt. It is remarkable to read in the memoirs of the commander of an RAF ground-radar unit deployed in France after the Invasion for Army Cooperation Command:

> Generally we did not know the squadrons by their number but by their R/T call signs Potter, Station, Jamjar, Iceberg or Wonder and so on. We answered with our call sign of Bazar but we did not know the pilots nor they us, they did not know of the capabilities of our radar or that we had one, we were just a control voice on the radio. I just hoped always that the orders and information that I gave over the radio had enough ring of confidence for the pilots to believe in what I was telling them. Their lives could depend on it [11].

The free exchange that marked TRE's Sunday Soviets became strangled by regulation and fear as one descended the chain of command, but as noted in Chapter 5.1 this attitude seems to have been common to Army Cooperation Command and may have resulted from policy decisions unrelated to secrecy. By contrast Fighter Command continued its free interaction between GCI directors and AI operators in night fighters [12].

The personnel at CH stations were strictly segregated according to their duties. The radar equipment was within guarded barbed-wire enclosures, and the administrative staff of the station, some of whom were superior in rank to the operators and mechanics and were responsible for the station, were not allowed inside. If there were two kinds of equipment being operated at the same station, crews were restricted to their own set and not permitted to learn about the other. At mess, conversation could not turn to technical matters [13].

It is not surprising to read of even worse restrictions on the other side:

> Even more astonishing, Major Schulze was forbidden in March 1942 by the Commanding General of the XII. Fliegerkorps to speak to the commanders of the air warning units about night fighting techniques. ... This 'Geheimniskrämerei' (keeping shop with secrets) took on such forms that officers, who should have been giving thought to ways of improving the air-warning service, were insufficiently oriented about night fighter use of radar [14].

Things were not much better in the Royal Navy. Derek Howse writes in *Radar at Sea*:

> While the very high level of secrecy about radar in the early days had the virtue of denying information to the enemy, its continuance once radar became operational—and particularly after the outbreak of war—was a great hindrance to the proper use of radar in the fleet. During the first year of the war, the

majority of those at sea were ignorant even of radar's very existence, let alone its capabilities and limitations. And this applied not only to the junior ranks, but to senior officers as well, unless it so happened that they had recently served in a staff appointment where there was a need to know about it. For example, when Rear-Admiral J G P Vivian hoisted his flag in *Carlisle* as Rear-Admiral AA Ships before the landings at Aandalsnes in April 1940, he knew nothing about radar, although it was fitted in three of the AA cruisers he commanded [15].

The extraordinary ineptitude that marked German and British use of radar in the Dieppe Raid was dealt with at length in Chapter 5.3; it had much—although certainly not all—of its origins in the secrecy that enveloped the technique. Command and staff at the levels of Lord Mountbatten and General Montgomery and at comparable levels among the Germans were either ignorant of radar or ignored its potential. Fighting men never realized that they had an enemy in their own ranks named secrecy.

The restraints of secrecy among Allied personnel were much more relaxed in the Pacific than anywhere in Europe. Other than the air war over Germany, no theater made such normal, daily use of radar, whether for navigation, defense against air attack, fire control or in the struggle with or against submarines. Everyone seemed to know about radar and appreciate it, as indicated in a piece of scurrilous doggerel written and quoted in Chapter 5.5 about Admiral Scott. The common feeling was certainly that lots of people 'needed to know'. This contributed greatly to the ever increasing efficiency and ingenuity of its deployment in that theater. This relaxation did not happen from a command decision; it simply grew among the personnel, to a large degree as a consequence of their complete isolation from possible eavesdroppers, but it was not repressed. Indeed by July 1944 this attitude had infiltrated official publications. The introduction of a Navy intelligence document concerning Japanese radar admonishes security officers:

> The widest possible use should be made of the <u>TECHNICAL DATA ON JAPANESE RADIO AND RADAR EQUIPMENT</u>. Although the material is <u>CONFIDENTIAL</u>, that classification should not be a bar to proper accessibility to this bulletin. Measures should be taken to insure availability of the report to Service personnel who may profit by reading it [16].

Secrecy provided a potent weapon for the bureaucratic wars between various services. We have seen how the Kriegsmarine kept DeTe II from the Luftwaffe and even tried to prevent the Luftwaffe from purchasing equipment from GEMA. In Japan the Imperial Army gave specific orders to those working for Ikuta Research Office of NEC not to communicate any radar information to the Imperial Navy. Those responsible for such stupidities no doubt had reasons that seemed valid at the time but they certainly puzzle an outside observer today.

Much was made in the United States by scientists and administrators in the National Defense Research Committee about the Army and Navy being ignorant of each other's radar activities before the war [17], statements based solely on administrative knowledge and that these pages have shown not to be true for the designing engineers, but it was true at command level.

> Colton heard that the Navy was detecting airplanes by radio, but was advised by his colleagues that the project was secret and the Navy would not talk to the Signal Corps about it. Colton apparently did not learn of the work that Hershberger was doing at Fort Monmouth until after the first part of 1935 [18].
>
> At Hershberger's request, Page agreed to forward the brief NRL monthly reports to the Signal Corps Laboratories. The first indication that NRL had achieved success with the pulse method seems to have been provided to the Army by the sudden and unexplained cessation of these reports. This occurred when the project was upgraded from a confidential to a secret classification by the Navy. Apparently the Army was not permitted to receive secret data from the Navy [19].

As demonstrated here, the engineers at the two labs managed to get around these restrictions when the occasion arose. A striking example of this is the Signal Corps use of Page's ring oscillator for the SCR-268 [20]. This clever design allowed several commercially available Eimac 100TH triodes to be formed into a very powerful 1.5 m transmitter. Other designers, world-wide, clung to the push–pull circuit, hence demanding more powerful pairs of output tubes. Page alludes to the transfer as having taken place in 1937 [21]. In another case, when Bell Labs demonstrated the first CXAS, the NRL people wanted it equipped with the lobe switching they knew was used on the SCR-268 but which they had not incorporated into NRL's XAF [22].

With authorities keen on preventing knowledge of even the very existence of radar from escaping, an incident in 1939 is amusing. A German publisher issued annually a pocketbook describing the ships of the world's navies. When the 1939 volume appeared it caused consternation in German radar circles [23] because of a photograph dated 1938 of the Torpedo School Ship G 10 displaying a prominent Seetakt antenna just forward of the foremast [24]. The photograph was passed for publication by various naval authorities, all kept in the dark about the new technique and, of course, unable to recognize the apparent mattress as the mark of a secret weapon. There is every reason to assume that the British naval attaché in Berlin purchased the book and that naval intelligence studied it, but there is no record of them having grasped the significance of the antenna either—very likely for the same reason that the picture had escaped in the first place. R V Jones was unaware of the matter until informed of it by the author [25].

This was followed, as we have noted in Chapter 3.1, about a year later when L H Bainbridge-Bell, a competent radar man, was sent by the Admiralty to examine the wreck of the *Admiral Graf Spee* in Montevideo. He sent a description of the Seetakt antenna only to have its significance lost on naval intelligence in London [26], very likely because of some convoluted demand of security.

Such failings did not mark the scientific intelligence section started by Jones in September 1939. Although immersed in Britain's radar very early, he had not incumbered himself with the idea that radar was theirs alone and began hunting for evidence of it across the Channel long before anyone else thought it remotely possible, supported in his quest by the disclosures of the Oslo report. He finally convinced British officialdom in early 1941 when photographs of a Freya were at hand. When a photograph of a Würzburg-Riese on a Flakturm at the Berlin Tiergarten came to him through the help of the American Embassy in Berlin, he was not long in putting the correct interpretation on it. In this he had the help of a Chinese physicist who had seen it and reported it to Jones—after suffering indignities at the hands of suspicious British officials [27].

The cavity magnetron was one technical secret for which every conceivable reason combined to demand secrecy, yet, as we have seen in Chapter 4.1, it was following the technical imperative just as did other components, although less uniformly with time. It probably eluded American microwave designers because of the hold that the klystron had taken on their thoughts. It probably eluded German designers because of their disappointment in the early years with the poor frequency stability of the split-anode magnetrons—and their failure to read German language engineering journals where the Brown–Boveri work was reported. Randall and Boot had the advantage of knowing nothing about magnetrons—a significant gain inadvertently resulting from the Bawdsey work having initially ignored microwaves!—and going back to fundamental papers by Hertz in their reading.

Concern for the magnetron unnecessarily cost the lives of British fliers. An effective way of reducing the number of Luftwaffe night fighters attacking Bomber Command was to have British night fighters attack their German counterparts as they left or returned to their bases, which could only be done with 10 cm AI because of its reduced ground returns at low altitude. This was foiled through the prohibition of fighters with microwave AI from supporting Bomber Command until May 1944 [28] even though H2S had been in use over enemy lines since December 1942. (Night-fighter support using 1.5 m AI for Bomber Command had been forbidden until August 1943 [29] even though the German use of 50 cm Lichtenstein was known to the Air Force well before.) In fact, although unknown at the time of course, concern for the loss of the magnetron was misplaced because the Germans made remarkably little use of the device when they did obtain it, despite their immediate recognition of its significance and value. On the contrary, it contributed significantly to the splintering of their efforts.

The Germans were just as convinced as the British of the value of their radar secrets. An instructive example of the way an obsession with secrecy can hurt were the measures taken at coastal radar stations after the Bruneval raid by constructing fortifications around them for protection against future commando action. This made every station stand out in aerial photographs, which greatly simplified their destruction before the Normandy invasion. Prior to the emplacement of barbed wire and entrenchments it had been difficult to locate these stations exactly, although approximate positions came from direction-finding techniques. It is a monstrous example of locking the barn after the horse was gone, as there was no need for the Allies to take another set. It is not known whether radar engineers were consulted on taking this step. Not very likely—it was secret!

Given the enormous amount of espionage material shipped from the United States to the Soviet Union during World War II [30], it would be surprising if radar had not had a sizeable component, but only one specific instance seems to be recorded. Julius Rosenberg is reported to have told David Greenglass, his brother-in-law and accomplice spy, that he had stolen a proximity fuze while working briefly at Emerson Radio in 1945 after having left the employ of the Signal Corps [31]. Inasmuch as he was bragging, it seems doubtful that he would have left off something he had obtained on other forms of radar. The absence of evidence of any other covert espionage about radar leads one to think it was minor, to Germany probably zero. It is unclear what the stolen fuze yielded. A most important component of the secret of its manufacture was in quality control, which could not be so easily transferred. Radar did not have high priority in the Soviet Union after the original flowering in the early 1930s, so there were probably no specific demands for radar information as there were for the atomic bomb. And there was little need when the SCR-584 was finally sent as Lend Lease, for it contained about every important radar secret that existed. The loving copy of it as SON-4 and long use tells of Russian evaluation [32].

It is idle yet unavoidable to speculate what secret agents of the radar powers might have procured had they been able to penetrate the electronics research laboratories at the beginning of the war, an occurrence that has left no record in a huge amount of published material, if it did take place. For those using and countering radar, the radiative characteristics of frequency, pulse repetition rate, polarization, beam width and power were sought urgently. These, however, could be obtained by listening, although not always an easy or safe procedure. The ability to resist jamming was the principal design parameter that might have been obtained by a spy.

Designers would have naturally been curious about what the people on the other side of the hill were doing, but it is questionable whether possession of detailed circuit diagrams would have made more than a perturbation on the work being done at any given time of disclosure. Consider what the consequences might have been, if a British master spy had obtained in late 1940 or early 1941 detailed plans for the Würzburg. Bear in mind that its superiority was only beginning to be recognized by the Germans. Months of adaptation by troops and engineers were indicating that it was going to be an excellent set for both Flak and Luftwaffe, but these were properties not yet felt by Bomber Command. In Britain the drawings would have drawn from the many who had little confidence in AA artillery a perfunctory dismissal, overruling General Pile's instinctive positive reaction to it. The Admiralty would have pointed out that they already had decimeter triodes and were well on the way to making their own 50 cm fire-direction set, the type 284, with them. In America, informed through the exchanges established through Tizard, the Naval Research Laboratory would have said that the Bell Labs CXAS, rapidly becoming the FD, was as good or better; the Signal Corps would have still preferred SCR-268 for the same reasons they had declined CXAS except for limited use in coast defense. Thus a tremendous espionage coup might well have been received by design engineers with 'Why do you suppose they did it that way?'. Those planning countermeasures would have noted with interest that the Würzburg—like nearly all radars of the time—had too small a frequency range to resist jamming, and such were the reactions to the Würzburg components seized in the famous commando raid.

The attentive reader will counter that the Japanese valued highly the notes of Corporal Newman describing the SLC radar captured at Singapore. The extent to which they helped the engineers transform it into the Army's Tachi 1, 2 and 4 searchlight and gun-laying radars is not clear. No such manual was obtained for the GL mark II, which was the basis for a somewhat improved copy in Tachi 3.

The most difficult task in dealing with secrecy is the official status that it assumes quite independent of its original purpose. Elaborate procedures are required for the control of secret documents, for the alteration of the levels of classification and for the grading of personnel to limit access; the conduct of this business becomes the concern of a special class of official who guards his tasks and secrets jealously. When the decision to label something secret is left entirely to him the results can approach the absurd. This is self-destructive, but a cure addressed to a national requirement does not allow brash treatment.

Without question the most profitable use of putting secrecy aside came from the Tizard Mission [33]. This extraordinary and unprecedented exchange of information between the United Kingdom and the United States in September 1940, strongly opposed by high-level parties on both sides of the Atlantic, came about through the keen insight of A V Hill and Henry Tizard reinforced by an understanding support from Prime Minister and President, both of whom realized the importance of science to modern warfare and were not timid. America gained the long-sought microwave generator; Britain gained microwave techniques and access to American electronic industry. Perhaps more important was the habit of exchanging secret information even before America was a belligerent. Had there not been a Tizard Mission, the Allies would have won the war but only after a longer struggle, spawning who knows what postwar complications.

The exchange of technical material during the missions of the Imperial Japanese Army and Navy to Germany in 1941 present a barren contrast to the openness of Britain and America, for this Axis exchange was no Tizard Mission. Although the Japanese were pleased with what they learned, their ally told them essentially nothing about radar, although the knowing eyes of Yoji Ito grasped much just from observation, and Japanese radar did not start in earnest until his reports of its importance reached Tokyo. The Japanese, on the other hand, told the Germans nothing about the resonant magnetron nor about how to make good torpedoes, critical knowledge the Germans needed.

A surprising and highly beneficial decision on secrecy came at the end of the war. In a particularly progressive decision the Radiation Laboratory ceased to exist by the end of 1945. This dispersed its great store of talent widespread across the land, filling universities and industry with the latest in electronic knowledge. I I Rabi had insisted that they produce a final technical report and had set grumbling people to work on it. Louis Ridenoir became editor and succeeded in getting the 28 volumes of a microwave-radar encyclopedia declassified [34], making it a technical best seller that went a long way in transforming the national scientific base. Selected volumes were to be found near the workbench or desk of nearly every Rad Lab veteran to remind him of how a problem similar to the one vexing him at the moment had been solved in Cambridge.

All this will bring a wan smile to the face of an official charged with deciding classifications, for it is easy to pontificate when in possession of the knowledge of how everything has turned out. In order to make an administratively wise decision about what to place on or remove from the secret list requires knowledge of what the enemy knows. This is uncertain at best, so when in doubt, clamp down. Two comments are needed as a counter-argument. First, technical secrets differ from other military secrets and should be treated differently than such matters as war plans, order of battle, status of supply or decryption capability. Second, successful military operations have always called for taking risks.

Technical secrets differ from other kinds in two ways. Foremost is the inescapable technical imperative. If you can do it, a competent adversary can too and may even be ahead. Second is the inevitable time delay in putting into practice what is learned from the enemy. A commander may act immediately on learning of the enemy's plans, but a technical chief may require months, perhaps years to duplicate some clever new device—if he thinks it worth doing. During that time one's own secrecy may have held back effective use of the device being so protected.

The prime example of this kind of thing is the resonant magnetron. The Germans made very little use of this super-secret device in the two years they had the opportunity. On the contrary it splintered their already over-burdened resources. GEMA recognized this by ignoring it completely; they had vital air-warning sets to deliver and improve even though their close collaborator, Rudolf Kühnhold, had continued microwave research for a long time. Yet during those years Allied decisions were made to protect AI-10 that cost airmen's lives, even though there was every reason to believe that a magnetron was in German hands. Allied exploitation of the magnetron had required a development and industrial effort that took two years before equipment began to have any tactical effects. Given this experience they must have thought German engineers had super-brains and unlimited resources.

At this point remarks should be made about the atomic bomb, as the degree of Soviet espionage pertaining to it is now widely known. The atomic bomb is, perhaps, one of the best examples of the technical imperative. Physicists everywhere made rough calculations of a uranium bomb almost as soon as they learned about fission. The experiments required to decide whether such a device was possible were obvious. The time between the discovery of fission and the construction of a bomb depended on the effort put into the project and the industrial and scientific resources available. The United States and Britain gave the project all possible support and completed it in less than five years. What espionage unquestionably did for the Soviet Union was reduce the development time, compensating thereby for a later start and for a substantially weaker scientific and industrial base.

The worst effects of technical secrecy are the restrictions on knowledge imposed on one's own forces through strict 'need to know' rules. Unauthorized knowledge of radar was widespread among personnel of the US Navy in the Pacific, and this contributed markedly to the efficient operation of the fleet. This is the form of risk taking that should be borne in mind by those responsible for classification. In this case it was a decision made from the bottom without official sanction. It is well to prevent technical secrets from being gained by the enemy, but when these decisions handicap one's own forces then these restriction are themselves the enemy. Better to let the enemy learn a technical secret from time to time than keep knowledge from those who may be able to put it to use, even the ones some official deems 'do not need to know'. Let the enemy puzzle over what it means, or wonder why it was done that way, or debate whether it was a plant, or argue about what to do with it. Remember the magnetron!

10.4. AN EVALUATION

And what have we learned from this long, detailed story of grand science, dramatic battles and wanton destruction? We have acquired some clarity about the origins of radar, a matter where a substantial amount of error and misconstruction, dare we even say mythology, can be found. We have encountered repeatedly the simultaneous occurrence of an idea independently at various laboratories when a common intellectual basis had been laid; this is hardly a new observation and carries the name technical imperative, but because of the secrecy imposed, radar probably has the richest supply of such manifestations that can be documented. We have also noted that radar owes its beginnings to the large development effort devoted to television by the broadcasting industry. We found differences in approach to prewar radar development but found them not in national or democratic–totalitarian characteristics, not even in widely different cultural backgrounds, but rather in whether the impetus came from engineers at the bottom or from officials at the top. We have encountered many examples that allow us to examine the sterile question about the relative merits of physicists or engineers. We have seen the deformities placed on rational actions and planning by the demands of secrecy. We have examined radar's status as a determining element in the greatest war of history. Finally we have looked briefly at the revolutionary change that radar and radar techniques made in navigation, the most important since the invention of the ship's chronometer. Paired with this remarkable civilian application came a burst of scientific activity based on the electronic skills that the engineers and scientists took back to their laboratories. Let us examine these matters one at a time.

Who invented radar? This question appeared frequently in the years following the war and did not lack for voices giving an answer. It was the subject of many articles, scholarly or otherwise, and was discussed at formal meetings and in countless bench-top discussions. Watson-Watt awarded the title to himself with complete assurance and disputed in print any challengers. Kühnhold certainly thought the title was his. Taylor, Young and Page had equally good claims. But it should be obvious to anyone who has persevered this far that the question really has no meaning; it is a question much like 'which vote decided the election?'. We generally imply by the word 'inventor' one or more persons who by study and experiment discover or produce for the first time a new device. Radar certainly qualified in the 1930s as a new device, but which one of the sets that appeared qualified as first and, if we select one, when did it actually become a radar?

If you begin with the question of who first thought of using echoes of radio waves for locating an airplane or ship, you will find, depending on your outlook, either no satisfactory answer or dozens. Christian Hülsmeyer may not have been the first to have thought of it but he was definitely the first to attempt to do something beyond dreaming. The patents listed by Tuska and Kern [1] alone demonstrate the fertility of imagination during the prewar years in the matter of radio location. The hundreds working in technical capacities on television of the 1930s had seen reflections from aircraft from the beginning of their work. They were impossible to miss with the amplitude-modulated equipment of the time and were as common as airplanes. It would have been a dull engineer who did not connect such observations with the idea that became radar. Had the realization been within the means of a radio amateur, there would have been hundreds of radars developed by 1939, but radar required moderate financial support, so there were only a dozen or so.

Leave aside for the moment the resonant magnetron and consider the meter-wave sets of the 1930s. Each was a composite of standard elements of communication engineering. It only required that a problem be formulated in order to receive the solution, if the solution lay within the capabilities of the time. All discovered within months of one another that transmitter tubes could be driven in pulsed operation to powers far beyond their normal maxima. That an array of dipoles could be made to produce a narrow beam with consequently high antenna gain for meter waves was something already entering textbooks [2]. The basic idea for lobe switching had already been used for methods of radio devices that brought fliers to their desired destination by 'flying the beam'; it required only slight adjustment to provide accurate direction capabilities for locating an aircraft with the primitive radar beams. Methods of protecting a sensitive receiver from the transmitter when using a common antenna depended on knowledge of the position of nodes on a transmission line and proved easy enough in application. The PPI indicator has a long trail of inventors in thought; practice came to TRE first because of the acute need for it in their first GCI equipment. Telefunken's fighter-control method did not require it, so German use came later. Rad Lab saw the immediate need for it in the SG. When circuit designers demanded ever shorter wavelengths, the tube designers delivered transmitter triodes with leads located to minimize the input reactance; Telefunken, GEMA, Lorenz, GEC and Bell Labs all delivered the ultimate tubes within months of each other. Some were better than others, but all provided the hearts of the 50 cm sets that appeared in 1939 and 1940.

The powerful generator of 10 cm waves does not fit so well into this pattern, yet its exception, now known to the attentive reader, actually emphasizes the rule being propounded here. Certainly all who knew of the invention by Randall and Boot of the cavity magnetron cannot think of it in any other way than as their great discovery. Only pedants would deny them the distinction of having invented the cavity magnetron. And yet the Japanese deployed in mid-1942 operational 10 cm radars using it. The Russians published a description of it in the open literature in 1940, and the Swiss company of Brown–Boveri had been working on it since the mid-1930s and first published in 1937. This device was indeed following the technical imperative as did all the other key elements of radar, just not quite so smoothly. Arthur Samuel at Bell Labs and A B Wood at His Majesty's Signal School both recorded designs that might well have led them to the cavity resonator, had their work been pushed.

Phased-array radars appeared whenever designers wanted electric scanning. The technique was first used at about the same time by GEMA and Bell Labs, the former for the 2.4 m Wassermann and Mammut long-range air-warning sets and the latter for the 10 cm mark 8 fire-direction set. Alvarez used the method in a more clever way slightly later.

And so it goes. One could tell similar stories about less significant parts of radar design, but the important part of all this is the independence of nearly simultaneous invention. There is no evidence that any of the parallel work just described was not, of itself, original. Furthermore, if one permits himself the thought experiment of removing any of the principals from our drama—Watt, Kühnhold, Taylor, Runge—or even the lesser characters, does he seriously believe that radar would not have moved forward just as fast? The answer must surely be that the pace would have been the same. The equipment would have been different, perhaps worse, perhaps better, but there would have been plenty of radar prototypes in 1940. That such equipment would have been tactically ready for the crucial battles of 1940–1942 cannot be said with the same assurance.

Radar is unusual in having sprung from civilian roots. Funding for electronic research at military service laboratories during the inter-war decades was grudgingly given until 1935, not even enough to provide the best communication equipment. One encounters careless statements about the negligence of various armed forces in not providing radar in the decade right after World War I. These criticisms are based on imperfect technical and historical knowledge. Until the early 1930s radar equipment was impossible because the components required for ranging did not exist, and radar without ranging was of no significance in World War II. The lack of range information was the reason that the collision-avoidance equipment of Christian Hülsmeyer and Henri Gutton was summarily dismissed by seamen.

Ranging in a useful form required the timing of the echo pulses, and this needed a receiver capable of delivering an output pulse of only a microsecond duration. Such receivers could not be constructed using triodes as amplifying elements; multiple-grid tubes were essential. Ranging also required a method of determining, generally in the presence of extraneous signals and noise, the time between transmission and return of the echo. This required a cathode-ray tube capable of displaying the signal obtained from the receiver as a function of time and fast enough to follow it. Although cathode-ray tubes had existed since the turn of the century, it required the electron-optically focused, high-vacuum tube to do this.

Both of these circuit elements, the pentode, which became the most important multiple-grid tube, and the focused cathode-ray tube, required development beyond the means of the military service laboratories of the between-war decades. They were developed because the requirements of radar were technically the same as those of high-definition television. During the 1920s subscribers to popular magazines about science and engineering read avidly about attempts to provide television. All were based on electro-mechanical optical scanners or in rare cases on the extremely slow gas-focused cathode-ray tubes. Capital flowed to the developers because radio broadcasting and the movies were clearly financial winners, even during the depression, and television was rightly seen to be just what the public wanted. By 1930 pentodes and suitable cathode-ray tubes were available, and by 1935 broadcast television was a reality. The radar men had the components they needed, and radar was also a reality.

(The determination of the height of the ionosphere with pulsed radio waves in 1925 has often been put forward as the basis for a radar set. Those experiments used pulse widths a hundred times longer than those necessary for radar and consequently were not so severely affected by the slow response of the electronics of the time.)

Radar may thus be unique in being a military usage having sprung from a technology developed for civilian purpose. It may also be the only military technology that has been more beneficial to mankind than the civilian use from which it came.

A comparison of the American, German and British approaches to radar is instructive. The similarities of the American and German approaches were emphasized early and have been noted by others [3]. In both nations radio engineers observed the phenomena of echoes and instituted research that culminated in prototype equipment constructed of commercially available components that they built out of meager funds. The first American work was done at the Naval Research and the Signal Corps Laboratories, and both had to use (illegally) funds appropriated for other purposes in their first experiments; their directors had the foresight to recognize radar's value and the courage to take a risk. Both labs worked early with private corporations to widen design and provide for later production. The first German work, although stimulated by Rudolph Kühnhold, a naval research scientist, was funded by private capital in the correct expectation that the government would contribute to the development and purchase the equipment they built. In both countries radar came from the workbenches of the engineers and had to be sold to the high levels of the war apparatus. The American effort found enthusiasm from a navy greatly worried about air attack and an army concerned about its inability to shoot down bombers at night. The Army Air Corps quickly saw a device to protect its bases from surprise attack. The German government's response was positive but with less enthusiasm; radar was and is a superb defensive weapon, and defense was not foremost in the minds of those planning the coming war. In both American and German examples the research costs of the first prototype sets were comparable to the price of a bombing plane.

Britain's approach was the inverse of this. By 1934 some political leaders and many officials of the Air Ministry had become concerned about Hitler's rapid rearmament and especially its emphasis on the air arm, concerns amplified by the knowledge of Britain's woeful weakness and the Air Force's stated inability to stop bombers. With the object of clearing away the numerous naive suggestions of a defense with 'death rays' they went to the Radio Research section of the National Physical Laboratory for advice. They were presented instead within days with a clever plan for radio location. The plan was evaluated by an Air Force officer of remarkable insight, who saw it providing the warning necessary to position defending fighters to break up the formations of attacking bombers; he provided the funds needed immediately and paved the way for all that was required for a complete air-defense system. The result was a national effort that produced the heroic engineering, CH, the only part of Britain's— indeed of democracy's—defense that was ready that fateful September of 1939.

CH was a technical anomaly. It was designed by physicists specializing in long waves, not electrical engineers experienced in the meter waves of television, engineers of whom Britain had many of the very best. But CH worked, and it saved Britain and possibly western civilization in the Battle of Britain, but it was a dead-end design. Its ability to track bombers once they passed the coastline was quite limited, which became particularly onerous when it forced the German attacks into the night. For this the methods already in use in America and Germany had to be independently learned and adapted, and CH had little to offer those designing them. It is idle and mean spirited to criticize something so obviously successful as the CH radar in accomplishing its original goals; nevertheless, it froze the Bawdsey designs for far too long as long-wave solutions for problems that demanded shorter wavelengths. The result was a scramble in 1939 and 1940. In his analysis of structural engineering David Billington noted that providing an engineer with unlimited funds is a guarantee of obtaining an inferior product; the best designs have invariably had tight budgets [4]. A basic fact remains that must silence such criticism: Watson-Watt and Wilkens conceived an effective defense quickly from their knowledge of radio. The plan was adopted and carried through with a will. Would someone else with a technically better plan have succeeded? Billington's criterion does not have a life-or-death parameter.

Much has been written about the dominant role played by physicists in the radar projects. Watt did not want electrical engineers introduced into the work early because he feared their judgement would be tied to the conventional, and he was not alone in this, but there is little to be found in these pages to justify this prejudice. One can say that physicists became first-rate radio engineers speedily and with an enthusiastic verve, but Hanbury Brown recalls in his memoirs how startled he was to encounter the much better techniques and tools of the EMI engineers with whom he worked in 1939 [5]. Until 1940 the American and German radars were designed almost entirely by engineers, who showed themselves in no way devoid of imaginative ideas.

The resonance magnetron of Randall and Boot is a prime example of a physicist's design—complete with sealing wax—and it was certainly not conventional. Another great radar-designing physicist, one active in the field for only two and a half years, was Luis Alvarez. To Alvarez must be applied the term 'radar scientist' rather than radar engineer. Whereas other scientists became superb engineers—one thinks immediately of Ivan Getting—Alvarez remained a scientist. He conceived three startlingly original projects, found for each a capable engineer to get things moving, and left for Los Alamos. But he was a distinct exception. One need only examine the pages of the Rad Lab's 28-volume encyclopedic final report. Those loving descriptions of circuits were written by engineers; they may have been physicists before or after, but at Rad Lab they were engineers. What was true there was also true at TRE.

The doleful price often paid for secrecy does not need to be reiterated, but closely associated with secrecy is the matter of administration. One of the most remarkable properties of the British–American alliance was the free flow of information and the collaboration in planning. If secrecy was baneful in its preventing serving military personnel from learning about the valuable weapon that was at hand, it did not inhibit work at the design and planning level. Top radar engineers gained access to the knowledge of colleagues at other laboratories with moderate ease. Similarly, there was a reasonable attempt to provide a coherent overall development program. True, there was duplication, which with the assurance of hindsight one can say was unnecessary. It is extremely doubtful whether the Allies needed four different groups working on 10 cm gun-laying radars: SCR-584, SCR-545 and GL marks IIIB and IIIC. Yet who could have selected in early 1941 the group that would deliver a 584? Even before America's entrance as an active belligerent there were important Allied technical agreements, the adoption of IFF mark III is an important example; disregard that it was the poorer of the choices available, one that led to its use by the German air-warning service as a range-enhancing secondary radar. At least it was a good try to take control of this horrible and ever unsolved problem.

When the question came up during the war, it was generally assumed that the totalitarian powers had a significant advantage over the democracies in their ability to control their military, scientific and economic apparatus with great efficiency from the top. Radar shows the very opposite. In the beginning German radar outdistanced both Britain and America in advancing excellent prototypes, well engineered for production. This was the result of a small group of designers employed at three electronics firms. These men found themselves relatively unhindered in what they undertook, but once their projects became national ones, the stifling hand of bureaucracy soon 'controlled' everything.

The German administrative organizations that had something to do with the planning and production of radar continued to grow and change throughout the war. It is difficult, very likely impossible, to understand their structure now. The Nazi bureaucracies were jealous of one another and fought—even as disaster was obvious—for control. The result was serious duplication and fragmentation. Germany never had a common IFF standard between the services. The muddle created by finding a cavity magnetron in the ruins of a Stirling bomber in February 1943 worked almost as a secret weapon for the Allies, for no coherent plan of exploitation can be extracted from the Rotterdam Protocols or anywhere else, only greater fragmentation and wasted effort.

Did radar win the war for the Allies? In a way this question is just as meaningless as 'who invented radar?'. Wars are won by determination, resources, manpower, leadership, courage and luck. Among resources are weapons, and superior weapons have had remarkable effects in many wars. The democratic states, Britain and France, certainly lacked determination during the years Hitler built his war machine. The appeasement at Munich left France relieved that war had not come but impressed the British that they had suffered a serious and humiliating defeat that transformed the whole nation to the same spirit that the radar men had held for years. Thus when the island later faced the monstrous tyranny alone, they had more than just spirit. They had a working air-defense system of the kind never seen before and never even imagined by other air powers. Did radar win the Battle of Britain? No, but it would have been lost without CH. This statement is as categorical as any one can make in historical matters. The battle was extremely close; radar made more than the difference. What would have been the fate of western civilization had Hitler won? At that point thoughts become decidedly morbid.

The crucial value of radar for Britain in the Mediterranean during 1941–1942 has been discussed at length in Chapter 5.1 with the conclusion that it provided the method by which supplies to Rommel's forces were seriously restricted. This prevented him from taking Suez and putting an end to British naval control of the eastern Mediterranean. German control of the Middle East meant control of Britain's oil and the interdiction of supplies to the Soviet Union. Japan's leaders might even have considered pushing on to link forces with their ally. The consequences of this are less dramatic than the invasion and defeat of Britain but nevertheless dire. It probably would have happened had the Royal Navy and its Fleet Air Arm not had radar.

Similarly, the importance of radar to the American fleet in 1942 has been treated in Chapters 5.4 and 5.5 with the conclusions that the American carriers would have been highly vulnerable to air attack without the CXAM that gave them the warning needed to clear planes from the deck and secure explosives and fuel. Holding Guadalcanal required the continued functioning of the Cactus Air Force, possible only through the timely warnings of their SCR-270 and 268. Had the United States been defeated in the Pacific in 1942—and it was a close thing—it is doubtful that America's Germany-first policy could have been retained. The reader can supply the plot for his own drama.

These are three major engagements in which radar kept the Allies from defeat. Such a statement can always be contested because there is no way to prove it, but it is as near to a 'truth' as history allows. It was certainly radar's heroic period. It was also the meter-wave period.

After 1942 radar became increasingly routine for all the contending forces, taking on an importance shared with other weapons. To have fought without it would have been the same as removing quick-firing artillery or machine guns, but it did figure critically at times.

The Battle of the Atlantic was certainly crucial for the Allies, and the defeat of the U-boats has frequently been credited to radar, often in statements that allow no contradiction, but the conclusion reached in Chapter 7.1 does indeed contradict them. The submarine lost when merchant ships traveled in convoys protected by adequately equipped and trained escorts with air cover over their entire routes. Radar was useful, but useful only to the same extent as were other Allied technical advances and probably less valuable than ship-borne high-frequency direction finding. The basis for the decisive importance ascribed by some to microwave radar lies in graphical displays of merchant-ship sinkings as functions of time. The decline of these losses after the introduction of the equipment in patrol bombers is impressive, as is the counter-diagram showing the increased sinkings of U-boats, but these plots are deceptive: correlation is not causality.

The failure of the Allied invasion of Normandy would have certainly stopped the Allied path to victory, which seemed so steady after 1942, and failure was a definite possibility. Two key elements in providing success were the remarkable experience the Allies had gained by spring 1944 in that most difficult of military operations, an amphibious landing on a hostile shore, and the near absolute command of the air over the beaches. Radar's part in this unprecedented operation was important, but declaring it vital or not, as has just been done for the other engagements, is more difficult. The radar-like devices, Gee and Decca, that allowed the entire fleet to navigate under difficult conditions can be given a substantial credit for success. For boats to arrive more or less on time and at the right place would have been one of the most difficult aspects but for Gee. The extremes of countermeasures plus the destruction of Seetakt and Freya sets in large numbers certainly made things easier for the attackers and confused the defenders during crucial hours. Radar beacons made air drops at night decidedly more accurate. Would the invasion have failed without radar? It is not so easy to say. There was a big force off shore with a strong determination not to fail.

One of radar's great moments came in the defense of London against the flying bombs. The Allies would not have lost the war had the SCR-584, the M-9 director and the proximity fuze not been ready, but London would have suffered great destruction.

The ferocity of the Pacific War reached a peak at Okinawa with the attacks by suicide pilots on the fleet supporting the invasion. This became an extreme test for radar fighter direction and radar-directed AA guns. It required a complete new concept for the former and proved difficult for the latter. It is conceivable that the invasion might have been repulsed had the fleet not had radar. It would not have cost the Allies the war, but it would have made the last year much more bloody.

The air offensive against Germany has been called here the Great Radar War. It has taken up more words than any other portion of this book because its radar activities were so varied and extensive. In the wide use of radar only the Pacific War can be compared with it, but the Pacific War, although radar became a standard component of the two contending forces, was fought without a countermeasure struggle comparable with what was found in the air over Germany, primarily because Japan was so completely outclassed. Radar shaped the struggle over the Reich, essentially determining its course.

We have seen how initially the Luftwaffe was unable to down attacking bombers at night and the RAF was incapable of even finding the targeted towns. Ground radar continually improved the defense and the radar-like device, Gee, allowed target location, although not accurate blind bombing. Airborne radar began to make night fighters really dangerous, and Gee was replaced by Oboe, a blind-bombing device as accurate as the best visual. Gee and Oboe were both limited by the distance to the horizon and so could not guide deep flights, a failure for which the microwave H2S was to provide a solution. The solution allowed by H2S was saturation bombing.

With the introduction of H2S at the beginning of 1943 Allied radar became technically superior to German, but it was a superiority that the bomber crews could hardly appreciate because the Luftwaffe made good use of radar's strong bias toward the defender. By March 1944 deep night attacks could no longer be sustained. Radar had helped the Luftwaffe dispel the delusion of penetration through the cover of darkness as months earlier they had dispelled the American delusion of penetration with self-protecting formations. Germany won the first round of the air offensive; when the second round began in late 1944, dominated by the long-range fighter, radar was much less important.

In some theaters radar bordered on being irrelevant. It figured not at all in the Blitzkriegs east and west that initiated the European war. Its importance in the Great Patriotic War fought by the Soviet Union was minor, except perhaps in the air defense of Leningrad and Moscow. The Japanese used radar to observe the swarms of attacking bombers that systematically destroyed their cities, but they no longer disposed of an effective air-defense force. Radar bombing became the rule in destroying Japanese cities and was much more effective than when used against Germany because the weak air defense allowed bombing from half the altitudes to which the RAF had been forced by German flak. The location of so many targets on the coast, just where H2X provided the best PPI views, and the inflammability of the cities made a terrible difference.

By the end of the war in Europe radar was transforming the nature of ground warfare. The SCR-584 was guiding dive bombers onto targets with the efficiency earlier found only in fighter control. Its automatic tracking, something a small group at Rad Lab had to fight for during initial planning stages, was even found to determine the positions of artillery by tracking the projectiles. Its ability to locate those deadly but elusive infantry mortars gave rise to a special set for that purpose that was better employed near the front, although deployed very late in the war. The 584 was even used to observe the movement of vehicles covered by darkness and fog.

It is obvious that radar transformed the nature of war more than has any other single invention. It turned the entire concept of strategic bombing on its head, replacing the dominance of the bomber with the dominance of the fighter and the AA gun. Reality and radar had quickly disposed of the air power fantasy of a short, decisive 'knockout blow', and a war of attrition was fought not in the trenches of Flanders but in the air over Europe with the same result as in 1918—the end came when Germany was exhausted. In forcing this transformation radar also helped move the conflict to a more barbaric level: trench warfare for civilians.

The changes imposed on fleet action were equally drastic. By providing aircraft carriers with a 20 to 30 minute warning it removed the fragility that had restrained plans for their employment and allowed them to participate in, indeed dominate major engagements. Without radar the Pacific War would have been a battleship war. Carriers and radar-directed guns with proximity fuzes had reduced to nothing Mitchell's prediction of the demise of surface ships to land-based air power.

In August 1945 the atomic bomb upstaged an announcement planned of Rad Lab's contribution to the Allied war effort, displacing the cover-page article in Time magazine to page 78 and prompting Lee DuBridge to make the oft quoted remark: 'The bomb may have ended the war, but radar won the war' [6]. Hoyt Taylor expressed it slightly differently, perhaps more accurately and in keeping with his personality: 'The bomb finished the war, radar fought the war' [7].

APPENDIX A

A FEW RADAR ESSENTIALS

Radar is the technique whereby the position of some object is determined by illuminating it with radio waves and observing the reflections. It is a highly technical subject requiring a large dose of electronics, physics and mathematics for a thorough understanding, but the basic ideas and conceptions are not all that hard to grasp and can be understood by a layman having a science education of the kind that should be gained in a good high school. The purpose of this appendix is to introduce a number of ideas necessary to the key elements. It can be skipped by many without loss, read selectively by others or studied carefully by those with little electronic knowledge. There will be brief portions of the text outside the explanation of this little essay. They are there because their omission would irritate the technically trained (for whom the book is also intended) and may be skipped without significant loss.

A.1. ELECTROMAGNETIC WAVES

Radio waves generated by various activities of civilization impinge upon us wherever we go and affect our lives to a degree that their discoverer, Heinrich Hertz, would have found incomprehensible. They are waves of co-dependent electric and magnetic fields whose motion shares some of the propagation characteristics of mechanical waves such as sound, seismic waves or waves on the surface of a liquid. They differ in requiring no material medium for their propagation and in this were the cause of a major crisis in physics at the turn of the century, resolved by the special theory of relativity. They are the same as mechanical waves in having a simple relationship between wavelength (the distance between two adjacent crests of the wave), frequency (the number of crests that pass a given point per second) and the velocity of propagation. For radio waves this velocity is the velocity of light; they differ from light in wavelength and in the inability of human senses to perceive them. Electromagnetic wavelengths extend from the extraordinarily short (gamma rays) to radio waves of hundreds of kilometers—the spectrum reproduced in countless textbooks.

The velocity of electromagnetic waves in a vacuum is constant and slightly slower when traveling through matter, such as the atmosphere. The frequency is not so affected, being the number of crests passing a given point per second, but the wavelength is consequently shorter in matter, and for this reason electrical engineers adopted frequency as the method of specifying a wave. Units of frequency were originally given as cycles per second with the standard prefixes of the metric system for thousand cycles per second (kc) or million cycles per second (Mc). The German practice of naming the cycle per second after Hertz has been adopted internationally, so we now use kilohertz (kHz) and megahertz (MHz). Wave velocity is the product of wavelength and frequency just as a person's velocity is the product of the length of stride times the number of strides per unit time. The velocity of light is approximately 300 million meters per second, so to obtain the approximate wavelength in meters from the frequency divide 300 by the frequency in MHz; to obtain frequency divide 300 by the wavelength in meters.

Wavelengths have from the beginning been measured in metric, not English units. For radar the wavelength is more useful to us than frequency because the interaction of waves with antennas and targets, which is strongly dependent on it, allows more convenient comparison and was the preference of most World War II radar men. For this reason wavelength is used rather than frequency throughout the book with two units of length sufficing for our purpose: meter (m) and centimeter (cm).

All waves transport energy, mechanical waves by the motion of particles, electromagnetic waves through the electric and magnetic fields that constitute them. The power radiated by a radio transmitter is measured in watts (W); power also uses metric prefixes to allow convenient sized units; MW, kW and μW (microwatt) will suffice.

When any wave is incident on a boundary between two different materials, e.g. air and water, part of the wave's energy is reflected and the rest transmitted. Sometimes nearly all is reflected, as with light on a mirror; sometimes most of it is transmitted, as with light on a pane of glass. For the waves used in radar, conducting surfaces are good reflectors; other materials also reflect but not to such a high degree and according to material characteristics that are not important to us. If the wavelength is much shorter than the size of the reflecting surface, reflection will take place as commonly observed with mirrors, called specular reflection; if the wavelength is comparable to the size of the reflector, the waves will be emitted in a wide distribution of directions. In the early days of radar some theorists argued that specular reflection would prevent wavelengths of a few centimeters from being useful in radio location, but this did not prove to be the case because of the irregularities of airplane and ship surfaces. When jet aircraft, which had neither propellers, large external motors nor radiators, were introduced, they produced much smaller radar signals for microwaves (10 cm or shorter) than for meter waves.

When separate trains of electromagnetic waves encounter one another their electric and magnetic fields add vectorially. Two waves of the same wavelength that are in phase, i.e. that have their wave crests moving together, increase their combined amplitude; if they are out of phase, they decrease their combined amplitude. This property, called interference, is particularly important in shaping the radiation patterns of radar equipment.

A.2. REFLECTED SIGNALS

All radar sets function by sending out a signal to encounter a target. This radiated signal loses intensity just as does light leaving an incandescent lamp, in proportion to the square of the distance. The signal reflected from a target is, of course, a very small fraction of the transmitted signal, but worse yet, the reflected signal is subject to the same law of propagation as the transmitter. Put together, this means that the reflected signal received by the radar receiver decreases as the fourth power of the distance. The only reason that radar is possible for any reasonable range given such diminution, is that transmitters can be made that radiate hundreds of kW and receivers that function with small fractions of a μW.

World War II radar sets operated in three modes: pulsed, continuous wave and frequency modulated. In the former the radio frequency signal produced by the transmitter has a duration that is short enough to form a wave train tens or at most hundreds of meters long. The speed of light is such that a train 300 m long is formed during one microsecond of transmission (μs). When operating in pulsed mode the time between pulses is usually set so that the round trip time is enough for the maximum range desired. The higher the pulse repetition rate the more reflected signals will be available for analysis.

Continuous-wave radar is much less often used than pulsed radar because distance information is more difficult to extract. It is excellent for measuring the speed of the target and as such has become the favorite of traffic police. If the target is in motion the wavelength of the reflected wave will be altered by the well known Doppler effect, noted acoustically by the change in pitch of the whistle of a passing locomotive or optically by the shifting to the red in the spectra taken of distant receding galaxies. When the reflected radio wave is mixed with the wave emitted by the transmitter, a new wave is formed, the familiar beat note of music. This is the easiest kind of radar set to build. Speeds of automobiles were measured this way in 1932.

If one modulates a transmitted signal so that its frequency changes with time, the reflected signal will return with a different frequency than the one being dispatched at that moment. Mixing the two frequencies will produce a beat frequency equal to the difference between the reflected and the momentarily transmitted frequencies, and this beat frequency is easily related to the distance of the target. This kind of operation is continuous, so separate antennas for transmission and reception are required. Obviously, motion of the target will produce a Doppler shift, which has generally restricted the use of frequency-modulated radar primarily to radio altimeters.

Some radar targets do not wish to escape detection and furthermore want the radar operator to know their identity. In war it is vital to recognize your own ships and aircraft. In peacetime a controller wishes to know the identity of the blip as well as other important pieces of information, such as its precise altitude (generally difficult for ground radar to determine), speed, remaining fuel and the like. This is accomplished by equipping the plane with a radio transmitter that is actuated by the interrogating radar signal or auxiliary signal. It responds on yet a different wavelength for which the radar set has a separate receiver and to which it sends the desired information in coded form. In war when enemy targets show no cooperation whatsoever, and when enemy intelligence becomes overly inquisitive about these signals, this gets to be a very messy problem. The signal transmitted by the responding set, the secondary radar, is much larger than the unaided reflected signal, allowing greater range.

A.3. ANTENNAS

An antenna is a metal structure that either transmits or receives electromagnetic waves. An antenna intended for transmission is equally good for reception. Hertz invented the first antenna, which is widely used in the radar equipment described here. It is called the oscillating dipole. Two metal rods, each a quarter-wavelength long, lying on the same line are connected at their common center to a high-frequency alternating current generator, called an oscillator. The oscillator forces currents of opposite polarity to flow in the two halves of the dipole, and these currents cause it to radiate. A wave that is incident on the dipole will cause currents to flow to the receiver that is connected to it. There is a glut of antenna designs, but the oscillating dipole suffices for most of the radar equipment we shall be discussing.

The radiation pattern of a dipole is at a maximum in the plane perpendicular to the axis and evenly distributed around it. The pattern reaches zero at the poles. This is satisfactory for broadcasting but not for a radar set, which needs a directed beam. A radio analog to a searchlight was the goal of many, although not all designers.

The first step in producing a directed radiation pattern is to place a metal surface on one side of the dipole and parallel to it. This produces a radiation pattern only on one side of the dipole but one that is distributed over 180 degrees. Placing dipoles in an array side by side allows them to interfere in such a way as to build up the radiation in the forward direction and to diminish it on the flanks. The greater the number of dipoles in the

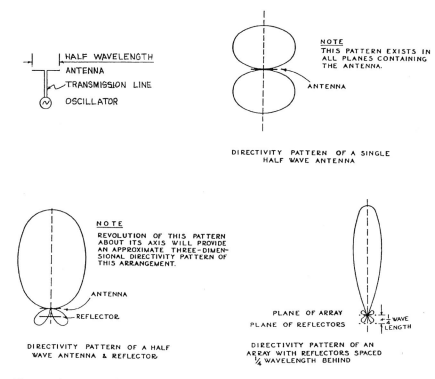

Figure A.1. *Dipole radiation pattern. The drawing on the upper left shows schematically the basic elements of a dipole antenna. It has two halves, each one quarter wavelength long that are connected at the center to a high-frequency alternating current generator. In practice the dipole is connected to the generator by means of a transmission line. The drawing on the upper right shows the radiation pattern of a dipole radiating into free space. The solid curve shows a constant value of the intensity and merely indicates the shape; the pattern extends out indefinitely. The maximum intensity is in the plane perpendicular to the dipole. The drawing on the lower left shows the radiation pattern of a dipole that has a reflector on one side. The drawing on the lower right shows the radiation pattern of an array of four dipoles placed side by side with reflectors. These kinds of pattern are called lobes.*

array, the tighter the radiation pattern.

Another method of producing a narrow beam is to place the dipole at the focal point of a curved reflecting surface, such as a paraboloid symmetric about the beam axis. This is the structure of a searchlight: the metallic paraboloid being replaced by a mirror of the same shape, and the dipole replaced by an electric arc. This is the form of antenna that is often used for microwaves, waves of a few centimeter wavelength. It used to be commonly seen at airports but with only a horizontal strip of the paraboloid, an arrangement that produces a vertical fan-shaped beam that will be able to interrogate aircraft at a wide range of altitudes. The same was long seen aboard ships, where the function was to locate other vessels or land forms regardless of roll and pitch. The paraboloid is generally being replaced by a linear array of microwave radiators, as is discussed in Chapter 4.5.

A.4. LOBES

The radiation pattern produced by the tricks just described is called a lobe. It determines the direction to a target with an accuracy limited by its angular dimension. One notices that the forward end of the lobe is rounded, which means that slight changes in the direction one points the beam make small changes in the energy incident on or received from the target. This is even more serious than it seems because aircraft targets change the amount of energy reflected with changes in the aircraft's orientation—they twinkle—making the determination of the direction of maximum signal difficult.

If one places two arrays of dipoles side by side but with slightly different directional orientations, two slightly overlapping lobes will result. The fronts of the lobes are blunt but the sides are steep. If the target is illuminated alternately with one lobe or the other, each lobe will yield a reflected signal, generally of different amplitude because one originates nearer the center of its lobe than the other. If the antenna is positioned so that the reflected signals are equal, then the target will lie on a line bisecting the angle that orients the two arrays. Remarkable angular accuracy can be had this way. In a radar set this technique is called lobe switching. The same idea was used during the 1930s for air navigation.

If a parabolic reflector is used, the same effect can be had by mounting the dipole slightly off axis and causing its position to rotate about the beam axis, producing what is called a conical scan.

A.5. VERTICAL LOBE STRUCTURE

The examples of lobes just reviewed assume that the antenna radiates into free space, which may be the case for a radar pointing upward to direct a searchlight or an antiaircraft gun or for airborne radar at high altitude, but if the radiation pattern is projected horizontally over land or sea a complication arises. The lower half of the radiation pattern will be reflected off the surface, which is an excellent conductor, and will interfere with the waves that come directly from the antenna to produce a vertical lobe pattern.

The consequences of this vertical lobe pattern are both detrimental and beneficial. Consider an airplane flying horizontally toward the radar set. At some point it will encounter the lowest lobe, from which it returns a signal. As it flies on it enters the region between lobes and can disappear completely from the radar screen only to reappear when it enters the next

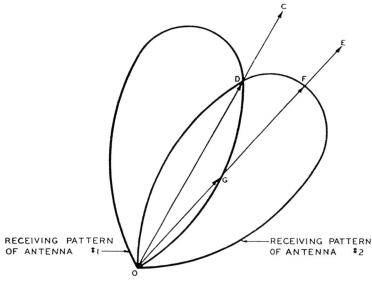

RADIO DIRECTION FINDING WITH TWO
ANTENNAS DISPLACED WITH RESPECT TO EACH
OTHER.

Figure A.2. *Lobe switching. Accurate direction can be obtained with the blunt lobes illustrated in figure A.1 by making use of the rapid change in intensity on the flanks of the lobes. The drawing shows the effect of a transmitter that radiates alternately pattern 1 and then 2. A target in the direction C will produce signals of equal amplitude (D) for either lobe, whereas a target in direction E will generate a larger amplitude (F) in pattern 2 than in pattern 1 (G). The radar operator adjusts the direction of the antenna so as to equalize the two signals.*

lobe, and so on according to the number of lobes. This is naturally confusing to an inexperienced operator. In principle it can be used—and was used successfully as operators became skillful—to estimate the height of the plane, which is almost as important a coordinate for an air controller as the plane's horizontal location. But over land this is complicated by the ground not being uniform. Even for the important case of observing the arrival of aircraft approaching over the sea, height estimation requires many calibration flights by test aircraft flying fixed courses. The interpretation of such data makes as much use of intuition as of science. But for many sets during the war it was the only way of determining the height of the target. Some long-wave sets were equipped with alternate antennas placed at various levels in order to switch the vertical lobe pattern to help the operator observe the target between lobes. The problem of the vertical lobe pattern is primarily one for long-wave equipment.

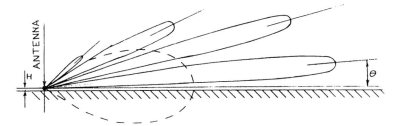

Figure A.3. *Vertical lobe patterns. The radiation patterns of figure A.1 are for antennas radiating into free space. If an antenna has its beam pointed parallel to a conducting surface, such as the ocean or very flat ground, and is located at a height H above it, the lower half of the pattern is reflected and interferes with the direct radiation, producing a pattern of vertical lobes. In the diagram the dashed curve represents the pattern in free space; the solid curves the first four lobes of the resulting radiation. The reader can easily imagine the confusion for an untrained observer when an aerial target approaching from the right disappears momentarily as it leaves the first lobe and proceeds to the second. The angle of the first lobe relative to the antenna axis (in radians), Θ, is approximately equal to the wavelength divided by four times the antenna height above the surface.*

A.6. THE YAGI ANTENNA

If one places a conductor that is slightly longer next to a radiating dipole, this parasitic electrode will function as a reflector. If the parasitic electrode is slightly shorter than the dipole, the effect will be to enhance or, as generally termed, direct the signal on that side. Hidetsugu Yagi utilized this in 1928 in a combination of elements that bears his name in America but is called the Yagi–Uda in Japan. In it an oscillating dipole, called the driver, has a reflector on one side and one or more directors, each somewhat shorter than its predecessor, on the other side. The result is a directional radiation pattern. The voltage at the center of all these elements is zero, so the reflector and directors can all be mounted on the same metal support at their centers, a matter of great convenience. Unlike the dipole arrays described, the theory of the Yagi is difficult to apply and inaccurate, so empirical design is the rule. It has come to dominate long-range, very-high-frequency communication and is almost the only exterior antenna for television. Many varieties are encountered.

A.7. ELECTRON TUBES

For anyone with an electronics knowledge acquired since 1970 the electron tube, also called the vacuum tube, is an archaic device that has been almost completely replaced by the transistor and its multifold descendant, the integrated circuit or chip. The radar story of World War II requires some understanding of tubes, or valves as they are called by the British, for the

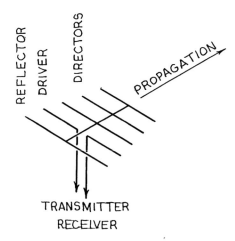

Figure A.4. *Schematic of a typical Yagi antenna. The driver is connected to the transmitter by a balanced transmission line. To its left is a metal rod somewhat greater in length than half a wavelength; to the right are directors that are somewhat shorter. The exact lengths and spacings, which generally lie between 0.1 and 0.25 wavelengths, and the number of directors are determined by experiment. The radiation pattern is strongly directed to the right.*

transistor was a postwar invention, one which had origins in radar work.

In experimenting with incandescent lamps Edison noted that current flowed to the hot filament from an adjacent electrode, otherwise stated, electrons flowed from filament to anode. John Fleming noted that the application of an electric potential between the filament and this second electrode resulted in current for one polarity of applied voltage, none for the reverse. He used this as a sensitive detector for early wireless and named it the diode. By passing current only in one direction it formed a variable direct current that could be heard in the operator's head phones. It took the form of a hot central filament, the cathode, surrounded by a sheet-metal second electrode, called the anode or plate. Lee De Forest inserted a grid between the two and found he could control the stream of electrons from the cathode to the anode, which allowed signals to be amplified. His invention, the triode, transformed radio. Triodes presented circuit designers with problems as they pushed to ever higher frequencies and shorter wavelengths. These problems were solved to a great degree by the introduction of one or two more grids between the control grid and the anode, tubes called tetrodes and pentodes. Tubes, along with components already in use, such as the resistor, capacitor, inductor and transformer, allowed the design of a huge number of electronic devices by 1939, generally for radio and audio amplification.

A.8. TRANSMITTERS

Transmitters are devices for producing an alternating current, such as one uses for household electric power, but at much higher frequencies than the 60 Hz of America or the 50 Hz of Europe. The wavelength of a 60 Hz oscillator is 5000 kilometers (km), hardly suitable for radar. A transmitter for communication by Morse code, a common application at the beginning of the radio age, generates power at the desired wavelength whenever the operator depresses the telegraph key. In early radio broadcasting the amplitude of the radio frequency signal varied according to the magnitude of the audio signal, the amplitude modulation that remains today on the AM band of radios. Television of that era also used amplitude modulation, so that the signal intensity varied according to whether the picture was to be bright or dark. The radar transmitter for the pulsed mode requires abrupt changes from zero to maximum signal and to zero again, design requirements in common with television transmitters of the time.

The final tubes in a transmitter, the output stage, must be the most robust, as high voltages and large currents are demanded of them. The early radar designers all quickly learned that these tubes could be driven to much more than their nominal power for the few microseconds required to form a pulse, because the limit of a tube was its ability to dispose of the heat formed when the electrons struck the anode, energy not radiated by the antenna. During the short period of the radar pulse the anode rose in temperature rapidly but had a relatively long period to cool by thermal radiation.

A.9. RECEIVERS

A radio receiver is a sophisticated instrument, but what we need to know here is relatively simple. Its function is to select the signal of interest from the jumble of wavelengths picked up by the antenna and to amplify it to a useful level. The input of a receiver is some device that resonates at the frequency of the signal to be amplified. For Morse code a very selectively tuned detector is used because the sharpness of its tuning suppresses the amplifier and antenna noise that is not at the resonant frequency, and the sharp tuning presents no problems. There is a problem in using such a receiver in radar, one that was discovered quickly by all the early experimenters who were not experienced enough to avoid it.

Consider this analogy. A guitar is a resonant device, so is a banjo. Pluck a guitar string and the tone continues for a some time; pluck a banjo string and the tone quickly disappears. The technical term for this difference is damping: the guitar has low damping, the banjo a great deal. There is a lot of wonderful theory about all this that is based on the mathematics of Joseph Fourier, whose work is thoroughly studied by electrical engineers. We shall, however, be satisfied with mentioning the concepts and deal with the results qualitatively.

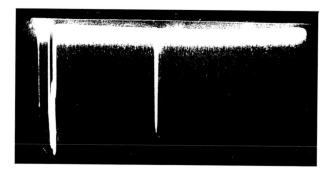

Figure A.5. *The simplest form of radar indicator, the A-scope. The drawing shows the appearance of an oscilloscope screen in which the horizontal trace begins at the left when the transmitter is pulsed. The spot moves across the screen with the output of the receiver applied in the vertical direction. Generally a large pulse is recorded simultaneously with transmitter output and target pulses occur at various times along the trace. The pattern repeats at the pulse repetition rate of the radar set, usually hundreds or thousands of times a second. This repetition is very useful in allowing the operator to examine carefully the very weak signals.*

When a short wave train is incident on a sharply tuned receiver, it responds much as does a plucked guitar, setting up vibrations that can hide subsequent plucks that are weak. Because of their damping banjos are better suited for fast tunes than guitars. For radar one needs a receiver that has the right amount of damping. It must resonate enough to select the radar frequency but damp the vibrations so they will not obscure the next signal. Engineers prefer to speak of 'pass band' rather than 'damping'. A communication receiver has a narrow pass band (sharp tuning), whereas a radar receiver requires a wide pass band, as does a television receiver, in order to follow the rapid changes of amplitude. The shorter the radar pulse, the wider must be the receiver pass band.

A wide pass band comes with a price: the wider the pass band, the larger the receiver noise, with consequent difficulty in observing small signals. This leads to the curious result that the maximum range of a radar set is not enhanced by decreasing the pulse width, thereby utilizing the increased amplitude of the transmitter that accrues from shorter pulses, because it requires a wider receiver pass band and with that comes enhanced receiver noise; the two effects work against each other.

A.10. INDICATORS

The information gained with a radar set must be made available to its operators. The crucial element in this is the cathode ray tube because it allows one to measure microsecond time differences. Such tubes are the display elements of television receivers. In one a highly mobile electron beam

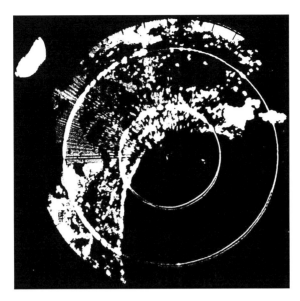

Figure A.6. *The plan position indicator (PPI). This is the most widely recognized radar indicator. It is a maplike display of panoramic sweeps of the radar beam. The intensity of the cathode-ray beam is modulated to show the presence of a target in the same manner that a television image results from beam modulation. The azimuthal direction of the outward trace of the spot follows the orientation of the antenna. The example is of Palermo harbor made with a 3 cm SU-2 surface-search shipborne radar. Rings for range are shown at 2 and 4 nautical miles. The dark region to the lower right indicates no reflected signal because of the signal's reflection from water; the dark splotches over the land indicate regions hidden from the line of sight.*

produces a spot of light on a phosphorescent screen that can be moved according to the amplitude of electric signals applied to electrodes of the tube. In the simplest radar application, a voltage that increases linearly with time (a time base) deflects the spot horizontally and the output of the radar receiver deflects the spot vertically. The screen shows a horizontal line trace with little vertical blips when the receiver picks up reflections. The spot begins its motion across the screen at the time the transmitter pulse is emitted. By knowing the speed of the time base, which can be determined by electronic means, the range to any of the blips can be determined. The size of the blips depends on the reflecting power of the target and its range. This form of indicator is called the A-scope.

Various combinations came into use for displaying a radar set's output. Some used cathode-ray tubes with horizontal and vertical inputs, others with circular traces in which one signal moves the spot along a radius from the center and the other moves it in a circular path around the center. The form of display that is most readily identified by the layman is the plan position indicator (PPI) in which the screen gives a maplike representation of the region interrogated by the radar. In it the time base moves the spot outward from the center of the tube face and a circular motion follows the rotation of the radar antenna. The tube's electron beam is suppressed when there is no signal from the receiver and enhanced when there is. This results in a glowing phosphor when there is a target and darkness otherwise.

NOTES AND SOURCES

Citations given in capital letters are from the list of abbreviated references; those in upper and lower case are from the general references.

Chapter 1.2 Electromagnetic waves

[1] James Clerk Maxwell *A Treatise on Electricity and Magnetism* article 771. Oxford: The Clarendon Press, 1892

[2] C W F Everitt *James Clerk Maxwell: Physical and Natural Philosopher* p 99. New York: Charles Scribner's Sons, 1975

[3] GUERLAC 2, pp 290–291

Chapter 1.3 Perceptions of air power, 1919–1939

[1] JONES H, Vol 5, pp 26–32

[2] Ibid., Vol 6, pp 1–27; appendices, pp 8–14

[3] WEBSTER & FRANKLIN 1, p 45

[4] JONES H, Vol 5, pp 153–154

[5] Beverley Nichols *Cry Havoc!* p 63. London: Jonathan Cape, 1933

[6] Douhet op. cit. R Ernest Dupuy and George Fielding Eliot introduced Douhet's ideas, which they encountered in a French military journal, to their English speaking readers in their book *If War Comes* pp 53–60, New York: The Macmillan Co., 1937. This book presents a calm evaluation of military technology on the obviously expected war

[7] James S Corum *The Roots of the Blitzkrieg: Hans von Seekt and German Military Reform* pp 144–146, 155. Lawrence, Kansas: University Press of Kansas, 1992

[8] Much uncertainty surrounds the pre-war German decisions concerning the four-motor bomber. The outcome probably turned on a combination of personal preferences, economics, experience in Spain and a flawed strategic outlook. See SUCHENWIRTH 1, pp 55–59

[9] Feldmarshall Albert Kesselring *Soldat bis zum letzten Tag* pp 32–33. Bonn: Athenäum Verlag, 1953

[10] Williamson Murray, Force Strategy, Blitzkrieg Strategy and the Economic Difficulties: Nazi Grand Strategy in the 1930s *Journal of the Royal United Services Institute in Defence Studies* Vol 128, pp 39–43, March 1983

[11] Saundby op. cit., p 28

[12] Peter C Smith *The History of Dive Bombing* Annapolis: The Nautical and Aviation Publishing Co., 1981

[13] SUCHENWIRTH 2, pp 7, 28, 79

[14] Colonel Roy M Stanley *Prelude to Pearl Harbor* pp 133–148. New York: Charles Scribner's Sons, 1982

[15] Schaffer op. cit., pp 107–108

[16] Greer op. cit., pp 67–69, 91–92

[17] Spaight op. cit.

[18] Ibid., p 35

[19] JONES H, Vol 6, pp 439–477, 465

[20] J C Slessor *Air Power and Armies* pp 163–164. London: Oxford University Press, 1936

[21] B H Liddell Hart *Paris and the Future of War*. New York: E P Dutton and Company, 1925

[22] Heinz Guderian *Panzer Leader* p 25. New York: E P Dutton and Co., 1952

[23] Roscoe op. cit., pp 141–174

[24] Mitchell op. cit. p 206

[25] Ibid., pp 125–126

[26] ROSKILL 5, pp 256–399

[27] ROSKILL 6, pp 392–405

Chapter 1.4 Navigation in 1939

[1] For a readable telling of the chronometer story see Dava Sobel *Longitude: the True Story of a Lone Genius Who Solved the Greatest Scientific Problem of His Time* New York: Walker Publishing Co., 1995

[2] Commander A E Fanning, Astronomical Navigation Since 1884 *Royal Institute of Navigation* Vol 38, pp 209–215, 1985

[3] TERRAINE 1, p 85

[4] William E Jackson *The Federal Airways System* pp 219–227. Institute of Electrical and Electronic Engineers, 1970; Robert I Colin, Otto Scheller: the Radio Range Principle *AES* Vol 2, pp 481–487, 1966

[5] JONES H, Vol 5, pp 8–18

[6] TUSKA 1

[7] SIG CORPS 1, pp 185–188

Chapter 1.5 Anti-aircraft artillery, 1914–1939

[1] KOCH, p 9

[2] PILE, p 173

[3] KOCH, pp 14–15. Anti-aircraft units generally follow the rule of about four heavy guns per battery with two or more batteries forming a battalion and two or more battalions forming a regiment

[4] Ibid., pp 10–15

[5] Ibid., p 19

[6] Ibid., pp 20–21

[7] Ibid., p 28

[8] Major Charles Edward Kirkpatrick *Archie in the AEF: the Creation of the Antiaircraft Service of the United States Army, 1917–1918* pp 5–11. Fort Bliss, Texas: US Army Air Defense School, 1984; James A Sawicki *Antiaircraft Artillery Battalions of the US Army* Vol 1, pp 1–2. Dumfries, Virginia: Wyvern Publications, 1991

[9] Coast Artillery units on the west coast and in the Philippines were not drawn for service in France until 1918, presumably through a mistrust of Japanese intentions

[10] Kirkpatrick [8], pp 20–24

[11] Ibid., pp 181–182

[12] Sawicki [8], pp 6–8

[13] John C Reilly Jr *United States Navy Destroyers of World War II* pp 68–79. Poole: Blandford Press, 1983

[14] PILE, pp 43–51

[15] Ibid., pp 52–60

[16] Liddell Hart *Memoirs* op. cit., Vol 1, p 127

[17] PILE, pp 41–42

[18] ROSKILL 6, p 420

Chapter 2.1 Electronic Component Development

[1] KURYLO & SUSSKIND

[2] Peter A Keller *The Cathode-Ray Tube: Technology, History and Applications* pp 45–54. New York: Palisades Institute for Research Service, 1991

[3] J B Johnson, A Low Voltage Cathode Ray Oscillograph *Journal of the Optical Society of America and Review of Scientific Instruments* Vol 6, pp 701–712, 1922

[4] V K Zworykin, Description of an Experimental Television System and the Kinescope *Proc. IRE* Vol 21, pp 1655–1673, 1933

[5] Manfred von Ardenne, Die Braunsche Röhre als Fernsehemfänger *Fernsehen* Vol 1, pp 193–202, 1930; Ueber neue Fernsehsender und Fernsehempfänger mit Kathodenstrahlröhren *Fernsehen* Vol 2, pp 65–80, 1931

[6] Manfred von Ardenne *Ein glückliches Leben für Technik und Forschung* pp 75–77, 82–90. Berlin: Verlag der Nation, 1973

[7] Ryoka Sawada *History of Electron Tubes* Sogo Okamura, editor, pp 23–24. Tokyo: Ohmsha Ltd, 1994

[8] Albert W Hull and N H Williams, Characteristics of Shielded-grid Pliotrons *Phys. Rev.* Vol 27, pp 432–438, 1926

[9] Albert W Hull, Measurements of High-frequency Amplification with Shielded-grid Pliotrons *Phys. Rev.* Vol 27, pp 439–454, 1926

[10] STOKES, pp 34–39. James Brittain in a letter to the author dated 24 January 1994 cautions that the entire tetrode–pentode invention may be more complicated than it appears here because of the intense activity in vacuum tube design during those years and the rather long time that four-element tubes had been available for experiment

[11] Ibid., pp 54–56; see also W I G Page, The Working Principles of a New Screened Grid Power Output Valve Explained *Wireless World* July, pp 7–9, 1928

[12] Zworykin, [4]

[13] H Barkhausen and K Kurz, Die kürzesten, mit Vakuumröhren herstellbaren Wellen *Phys. Zeit.* Vol 21, pp 1–6, 1920. This is a beautiful little paper. It reports an encounter with an unsuspected form of oscillation, describes the experiments that revealed its nature and derives a formula that gives good approximation for the wavelength as a function of the one adjustable parameter

[14] For a review of early magnetron work see James E Brittain, The Magnetron and the Beginnings of the Microwave Age *Phys. Today* Vol 38, pp 60–67, 1985

[15] August Zacek, Über eine Methode zur Erzeugung von sehr kurzen elektromagnetischen Wellen *Zeit. Hf.* Vol 32, p 172, 1928

[16] STOKES, p 137

[17] CALLICK, pp 41–42

[18] ALLISON, p 102

[19] Ed Simmonds and Norm Smith *Radar Yarns: Being Memories and Stories Collected from RAAF Personnel Who Served in Ground Based Radar During World War II* p 219. Published privately by E W & E Simmonds, 15 Blair Street, Port Macquarie, NSW 2444, Australia, 1991

[20] HOWSE, p 8

[21] CALLICK, pp 37–38

[22] Raymond R Myers, Plastics and Resins *Encylopaedia Britannica* Vol 21, p 339, 1993

[23] Dipl-Phys. Hans Ulrich Widdel in a letter to the author dated 18 November 1994

[24] J C Swallow, The History of Polythene *Polythene: the Technology and Uses of Ethylene Polymers* A Renfrew and Phillip Morgan, editors, pp 1–10. London: Iliffe and Sons Ltd, 1960

[25] Paul Kokulis of Washington, patent attorney and friend of Fawcett, reports the disappointment on finding a solid

[26] Maurice V Wilkes *Memoirs of a Computer Pioneer* pp 45–46. Cambridge, Massachusetts: The MIT Press, 1985

[27] John Harry DuBois and Frederick W John *Plastics* p 47. New York: Reinhold Publishing Company, 1967

[28] ROTTERDAM, 23 February 1943, p 3 and 17 March 1943, p 8

[29] Gordon M Kline, Plastics in Germany, 1939–1945 *Modern Plastics* Vol 23, pp 152a–152p, October 1945

[30] Gordon M Kline, interviewed by Jeffrey L Meikle at Lake Worth, Florida on 15 and 16 May 1987

Chapter 2.2 Beginnings, 1902–1934

[1] KERN, pp 34–40

[2] The superheterodyne principle was discovered independently in Germany with a patent having been filed on 18 June 1918, predating Armstrong's by six months. See Walter Schottky, On the Origin of the Super-heterodyne Method *Proc. IRE* Vol 14, pp 695–698, 1926

[3] Guglielmo Marconi, Radio Telegraphy *Proc. IRE* Vol 10, pp 215–238, 1922

[4] WATT, p 62

[5] L S Alder, Provisional Patent Specification No 6433/28, 1 March 1928

[6] TUSKA 2, pp 13–20, 95–115, identifies nine, and KERN, pp 44–63, ten patents that present the idea of radar

[7] Sean Swords, The significance of radio wave propagation studies in the evolution of radar, BLUMTRITT, pp 185–197. This experiment became transformed in *Popular Science* Vol 127, November 1935, p 18, as 'a way of halting airplane motors at a distance'

[8] Leo C Young in the Stanford Caldwell Hooper audio collection History of Radio–Radar–Sonar at the Library of Congress, Reel 23

[9] Gregory Breit and Merle A Tuve, A Radio Method of Estimating the Height of the Conducting Layer *Nature* Vol 116, p 357, 1925; Gregory Breit and Merle A Tuve, A Test of the Existence of the Conducting Layer *Phys. Rev.* Vol 28, pp 554–575, 1926; Merle A Tuve, Early Days of Pulse Radio at the Carnegie Institution *Journal of Atmospheric and Terrestrial Physics* Vol 36, pp 2079–2083, 1974

[10] MCKINNEY, pp 84–87

[11] Ibid., p 132

[12] SOUTHWORTH, pp 79–81

[13] L F Jones, A Study of the Propagation of Wavelengths between Three and Eight Meters *Proc. IRE* Vol 21, pp 349–386, 1933

[14] William H Wenstrom, Notes on Television Definition *Proc. IRE* Vol 21, pp 1317–1327, 1933

[15] Carl R Englund, Arthur B Crawford and William W Mumford, Some Results of a Study of Ultra-short-wave Transmission Phenomena *Proc. IRE* Vol 21, pp 464–492, 1933

[16] TUSKA 2, p 13

[17] E Giboin, L'Évolution de la Détection Électromagnétique dans la Marine Nationale *Onde* Vol 31, pp 53–64, 1951

[18] J Bion, Le Radar *La Revue Maritime* July–August, pp 330–346, 456–471, 1946; R B Molyneux-Berry, Dr Henri Gutton, French radar pioneer, BURNS, pp 45–52

[19] J B Johnson, A Low Voltage Cathode Ray Oscillograph *Journal of the Optical Society America and Review of Scientific Instruments* Vol 6, pp 701–712, 1922

[20] Heinrich Löwy, Die Fizeausche Methode zur Erforschung des Erdinnern *Phys. Zeit.* Vol 12, pp 1001–1004, 1911

[21] SWORDS, pp 46–47, 59. For detail about Löwy see Ulrich Kern, Review concerning the history of German radar technology up to 1945, BLUMTRITT, pp 171–183

[22] KROGE, p 16; TRENKLE 1, pp 23–24

[23] Ibid., pp 12–18; REUTER, pp 15–22

[24] LOBANOV, pp 25–33

[25] ERICKSON, pp 247–252; LOBANOV, pp 101–109

[26] LOBANOV, p 103

[27] Cited in SWORDS, pp 45–46

[28] Harold A Zahl, From an Early Radar Diary *Coast Artillery Journal* Vol 91, no 4 (March–April), pp 8–15, 1948. A rather accurate description of the incident was found in 'Mystery Ray Locates "Enemy": U.S. Army Tests Detector for Hostile Ships and Planes' *Popular Science* Vol 127, October 1935, p 25

[29] News item *Elektrisches Nachrichtenwesen* Vol 10, pp 24–25, 1932

Chapter 2.3 Britain builds an air defense system

[1] For a survey of the politics of the major powers up to the outbreak of war see Richard Overy and Andrew Wheatcroft *The Road to War* London: Macmillan Co., 1989

[2] For a glimpse into some of the ideas that had been studied see Russell Burns, Aspects of UK Air Defence from 1914 to 1935 *Proc. IEE* Vol 136, pp 267–278, 1989

[3] Ronald W Clark *Sir Edward Appleton* pp x, 90–91, 96–102. Oxford: Pergamon Press, 1971. The radio section of the Post Office (British) had noticed reflections from aircraft the year before. A H Mumford and R F J Jarvis, Radio Report No 273, 18 July 1934

[4] Wright op. cit., pp 26–36

[5] Ibid., pp 53–56

[6] William Roy Piggott, who worked closely with both Appleton and Watt, in a letter to the author dated 13 March 1997

[7] Gordon Kinsey *Orfordness: Secret Site*. Lavenham: Terence Dalton Ltd, 1981. A book about the place, primarily dealing with its World War I functions. Two chapters about the radar work there were written by A F Wilkens

[8] WATT, pp 113–115

[9] CALLICK, pp 30–38

[10] SWORDS, pp 186–236; B T Neale, CH—The First Operational Radar *GEC Journal of Research* Vol 3, pp 73–83, 1985

[11] ROWE, pp 25–26

[12] Ibid., p 23

[13] WOOD & DEMPSTER, p 83

[14] WOOD & DEMPSTER, p 88; WATT, p 208

[15] The Germans made a few copies of CH, called Elefant. This was equipment that originated in a limited excursion of the Reichspost Zentrale into radar. I have not been able to determine whether this was to test the CH design or, as suggested by Trenkle, experiment with over-the-horizon capability. The transmitter array was mounted on a tower, electrically very similar to CH, although much more compact. The receiver was improved by using a steerable high-gain dipole array. TRENKLE 1, pp 101–102. The Japanese Army used a design, Tachi-6, with broadcast transmission and multiple receivers having dipole arrays for direction (see Chapter 9.2), and the Russians made a similar broadcast-transmission radar, RUS-2 (see Chapter 5.6).

[16] *Radar Bulletin* (published by RAF 60 Group beginning April 1941) October 1945, p 9

[17] C P Snow *Science and Government*. Cambridge, Massachusetts: Harvard University Press, 1961, appendix 1962

[18] The Earl of Birkenhead *The Professor and the Prime Minister: The Official Life of Professor F A Lindemann, Viscount Cherwell*. Boston: Houghton Mifflin and Co., 1962

[19] ROWE, p 22

[20] BROWN, p 17

[21] Ibid., p 13

[22] R W Burns, A D Blumlein—engineer extraordinary *Engineering Science & Education Journal* February, pp 19–33, 1992

[23] JAY, part 1, para 5

[24] Sidney Jefferson, LATHAM & STOBBS, p 215. Young readers may be puzzled by the meaning of 'dress for dinner'. They can look for instruction in Noël Coward's plays.

[25] SAYER, pp 20–26

[26] Ibid., p 117; BOWEN, p 50

[27] Michael Pearson, Coast Defence Radar *Fort* Vol 19, pp 93–105, 1991

[28] Ibid., p 96; SAYER, pp 120–125

[29] Alfred Price *Cossor Radar: The First Fifty Years* pp 10–11. Harlow: Cossor Electronics, 1985; SAYER, p 46–50

[30] SAYER, p 319

[31] P M S Blackett, Operational Research: Recollections of Problems Studied, 1940–45 *Brassey's Annual: The Armed Forces Yearbook 1953* pp 88–106

[32] JAY, part 2, paras 312–313

[33] Russell Burns, Early History of the Proximity Fuze (1937–1940) *Proc. IEE* Vol 140A, pp 224–236, 1993

[34] JAY, part 1, paras 186–188

[35] A straight-vision receiver carried only the video portion of the broadcast and in amplitude, not frequency modulation; the audio was in a separate radio-frequency channel. This allowed the wide-band design characteristics to be dealt with without the complication of an intermediate-frequency amplifier or the need to separate the audio and video portions of the signal in a common carrier

[36] BOWEN, pp 30–64

[37] J F Coales, The Origins and Development of Radar in the Royal Navy, 1935–1945, KINGSLEY 1, pp 11–29

[38] HOWSE, p 12

[39] Ibid., pp 18–22

[40] J F Coales and J D S Rawlinson, The Development of UK Naval Radar, BURNS, pp 53–59

[41] ROWE, p 27

[42] Ed Simmonds and Norm Smith *Radar Yarns* Ed Simmonds editor, p 35. Published privately by E W & E Simmonds, 15 Blair Street, Port Macquarie NSW 2444, Australia, 1991

[43] WATT, p 46

Chapter 2.4 Americans and Germans build prototypes

[1] Carl R Englund, Arthur B Crawford and William W Mumford, Some Results of a Study of Ultra-short-wave Transmission Phenomena *Proc. IRE* Vol 21, pp 464–492, 1933

[2] ALLISON, pp 68–78

[3] L A Hyland, A personal reminiscence: the beginnings of radar 1930–1934, BURNS, pp 29–34

[4] PAGE, pp 64–66

[5] René Mesny, Constantes de Temps, Durées d'Établissement, Décréments *Onde* Vol 13, pp 237–243, 1934. This approach to the problem differs entirely from that taken by electrical engineers of the time or subsequently, but it must have looked familiar to Page as the analysis of a galvanometer widely employed in undergraduate laboratories

[6] G L Beers, Description of Experimental Television Receivers, *Proc. IRE* Vol 21, pp 1692–1706, 1933

[7] ALLISON, pp 120–124

[8] TAYLOR, p 192

[9] ALLISON, p 124; BELL, pp 24–25

[10] A L Samuel, A Negative Grid Triode Oscillator and Amplifier for Ultra-High Frequencies *Proc. IRE* Vol 25, pp 1243–1252, 1937

[11] W Tinus and W H C Higgins, Early Fire Control Radars for Naval Vessels *Bell Tech. J* Vol 25, pp 1–47, 1946

[12] FRIEDMAN, p 172

[13] SIG CORPS 2, pp 256–257

[14] RADAR SURVEY 2,, pp 33–38

[15] Interview with Blair and other sources in MCKINNEY, pp 64–68

[16] MCKINNEY, pp 97–103

[17] Interview with Colton in MCKINNEY, pp 130–131

[18] MCKINNEY, pp 253–254

[19] GETTING, p 113

[20] SIG CORPS 1, pp 43–45; MCKINNEY, pp 132–136

[21] Arthur L Vieweger, Radar in the Signal Corps *Institute of Radio Engineers Transactions* Vol 1-MIL, pp 555–561, 1960

[22] Interviews with Blair and Colton in MCKINNEY, pp 157–158

[23] Interview with Colton in MCKINNEY, pp 165–166

[24] SIG CORPS 1, p 123

[25] Colton op. cit., p 746; MCKINNEY, pp 168–172; SIG CORPS 1, pp 125–127; Harold A Zahl, From an Early Radar Diary *Coast Artillery Journal* Vol 91, no.4 (March–April), pp 8–15, 1948

[26] Vieweger [21], p 556

[27] Ibid., pp 558–559

[28] Frank Voltaggio, The SCR-270 in Japan *AES* Vol 3, pp 7–14, 1988

[29] REUTER, pp 18–19

[30] KROGE, pp 17–26

[31] Ibid., pp 24–27

[32] Ibid., pp 22–23

[33] HOLLMANN

[34] KROGE, p 44; TRENKLE 1, pp 26–30; REUTER, p 23

[35] KROGE, pp 55–56; P-G Erbslöh and H-K Freiherr von Willisen, patent declaration dated 27 April 1935

[36] KROGE, p 77

[37] REUTER, pp 24–25

[38] KROGE, p 136; TRENKLE 1, pp 151–152

[39] F A Kingsley, who interrogated German naval personnel after the surrender, in a letter to the author dated 20 November 1995

[40] KROGE, p 110; REUTER, p 43(n)

[41] KROGE, p 94

[42] Ibid., p 118

[43] Ibid., pp 93–94

[44] WATT, p 405

[45] RUNGE, p 30. It should be noted that Marconi was excluded from early radar work because of their Italian connection

[46] HOFFMANN 2, pp 329–440

[47] Iris Runge *Carl Runge und sein wissenschaftliches Werk* p 156. Göttingen: Vandenhoeck und Ruprecht, 1948

[48] Ibid., p 171

[49] Kurt Fränz, Wer war Wilhelm Tolmé Runge, address to the Direktionsbereich der Forschungs-Institut-Telefunken, November 1990

[50] *Electronics* Vol 8, pp 284–286 (September, pp 18–19), 1935; *Popular Science* Vol 127, October 1935, p 25. These articles describe transmitters, receivers and their antennas of equipment supposedly for 10 and 15 cm waves but the photographs indicate that they are almost certainly the 50 cm equipment that Runge employed. The drawing of an array of transmitters and receivers would seem to show the first idea for a radar system that occurred after the 50 cm reflections

[51] RUNGE, pp 42–44

[52] Dr-Ing Gotthard Müller, Funkmessgeräte-Entwicklung bei der C Lorenz AG, 1935–1945, pp 2–4. Stuttgart: Standard Elektrik Lorenz AG (Technisch-wissenschaftliches Schriftum), 1983

[53] Robert I Colin, Otto Scheller: the Radio Range Principle *IEEE Transactions on Aerospace and Electronic Systems* Vol AES-2, pp 481–487, 1966

[54] KROGE, p 50

[55] TRENKLE 1, pp 30–31; PRITCHARD, pp 42–47

[56] HOFFMANN-HEYDEN, pp 31–34

[57] REUTER, p 47

[58] Office of the Chief of Naval Operations, order dated 18 November 1940 (Op-20-E/AB, (SC)A6/A1, serial 069120)

[59] HOWSE, p 162

Chapter 2.5 Five other nations

[1] NAKAGAWA, p 7

[2] TAYLOR, p 188

[3] Shigeru Nakajima on receiving the Barkhausen-Medaille der Akademie der Wissenschaft der DDR, Wissenschaft: eine Herausforschung an den erfinderischen Geist, reprinted in *Japan Radio Company Review* No 16, 1981

[4] NAKAGAWA, pp 6–7

[5] Ibid., pp 7–8

[6] Ibid., p 11

[7] Ibid., pp 15–16, 82

[8] PRICE 2, p 291

[9] NAKAJIMA, pp 243–246

[10] ERICKSON, p 248

[11] Ibid., pp 255–257

[12] LOBANOV, pp 107–108

[13] Ibid., p 57

[14] Ibid., p 59

[15] ERICKSON, p 255

[16] Ibid., pp 256–258

[17] Steven J Zaloga, Soviet Air Defense Radar in the Second World War *The Journal of Soviet Military Studies* Vol 2, pp 104–116, 1989

[18] LOBANOV, p 139

[19] Ibid., p 129

[20] Ibid., pp 240–241

[21] Ibid., pp 92–94

[22] ERICKSON, pp 258–259

[23] N F Alekseev and D D Malairov, Generation of High-Power Oscillations with a Magnetron in the Centimeter Band *Zhurnal Tekhnicheskoi Fiziki* Vol 10, pp 1297–1300, 1940

[24] E Giboin, L'Évolution de la Détection Électromagnétique dans la Marine Nationale *Onde* Vol 31, pp 53–64, 1951

[25] J Oger, Pré-histoire du Radar *La Revue Maritime* Vol 108, pp 433–469, 1955

[26] Giboin, [24]

[27] Maurice Ponte, Sur des Apports Français a la Technique de la Détection Électromagnétique *Revue Technique Thomson-CSF* Vol 1, pp 171–180, 1946

[28] M Staal and J L C Weiller, Radar Development in the Netherlands before the War, BURNS, pp 235–242

[29] M Calamia and R Palandri, The History of the Italian Radio Detector Telemetro, BURNS, pp 97–105

Chapter 3.1 War in Europe

[1] Manfred Bauer and John Duggan *LZ 130 'Graf Zeppelin' und das Ende der Verkehrsluftschiffahrt* pp 136–140, 153–166. Friedrichshafen: Zeppelin-Museum, 1994

[2] These positions are based on the ship's log. There is reason to believe they were inaccurate between fixes. A photograph that purports to be of the airship straying over British terra firma is one of LZ-127 during a pre-Nazi flight, as the shape of the tail fins and the absence of swastikas on them attest

[3] Sir Edward Fennessy, The Zeppelin Incident and the Battle of Britain, unpublished manuscript, February 1994. The map shown in Bauer and Duggan's book does not show the ship straying over land. Political pressures may have intervened to require the navigator to cook his log

[4] Len Dobson, LATHAM & STOBBS, pp 1–3

[5] Letter to the authors (Bauer and Duggan [1], p 156.) from Keith Wood, a scientist then employed at Bawdsey

[6] This interpretation of the Zeppelin incident is based on Colin Latham's, I see the cat but he can't see me! *News and Views* (Newspaper of Marconi Radar and Control Systems Limited), July 1992, pp 8–9; also LATHAM & STOBBS, pp 3–6. Latham retired as Chief Engineer, Airspace Control Division, Marconi Radar Systems

[7] Robert Watson Watt, memorandum to the Committee for the Scientific Survey of Air Defence dated 27 February 1935, Detection of Aircraft by Radio Methods, paragraph 9, cited in WATT, p 431

[8] Fennessy, [3]

[9] Interview with Dr Ernst Breuning on 21 February 1969 by Alfred Price

[10] WOOD & DEMPSTER, pp 98–99

[11] BOWEN, pp 83–91; GUERLAC 1, p 153

[12] Sir Edward Fennessy in a letter to the author dated 7 September 1995

[13] BOWEN, pp 83–91; BROWN, pp 45–46

[14] LOVELL, p 21

[15] BOWEN, p 84

[16] LOVELL, p 25; CALLICK, pp 38–40

[17] BOWEN, pp 117–119

[18] Ibid., pp 58–59

[19] R W Burns, A D Blumlein—Engineer Extraordinary *Engineering and Science Education Journal* February 1992, pp 19–33

[20] Derek Martin *THORN EMI: 50 Years of Radar* pp 4–5. Hayes, Middlesex: THORN EMI Electronics Ltd, 1986

[21] BROWN, pp 48–51; BOWEN, p 99

[22] PRICE 4, pp 54–55

[23] LOVELL, p 25

[24] Ibid., pp 55–56

[25] TERRAINE 1, pp 95–106

[26] HOFFMANN 1, p 11

[27] Ibid., pp 12–13

[28] TERRAINE 1, pp 103–104

[29] HOFFMANN 1, pp 95–96

[30] TRENKLE 1, pp 76–78

[31] BRANDT, p 19

[32] HOFFMANN 1, p 314

[33] GIESSLER, p 66

[34] GIESSLER, pp 16–17

[35] JONES 1, pp 190–192

[36] Ibid., pp 67–71

[37] JONES 2, pp 265–332

[38] TERRAINE 1, pp 95–115

[39] The Germans designated these vessels as battleships despite their 11 inch artillery. The British terminology is followed here to distinguish them from the 15 inch *Bismark* and *Tirpitz*

[40] REUTER, pp 44, 52

[41] ROSKILL 1, pp 82–87

[42] REUTER, pp 51–52

[43] Ibid., pp 111–121

[44] JONES 1, p 93

[45] Mr F A Kingsley in a letter to the author dated 16 January 1996

[46] Liddell Hart *History* op. cit., pp 51–53

[47] KROGE, pp 108–109

[48] HEZLET, p 193

[49] REUTER, pp 53–54

[50] HEZLET, p 193

[51] RAF SIGNALS 5, pp 56–58

Chapter 3.2 The Battle of Britain and the Blitz

[1] I am indebted to Harry von Kroge for locating the specific directive, which is mentioned without citation in German papers; it was given by Göring on 3 February 1940. Der Führer legt entscheidenen Wert darauf, dass die Rüstung im Jahre 1940 zur grösstmöglichen Höhe gebracht wird. Es müssen daher mit allen Mitteln die Vorhaben gefördert werden, die im Jahre 1940 bezw. 1941 zur Auswirkung kommen können. Alle anderen Programme, die sich erst später auswirken müssen, falls es die Belebung der Wirtschaft erfordert, zugunsten der obigen Vorhaben zurückgestellt werden, Percy E Schramm *Kriegstagebuch*

des Oberkommandos der Wehrmacht Vol 1, p 962. Bonn: Bernard & Graefe Verlag. For other discussions see GIESSLER, p 74; TRENKLE 1, pp 41–42. For the Allied understanding see OSRD 5, p 117. For a complete discussion of this question, which is more involved than the Göring quotation seems to imply, see SUCHENWIRTH 2, pp 49–54

[2] TRENKLE 2, p 16

[3] WOOD & DEMPSTER, pp 66–67

[4] TERRAINE 1, p 186

[5] Alfred Price has grippingly described this day through official records and from interviews with pilots from both sides who were present at an RAF–Luftwaffe reunion. Read his *Battle of Britain: The Hardest Day, 18 August 1940*. New York: Charles Scribner's Sons, 1979

[6] Details about the front-line stations of Chain Home are covered by Mike Dean *Radar on the Isle of Wight*. Scampton: Historical Radar Archive, 1994. This attack was the cause of the first Military Medal to be awarded to a woman, Avis Parsons (née Hearn) who remained at her post relaying messages while Stukas dropped 90 bombs on the station. LATHAM & STOBBS, pp 25–29

[7] WEBSTER & FRANKLIN Vol 1, p 152

[8] NIEHAUS, pp 37–40

[9] D V Pritchard, The Battle of the Beams *Ham Radio* (published by Communications Technology, Greenville, New Hampshire) pp 29–39, June 1989; pp 20–29, August 1989; pp 53–61, October 1989

[10] W D Hershberger, Seventy-five Centimeter Radio Communication Test *Proc. IRE* Vol 22, pp 870–877, 1933

[11] TRENKLE 3, pp 119–120

[12] Pritchard *Ham Radio* [9]

[13] TRENKLE 3, pp 137–144

[14] NIEHAUS, pp 39–40

[15] Ibid., p 49

[16] Winston S Churchill *The Second World War: Vol 2, Their Finest Hour* pp 384–385. Houghton Mifflin Company, Boston, 1949

[17] JONES 1, p 169

[18] Members of KGr 100 recalled in interviews with Dr Alfred Price that they were generally more aware of the deficiencies of the navigation systems than of British interference. Specifically they said the clear, moonlit night plus the initial fires contributed most to the accuracy of the Coventry attack. TRENKLE 3, p 126, reports that KGr 100 was not troubled by interference until May 1941

[19] LOVELL, p 14

[20] KEMP, pp 20–22, 28

[21] Ernest Putley, Ground Control Interception, BURNS, pp 162–176

[22] BROWN, p 63

[23] SIG CORPS 2, pp 86–87, 96

[24] PILE, p 172

[25] Maurice V Wilkes *Memoirs of a Computer Pioneer* pp 64–65. The MIT Press: Cambridge, Massachusetts, 1985

[26] P M S Blackett, Operational Research: Recollections of Problems Studied, 1940–45 *Brassey's Annual: The Armed Forces Yearbook, 1953* pp 88–106

[27] PILE, p 173

[28] Ibid., p 287

[29] Viktor Reimann *Goebbels* p 245. Garden City, New York: Doubleday and Company, 1976. NIEHAUS, p 54
[30] ROWE, pp 73–74
[31] Churchill [16], p 333
[32] Ibid., pp 383–391
[33] Editor, Radiolocators *J. Ap. Phys.* Vol 12, p 511, 1941
[34] For example, Larry Wolters, Radio Locator Called a Child of Television *Chicago Sunday Tribune* 13 July 1941

Chapter 3.3 The Atlantic, 1941

[1] ROSKILL 1, pp 279–282, 604–605
[2] REUTER, pp 55–57; ROSKILL 1, pp 291–292
[3] REUTER, pp 61–63; ROSKILL 1, pp 373–379
[4] REUTER, pp 63–65; ROSKILL 1, pp 371–373
[5] REUTER, pp 65–68; ROSKILL 1, pp 287–291, 367–372
[6] HOWSE, p 91
[7] TRENKLE 1, pp 116–120
[8] Burkard Baron von Müllenheim-Rechberg *Battleship Bismarck: A Survivor's Story* pp 130–131. Annapolis: Naval Institute Press, 1990
[9] GIESSLER, p 72
[10] ROSKILL 1, pp 403–404
[11] HOWSE, pp 95–96
[12] Müllenheim-Rechberg [8], 130–131
[13] Ibid., pp 177–178
[14] GIESSLER, p 73
[15] HEZLET, pp 207–208
[16] Ibid., p 207
[17] Müllenheim-Rechberg [8]
[18] Robert D Ballard *The Discovery of the Bismarck*. New York: Warner Books, 1990
[19] REUTER, p 47
[20] GIESSLER, p 75
[21] HOWSE, pp 341–343
[22] Ibid., pp 84–86

Chapter 3.4 Friend, foe or home?

[1] Robert Watson Watt, memorandum to the Committee for the Scientific Survey of Air Defence dated 27 February 1935, Detection of Aircraft by Radio Methods, paragraph 19, cited in WATT, p 434
[2] WATT, p 138
[3] GEBHARD, pp 251–256
[4] Lord Bowden of Chesterfield, The story of IFF (identification friend or foe) *Proc. IEE* Vol A132, 1985, pp 435–437
[5] RAF SIGNALS 5, pp 75–78
[6] Bowden [4], p 435
[7] SIG CORPS 2, pp 242–243
[8] Harold Alden Wheeler *Hazeltine Corporation in World War II* Ventura, California: Pathfinder Publishing Company, 1993

[9] R M Trim, The development of IFF in the Period up to 1945, BURNS, pp 436–457

[10] RAF SIGNALS 5, pp 91–93

[11] SIG CORPS 1, pp 264–266

[12] RAF SIGNALS 5, pp 98–99

[13] For examples of radar-directed fire on friendly ships see MORISON 6, pp 188, 243, 317, 354; 7, p 154; 12, pp 222, 227

[14] REUTER, pp 34–35

[15] TRENKLE 1, p 172

[16] Ibid., p 173

[17] Harry von Kroge, letter dated 7 March 1995

[18] HOFFMANN-HEYDEN, pp 129–131

[19] Ibid., pp 133–135

[20] REUTER, pp 35–36

[21] Trim [9], pp 442–445

[22] Wheeler [6], pp 215–238

[23] TRENKLE 1, p 178

[24] Bowden [4], p 436

[25] RADAR, No 2, May 1944, pp 21–27

[26] ROWE, p 126

[27] TRENKLE 2, p 16

[28] WATT, p 147

[29] Ibid.

Chapter 3.5 The Japanese realize they are behind

[1] NAKAGAWA, pp 21–22. Nakagawa's treatment of history differs from that of Wilkinson (Roger I Wilkinson, Short Survey of Japanese Radar *Transactions of the American Institute of Electrical Engineers* Vol 65, pp 370–377, 455–463, 1946), although not with his technical descriptions of the equipment. This difference is repeated in appendix G of PRICE 2. Consultation with a number of Japanese sources bears out Nakagawa's presentation, which draws on a wider base, was researched at a more leisurely pace and was not veiled by greatly dissimilar tongues

[2] Ibid., pp 22–24

[3] Ibid., pp 25–26

[4] Ibid., p 25

[5] Ibid., p 26

[6] Ibid., pp 26–27

[7] Ibid., pp 27–28

[8] Ibid., pp 28–29, 87

[9] Ibid., pp 32–33

[10] Ibid., p 83; PRICE 2, p 289

[11] NAKAGAWA, pp 34, 84–85

[12] PRICE 2, p 293

[13] NAKAGAWA, pp 33–35

[14] Genzo Sato, The Secret Story of the Yagi Antenna in World War II *The Radioscientist* Vol 2, No 4, pp 71–74, 1991. Colin MacKinnon in a letter to the author dated 31 May 1995

[15] PRICE 2, pp 290–291. Price attributes the manufacture of Tachi-1 to Sumitomo and Tachi-2 to Shibaura. Sumitomo was the name NEC adopted briefly during

the war, and Shibaura was Tokyo Shibaura Electric, later Toshiba. Japanese Wartime Military Electronics and Communications, Section VI, Japanese Army Radar, Technical Liaison and Investigation Division, Office of the Chief Signal Officer, GHQ, US Army Forces Pacific found at US National Archives, SCAP Box 7428

[16] TRENKLE 1, p 46

[17] PRICE 2, p 291. Wilkinson [1], p 375

[18] Air Technical Intelligence Group, Advanced Echelon FEAF, Report No 261, 12 December 1945

[19] SCIENTIFIC, Vol 1, pp 12–14

[20] NAKAJIMA, p 255

[21] Ibid., p 254–255

Chapter 4.1 Microwaves

[1] GIESSLER, p 63

[2] Ben R Rich and Leo Janos *Skunk Works: A Personal Memoir of My Years at Lockheed* pp 19–27. Boston: Little, Brown and Co., 1994

[3] W W Hansen, A Type of Electric Resonator *J. Ap. Phys.* Vol 9, pp 654–663, 1938

[4] Arthur L Norberg and Robert W Seidel, The Contexts for the Development of Radar, BLUMTRITT, pp 199–216

[5] The term 'ionosphere' seems to have been coined independently by Robert Watson Watt and E V Appleton in 1926. C S Gillmor, The history of the term 'ionosphere' *Nature* Vol 262, pp 347–348, 1976. The term Ionosphäre was used by Hans Plendl and was apparently introduced to American researchers through his papers. Wilbert F Snyder and Charles L Bragaw *Achievement in Radio: Seventy Years of Radio Science, Technology, Standards, and Measurement of the National Bureau of Standards* footnote, p 172. Washington: US Government Printing Office, 1986

[6] Lord Rayleigh, On the passage of electric waves through tubes or the vibrations of dielectric cylinders *Phil. Mag.* Vol 43, pp 125–132, 1897

[7] SOUTHWORTH, pp 60–65

[8] O Schriever, Elektromagnetishe Wellen an dielektrischen Drähten *Ann. Phys.* Vol 63, pp 645–673, 1920

[9] Karle S Packard, The Origin of Waveguides: a Case of Multiple Rediscovery *IEEE Transactions on Microwave Theory and Techniques* Vol MTT-32, pp 961–969, 1984

[10] Ibid., pp 966–968

[11] A Arsenjewa-Heil and O Heil, Eine neue Methode zur Erzeugung kurzer, ungedämpfter, elektromagnetischer Wellen grösser Intensität *Zeit. Phy.* Vol 95, pp 752–762, 1935

[12] Karl R Spangenberg *Vacuum Tubes* p 616. New York: McGraw-Hill Book Company, 1948

[13] GUERLAC 1, p 194

[14] Russell H Varian and Sigurd F Varian, A High Frequency Oscillator and Amplifier *J. Ap. Phys.* Vol 10, pp 321–327, 1939; W W Hansen and R D Richtmyer, On Resonators Suitable for Klystron Oscillators *J. Ap. Phys.* Vol 10, pp 189–199, 1939

[15] Edward L Ginzton, The $100 idea: How Russell and Sigurd Varian with the help of William Hansen and a $100 appropriation, invented the klystron *IEEE Spectrum* Vol 12, pp 30–39, 1975

[16] GUERLAC 1, p 213

[17] Luis W Alvarez, Alfred Lee Loomis—last great amateur of science *Phys. Today* Vol 36, 1983, pp 25–34. Adapted from *Biographical Memoirs* Vol 51, National Academy of Sciences, 1980

[18] GUERLAC 1, p 221

[19] Ibid., pp 247–250

[20] David H Sloan and Lauritsen C Marshall, Ultra-High Frequency Power, (abstract) *Phys. Rev.* Vol 58, p 193, 1940 Winfield W Salisbury, The Resnatron *Electronics* February 1946, pp 92–97

[21] BOWEN, p 143

[22] Russell W Burns, The Early History of Centimetric Radar: the Contributions of the General Electric Company, manuscript intended for publication, 1998

[23] GUERLAC 1, pp 225–226

[24] BATT, p 43

[25] The space-charge limitation of klystrons was overcome after the war by imposing an axial magnetic field on the electron beam, thereby allowing much larger currents. These were the generators that made possible the high-energy electron accelerators at Stanford. The excellent frequency stability of the klystron was the determining factor in choosing klystron over magnetron

[26] J T Randall, The Cavity Magnetron *Proc. Phys. Soc.* Vol 58, pp 247–252, 1946

[27] CALLICK, pp 55–57

[28] GUERLAC 1, pp 228–231

[29] Burns, [22]

[30] CALLICK, pp 78–80

[31] HOLLMANN, Vol 2, pp 2–4

[32] SOUTHWORTH, pp 153–157

[33] Reg Batt in MAGNETRON, p 34

[34] LOVELL, pp 1–43

[35] Ibid., pp 41–42

[36] C A Cochrane, Development of Naval Warning and Tactical Radar, KINGSLEY 1, pp 189–203

[37] HOWSE, pp 83–84

[38] James Sayers, MAGNETRON, pp 12–14

[39] Nakajima, BURNS, pp 243–258. In Chapter 2.5 the 1938 visit of Professor Barkhausen to Japan is described. Although it was a highly technical visit lasting two months during which he went to the principal laboratories and industries, he did not learn about the new centimeter-wave generator

[40] Sogo Okamura, editor *History of Electron Tubes* p 29. Washington: IOS Press (republished for Ohmsha, Tokyo), 1994. The cavity magnetron was refered to at Bell Labs for a while as the 'Samuel oscillator', letter from J R Wilson to L A DuBridge, 30 April 1940

[41] Arthur L Samuel, Electron Discharge Device, US Patent No 2 063 342, 6 December 1936

[42] Kinjiro Okabe *Magnetron-Oscillations of Ultra-Short Wavelengths and Electron Oscillations in General* pp 30–31. Tokyo: Shokendo, 1937

[43] NAKAJIMA; Marvin Hobbs, Japanese Magnetrons *Electronics* May, 1946, pp

114–115

[44] NAKAGAWA, pp 30–31

[45] Ibid., pp 32, 89

[46] NAKAJIMA

[47] James Phinney Baxter *Scientists Against Time* p 142. Boston: Little, Brown and Company, 1946

[48] N F Alekseev and D D Malairov, Generation of High-Power Oscillations with a Magnetron in the Centimeter Band *Zhurnal Tekhnicheskoi Fiziki* Vol 10, pp 1297–1300, 1940; transliteration and translation by I B Benson for *Proc. IRE* Vol 32, pp 136–139, 1944

[49] LOBANOV, pp 65–66

[50] Ibid., pp 92–94

[51] Ibid., pp 67–69

[52] F Fischer and F Lüdi, Die Posthumus-Schwingungen im Magnetron *Schweizerischer Elektrotechnischer Verein Bulletin* Vol 28, pp 277–283, 1937

[53] F Lüdi, Zur Theorie der geschlitzten Magnetfeldröhre *Helvetica Physica Acta* Vol 16, pp 59–82, 1942

[54] Hans Paul, Neuere Entwicklungen auf dem Gebiet der Zentimeterwellen *Elektrotechnische Zeitschrift* Vol 77, pp 849–854, 1956

[55] Hans H Jucker, who is preparing a study of Swiss radar history, in a letter to the author dated 21 April 1995

Chapter 4.2 The Tizard Mission

[1] Walter Millis *Road to War: America 1914–1917*. Boston: Houghton Mifflin Co., 1935

[2] Clark op. cit., pp 251–252

[3] Churchill vacillated in his support, withdrawing it even after preparations were well under way. Lindemann favored it, despite it being Tizard's idea. ZIMMERMAN, pp, 61–89

[4] BOWEN, p 151

[5] Ibid., pp 157–158

[6] Wallace is reported to have said that GL mark I was responsible for having brought down 400 German planes with the British Expeditionary Force in France and Belgium. GUERLAC 1, p 170. Unknown to those participating in these discussions at the time, the Battle of Britain was demonstrating the fantasy of these claims, as GL mark I at the time of the Tizard meetings had not been instrumental in bringing down a single plane. PILE, p 173

[7] Interview with A E Cassevant and John J Slattery cited in MCKINNEY, p 213

[8] SIG CORPS 1, p 193

[9] Ibid., p 200

[10] Interview with Karl T Compton on 20 October 1943 by Henry Guerlac

[11] BOWEN, p 180

[12] SIG CORPS 1, pp 198–199

[13] BOWEN, p 158

[14] SIG CORPS 1, pp 288–291. Ruth F Sadler with Lt Col Herbert H Butler, History of the Electronics Training Group in the United Kingdom, March 1944, manuscript at US Army Center for Military History, Washington, DC

[15] PILE, p 114

[16] Ibid., p 166

[17] Ibid., pp 171–174

[18] Ibid., p 215

[19] GUERLAC 1, p 258

[20] BELL, pp 25–26. The Bureau of Ordnance assumed responsibility for fire-control radar and replaced the Bureau of Engineering's FA designation with mark 1 and so on

[21] Wilfred Eggleston *Scientists at War* pp 28–40. London: Oxford University Press, 1950

[22] OSRD 6, p 3

Chapter 4.3 The Radiation Laboratory

[1] GUERLAC 1, pp 259–261; RABI, p 134

[2] ALVAREZ, p 87

[3] GUERLAC 1, pp 262–263

[4] ALVAREZ, pp 88–92

[5] CALLICK, pp 98–99

[6] BELL, pp 95–100. The comparison of British and American microwave AI sets is made nicely in BUDERI, pp 116–119

[7] GUERLAC 1, p 324

[8] GETTING, pp 3–80

[9] Lee L Davenport, IEEE, pp 61–62

[10] GETTING, pp 108–110

[11] Monopulse radar later removed this jitter much more effectively than the smoothing circuits, but such accuracy was not required for the fire-control problems of World War II

[12] GETTING, p 113; GUERLAC 1, p 278

[13] GETTING, pp 120–121. Davenport remembers that the mechanical M-7 director was used for this test, IEEE, p 66. The Bell Labs history does not mention this test but states that the T-10 prototype was functioning in December 1941. BELL, pp 145–146

[14] GUERLAC 1, p 481

[15] SIG CORPS 2, pp 268–274

[16] GETTING, pp 133–142

[17] Perhaps the best judgment on the value of SCR-584 comes from the 1995 catalogue of Radio-Research Instrument Co., Waterbury, Connecticut: 'Designed at MIT Radiation Labs and still considered one of the finest automatic tracking radars ever built, it is now being used in hundreds of installations in its original form and in various modifications. We have them in stock for immediate delivery complete in their own 20 ft trailer van containing the entire system'

[18] MORISON 10, p 154

[19] POLLARD, pp 53–60

[20] GUERLAC 1, pp 273–274, 399–400

[21] The administrative history is entirely from GUERLAC 1, pp 647–692

[22] Britton Chance, IEEE, p 50

[23] Ernest C Pollard, IEEE, pp 214–215

[24] Louis N Ridenour, Editor in Chief *Radiation Laboratory Series* New York: McGraw-Hill Book Company, 1947

[25] RABI, pp 164–165

Chapter 4.4 The proximity fuze—the smallest radar

[1] The Luftwaffe tried without notable success the 'bomb the bombers' technique when the air defense of Germany began to call for desperate measures. BEKKER 3, pp 404–405
[2] Russell Burns, Early History of the Proximity Fuze (1937–1940) *Proc. IEE* Vol 140A, pp 224–236, 1993
[3] Dennis Newton, The Remarkable Work of Alan Butement *Despatch: Journal of the New South Wales Military Historical Society* Vol 27, pp 70–82, 1992; Betty Williams *Dr W A S Butement, the First Chief Scientist for Defence* Canberra: Australian Government Publishing Service, 1991
[4] JONES 2, pp 307–313
[5] The Proximity Fuze *US Naval Administrative Histories of World War II* part II, Vol II, pp 192–193. Washington: Bureau of Ordnance
[6] Section T Final Summary Report, 19 April 1942, Archives of the Department of Terrestrial Magnetism. Applied Physics Laboratory, The 'VT' or Radio Proximity Fuze, with confidential supplement, 20 September 1945; Merle A Tuve and Richard B Roberts, US Patent 3 166 015, Radio Frequency Proximity Fuze, 6 January 1943; M A Tuve, Affidavit on behalf of Butement *et al* v. Varian before the Board of Patent Interferences, No 86 648, 4 September 1964
[7] R B Roberts, extract from a manuscript autobiography and reminiscences, Archives of the Department of Terrestrial Magnetism, Carnegie Institution of Washington, 1978
[8] Statement to the author circa 1965
[9] Roberts, [7]
[10] Either Roberts's memory was in error or he was misinformed because the *Cleveland* was soon to take part in the invasion of North Africa
[11] *Administrative Histories* [5], p 225
[12] Ibid., pp 233–236
[13] MORISON 5, p 107
[14] BALDWIN, pp 181–189
[15] *Administrative Histories* [5], pp 253–255
[16] SIG CORPS 3, pp 297–298
[17] Botho Stüve *Peenemünde West: Die Erprobungsstelle der Luftwaffe für geheime Fernlenkwaffen und deren Entwichlungsgeschichte* pp 761–777. Munich: Bechtle Verlag, 1995
[18] R W Burns, Factors Affecting the Development of the Radio Proximity Fuze 1940–1944 *IEE Proc. Sci. Meas. Technol.* Vol 143, pp 1–9, 1996

Chapter 4.5 Greater and lesser microwave sets

[1] LOVELL, p 45
[2] RAWNSLEY, pp 185–189
[3] James Sayers, MAGNETRON, pp 12–14
[4] LOVELL, p 52
[5] Ibid., pp 65–84
[6] BROWN, p 33

[7] Sir Bernard Lovell, H2S/ASV, MAGNETRON, pp 38–40

[8] Scope Distortion, RADAR, No 11, pp 29–32, 1945

[9] In allocating electron tubes in April 1942 Cherwell favored elimination of GL mark III as the best means of reconciling production with need. JAY, part 3, para 69

[10] BELL, pp 83–84

[11] D H Tomlin, The Origins and Development of UK Army Radar to 1946, BURNS, pp 284–295; D H Tomlin, Army Radar 1939–1945, MAGNETRON, pp 53–58

[12] ALVAREZ

[13] RADAR, No 10, pp 28–33, 1945

[14] ALVAREZ, pp 101–103

[15] GUERLAC 1, pp 386–394

[16] RADAR, [13]

[17] RADAR SURVEY 1, pp 127–132

[18] RADAR, [13], p 33

[19] For a description of some of these systems and the problems of implementing them see William L Leary, The Search for an Instrument Landing System, 1918–48 *Innovation and the Development of Flight* Roger D Saunius, editor, pp 80–99. College Station, Texas: Texas A&M University Press, 1999

[20] GUERLAC 1, pp 497–505

[21] Luis W Alvarez, Alfred Lee Loomis—Last Great Amateur of Science *Phys. Today* Vol 36, No 1, pp 25–34, 1983

[22] Captain C W Watson, Ground-Controlled Approach for Aircraft *Electronics* pp 112–115, 1945; Charles Fowler, IEEE, pp 101–114

[23] ALVAREZ, pp 98–101

[24] RADAR, No 7, pp 19–25, 1945. Details of the operation of GCA draw in part from a letter to the author from Squadron Leader (ret.) T Winchcombe dated 31 December 1997. The trials of mark I are the basis of an historically and technically correct autobiographical novel by Sir Arthur C Clarke *Glide Path* London: Sidgwick and Jackson, Ltd, 1963. Sir Edward Fennessy wrote that he had been the Wing Commander who selected Clarke as RAF Technical Officer to join the team working with Alvarez. The decision was influenced by the marginal comment on Clarke's résumé: 'brilliant but mad'.

[25] GUERLAC 1, pp 448–459

[26] RADAR SURVEY 3, pp 5–10

[27] Milton A Chaffee, IEEE, pp 42–46

[28] KEMP, pp 120–121

[29] GUERLAC 1, pp 1025–1030

[30] Ibid., pp 537–550

Chapter 5.1 The Mediterranean, 1940–1942

[1] HOWSE, pp 20–21

[2] Ibid., pp 63–64

[3] RAF SIGNALS 4, p 70

[4] HEZLET, p 194

[5] Macintyre *The Battle for the Mediterranean*, p 36

[6] Charles Lamb *War in a Stringbag* pp 105–113. London: Cassell and Company, 1977; Vice-Admiral Brian Betham Schofield *The Attack on Taranto* pp 40–52.

Annapolis: US Naval Institute, 1973. Imitation is often said to be the highest form of flattery. The Kriegsmarine built a number of multi-purpose aircraft for their carrier that was never finished, the *Graf Zeppelin* that had a design very similar to the Swordfish, the Fiessler Fi 167. It had a more powerful engine than the Swordfish, which provided higher speed and greater range

[7] Lamb [6], pp 40–44

[8] Tony Spooner, Goofingtons and Malta's War *Aeroplane Monthly* July, pp 409–412, 1988

[9] S W C Pack *Night Action off Cape Matapan* Annapolis: US Naval Institute, 1972

[10] Lutton op. cit., p 84

[11] Peter Brain, Sheilah Lloyd and F J Hewitt *South African Radar in World War II* Cape Town: The SSS Radar Book Group, 1993; B A Austin, Radar in World War II: The South African Contribution *Engineering Science and Education Journal* June, pp 121–130, 1992

[12] Ralph Bennett *Ultra and the Mediterranean Strategy* pp 72–74. New York: William Morrow and Company, 1986

[13] Ibid., pp 38–39

[14] RAF SIGNALS 4, pp 165–167

[15] Playfair op. cit., Vol 2, p 45

[16] RAF SIGNALS 4, pp 169–172

[17] Flying Officer John W Findlay, 202 Group, RAF, in a letter to the author dated 11 February 1994

[18] KEMP, pp 43, 52, 93, 109, 121

[19] HOFFMANN 1, pp 201–204

[20] Ibid., pp 206–207

[21] TERRAINE 2, p 651

[22] Flying Officer George A Emery, 230 Squadron, RAF, in a letter to the author dated 16 December 1994

[23] Macintyre *The Battle for the Mediterranean*, p 141

[24] JONES 1, p 256

[25] C Powell, A personal reminiscence: GL radar, an elementary ECCM technique, BURNS, pp 503–505

[26] German aerial photographs in the US National Archives, Record Group 242

[27] TEDDER, p 245

[28] Adolf Galland *Die Ersten und die Letzten: Die Jagdflieger im zweiten Weltkrieg* p 185. Darmstadt: Franz Schneekluth, 1953

[29] HOWSE, p 154

[30] PRICE 1, p 94; REUTER, p 109

[31] HOFFMANN 1, p 208

[32] BOWEN, pp 113–114. This brings up again the question of when Germany got an ASV According to Leo Brandt it was May 1941. For further discussion see Chapter 5.3, The Channel, 1942

Chapter 5.2 War in the Pacific

[1] Harold A Zahl *Radar Spelled Backwards* p 74. New York: Vantage Press, 1972

[2] Myron J Smith *Pearl Harbor, 1941: a Bibliography* New York: Greenwood Press, 1991

[3] US Congress *Pearl Harbor Attack: Hearings Before the Joint Committee on the Investigation of the Pearl Harbor Attack* part 26, pp 367–375. Washington: Government Printing Office, 1946

[4] The Army Air Corps changed into the Army Air Forces on 20 June 1941

[5] CRAVEN & CATE 1, pp 289–291

[6] Congress [3], part 26, pp 379–386; part 27, pp 615–632

[7] Ibid., part 10, pp 5027–5080; part 26, pp 517–536; part 32, pp 341–351

[8] SIG CORPS 2, pp 10–15

[9] HOWSE, pp 122–126

[10] Sir John Rupert Colville *The Churchillians* p 140. London: Weidenfeld and Nicholson, 1981

[11] Ed Simmonds, Historic Background *More Radar Yarns* Ed Simmonds, editor, pp 2–29. Published privately by E W & E Simmonds, 15 Blair Street, Port Macquarie NSW 2444, Australia, 1992

[12] Ed Simmonds in a letter to the author dated 19 November 1994

[13] Ed Simmonds in a letter to the author dated 6 August 1994

[14] WATT, pp 308–310

[15] Elting E Morison *Turmoil and Tradition: a Study of the Life and Times of Henry L Stimson* p 562. Boston: Houghton Mifflin Company, 1960

[16] SIG CORPS 2, pp 95–102

[17] Ibid., p 97

[18] CRAVEN & CATE 1, pp 291–293

[19] Elting Morison [15], 563

Chapter 5.3 The Channel, 1942

[1] ROSKILL 2, pp 147–150

[2] REUTER, p 76

[3] Apparently the ships were able to obtain bearings on shore Seetakt stations by using the directional capability of their receivers, but the wording of the only source for this, REUTER, is ambiguous

[4] John Deane Potter *Fiasco: the Break-out of the German Battleships* pp 62–63. New York: Stein and Day, 1970

[5] JONES 1, pp 233–234

[6] HOWSE, pp 129–131

[7] ROSKILL 2, p 154

[8] Captain H J Reinicke, The German Side of the Channel Dash *US Naval Institute Proceedings* Vol 81, pp 636–646, 1955

[9] BRANDT, pp 39–40. Brandt is quite specific in giving the time of a German examination of an ASV set: 'Am Tage der Versenkung der "Bismarck" befand sich ein "ASV"-Gerät im Laboratorium der Firma Telefunken'. This together with Brandt's extraordinary competence and high position would seem to exclude the possibility of the date being a typographical error, yet Metox radar receivers to counter ASV and designed from a knowledge of it were not issued to U-boats until August 1942. REUTER, p 153

[10] WATT, pp 358–378

[11] Sir Robert Cockburn, The Radio War BURNS, p 337

[12] F A Kingsley, Electronic Countermeasures in the Royal Navy KINGSLEY 2, pp 196–199

[13] JONES 1, pp 121–125; for a detailed description of the raid see Millar, op. cit.

[14] Cockburn [11], pp 337–338

[15] JONES 1, pp 130–134; Millar, op. cit., pp 109–133

[16] Ibid., p 192

[17] PRICE 1, p 78

[18] Cockburn, loc. cit.

[19] JONES 1, pp 233–249

[20] Ibid., pp 244–246

[21] ROWE, pp 128–134

[22] Terence Robertson *Dieppe: the Shame and the Glory* Boston: Little, Brown and Company, 1962

[23] John P Campbell *Dieppe Revisited: a Documentary Investigation* pp 130–132. London: Frank Cass and Co., 1993

[24] Ibid., p 137

[25] ROSKILL 2, p 241

[26] Campbell [23], p 131

[27] NIEHAUS, pp 93–94; HOFFMANN 1, pp 268–269; Campbell, [23], p 141

[28] J R Robinson, Radar Intelligence and the Dieppe Raid *Canadian Defence Quarterly* Vol 20, pp 37–43, 1991

[29] Jack Nissen and A W Cockerill *Winning the Radar War: a Memoir* pp 162–191. New York: St. Martin's Press, 1987

[30] Robinson [28], p 41

[31] JONES 1, pp 195–198; Robinson [28], p 42

[32] JONES 1, p 402

[33] Campbell [23], pp 147, 168

[34] ROSKILL 2, pp 246–247

[35] Campbell [23], p 148

[36] Derek Howse, Type Number of Radar Sets, Operational or Designed, KINGSLEY 1, pp 372–373

[37] Hugh G Henry III, draft of a dissertation intended for submission to St John's College, University of Cambridge

[38] Robinson [28], p 42

Chapter 5.4 Carrier warfare defined

[1] FRIEDMAN, p 149

[2] LUNDSTROM 1, p 63

[3] Ibid., pp 91–93

[4] Ibid., pp 96–98, 101–102

[5] Ibid., pp 118–119

[6] Ibid., pp 148–149

[7] Ibid., p 169

[8] Ibid., p 196

[9] Ibid., pp 210–211

[10] Ibid., pp 243–248

[11] Ibid., p 293

[12] MORISON 4, pp 141–159

[13] Walter Lord *Incredible Victory* New York: Harper and Row, 1967

[14] Gordon Prange *Miracle at Midway* New York: McGraw-Hill, 1982

[15] LUNDSTROM 1, pp 309–449

[16] MORISON 4, p 104

[17] Mitsuo Fuchida and Masatake Okumiya *Midway: the Battle That Doomed Japan: The Japanese Story* pp 243–244. Annapolis, Maryland: US Naval Institute, 1955

[18] NAKAGAWA, pp 32, 89

[19] LUNDSTROM 1, p 374

Chapter 5.5 The South Pacific, 1942

[1] The determination of when vessels received radar was made from the collection of photographs in the Still Picture Division of the US National Archives. Photographs were taken on completion of construction, repair or modification and generally allowed the radar antennas to be recognized

[2] Letter dated 16 January 1995 to the author by Captain Russell S Crenshaw, Jr, who was Gunnery Officer aboard USS *Maury* during these actions

[3] FRIEDMAN, pp 147–148. The important information, that the *San Juan* carried an SG radar at the time of the Battle of Savo Island on 8/9 August 1942, was confirmed at the National Archives by examination of photographs taken on 31 May 1942 in Boston Harbor. She also had an SC and two FDs

[4] MORISON 5, p 154

[5] MCNALLY, pp 13a–13c

[6] LUNDSTROM 2, p 39

[7] PRICE 2, pp 47–48, 290

[8] LUNDSTROM 2, p 89

[9] Lieutenant Lewis C Mattison, USNR, and Master Sergeant Dermott H MacDonnell, USMC, Report on Fighter Direction at Cactus, October 8, 1942 to January 1, 1943. This document obtained through the courtesy of Dr John B Lundstrom, Milwaukee Public Museum

[10] Frank op. cit., p 111

[11] Ibid., p 117

[12] LUNDSTROM 2, p 93

[13] Frank op. cit. p 178

[14] LUNDSTROM 2, p 117

[15] Jennings B Dow, Navy Radio and Electronics During World War II *Proc. IRE* Vol 34, pp 284–287, 1946

[16] Robert C Rasmussen, It Helped Sink Six Jap Warships *Bell Laboratories Record* Vol 24, pp 201–202, 1946

[17] NAKAGAWA, pp 43–44

[18] LUNDSTROM 2, pp 467–471

[19] Frank op. cit., pp 402–403

[20] Ibid., p 379

[21] LUNDSTROM 2, pp 339–340

[22] Ibid., pp 384, 458

[23] Ibid., pp 454–455

[24] C W Kilpatrick *The Night Naval Battles in the Solomons* p 81. Pompano Beach, Florida: Exposition Press of Florida, Inc., 1987

[25] The gun flashes were only on American ships, as the Japanese used flashless powder

[26] LUNDSTROM 2, pp 478–483

[27] ROSKILL 2, p 232

[28] Undersea exploration by Dr Robert D Ballard, Woods Hole Oceanographic Institution presented at the conference 'World War II in the Pacific', 10–12 August 1994, Crystal City, Virginia

[29] Frank op. cit., pp 486–487

[30] Ibid., p 207

[31] SIMMONDS & SMITH, pp 97–108; see also [32]

[32] Samuel Milner *United States Army in World War II The War in the Pacific: Victory in Papua* Washington: Office of the Chief of Military History, 1957

[33] SIG CORPS 2, pp 261–265. Lt Col Harold A Zahl and Maj John W Marchetti, Radar on 50 Centimeters *Electronics* January 1946, pp 98–104; February 1946, pp 98–103

[34] SIMMONDS & SMITH, pp 39–53, 97–108

Chapter 5.6 The Eastern Front

[1] Steven J Zaloga, Soviet Air Defense Radar in the Second World War *Journal of Soviet Military Studies* Vol 2, No 1 (March), pp 104–116, 1989

[2] LOBANOV, pp 238, 245, 80

[3] Zaloga [1], p 105

[4] Generalleutnant a.D Klaus Uebe *Russian Reactions to German Airpower in World War II* p 10. New York: US Air Force Historical Division, Aerospace Studies Institute, Air University, Arno Press, 1964

[5] PLOCHER 1, pp 239, 156

[6] LOBANOV, pp 184–185

[7] Ibid., pp 228–235

[8] Pekka Eskelinen, View of Military Radio Systems and Electronic Warfare in Finland During WWII *AES* Vol 11, pp 3–7, 1996

[9] Kurt Petsch *Nachtjagdleitschiff ¡Togo'* pp 58–100. Reutlingen: Preussischer Militär-Verlag, 1988

[10] LOBANOV, pp 81–82

[11] Ibid., pp 191–197

[12] PRICE 3, p 338

[13] LOBANOV, p 199

[14] Zaloga [1], p 111

[15] Ibid., pp 111–112

[16] HOFFMANN 1, p 117

[17] PLOCHER 1, p 163

[18] HOFFMANN 1, p 127

[19] Walter Schwabedissen *The Russian Air Force in the Eyes of the German Commanders* p 360. New York: US Air Force Historical Division, Aerospace Studies Institute, Air University, Arno Press, 1960

[20] HOFFMANN 1, pp 169–170

[21] Ibid., pp 303–305

[22] PLOCHER 3, p 64

[23] Walter Bartig, Geschichte einer Funkkompanie und ihrer Männer, p 34. Manuscript, Berlin, September 1987

[24] Schwabedissen [19], pp 318–323

[25] PLOCHER 3, p 136

[26] HOFFMANN 1, pp 163–165

[27] Schwabedissen op. cit., p 376

[28] HOFFMANN 1, pp 184–188

Chapter 6.1 The destruction of German cities initiated

[1] WEBSTER & FRANKLIN 1, pp 155–166

[2] Ibid., p 459

[3] The term 'AN-Verfahren' is traced back to the Lorenz beam-navigation system in which the pilot heard dots, if he were on one side of the correct course (in one of the two radiation lobes), dashes if he were on the other side (in the second lobe) and long dashes if on course (equal amplitudes from both lobes). Straying a little to the dot side generated from the unequal amplitudes an audible signal that fooled the ear into recognizing a Morse letter A (dit–dah); straying to the dash side a Morse letter N (dah–dit)

[4] KROGE, p 121

[5] REUTER, p 244; TRENKLE 1, pp 91–93

[6] REUTER, p 243; TRENKLE 1, pp 93–94

[7] One version of Wassermann shows a cabin half way up the structure that is frequently taken to be the operating room. The dipoles had to be fed in phase, which was done by using the same lengths of cable to connect each dipole with the generator. The cabin stored excess cable on reels

[8] FRIEDMAN, p 173

[9] See Chapter 4.5

[10] REUTER, p 84; TRENKLE 1, pp 74–76

[11] BRANDT, pp 32–33

[12] RUNGE, p 47

[13] REUTER, p 34

[14] Ibid., p 86

[15] Udet's genial nature is demonstrated by his long friendship with Carl Zuckmayer, a strong anti-Nazi playwright, who was forced to flee Europe minutes ahead of the Gestapo with each new advance of the dictator's power. His play 'Des Teufels General' attempts to salvage the memory of his friend

[16] PRICE 1, pp 64–65

[17] REUTER, p 87; TRENKLE 1, pp 46–49

[18] STREETLY, pp 214–215

[19] TRENKLE 3, pp 182–192

[20] ADERS, pp 76–77

[21] R Cockburn, The Radio War, BURNS, pp 330–356

[22] HOFFMANN 1, pp 330–332; TRENKLE 1, pp 102–104

[23] R V Jones in a letter to the author dated 18 April 1995

[24] TRENKLE 1, pp 104–106; Theodor Schultes, Funkmess- Übersichtsverfahren *Die Funkortung der deutschen Flugsicherung* pp 56–94. Dortmund: Verkehrs- und Wirtschafts-Verlag, GmbH, 1953

[25] STREETLY, p 215

[26] TRENKLE 1, pp 51–54

[27] A E Hoffman-Heyden, German World War II Anti-jamming Techniques, BURNS, pp 374–396

[28] RUNGE, pp 54–55

[29] REUTER, pp 87–88
[30] PRICE 1, pp 69–70
[31] RUNGE, p 49
[32] ADERS, pp 77–79
[33] HOFFMANN 1, pp 32–34
[34] PRICE 1, p 70
[35] RUNGE, pp 55–57, 76–82
[36] Manfred von Ardenne *Ein glückliches Leben für Technik und Forschung* pp 129–130. Berlin: Verlag der Nation, 1973
[37] V Ardenne [36], pp 105, 101, 124–128, 67–68
[38] Radar attracted others who had little stomach for some kinds of war research. George Valley, who pushed for the construction and designed in the late 1940s the radar system to guard the North American continent from Soviet attack, was a Rad Lab veteran. His graduate research had been in nuclear physics, and when demands for radar work slacked DuBridge asked him to transfer to Los Alamos for bomb work. Valley refused, saying 'No, I think that's filthy, I won't do it'. BUDERI, p 358
[39] ROWE, p 108
[40] There are numerous descriptions of Gee. Colin Latham provided this one
[41] ROWE, p 109
[42] The frequencies used were the same used by Knickebein, as these were those of the pre-war Lorenz blind approach system that the RAF used under license and for which their planes were equipped
[43] PRICE 1, pp 98–104

Chapter 6.2 Countermeasures

[1] PRICE 2, p 11
[2] STREETLY, pp 17–18, 160; RAF SIGNALS 7, pp 191–194
[3] STREETLY, p 18–19; RAF SIGNALS 7, pp 76–78
[4] R Cockburn, The radio war, BURNS, pp 330–356
[5] REUTER, p 142
[6] OSRD 5, pp 12–13
[7] Ibid., p 18
[8] PRICE 2, pp 11–33
[9] Ibid., p 59
[10] OSRD 5, p 28
[11] RAF SIGNALS 7, pp 153–154
[12] Ibid., p 41
[13] PRICE 2, pp 86–89
[14] STEPP, p 72
[15] J S Hey and G S Stewart, Radar Observations of Meteors *Proc. Phys. Soc.* Vol 59, pp 858–883, 1947
[16] J S Hey, Solar Radiations in the 4–6 Metre Radio Wave-Length Band *Nature* Vol 157, pp 47–48, 1946
[17] M J B Scanlan, Chain Home Radar—a Personal Reminiscence *GEC Review* Vol 8, No 3, pp 171–183, 1993. Scanlan points out that the CH signal/noise ratio was established at the antenna, which made imperfections in the long transmission lines to the receivers less troublesome

[18] PRICE 1, pp 112–114

[19] WATT, p 395

[20] PRICE 1, p 114

[21] HOFFMANN 1, pp 201–202

[22] RAF SIGNALS 5, pp 149–150, 162

[23] The Würzburg's rotating polarization allowed a blanking circuit to suppress the plane-polarized jammer, Carpet, which then had to be provided with a rotating field antenna. OSRD 5, p 77

[24] REUTER, pp 106–107

[25] There are reports that the name comes from the Danish village Dybböl (German Düppel) on the Baltic Sea where tests were presumably made

[26] PRICE 2, pp 31, 62–63

[27] OSRD 5, p 25

[28] The Würzburg transmitter had to be altered for these techniques. The pass band of the receiver was much larger than typical Doppler shifts, and filters designed to pass the shifted radio frequency would have failed to generate an observable signal, if the phase of the radio frequency had not been coherent from pulse to pulse, as it was not in the self-exciting oscillator used. Coherence was attained by using a continuously-running, low-power auxiliary oscillator to stimulate the onset of oscillation (demanded by the modulator pulse) to be in phase with this reference signal, a process called phase locking or injection locking

[29] PRICE 2, p 284

[30] HOFFMANN-HEYDEN, p 382

[31] Ibid., pp 194–202

[32] Ibid., pp 386–387

[33] DAVIS, p 527

[34] Memorandum of the Reichsminister der Luftfahrt und Oberbefehlshaber der Luftwaffe, Berlin, 5 January 1944

[35] PRICE 1, p 142

[36] KINGSLEY 2, pp 197–201

[37] JONES 1, pp 284–286

[38] PETERSEN, pp 18–20

[39] PRICE 2, pp 47–56

[40] E H Cooke-Yarborough, Countermeasures Receiver Techniques, BURNS, pp 365–373,

Chapter 6.3 An air war of attrition

[1] WEBSTER & FRANKLIN 4, p 6

[2] JONES 1, pp 258–259. For a technical description that illustrates the extent of Allied interest, see J H Buck and J A Pierce, Nonradar Navigational Methods *Radar Aids to Navigation* John S Hall, editor *Radiation Laboratory Series* Vol 2, pp 47–50. New York: McGraw-Hill Book Company, 1947

[3] F E Jones, OBOE—a precision ground controlled blind bombing system, BURNS, pp 319–329

[4] ROWE, pp 144–147

[5] F E Jones [3], pp 324–325

[6] Sir Edward Fennessy in letters to the author dated 7 September and 23 November 1995

[7] DAHL, p 10

[8] WEBSTER & FRANKLIN 2, pp 108–137

[9] REUTER, pp 96–97

[10] Ibid., p 123

[11] WEBSTER & FRANKLIN 2, p 12

[12] REUTER, p 250

[13] RAWNSLEY, pp 278–292

[14] WEBSTER & FRANKLIN 2, p 143

[15] RAF SIGNALS 7, p 111

[16] This belief came full blown from the pages of the air prophet. Giulio Douhet *The Command of the Air* pp 371–389. New York: Coward-McCann, Inc., 1942

[17] GUERLAC 1, p 773

[18] Ibid., pp 766–772

[19] The name H2S, the origin of which has various traditions, caused 10 cm microwave equipment to be referred to as S-band. H2X, whose nominal origin is obscure, resulted in 3 cm microwaves being referred to as X-band

[20] GUERLAC 1, pp 776–783

[21] Just How Accurate is H2X Bombing? RADAR, No 9, pp 43–44, 1945

[22] GUERLAC 1, pp 788, 800

[23] Ibid., pp 746–753

[24] CRAVEN & CATE 2, pp 681–706, 848–850

[25] REUTER, p 134

[26] WEBSTER & FRANKLIN 2, pp 190–211

Chapter 6.4 Arbeitsgemeinschaft Rotterdam

[1] The number has been recorded ranging from 1 500 to 9 000

[2] TRENKLE 1, pp 62–63

[3] REUTER, pp 48–49, 113–114. It is useful to note the German names for the various bureaus and offices just given. Air Ministry = Reichs-Luftfahrts-Ministerium; Air Signals = Luftnachrichtentruppe; Chief of Air Signals (Martini) = Chef des Nachrichtenverbindungswesens der Luftwaffe; Special Commissioner for Radar (Martini) = Sonderbeauftragter für Funkmessgeräte; Supervisor of Technical Communications (Fellgiebel) = Generalbevollmächtiger für technische Nachrichtenmittel; Plenipotentiary of High-frequency Research (Plendl, Esau) = Bevollmächtiger für die Hochfrequenzforschung

[4] Arbeitsgemeinschaft does not translate into committee, although it functioned as such. Working group is perhaps better

[5] HOLLMANN Denis M Robinson cites a German book by Thoma in an interview published in IEEE Frederick Seitz has shown this to have certainly been Hollmann's book

[6] ROTTERDAM, 23 February and 17 March 1943

[7] A technical report based on a meeting about antennas held 24–26 March 1943 and issued by Zentrale für wissenschaftliches Berichtswesen der Luftfahrtforschung, pp 153–161

[8] ROTTERDAM, 23 February, p 3; 17 March 1943, p 8

[9] Ibid., 8 April 1943, p 1

[10] Ibid., 1 and 22 June and 14 December 1943

[11] Ibid., 14 December 1943, p 6

[12] A succinct statement about German radio amateurs can be found in *Die geheimen Konferenzen des Generalluftzeugmeisters* Georg Hentschel, editor, pp 124–125. Koblenz: Bernard und Graefe Verlag, 1989. The following dialogue discloses not only the difficulties faced by radio amateurs but the extraordinary mistrust by the Nazis for a large part of the German population. Pasewaldt: 'I think the progress of this field in England and America is essentially the result of the unheard of importance of their radio amateurs, while in Germany they have been unyieldingly suppressed . . .' Feldmarschall Milch: 'It was done by the offices responsible for security. The whole German radio group before the war had been trained for pure communist espionage. The amateurs were up to 99% Moscow boys, so we said that now we are going to cut the wires of those fellows, which was just as it should have been. . . . After the war we must introduce amateur radio into the Hitler Youth and let it bloom. . . .' (In answer to the problem that some were still communicating illegally with radio, Milch discloses a less endearing side of his personality.) 'Those scoundrels are lucky that I am not Chief of the Gestapo for there would be far more executions. They are far too mild and humane; they cannot compare themselves with the Russian GPU'

[13] STEGSKOPFER The search for qualified technicians continued by other means. Public notices calling for soldiers engaged in non-electronic duties to apply for transfer, even recommending that family members apply for those at the front who might miss the announcements. There was no mention of radar. News item, Hochfrequenz-Fachkräfte für die Luftwaffe *Funkschau* p 295, Vol 16, October–December, 1943

[14] JONES 2, pp 325–326

[15] Letter from Staatsrat Dr-Ing H Plendl to Reichsführer SS Himmler dated 7 January 1944

[16] JONES 2, loc. cit.

[17] Letter from Hans Plendl, Jr., dated 14 March 1995

[18] Fritz Trenkle, Zum 90. Geburtstag von Hans Plendl *Funkgeschichte* No 78, pp 3–5, 1991. See Chapter 10.3, Secrecy and the Technical Imperative.

[19] H Frühauf, H E Hollmann zum 60. Geburtstag *Hochfrequenztechnik und Elektroakustik* Vol 68, pp 141–143, 1959

[20] REUTER, pp 198–200

[21] Personal communication dated 10 April 1996 from Frau Anna Maria Elstner, Runge's daughter

[22] Professor Kurt Fränz, Als Student und Doktorand in der Weimarer Republik und im Dritten Reich, unpublished manuscript communicated in a letter to the author dated 15 July 1996

[23] Professor Hans Plendl, Jr, in a statement to the author, 26 August 1996

[24] DAHL, pp 16–24; KAUFMANN, pp 43–46

[25] REUTER, pp 135, 141

[26] HOFFMANN-HEYDEN, pp 239–242

[27] Ibid., pp 248–264, 293

[28] REUTER, pp 147–148

[29] Ibid., p 195

Chapter 6.5 The destruction of German cities completed

[1] For a study of how Germany coped with the air attacks see Earl R Beck *Under the Bombs: the German Home Front, 1942–1945* Lexington, Kentucky: The University Press of Kentucky, 1986

[2] PRICE 1, p 225

[3] Ibid., pp 213–214

[4] Ibid., pp 230–233

[5] Ibid., pp 221–222

[6] HOFFMANN 1, p 329; PETERSEN, pp 60–61. These books provide the most detailed account of how the Luftwaffe night fighters used radar and radar control in their ever changing struggle with Bomber Command

[7] PRICE 2, pp 280–288

[8] KAUFMANN, pp 101–114

[9] KOCH, p 75

[10] Sir Edward Fennessy in a letter to the author dated 23 November 1995

[11] PRICE 2, pp 189–190

[12] G Förster, German experiments in Jamming H2S airborne radar, BURNS, pp 397–404

[13] TRENKLE 2, pp 138–139; ROTTERDAM, 29 September 1943

[14] OSRD 5, p 107

[15] CRAVEN & CATE 3, p 723

[16] Ibid., pp 666–669

[17] GUERLAC 1, pp 772, 1076–1079

[18] RADAR SURVEY 1, pp 145–149, 213–216

[19] Ibid., p 189

[20] ROWE, p 117

Chapter 7.1 The Battle of the Atlantic, 1939–1945

[1] For a brief description of the code war see Brian Johnson *The Secret War* pp 305–349. New York: Methuen Inc., 1978

[2] HOWSE, pp 58, 100

[3] David Zimmerman, Technology and Tactics *The Battle of the Atlantic, 1939–1945: the 50th Anniversary International Naval Conference* pp 476–489, Stephen Howarth and Derek Law, editors. London: Greenhill Books, 1994

[4] Captain Donald Macintyre *U-Boat Killer* p 50. Annapolis: Naval Institute Press, 1976

[5] HOWSE, p 100

[6] Ibid., pp 108–110; C A Cochrane, Development of Naval Warning and Tactical Radar Operating in the 10-cm Band, 1940–5, KINGSLEY 1, pp 185–276

[7] Macintyre [4], p 63

[8] BOWEN, pp 102–103

[9] PRICE 4, pp 78–79

[10] Ibid., pp 60–65; Axel Niestlé in a letter to the author dated 15 April 1995

[11] PRICE 4, pp 87–91

[12] Blair, 1996 op. cit.

[13] TRENKLE 2, p 44. See also the historical survey presented by Dr Bode at a meeting of the Rotterdam Committee. ROTTERDAM, 26 April 1944, pp 30–40

[14] BRANDT, pp 39–40

[15] Axel Niestlé, German Technical and Electronic Development *The Battle of the Atlantic* pp 430–451, Stephen Howarth and Derek Law, editors. London: Greenhill Books, 1994.

[16] PRICE 4, pp 94–95

[17] P M S Blackett, Operational Research: Recollections of Problems Studied, 1940–45 *Brassey's Annual: the Armed Forces Yearbook, 1953* pp 88–106

[18] Marshal of the Royal Air Force Sir John Slessor *The Central Blue: Recollections and Reflections* pp 524–525. London: Cassell and Company Ltd, 1956

[19] The equipping of these planes with microwave radar had taken place with the active participation of Rad Lab personnel, even to the attacking of some of the raiders

[20] Max Schoenfeld *Stalking the U-Boat: USAAF Offensive Antisubmarine Operations in World War II* pp 3–6. Washington: Smithsonian Institution Press, 1995

[21] ROTTERDAM, 17 March 1943

[22] PRICE 4, p 168

[23] Niestlé [15], p 443

[24] This was the case for SCR-517A and C-SCR-517B had a PPI indicator

[25] Schoenfeld [20], pp 37–53. In letter to the author dated 15 April 1995 Niestlé asserts this one sinking was not justified

[26] Russell W Burns, Impact of Technology on the Defeat of the U-boat, September 1939–May 1943 *IEE Proceedings: Science, Measurement and Technology* Vol 141, pp 343–355, 1994

[27] LOVELL, p 161

[28] As late as April 1944 Brandt opened a meeting of the Rotterdam Committee with an admonition not to discuss matters with anyone not officially involved, clear evidence of a demand to tighten security. ROTTERDAM, 5 April 1944, p 1

[29] Based on a search of applicable records by Axel Niestlé reported in letters to the author dated 12 March and 16 April 1995

[30] REUTER, p 161

[31] Schoenfeld [20], pp 80–83

[32] Brian McCue *U-Boats in the Bay of Biscay: an Essay in Operations Analysis* p 65. Washington: National Defense University Press, 1990

[33] REUTER, p 161

[34] PRICE 4, pp 165–171; REUTER, pp 162–166

[35] PRICE 4, pp 165–171

[36] ROTTERDAM, 26 April 1944, pp 12–22

[37] RUNGE, pp 50–51

[38] ROSKILL 3, pp 365–366

[39] HOFFMANN 1, p 63

[40] Richard Natkiel, Maps *Battle of the Atlantic* see [15], p 23

[41] Slessor [18], p 518

[42] TERRAINE 2, pp 767–768

[43] R W Burns, The Background to the Development of the Cavity Magnetron, BURNS, pp 259–283

[44] The Allies lost 2 353 ships of which only 19 were from convoys with air cover. Macintyre, 1976 [4], p 173

[45] Air Commodore Henry A Probert, Head of the RAF Air Historical Branch, in a letter to the author dated 2 February 1995; Axel Niestlé in a letter to the author

dated 15 April 1995

[46] For a summary of such matters see Burns, [43]

[47] I base this statement on the extensive descriptions of the use of both by Donald Macintyre, an outstanding escort commander. See Macintyre [4]. For an excellent account of the HF/DF technique stripped of its secrecy, see P G Redgment, High-Frequency Direction Finding in the Royal Navy, KINGSLEY 2, pp 229–266. For the German perspective, see J Rohwer, Die Funkführung der deutschen U-Boote im zweiten Weltkrieg *Wehrtechnik* pp 324–328, 360–364, 1969. For an extensive review of American HF/DF see Kathleen Broome Williams *Secret Weapon: US High-Frequency Direction Finding in the Battle of the Atlantic* Washington: Naval Institute Press, 1996. For a very complete description of all aspects of U-boat wireless communication and British surface vessel HF/DF see Arthur O Bauer *Funkpeilung als aliierte Waffe gegen deutsche U-Boote 1939–1945* Rheinberg (D-47486), Germany: Herausgeber & Vertrieb (Postfach 301 217), 1997

[48] Francis Harry Hinsley *British Intelligence in the Second World War: its Influence on Strategy and Operations* Vol 2, p 177. New York: Cambridge University Press, 1981

[49] P G Redgment [47], p 255

[50] Axel Niestlé in a letter to the author dated 15 April 1995

[51] Y'Blood op. cit., pp 273–274

[52] Ibid. op. cit.

[53] MORISON 10, pp 171–177

[54] TRENKLE 1, pp 125–126

[55] Niestlé [15], pp 441–442

[56] TRENKLE 1, pp 139–142; REUTER, pp 167–168; Niestlé, [15], p 441

[57] GUERLAC 1, p 726

[58] TERRAINE 2, pp 626–627

[59] Niestlé [15], pp 445–446

[60] Y'Blood op. cit.

Chapter 7.2 Radar in arctic waters

[1] PLOCHER 2, pp 35–49

[2] Ibid., p 38

[3] ROSKILL 2, pp 138–144

[4] Ibid., pp 279–285

[5] Ibid., p 291

[6] HOWSE, pp 158–159; ROSKILL 2, loc. cit., reports that only two escort vessels had radar, of which the minesweeper was one. Given Howse's sources, this is clearly in error. Dudley Pope *73 North: the Battle of the Barents Sea* Annapolis: Naval Institute Press, 1989, gives detail of radar use by all British destroyers

[7] HOWSE, pp 158–159

[8] For details of the engagement, skillfully told, see Pope [6]

[9] Pope [6], p 185

[10] Walter Bartig, Geschichte einer Funkkompanie und ihrer Männer, p 38. Manuscript, Berlin, September 1987

[11] PLOCHER 3, p 209

[12] G Muller and R Bosse, German Primary Radar for Airborne and Ground-Based Surveillance, BURNS, pp 200–208

[13] BEKKER 2, p 240

[14] ROSKILL 3, pp 78–88

[15] The disposition with time of installation of radar in the Royal Navy is given in appendix 3 of KINGSLEY 1, pp 387–400

[16] There is a report (HOWSE, pp 187–188) of a naval adaptation of the 55 cm air-search set Hohentwiel on the mainmast, but none of the reports suggest that it made any contribution to the battle; it may have been knocked out by the same shot that disabled the forward Seetakt

[17] For details of all these attempts and those that follow see Ludovic Kennedy *The Death of the Tirpitz* Boston: Little, Brown and Company, 1979

[18] PRICE 1, pp 244–246

Chapter 7.3 The Mediterranean, 1943–1945

[1] CRAVEN & CATE 2, p 3

[2] SIG CORPS 2, pp 374–375

[3] RAF SIGNALS 4, pp 261–268

[4] War Diary, 62nd Coast Artillery, 12 November 1942 to 26 April 1946. Copy furnished by Colonel John M Godfrey, Adjutant

[5] John Manning, commander of Battery C, 62nd CA (AA), in a letter to the author dated 23 March 1994

[6] War Diary, 62nd Coast Artillery, 30 September 1943

[7] GUERLAC 1, p 110

[8] RAF SIGNALS 4, p 272

[9] GUERLAC 1, p 705

[10] SIG CORPS 2, pp 257–260, 377–378

[11] GUERLAC 1, pp 699–706

[12] HOFFMANN 1, pp 212–221

[13] TERRAINE 1, pp 390–393

[14] E Giboin, L'Evolution de la Detection Electromagnetique dans la Marine Nationale *Onde* Vol 31, pp 53–64, 1951; Maurice Ponte, Sur des Apports Français a la Technique de la Détection Électromagnétique *Revue technique Thomson-CSF* Vol 1, pp 171–180, 1946

[15] PRICE 2, pp 71–79

[16] RAF SIGNALS 4, pp 302–304

[17] Ibid., p 325

[18] WATT, pp 146–147, 332–334

[19] JAY, part 2, para. 30

[20] HOFFMANN 1, p 224; PRICE 2, p 301

[21] CRAVEN & CATE 2, pp 520–545

[22] RAF SIGNALS 4, p 347

[23] SIG CORPS 3, p 303

[24] CRAVEN & CATE 3, pp 346–352

[25] SIG CORPS 3, p 58

[26] RADAR, No 5, 30 September 1944, pp 10–12

Chapter 7.4 Japanese shipping destroyed

[1] FRIEDMAN, pp 146–147

[2] BELL, pp 69, 75–81; RADAR SURVEY 4, pp 135–136

[3] Blair, 1975 op. cit., p 113

[4] Ibid., pp 321–322

[5] Roscoe, 1949 op. cit., pp 170–172

[6] Blair, 1975 op. cit., p 530

[7] Vice Admiral Charles A Lockwood, Electronics in Submarine Warfare *Proc. IRE* Vol 35, pp 712–715, 1947

[8] Mochitsura Hashimoto (translated by Commander E H M Colegrave) *Sunk: the Story of the Japanese Submarine Fleet, 1941–1945* pp 200–205. New York: Henry Holt and Company, 1954

[9] See Chapter 4.2 for details

[10] Richard C Knott *Black Cat Raiders of WWII* Annapolis: The Nautical and Aviation Publishing Company, 1981

[11] GUERLAC 1, pp 997–1001

[12] BELL, pp 71, 100–102; RADAR SURVEY 1, pp 117–120

[13] Low Altitude, High Precision, RADAR, No 1, April 1944, pp 19–22

[14] LAB vs. Jap Shipping, RADAR, No 5, November 1944, pp 3- 9

[15] Ibid.

[16] GUERLAC 1, pp 1006–1013

Chapter 7.5 The wide Pacific

[1] FRIEDMAN, p 149; RADAR SURVEY 2, pp 137–138

[2] FRIEDMAN, p 150; RADAR SURVEY 2, pp 143–144; GUERLAC 1, pp 439–442

[3] FRIEDMAN, pp 151–152; RADAR SURVEY 2, pp 163–164; GUERLAC 1, pp 442–443

[4] MONSARRAT, p 46

[5] Ibid., pp 53–57

[6] MORISON 6, p 108

[7] MONSARRAT, pp 61–62

[8] PAGE, p 148

[9] MCNALLY

[10] A E Fanning, The Action Information Organization, KINGSLEY 2, p 168

[11] FRIEDMAN, pp 147–148; RADAR SURVEY 2, pp 131–132

[12] FRIEDMAN, p 173; RADAR SURVEY 2, pp 39–44

[13] SOUTHWORTH, pp 201–207

[14] H T Friis and W D Lewis, Radar Antennas *Bell Tech. J.* Vol 26, pp 219–317, 1947

[15] TRENKLE 1, pp 147–151

[16] Lloyd V Berkner, Naval Airborne Radar *Proc. IRE* Vol 34, pp 671–706, 1946

[17] BELOTE, pp 204–205. There is evidence that O'Hare was shot down by one of the Japanese aircraft; John Lundstrom and Steve Ewing *Fateful Rendezvous* Annapolis: Naval Institute Press, 1997

[18] Berkner [16], pp 692–695

[19] NAKAGAWA, p 49

[20] Ibid., pp 57–58

[21] NAKAGAWA, p 66

[22] Japanese Monograph No 118, Operational History of Naval Communications, December 1941—August 1945, pp 110–111. Washington: Office of the Chief of Military History, Department of the Army, 26 May 1953

[23] NAKAJIMA, p 255

[24] NAKAGAWA, p 36

[25] BELOTE, p 208; PRICE 2, p 293; NAKAGAWA, p 91

[26] NAKAGAWA, pp 54–55

[27] PRICE 2, p 143

[28] NAKAGAWA, pp 55–56

[29] Charles A Lockwood and Hans Christian Adamson *Battles of the Philippine Sea* pp 14, 57. New York: Thomas Y Crowell Company, 1967

[30] MORISON 6, p 394

[31] Lockwood and Adamson [29], pp 19–23

[32] MONSARRAT, p 77

[33] Brian Garfield *The Thousand-Mile War: World War II in Alaska and the Aleutians* pp 24–32. Garden City, New York: Doubleday and Company, 1969

[34] PRICE 2, pp 53–55

[35] Garfield [33], pp 717–179

[36] Guadalcanal preceded Attu but the landing was unopposed; the tough part came later

[37] NAKAGAWA, p 46

[38] Action Report of USS *Mississippi* 26 July 1943, Office of Naval Records and Library

[39] Garfield [33], pp 271–282; PAGE, p 151; MORISON 7, p 59–61, erroneously includes the *Idaho* and the *Monaghan* in this action

[40] *Mississippi* Action Report, loc. cit.

[41] MORISON 6, p 104

[42] John Miller, Jr *United States Army in World War II The War in the Pacific. Cartwheel: the Reduction of Rabaul* p 94. Washington: Office of the Chief of Military History, 1959

[43] GUERLAC 1, pp 994, 1020

[44] MORISON 8, pp 233, 321

[45] William A Klingaman *APL—Fifty Years of Service to the Nation: a History of the Johns Hopkins University Applied Physics Laboratory* pp 14–15. Laurel, Maryland: The Johns Hopkins University Press, 1993

[46] The details of the Battle of the Philippine Sea, including its radar component are well told in the general references given

[47] John Hightower, Most Secret Weapon, Radar *The Evening Star* (Washington), 21 June 1943, p A5; 22 June, p A6; 23 June, p A6. Radar Stories Are Released by US and Great Britain *Electronics* June, pp 274–282, 1943

[48] US Naval Historical Center, Washington Navy Yard, photograph NH72278-KN

Chapter 8.1 Invasion

[1] Mark A Stoler *D-Day 1944* Theodore A Wilson, editor, pp 298–317. Lawrence, Kansas: University of Kansas Press, 1994

[2] JONES 1, pp 400–401

[3] Ibid., pp 116–118

[4] HOFFMANN 1, p 276

[5] Friedrich Ruge *D-Day* [1], p 127

[6] PRICE 2, p 128

[7] STREETLY, pp 53–54; RAF SIGNALS 7, pp 232–233

[8] Ibid., loc. cit.

[9] F A Kingsley, Electronic Countermeasures in the Royal Navy, KINGSLEY 2, pp 214–221

[10] Professor Sir Martin Ryle, D-13: Some Personal Memories of 24–28th May 1944 *Proc. IEE* Vol 132A, pp 438–440, 1985

[11] RAF SIGNALS 7, pp 233–234

[12] Ron Colledge, LATHAM & STOBBS, pp 137–139

[13] PRICE 2, pp 123–127

[14] Kingsley [9], p 219

[15] PRICE 2, pp 128–130

[16] HOFFMANN, p 278

[17] JONES 1, pp 410–411

[18] HOWSE, pp 212–213

[19] GUERLAC 1, pp 837–838

[20] HOWSE, pp 213–217

[21] WATT, pp 331–334; RADAR SURVEY 4, pp 5–8, 99–101

[22] TERRAINE 1, pp 630–631

[23] It is possible that advanced forms of the light-weight air-warning equipment, AN/TPS-1 and TPS-3, were also on the beaches. Portable Radar Gets Better, RADAR, No 3, June, pp 22–23, 1944

[24] KEMP, pp 54–57

[25] HOWSE, pp 217–219

[26] RADAR SURVEY 3, pp 5–10

[27] Sixty-six Tons of MEW, RADAR, No 3, June, pp 6–7, 1944

[28] HOFFMANN 1, pp 278–285. The British record of this minor affair is noted in the situation map of 10 June. Major L F Ellis *Victory in the West* Vol 1 *The Battle of Normandy* map opp. p 248. London: Her Majesty's Stationery Office, 1962

[29] GUERLAC 1, pp 838, 842–844

[30] Gerhard L Weinberg *D-Day* [1], pp 324–326

Chapter 8.2 Flying bombs

[1] Lieutenant Colonel F M Rickard, Ordnance *Encyclopaedia Britannica* (12th edition), Vol 31, pp 1202–1203, 1922

[2] Michael J Neufeld *The Rocket and the Reich: Peenemünde and the Coming of the Ballistic Missile Era* p 50. New York: The Free Press, 1995

[3] JONES 1, pp 359–360

[4] Ibid., pp 414–415

[5] Neufeld [2], pp 197–200

[6] PILE, pp 329–330

[7] Ibid., p 303

[8] Air Chief Marshal Sir Roderic Hill, Air Operations by Air Defence of Great Britain and Fighter Command in Connection with the German Flying Bomb and Rocket Offensives, 1944–1945 *The London Gazette* (supplement), 19 October 1948, pp 5585–5617

[9] Ibid., pp 5596–5598

[10] PILE, pp 339–340

[11] Henry B Abajian and Lee Davenport, IEEE, pp 1–19, 72–73

[12] BALDWIN, pp 260–262

[13] Hill [8], p 5599

[14] BUDERI, pp 226–228

[15] GUERLAC 1, pp 448–459, 857–859

[16] Sir Edward Fennessy in a letter to the author dated 18 June 1996; M S Dean, The UK's First Ballistic Missile Early Warning System, unpublished manuscript, August 1996

[17] TRENKLE 1, pp 101–102

[18] Neufeld [2], p 247

[19] SIG CORPS 3, pp 319–321; Pat Hawker, Aspi 5, Task Z & Operation 'Silent Minute': the Secret Story of the Vain Attempt to Jam the V2 *Radio Bygones* August/September, pp 16–19, 1944, October/November, pp 9–13, 1944

[20] Neufeld [2], p 273

Chapter 8.3 The battlefield transformed

[1] The Mobile MEW, RADAR, No 5, September, pp 31–33, 1944

[2] Radar to Fighter to Target, RADAR, No 7, January 1945, pp 36–37

[3] GETTING, pp 136–137

[4] Radar at the Front, RADAR, No 8, February 1945, pp 9–15

[5] GUERLAC 1, pp 906–907

[6] KEMP, pp 109–121

[7] RAWNSLEY, pp 88–101

[8] KEMP, p 108

[9] GUERLAC 1, pp 891–892

[10] KEMP, p 118

[11] GUERLAC 1, p 488

[12] GETTING, p 138

[13] Countermeasure for the Mortar Menace, RADAR, No 10, June, pp 3–8, 1945

[14] GUERLAC 1, pp 883–885

[15] Ibid., pp 885–886

[16] DAVIS, pp 453–482

[17] BALDWIN, p 279

Chapter 8.4 Post Mortem

[1] HOFFMANN 1, pp 351–352

[2] Exercise 'Post Mortem': Report on an Investigation of a Portion of the German Raid Reporting and Control System, pp 3–5. Air Ministry Report, 1945

[3] Generalmajor (a.D.) Alfred Boner, Ein Kapitel Nachkreigsgeschichte, p 10. Unpublished manuscript dated 23 September 1985 provided the author by Harry von Kroge

[4] STREETLY, p 122

[5] Ibid., pp 15–20

[6] Ibid., pp 7–10

[7] Ibid., pp 43, appendix pp 1–2

[8] Post Mortem [2], appendix p 19

[9] Ibid., p 5; STREETLY, p 122

[10] PETERSEN, p 52

[11] STEGSKOPFER

[12] PRICE 1, p 242

[13] Post Mortem [2], p 43

[14] Boner [3], p 15

[15] Gunnar Krogsøfe as told to Niels Chr Bahnson and related in a letter to the author dated 12 August 1996

[16] Dipl-Phys. Hans Ulrich Widdel in a letter to the author dated 4 September 1994. Mr Widdel has been active for many years at the Max-Planck-Institut für Aeronomie, where his professional activities caused him to know and work with many of the top German radar engineers

[17] BEKKER 1, chap 1. The third edition of this book, BEKKER 2, omits the incident

[18] ROTTERDAM

[19] PRITCHARD, p 219

[20] Dr-Ing Gotthard Müller, Funkmessgeräte-Entwicklung bei der C Lorenz AG, 1935–1945, p 37. Stuttgart: Standard Elektrik Lorenz AG (Technisch-wissenschaftliches Schrifttum), 1983

[21] RUNGE, p 87

[22] KROGE, pp 191–192

[23] Fritz Trenkle, Zum 90. Geburtstag von Hans Plendl *Funkgeschichte* No 78, pp 3–5, 1991

[24] Kurt Mauel, Leo Brandt *Männer der Funktechnik* pp 25–27, Sigfrid von Weiher, editor. Berlin: VDE-Verlag, GmbH, 1983

Chapter 9.1 The Philippines and Okinawa

[1] Captain Rikihei Inoguchi and Commander Tadashi Nakajima *The Divine Wind* pp 90–99. Annapolis: US Naval Institute, 1958

[2] Inoguchi and Nakajima [1], p 64

[3] FRIEDMAN, p 150

[4] MORISON 12, p 206

[5] Ibid., p 191

[6] Ibid., p 224

[7] The guns of *Yamato* fired projectiles about 50 times heavier than those of the destroyers

[8] GUERLAC 1, pp 1020–1021

[9] RADAR SURVEY 3, pp 85–87

[10] Unit histories of 583rd and 597th Signal Air Warning Battalions, US National Archives

[11] GUERLAC 1, p 1022

[12] Initially five planes flew together: three kamikazes with two escorting fighters. The fighters had experienced pilots who were to report the effectiveness and recommend alterations in tactics. Inoguchi and Nakajima [1], p 62

[13] FRIEDMAN, pp 174–175

[14] GETTING, pp 173–185

[15] MONSARRAT, pp 102–107

[16] Ibid., pp 112–113, 133

[17] Ibid., pp 118–119

[18] Commander R H Maynard, USN, Radar and Weather *Journal of Meteorology* Vol 2, pp 214–225, 1945

[19] The Li'l Abners, RADAR, No 11, September 1945, pp 22–25

[20] BELOTE, p 39

[21] Colonel Hiromichi Yahara *The Battle for Okinawa* p xiii. New York: John Wiley and Sons, 1995

[22] BELOTE, p 267

[23] MORISON 14, pp 234–235

[24] Roy E Appleman, James M Burns, Russell A Gugeler and John Stevens *United States Army in World War II Okinawa: the Last Battle* p 102. Washington: Office of the Chief of Military History, 1948

[25] BELOTE, p 307

[26] Ibid., p 267

Chapter 9.2 The destruction of Japanese cities

[1] Barbara W Tuchman *Stilwell and the American Experience in China, 1911–45* pp 349–509. New York: The Macmillan Company, 1970; Eric Larrabee *Commander in Chief: Franklin Roosevelt, his Lieutenants, and Their War* pp 509–578, 606–614. New York: Harper and Row, 1987.

[2] *Report of Division 15* Vol 1 *Radio Countermeasures* pp 311–313. Washington: National Defense Research Committee, 1945. For a table of Japanese radars see NAKAGAWA, pp 83–91

[3] PRICE 2, pp 151–154, 292.

[4] CRAVEN & CATE 5, pp 94–102.

[5] PRICE 2, p 160.

[6] Ibid., p 226.

[7] This is based on the interrogation of a Japanese radar officer specializing in IFF at the Tama Laboratories, who produced a document outlining their investigation of a captured American IFF Air Technical Intelligence Group, Advanced Echelon FEAF, Report No 275, 14 December 1945

[8] PRICE 2, pp 227–231

[9] Gee and Loran, RADAR, No 9, April 1945, pp 14–23

[10] CRAVEN & CATE 5, pp 656–657

[11] GUERLAC 1, pp 1050–1053

[12] NAKAGAWA, pp 50, 67–68; SCIENTIFIC, Vol 1, p.6

[13] PRICE 2, pp 232–233

[14] For details of the destruction of these two cities see Richard Rhodes *The Making of the Atomic Bomb* New York: Simon and Schuster, 1986

[15] Edward S Miller *War Plan Orange: the US Strategy to Defeat Japan, 1897–1945* Annapolis: Naval Institute Press, 1991

Chapter 10.1 Navigation transformed

[1] A E Fanning, The Action Information Organization, KINGSLEY 2, pp 170–171

[2] The author was a passenger on an American-flag freighter in 1958 that had neither radar nor loran

[3] F J Wylie *The Use of Radar at Sea* pp 175–181. London: Hollis and Carter, 1952

[4] Philip Van Horn Weems *Air Navigation* pp 153–158. Annapolis: Weems System of Navigation, 1955

[5] R A Sheppard and M C Stevens, The Development of IFF and SSR in the Post War Years, BURNS, pp 458–461

[6] Jack Gough *Watching the Skies: a History of Ground Radar* pp K1-K7. London: Her Majesty's Stationery Office, 1993

[7] Richard M Trim, Secondary Surveillance Radar—Past, Present and Future, BLUMTRITT, pp 93–120

[8] Hall op. cit., pp 252–255; GUERLAC 1, pp 615–617

[9] Hall op. cit., pp 131–142

[10] TRENKLE 1, pp 136–142

[11] GUERLAC 1, pp 525–529

[12] Pierce *et al* op. cit., pp 1–34

[13] Ibid., p 97

[14] Ibid., p ix

[15] Elbert S Maloney *Dutton's Navigation and Piloting* pp 701–732. Annapolis: Naval Institute Press, 1978

[16] Hall op. cit., pp 76–77

[17] GUERLAC 1, pp 1082–1083

[18] Carl I Aslakson, Velocity of Electromagnetic Waves *Nature* Vol 164, pp 711–712, 1949

[19] A A Michelson, F G Pease and F Pearson, Measurement of the Velocity of Light in a Partial Vacuum *Ap. J.* Vol 82, pp 26–61, 1935

[20] Wilmer C Anderson *Rev. Sci. Inst.* Vol 8, pp 239–247, 1937

[21] L Essen, Velocity of Electromagnetic Waves *Nature* Vol 159, pp 611–612, 1947

Chapter 10.2 Science and the electronic age

[1] Karl G Jansky, Electrical Disturbances Apparently of Extraterrestrial Origin *Proc. IRE* Vol 21, pp 1387–1398, 1937

[2] Grote Reber, Cosmic Static *Ap. J.* Vol 100, pp 279–287, 1944

[3] J S Hey op. cit., p 20

[4] STEPP, pp 75–76

[5] Sir Edward Appleton and J S Hey, Solar Radio Noise *Phil. Mag.* Vol 37, p 73, 1946

[6] General Pile, chief of AA Command, writes that the Germans never attempted to jam GL mark II PILE, pp 288, 330

[7] Hey op. cit., p 15

[8] Hey op. cit., pp 19–23. J S Hey and G S Stewart, Radar Observations of Meteors *Proc. Phy. Soc.* Vol 59, pp 858–883, 1946

[9] Sir Bernard Lovell, Impact of World War II on Radio Astronomy *Serendipitous Discoveries in Radio Astronomy* K Kellermann and B Sheets, editors. Green Bank, West Virginia: National Radio Astronomy Observatory, 1983

[10] P M S Blackett and A C B Lovell, Radio Echoes and Cosmic Ray Showers *Proc. Roy. Soc.* Vol 177, pp 183–186, 1941

[11] Hey op. cit., p 124

[12] BOWEN, pp 196–199, 205–207

[13] J L Pawsey, R Payne-Scott and L L McCready, Radio-frequency Energy from the Sun *Nature* Vol 157, pp 158–159, 1946

[14] M Ryle and D D Vonberg, Solar Radiation on 175 Mc./s. *Nature* Vol 158, pp 339–340, 1946

[15] John H DeWitt and E K Stodola, Detection of Radio Signals Reflected from the Moon *Proc. IRE* Vol 37, pp 229–242, 1949

[16] Zoltán Bay, Reflections of Microwaves from the Moon *Hungarica Acta Physica* Vol 1, pp 1–22, 1947 (Note the 2.4 m wavelength described as microwaves.)

[17] Andrew J Butrica *To See the Unseen: a History of Planetary Radar Astronomy* Washington: National Aeronautics and Space Administration, 1996

[18] TRENKLE 4, pp 112–119

[19] TRENKLE 1, p 94

[20] STEPP, p 95. Hans Ulrich Widdel has pointed out the curious manner in which this sentence is incorporated into the typescript. It occupies the bottom of the page in space that is left blank on all other pages. Taking into account the circumstances of a military occupation, one cannot exclude the possibility that it was added after the thesis had been accepted by a nervous faculty committee

[21] This lecture was a tutorial on radar sensitivity in which Stepp discussed the Moon observations as an example. W Stepp, Die Reichweite von Funkmess-geräten *Hochfrequenztechnik und Weltraumfahrt* Dr-Ing R Merten, editor, pp 36–43. Stuttgart: S Hirzel Verlag, 1951

[22] Dr Hans J Albrecht in a letter to the author, dated 8 June 1995

[23] Mr Widdel has made a study of the propagation of various German radars to see what was possible and concluded that the Würzmann could have seen the Moon as enhanced noise with real-time signals. He also concluded that a very alert operator might have seen it with Wassermann or Mammut, but there are no reports of such

[24] Wilhelm Stepp, Ueber die erste Erfassung des Monds mit einem Funkmess-gerät (Radar-Gerät) in Deutschland *Der Seewart* Vol 35, No 2, p 71, 1974. This paper was the basis of another, more often cited paper. Hans Mogk, Die Mon-dentfernung 1943 funktechnisch vermessen *Funkgeschichte* No 87, pp 323–324, 1992

[25] G C Southworth, Microwave Radiation from the Sun *J. Frank. Inst.* Vol 239, pp 285–297, 1945

[26] Hey op. cit., pp 23–25. For details of the work by Purcell see BUDERI, pp 291–307

[27] H I Ewen and E M Purcell, Observation of a Line in the Galactic Radio Spectrum *Nature* Vol 168, p 356, 1951. C A Muller and J H Oort, The Interstellar Hydrogen Line at 1,420 Mc/sec., and an Estimate of Galactic Rotation *Nature* Vol 168, pp 357–358, 1951

[28] Robert H Dicke, The Measurement of Thermal Radiation at Microwave Fre-quencies *Rev. Sci. Inst.* Vol 17, pp 268–275, 1946

[29] Antenna temperature is the standard way radio astronomers report the inten-sity of received radiation. If one has a perfect receiver and connects a resistor across the input, the output will result from the amplification of the random motion of the free electrons in that resistor. The learned have shown that black-body radiation of temperature T incident on an antenna will generate the same noise signal as a resistor input of temperature T. Black-body radiation is rare in radio astronomy, but the Dicke comparison method makes temperature the natural unit for a given frequency band

[30] R E Burgess, Noise in Receiving Aerial Systems *Proc. Phys. Soc.* Vol 53, pp 293–304, 1941

[31] Robert H Dicke and Robert Beringer, Microwave Radiation from the Sun and Moon *Ap. J.* Vol 103, pp 375–376, 1946

[32] For references to this early work see Forman op. cit., pp 416- 419, and J C Slater, The Design of Linear Accelerators *Rev. Mod. Phys.* Vol 20, pp 473–518, 1948

[33] Luis W Alvarez, The Design of a Proton Linear Accelerator *Phys. Rev.* Vol 70,

pp 799–800, 1946; ALVAREZ, pp 153–160

[34] Forman op. cit., p 417. Luis W Alvarez *et al*, Berkeley Proton Linear Accelerator *Rev. Sci. Inst.* Vol 26, pp 111–133, 1955

[35] *TM 11–1106. Technical Manual, Radio Set SCR-268* pp 101–102. Washington: War Department, 26 August 1942

[36] For references to these first experiments see D W Fry and W Walkinshaw, Linear Accelerators *Reports on Progress in Physics* Vol 12, pp 102–132, 1948–49

[37] E G Bowen, O O Pulley and J S Gooden, Application of Pulse Technique to the Acceleration of Elementary Particles *Nature* Vol 157, p 840, 1946. W D Allen and J L Symonds, Experiments in Multiple-Gap Linear Acceleration of Electrons *Proc. Phys. Soc.* Vol 59, pp 622–628, 1947

[38] D W Fry, R B R-S Harvie, L B Mullett and W Walkinshaw, Travelling-Wave Linear Accelerator for Electrons *Nature* Vol 160, pp 351–353; A Travelling-Wave Linear Accelerator for 4 MeV Electrons *Nature* Vol 162, pp 859–861, 1948

[39] E L Ginzton, W W Hansen and W R Kennedy, A Linear Electron Accelerator *Rev. Sci. Inst.* Vol 19, pp 89–108, 1948

[40] Felix Bloch, William Webster Hansen *Biographical Memoirs* Vol 27, pp 121–137. Washington: National Academy of Sciences, 1952

[41] J E Nafe, E B Nelson and I I Rabi, The Hyperfine Structure of Atomic Hydrogen and Deuterium *Phys. Rev.* Vol 71, pp 914–915, 1947. J E Nafe and E B Nelson, The Hyperfine Structure of Hydrogen and Deuterium *Phys. Rev.* Vol 73, pp 718–728, 1948

[42] Willis E Lamb, Jr and Robert C Retherford, Fine Structure of the Hydrogen Atom by a Microwave Method *Phys. Rev.* Vol 72, pp 241–243, 1947

[43] N Bloembergen, E M Purcell and R V Pound, Relaxation Effects in Nuclear Magnetic Resonance Absorption *Phys. Rev.* Vol 73, pp 679–712, 1948

[44] Felix Bloch, W W Hansen and Martin Packard Nuclear Induction *Phys. Rev.* Vol 69, p 127, 1946; Felix Bloch, W W Hansen and Martin Packard The Nuclear Induction Experiment *Phys. Rev.* Vol 70, pp 474–485, 1946

[45] GUERLAC 1, pp 515–522

[46] Robert H Dicke, Robert Beringer, Robert L Kyhl and A B Vane, Atmospheric Absorption Measurements with a Microwave Radiometer *Phys. Rev.* Vol 70, pp 340–348, 1946

[47] John H Van Vleck and Victor F Weisskopf, On the Shape of Collision-broadened Lines *Rev. Mod. Phys.* Vol 17, pp 227–236, 1945

[48] B Bleaney and R P Penrose, Ammonia Spectrum in the 1 cm Wave-length Region *Nature* Vol 157, pp 339–340, 1946

[49] J P Gordon, H J Zeiger and C H Townes, Molecular Microwave Oscillator and New Hyperfine Structure in the Microwave Spectrum of Ammonia *Phys. Rev.* Vol 95, pp 282–284, 1954

[50] For a detailed report of this early work see Forman [32], pp 407–410

[51] Commander R H Maynard, USN, Radar and Weather *Journal of Meteorology* Vol 2, pp 214–225, 1945

[52] Arthur E Bent, Radar Detection of Precipitation *Journal of Meteorology* Vol 3, pp 78–84, 1946. GUERLAC 1, pp 641–642

[53] Horace E Byers and Richard D Coons, The 'Bright Line' in Radar Cloud Echoes and its Probable Explanation *Journal of Meteorology* Vol 4, pp 75–81, 1947

[54] J S Marshall, R C Langille and W McK Palmer, Measurement of Rainfall by Radar *Journal of Meteorology* Vol 4, pp 186–192, 1947

[55] E G Bowen, Radar Observations of Rain and Their Relation to Mechanisms of Rain Formation *Journal of Atmospheric and Terrestrial Physics* Vol 1, pp 125–140, 1950

[56] SIG CORPS 3, pp 465–466

[57] HOLLMANN

[58] G C Southworth and A P King, Metal Horns as Directive Receivers of Ultrashort Waves *Proc. IRE* Vol 27, pp 95–102, 1939

[59] CALLICK, p 92

[60] Henry C Torrey and Charles A Whitman *Crystal Rectifiers. Radiation Laboratory Series* Vol 15, pp 5–11. New York: McGraw-Hill Book Company, 1948

[61] Frederick Seitz, Research on Silicon and Germanium in World War II *Phys. Today* January, pp 22–27, 1995, The Tangled Prelude to the Age of Silicon Electronics *Proceedings of the American Philosophical Society* Vol 140, pp 289–337, 1996, and with Norman G Einspruch *Electronic Genie: the Tangled History of Silicon* Urbana, Illinois: University of Illinois Press, 1998. For details of diode development during the war see J H Scaff and R S Ohl, Development of Silicon Crystal Rectifiers for Microwave Radar Receivers *Bell Tech. J.* Vol 26, pp 1–30, 1947

Chapter 10.3 Secrecy and the Technical Imperative

[1] *Fuehrer Directives and Other Top-level Directives of the German Armed Forces, 1939–1941* p 81. Washington: Office of the Judge Advocate General, ca. 1946

[2] Letter from Hans Ulrich Widdel to the author dated 11 August 1994

[3] Dr-Ing Gotthard Müller, Funkmessgeräte-Entwicklung bei der C Lorenz AG, 1935–1945, pp 35–36. Stuttgart: Standard Elektrik Lorenz AG (Technisch-wissenschaftliches Schriftum), 1983

[4] KROGE, pp 174–175

[5] SIG CORPS 2, pp 268–274

[6] FRIEDMAN, pp 149–150

[7] SIG CORPS 2, p 481

[8] RAF SIGNALS 4, pp 67–69

[9] Annalee Saxenian, Lessons from Silicon Valley *Technology Review* July 1994, pp 42–51

[10] Akio Morita *Made in Japan: Akio Morita and Sony* pp 211–213. New York: E P Dutton, 1986

[11] KEMP, p 93

[12] RAWNSLEY, pp 354–357

[13] Anne Stobbs, LATHAM & STOBBS, p 198

[14] HOFFMANN 1, p 101

[15] HOWSE, p 54

[16] Technical Data on Japanese Radio and Radar Equipment, Naval Research Laboratory, Washington, July 1944

[17] GUERLAC 1, p 249

[18] MCKINNEY, p 130

[19] Ibid., p 156

[20] *TM 11–1106. Technical Manual, Radio Set SCR-268* p 103. Washington: War Department, 26 August 1942

[21] PAGE, p 129

[22] BELL, p 51

[23] KERN, pp 126–127

[24] *Weyers Taschenbuch der Kriegsflotten 1939* p 220. München/Berlin: J F Lehmanns Verlag, 1939

[25] R V Jones in a letter to the author dated 26 October 1995

[26] JONES 1, p 93

[27] Ibid., p 224

[28] STREETLY, p 46

[29] Ibid., pp 26–31

[30] Richard Rhodes *Dark Sun: the Making of the Hydrogen Bomb* pp 83–120. New York: Simon and Schuster, 1995

[31] Ronald Radosh and Joyce Milton *The Rosenberg File: a Search for the Truth* p 72. New York: Holt, Rinehart and Winston, 1983

[32] PRICE 3, pp 338–340

[33] To contrast this with the strangled attempts at exchange when attempted during the preceding few months see ZIMMERMAN, pp 25–48

[34] Britton Chance, IEEE, p 57

Chapter 10.4 An evaluation

[1] TUSKA 2; KERN, pp 44–63

[2] Frederick Emmons Terman *Radio Engineering* pp 663–671. New York: McGraw-Hill Book Co., 1937

[3] KERN, p 160

[4] David P Billington *The Tower and the Bridge: the New Art of Structural Engineering* pp 90–91. New York: Basic Books, Inc., 1983

[5] BROWN, p 13

[6] RABI, p 164

[7] TAYLOR, p 247

BIBLIOGRAPHY

General references

Hugh G Aitken *Syntony and Spark: the Origins of Radio* Princeton, New Jersey: Princeton University Press, 1979

Hugh G Aitken *The Continuous Wave: Technology and American Radio, 1900–1932* Princeton, New Jersey: Princeton University Press, 1985

Park Benjamin *A History of Electricity (The Intellectual Rise in Electricity) From Antiquity to the Days of Benjamin Franklin* New York: John Wiley and Sons, 1898

Clay Blair *Hitler's U-Boat War: the Hunters 1939–1942* New York: Random House, 1996

Clay Blair Jr *Silent Victory: the US Submarine War Against Japan* Philadelphia: J B Lippincott Company, 1975

John H Bryant, The First Century of Microwaves—1886 to 1986 *IEEE Transactions on Microwave Theory and Techniques* Vol 36, pp 830–58, 1988

Russell W Burns *Television: an International History of the Formative Years* London: The Institution of Electrical Engineers, 1998

Ronald William Clark *Tizard* Cambridge, Massachusetts: The MIT Press, 1965

Roger B Colton, Radar in the United States Army *Proc. IRE* Vol 33, pp 740–753, 1945

Giulio Douhet *The Command of the Air* New York: Coward-McCann, Inc., 1942

C W F Everitt *James Clerk Maxwell: Physical and Natural Philosopher* New York: Charles Scribner's Sons, 1975

John A Fleming, Electricity *Encyclopaedia Britannica* (11th edition) Vol 9, pp 179–193, 1911

John A Fleming, Telegraph. Part II—Wireless Telegraphy *Encyclopaedia Britannica* (11th edition) Vol 26, pp 529–541, 1911

Paul Forman, 'Swords into ploughshares': breaking new ground with radar hardware and technique in physical research after World War II *Rev. Mod. Phys.* Vol 67, pp 397–455, 1995

Richard B Frank *Guadalcanal: the Definitive Account of the Landmark Battle* New York: Random House, Inc., 1990

Noble Franklin *The Bombing Offensive Against Germany: Outlines and Perspectives* London: Faber and Faber, 1965

Thomas H Greer *The Development of Air Doctrine in the Army Air Corps, 1917–1941* Washington: Office of Air Force History, 1955

John S Hall, editor *Radar Aids to Navigation: Radiation Laboratory Series* Vol 2. New York: McGraw-Hill Book Company, 1947

Max Hastings *Bomber Command* New York: Dial Press, 1979

J S Hey *The Evolution of Radio Astronomy* New York: Science History Publications, 1973

Basil Henry Liddell Hart *History of the Second World War* New York: G P Putnam's Sons, 1970

B H Liddell Hart *The Liddell Hart Memoirs* Vol 1, *1895–1938*; Vol 2 *The Later Years* New York: G P Putnam's Sons, 1965

John B Lundstrom *The First South Pacific Campaign: Pacific Fleet Strategy December 1941–June 1942* Annapolis: Naval Institute Press, 1976

Wayne Charles Lutton *Malta and the Mediterranean: a Study in Allied and Axis Strategy, Planning and Intelligence During the Second World War* Dissertation, Southern Illinois University, 1983

Donald Macintyre *The Battle of the Atlantic* London: B T Batsford Ltd, 1961

Donald Macintyre *The Battle for the Mediterranean* New York: W W Norton and Co., 1964

George Millar *The Bruneval Raid: Flashpoint of the Radar War* Garden City, New York: Doubleday and Co., 1975

William Mitchell *Winged Defense: the Development and Possibilities of Modern Air Power—Economic and Military* New York: G P Putnam's Sons, 1925

Williamson Murray *Luftwaffe* Baltimore: The Nautical and Aviation Publishing Co., 1985

Paul J Nahin *Oliver Heaviside: Sage in Solitude: the Life, Work, and Times of an Electrical Genius of the Victorian Age* New York: The Institute of Electrical and Electronic Engineers, 1987

Vincent Orange *A Biography of Air Chief Marshal Sir Keith Park* London: Methuen & Co., Ltd, 1984

J A Pierce, A A McKenzie and R H Woodward, editors *Loran: Radiation Laboratory Series* Vol 4. New York: McGraw-Hill Book Company, 1948

Major General I S O Playfair *The Mediterranean and Middle East* Vols 1–3. London: Her Majesty's Stationery Office, 1954, 1956, 1960

Gordon W Prange with Donald M Goldstein and Katherine V Dillon *Pearl Harbor: the Verdict of History* New York: McGraw-Hill Company, 1986

R A F Air Historical Branch *The Rise and Fall of the German Air Force, 1933–1945* New York: St Martin's Press, 1983 (a reprint of the classified, 1948, Her Majesty's Stationery Office edition)

Theodore Roscoe *On the Seas and in the Skies: a History of the US Navy's Air Power* New York: Hawthorn Books, 1970

Theodore Roscoe *United States Submarine Operations in World War II* Annapolis: US Naval Institute, 1949

Friedrich Ruge (translated by M G Saunders, RN) *Der Seekrieg: the German Navy's Story, 1939–1945* Annapolis: US Naval Institute, 1957

Dudley Saward *Bomber Harris: the Story of Marshal of the Royal Air Force Sir Arthur Harris* Garden City, New York: Doubleday and Co., 1985

Air Marshal Sir Robert Saundby *Air Bombardment: the Story of Its Development* New York: Harper and Brothers, 1961

Ronald Schaffer *Wings of Judgment: American Bombing in World War II* New York: Oxford University Press, 1985

George C Southworth, Survey and History of the Progress of the Microwave Arts *Proc. IRE* Vol 50, pp 1199–1206, 1962

J M Spaight *Air Power and the Cities* New York: Longmans, Green and Co., 1930

Gerald F Tyne *Saga of the Vacuum Tube* Indianapolis: Howard W Sams, 1977

E T Whittaker *A History of the Theories of Aether and Electricity from the Age of Descartes to the Close of the Nineteenth Century* New York: Longmans, Green, and Co., 1910

Robert Wright *The Man Who Won the Battle of Britain* New York: Charles Scribner's Sons, 1969

William T Y'Blood *Hunter–Killer: US Escort Carriers in the Battle of the Atlantic* Annapolis: Naval Institute Press, 1983

William T Y'Blood *Red Sun Setting: the Battle of the Philippine Sea* Annapolis: Naval Institute Press, 1981

References cited through abbreviation. General references are cited as op. cit. in Notes.

ADERS = Gebhard Aders *History of the German Night Fighter Force, 1917–1945* (translation of *Geschichte der deutschen Nachtjagd 1917–1945)* London: Jane's Publishing Company, 1979

ALLISON = David Kite Allison *New Eye for the Navy: the Origin of Radar at the Naval Research Laboratory* NRL Report 8466. Washington: US Government Printing Office, 1981

ALVAREZ = Luis W Alvarez *Alvarez: Adventures of a Physicist* New York: Basic Books, Inc., 1987

BALDWIN = Ralph B Baldwin *The Deadly Fuze: Secret Weapon of World War II* San Rafael, California: Presidio Press, 1980

BATT = Reg Batt *The Radar Army: Winning the War of the Airwaves* London: Robert Hale, 1991

BEKKER 1 = Cajus Bekker (pseudonym of Hans Dieter Berenbrok) *Radar: Duell im Dunkel* Oldenburg/Hamburg: Gerhard Stalling Verlag, 1958

BEKKER 2 = Cajus Bekker *Augen durch Nacht und Nebel: Die Radar Story* Oldenburg/Hamburg: Gerhard Stalling Verlag, 1964

BEKKER 3 = Cajus Bekker *The Luftwaffe War Diaries: the German Air Force in World War II* (translation of *Angriffshöhe 4000*) pp 404–405. Garden City, New York: Doubleday and Co., 1968

BELL = *A History of Engineering and Science in the Bell System: National Service in War and Peace (1925–1975)* M D Fagen, editor. New York: Bell Telephone Laboratories, 1978

BELOTE = James H Belote and William M Belote *Titans of the Seas: the Development and Operations of Japanese and American Carrier Task Forces During World War II* New York: Harper and Row, Publishers, 1975

BLUMTRITT = *Tracking the History of Radar* Oskar Blumtritt, Hartmus Petzold and William Aspray, editors. Piscataway, New Jersey: IEEE Press, 1994

BOWEN = E G Bowen *Radar Days* Bristol: Adam Hilger, 1987

BRANDT = Leo Brandt *Zur Geschichte der Radartechnik in Deutschland und Gross Britannien* Genoa: Pubblicazioni dell'Istituto Internazionale dell Comunicazioni, 1967

BROWN = R Hanbury Brown *Boffin: a Personal Story of the Early Days of Radar, Radio Astronomy and Quantum Optics* Bristol: Adam Hilger, 1991

BUDERI = Robert Buderi *The Invention that Changed the World* New York: Simon and Schuster, 1996

BURNS = *Radar Development to 1945* Russell Burns, editor. London: Peter Peregrinus Ltd (for the IEE) 1988

CALLICK = E B Callick *Metres to Microwaves: British Development of Active Components for Radar Systems 1937 to 1944* London: Peter Peregrinus Ltd (for the IEE) 1990

CRAVEN & CATE 1 = Wesley Frank Craven and James Lea Cate, editors *The Army Air Forces in World War II* Vol 1 *Plans and Early Operations, January 1939 to August 1942* Chicago: University of Chicago Press, 1948

CRAVEN & CATE 2 = Vol 2 *Europe: Torch to Pointblank, August 1942 to December 1943* 1949

CRAVEN & CATE 3 = Vol 3 *Europe: Argument to V–E Day, January 1944 to May 1945* 1951

CRAVEN & CATE 5 = Vol 5 *The Pacific: Matterhorn to Nagasaki, June 1944 to August 1945* 1953

DAHL = Alexander Dahl *Bumerang: Ein Beitrag zum Hochfrequenzkrieg* München: J F Lehmanns Verlag, 1973

DAVIS = Richard G Davis *Carl A. Spaatz and the Air War in Europe* Washington: Center for Air Force History, 1993

ERICKSON = John Erickson, Radio-location and the Air Defence Problem: the Design and Development of Soviet Radar *Science Studies* Vol 2, pp 241–63, 1972

FRIEDMAN = Norman Friedman *Naval Radar* Annapolis: Naval Institute Press, 1981

GEBHARD = Louis A Gebhard *Evolution of Naval Radio-Electronics and Contributions of the Naval Research Laboratory* NRL Report 8300. Washington: US Government Printing Office, 1979

GETTING = Ivan Alexander Getting *All in a Lifetime: Science in the Defense of Democracy* New York: Vantage Press, 1989

GIESSLER = Helmuth Giessler *Der Marine–Nachrichten–und–Ortungsdienst* Munich: J F Lehmanns Verlag, 1971

GUERLAC 1 = Henry E Guerlac *Radar in World War II* New York: Tomash–American Institute of Physics Publishers, 1987

GUERLAC 2 = Henry Guerlac, The Radio Background of Radar *J. Frank. Inst.* Vol 250, pp 285–308, 1950

HEZLET = Vice-admiral Sir Arthur Richard Hezlet *Electronics and Sea Power* New York: Stein and Day Publishers, 1975

HOFFMANN 1 = Karl Otto Hoffmann *Ln—Die Geschichte der Luft–nachrichtentruppe* Vol 2, part 1 *Der Weltkrieg: Der Flugmelde–und Jägerleitdienst 1939–1945* Neckargemünd: Kurt Vowinckel Verlag, 1968

HOFFMANN 2 = Vol 2, part 2 *Der Weltkrieg: Drahtnachrichtenverbindungen Richtfunkverbindungen 1939–1945* 1973

HOFFMANN-HEYDEN = Adolf-Eckard Hoffmann-Heyden *Die Funkmessgeräte der deutschen Flakartillerie (1938–1945)* Dortmund: Verkehrs–und Wirtschafts–Verlag GmbH, 1953

HOLLMANN = Hans E Hollmann *Physik und Technik der ultrakurzen Wellen* Berlin: Springer Verlag, 1936

HOWSE = Derek Howse *Radar at Sea: the Royal Navy in World War 2* Annapolis: Naval Institute Press, 1993

IEEE = IEEE Center for the History of Electrical Engineering *RAD Lab: Oral Histories*

Documenting World War II Activities at the MIT Radiation Laboratory Piscataway, New Jersey: IEEE Press, 1993

JONES 1 = R V Jones *The Wizard War: British Scientific Intelligence, 1939–1945* (American edition of *Most Secret War*) New York: Coward, McCann and Geoghegan, Inc., 1978

JONES 2 = R V Jones *Reflections on Intelligence* London: William Heinemann, 1989

JONES H = H A Jones *The War in the Air* Oxford: The Clarendon Press, 1935

KEMP = John Kemp *Off to War with ¡054'* Braunton: Merlin Books, Ltd, 1989

KINGSLEY 1 = *The Development of Radar Equipments for the Royal Navy, 1935–45* F A Kingsley, editor. London: Macmillan Press, Ltd (for the Naval Radar Trust) 1995

KINGSLEY 2 = *The Application of Radar and Other Electronic Systems in the Royal Navy in World War 2* F A Kingsley, editor. London: Macmillan Press, Ltd (for the Naval Radar Trust) 1995

KOCH = Horst-Adalbert Koch *Flak: Die Geschichte der deutschen Flakartillerie* Bad Nauheim: Verlag Hans-Henning Podzun, 1954

KURYLO & SUSSKIND = Friedrich Kurylo and Charles Susskind *Ferdinand Braun: a Life of the Nobel Prizewinner and Inventor of the Cathode-Ray Oscilloscope* Cambridge, Massachusetts: The MIT Press, 1981

LATHAM & STOBBS = Colin Latham and Anne Stobbs with Foreword by Sir Edward Fennessy, CBE *Radar: a Wartime Miracle* Stroud: Alan Sutton Publishing Ltd, 1996

LOBANOV = M M Lobanov *Nachalo Sovetskoy Radiolokatsii* (*The Beginnings of Soviet Radar*) Moscow: Sovetskoye Radio, 1975

LOVELL = Sir Bernard Lovell *Echoes of War: the Story of H2S Radar* Bristol: Adam Hilger, 1991

LUNDSTROM 1 = John B Lundstrom *The First Team: Pacific Naval Air Combat from Pearl Harbor to Midway* Annapolis: Naval Institute Press, 1990

LUNDSTROM 2 = John B Lundstrom *The First Team and the Guadalcanal Campaign: Naval Fighter Combat from August to November 1942* Annapolis: Naval Institute Press, 1994

MAGNETRON = *Fifty Years of the Cavity Magnetron* P M Rolph, editor. Birmingham: School of Physics and Space Research, University of Birmingham, 1991

MONSARRAT = John Monsarrat *Angel on the Yardarm: the Beginnings of Fleet Radar Defense and the Kamikaze Threat* Newport, Rhode Island: Naval War College Press, 1985

MORISON 1 = Samuel Eliot Morison *History of United States Naval Operations in World War Two* Vol 1 *The Battle of the Atlantic, September 1939–May 1943* Boston: Little, Brown and Company, 1947

MORISON 3 = Vol 3 *The Rising Sun in the Pacific, 1931–April 1942* 1948

MORISON 4 = Vol 4 *Coral Sea, Midway and Submarine Actions, May 1942–August 1942* 1949

MORISON 5 = Vol 5 *The Struggle for Guadalcanal, August 1942–February 1943* 1948

MORISON 6 = Vol 6 *Breaking the Bismarcks Barrier, 22 July 1942–1 May 1944* 1950

MORISON 7 = Vol 7 *Aleutians, Gilberts and Marshalls, June 1942–April 1944* 1951

MORISON 8 = Vol 8 *New Guinea and the Marianas, March 1944–August 1944* 1953

MORISON 10 = Vol 10 *The Atlantic Battle Won, May 1943–May 1945* 1956

MORISON 12 = Vol 12 *Leyte, June 1944–January 1945* 1958

MORISON 13 = Vol 13 *The Liberation of the Philippines, Luzon, Mindanao, the Visayas,*

1944–August 1945 1959

MORISON 14 = Vol 14 *Victory in the Pacific, 1945* 1960

NAKAGAWA = Yasuzo Nakagawa *Japanese Radar and Related Weapons of World War II* Laguna Hills, California: Aegean Park Press, 1997; an English translation edited by Louis Brown, John Bryant and Naohiko Koizumi of relevant parts of the author's *Document: Kaigun Gijutsu Kenkyu-sho (The Naval Technical Research Department)* and *Jishu Gijutsu de Ute: Nippon Denki ni Miru Erekutoronikusu Hatten no Kiseki (Electronics Development at Nippon Electric)* Tokyo: Nihon Keizai Shinbunsha Inc., 1987

NAKAJIMA = Shigeru Nakajima, The history of Japanese radar development to 1945, BURNS, pp 243–258

NIEHAUS = Werner Niehaus *Die Radarschlacht 1939–1945: Die Geschichte des Hochfrequenzkriges* Stuttgart: Motorbuch Verlag, 1977

OSRD 5 = *Science in World War II* Vol 5 *Electronics: a History of Divisions 13 and 15 and the Committee on Propagation* C Guy Suits, editor. Boston: Little, Brown and Company (for Office of Scientific Research and Development) 1948

OSRD 6 = Vol 6 *The Administrative History of the Office of Scientific Research and Development* Irvin Stewart, editor, 1948

PAGE = Robert Morris Page *The Origin of Radar* New York: Doubleday and Company, 1962

PETERSEN = Carsten Petersen *Luftkrig over Danmark* Vol 3 *Jagerkontrol 1943–1945, Karup 1940–1945* Ringkøing, Denmark: Bollerrup Boghandels Forlag, 1988

PILE = General Sir Frederick Pile *Ack-Ack: Britain's Defence Against Air Attack During the Second World War* London: George G Harrap, 1949

PLOCHER 1 = Generalleutnant Hermann Plocher, edited by Harry R Fletcher *The German Air Force Versus Russia* Vol 1 *1941* New York: US Air Force Historical Division, Aerospace Studies Institute, Air University, Arno Press, 1965

PLOCHER 2 = Vol 2 *1942* 1966

PLOCHER 3 = Vol 3 *1943* 1967

POLLARD = Ernest C Pollard *Radiation: One Story of the MIT Radiation Laboratory* Durham, North Carolina: The Woodburn Press, 1982

PRICE 1 = Alfred Price *Instruments of Darkness: the History of Electronic Warfare* London: Macdonald and Jane's Publishers, 1977

PRICE 2 = Alfred Price *The History of US Electronic Warfare: the Years of Innovation-Beginnings to 1946* Washington: Association of Old Crows, 1984

PRICE 3 = Alfred Price *The History of US Electronic Warfare: the Renaissance Years, 1946 to 1964* Washington: Association of Old Crows, 1989

PRICE 4 = Alfred Price *Aircraft versus Submarine: the Evolution of the Antisubmarine Aircraft, 1912–1980* London: Jane's Publishing Co., Ltd, 1980

PRITCHARD = David Pritchard *The Radar War: Germany's Pioneering Achievement 1904-45* Wellingborough: Patrick Stephens Limited, 1989

RABI = John S Rigden *Rabi: Scientist and Citizen* New York: Basic Books, Inc., 1987

RADAR = *Radar* (magazine). Washington: Headquarters, Army Air Forces, 1944–1945 (a Ferranti limited edition reprint as a book)

RADAR SURVEY 1 = *US Radar Survey* Vol 1 *Airborne Radar* Washington: National Defense Research Committee, 1945

RADAR SURVEY 2 = Vol 2 *Shipborne Radar*

RADAR SURVEY 3 = Vol 3 *Ground Radar*

RADAR SURVEY 4 = Vol 4 *Navigational Radar*

RADAR SURVEY 5 = Vol 5 *Radar Definitions*

RADAR SURVEY 6 = Vol 6 *Test Equipment*

RADAR SURVEY 7 = Vol 7 *Nomenclature Index*

RAF SIGNALS 4 = *The Second World War, 1939–1945: Royal Air Force Signals* Vol 4 *Radar in Raid Reporting* London: The Air Ministry, 1952

RAF SIGNALS 5 = Vol 5 *Fighter Control and Interception*

RAF SIGNALS 7 = Vol 7 *Countermeasures*

RAWNSLEY = C F Rawnsley and Robert Wright *Night Fighter* New York: Henry Holt and Company, 1957

REUTER = Frank Reuter *Funkmess: Die Entwicklung und Einsatz des RADAR-Verfahrens in Deutschland bis zum Ende des Zweiten Weltkrieges* Opladen: West-deutscher Verlag GmbH, 1971

ROSKILL 1 = Captain Stephen Wentworth Roskill *The War at Sea, 1939–1945* Vol 1 *The Defensive* London: Her Majesty's Stationery Office, 1954

ROSKILL 2 = Vol 2 *The Period of Balance* 1956

ROSKILL 3 = Vol 3, part 1 *The Offensive* 1960

ROSKILL 4 = Vol 3, part 2 *The Offensive* 1961

ROSKILL 5 = Stephen Roskill *Naval Policy Between the Wars* Vol 1 *The Period of Anglo–American Antagonism, 1919–1929* London: Collins, 1968

ROSKILL 6 = *Policy* Vol 2 *The Period of Reluctant Rearmament* 1976

ROTTERDAM = *Sitzungsprotokolle der Arbeitsgemeinschaft Rotterdam (Minutes of the Rotterdam Committee)* Leo Brandt, editor. Düsseldorf: Sonderbücherei für Funkortung, 1953

ROWE = Albert Percival Rowe *One Story of Radar* Cambridge: The University Press, 1948

SAYER = Brigadier A P Sayer *The Second World War, 1939–1945: army Radar* London: The War Office, 1950

SCIENTIFIC = *Report on the Scientific Intelligence Survey in Japan* Scientific and Technical Advisory Section, General Headquarters United States Army Forces, Pacific, 1945. US National Archives

SIG CORPS 1 = Dulany Terrett *United States Army in World War II. The Signal Corps: the Emergency (to December 1941)* Washington: Office of the Chief of Military History, 1956

SIG CORPS 2 = George Raynor Thompson, Dixie R Harris, Pauline M Oakes and Dulany Terrett *United States Army in World War II. The Signal Corps: the Test (December 1941 to July 1943)* Washington: Office of the Chief of Military History, 1957

SIG CORPS 3 = George Raynor Thompson and Dixie R Harris, *United States Army in World War II. The Signal Corps: the Outcome (Mid-1943 through 1945)* Washington: Office of the Chief of Military History, 1966

SIMMONDS & SMITH = Ed Simmonds and Norm Smith *Echoes Over the Pacific: an Overview of Allied Air Warning Radar in the Pacific from Pearl Harbor to the Philippines Campaign* Murwillumbah: Murwillumbah Print Shop, 1995

SOUTHWORTH = George C Southworth *Forty Years of Radio Research: a Reportorial* New York: Gordon and Breach Science Publishers, 1962

STOKES = John W Stokes *70 Years of Radio Tubes and Valves* Vestal, New York: Vestal Press, 1982

STREETLY = Martin Streetly *Confound and Destroy: 100 Group and the Bomber Support Campaign* London: Jane's Publishing Company, 1978

SUCHENWIRTH 1 = Richard Suchenwirth *The Development of the German Air Force, 1919–1939* New York: US Air Force Historical Division, Aerospace Studies Institute, Air University, Arno Press, 1968

SUCHENWIRTH 2 = Richard Suchenwirth *Historical Turning Points in the German Air Force War Effort* New York: US Air Force Historical Division, Aerospace Studies Institute, Air University, Arno Press, 1968

SWORDS = Sean S Swords *Technical History of the Beginnings of Radar* London: Peter Peregrinus (for IEE) 1986

TAYLOR = Albert Hoyt Taylor *Radio Reminiscences: a Half Century* Washington: US Naval Research Laboratory, 1948, 1960

TECHNICAL = *Reports of the US Naval Technical Mission to Japan* 1945, 1946. Operational Archives, US Naval Historical Center

TEDDER = Lord Arthur William Tedder *With Prejudice: the War Memoirs of Marshal of the RAF Lord Tedder* Boston: Little, Brown and Company, 1966

TERRAINE 1 = John Terraine *A Time for Courage: the Royal Air Force in the European War, 1939–1945* (American edition of *The Right of the Line*). New York: Macmillan Publishing Co., 1985

TERRAINE 2 = John Terraine *The U-boat Wars, 1916–1945* (American edition of *Business in Great Waters: the U–Boat Wars, 1916–1945*) New York: G P Putnam's Sons, 1989

TRENKLE 1 = Fritz Trenkle *Die deutschen Funkmessverfahren bis 1945* Heidelberg: Alfred Hüthig Verlag, 1986

TRENKLE 2 = Fritz Trenkle *Die deutschen Funkstö rverfahren bis 1945* Ulm: AEG– Telefunken Aktiengesellschaft, 1982

TRENKLE 3 = Fritz Trenkle *Die deutschen Funkführungsverfahren bis 1945* Heidelberg: Alfred Hüthig Verlag, 1987

TRENKLE 4 = Fritz Trenkle *Die deutschen Funknachrichtenanlagen bis 1945* Vol 2 *Der Zweite Weltkrieg* Heidelberg: Alfred Hüthig Verlag, 1990

TUSKA 1 = C D Tuska, Historical Notes on the Determination of Distance by Timed Radio Waves *J. Frank. Inst.* Vol 237, pp 1–20, 83–102, 1944

TUSKA 2 = C D Tuska, Pictorial Radio *J. Frank. Inst.* Vol 253, pp 1–20, 95–124, 1952

WATT = Sir Robert Watson-Watt *The Pulse of Radar: the Autobiography of Robert Watson-Watt* (American edition of *Three Steps to Victory*) New York: Dial Press, 1959

WEBSTER & FRANKLIN 1 = Sir Charles Webster and Noble Franklin *The Strategic Air Offensive Against Germany, 1939–1945* Vol 1 *Preparation* London: Her Majesty's Stationery Office, 1961

WEBSTER & FRANKLIN 2 = Vol 2 *Endeavor* 1961

WEBSTER & FRANKLIN 3 = Vol 3 *Victory* 1961

WEBSTER & FRANKLIN 4 = Vol 4 *Annexes and Appendices* 1961

WOOD & DEMPSTER = Derek Wood and Derek Dempster *The Narrow Margin: the Battle of Britain and the Rise of Air Power, 1930–1940* Washington: Smithsonian Institution Press, 1990

ZIMMERMAN = David Zimmerman *Top Secret Exchange: the Tizard Mission and the Scientific War* Montreal: McGill–Queen's University Press, 1996

Principal dissertations, unpublished reports and private publications cited through abbreviation.

JAY = K E B Jay, History of the Development of Radio and Radar. Unpublished manuscript prepared for contribution to the *Official History of the Second World War* 1946, located at Library of the Defence Research Agency Malvern

KAUFMANN = Robert Kaufmann *Funk–Radar–Bumerang aktiv dabei* Florstadt: private publication, 1986, author's address Lauterbacher Strasse 36, D–61197 Florstadt/Paderborn

KERN = Ulrich Kern *Die Entstehung des Radarverfahrens: Zur Geschichte der Radartechnik bis 1945* (Dissertation, Universität Stuttgart) Historisches Institut der Universität Stuttgart, 1984

KROGE = Harry von Kroge *GEMA—Berlin: Geburtsstätte der deutschen aktiven Wasseschall–und Funkortungstechnik* 1998. Private printing by Lühmanndruck, Hamburg. Author's address: Sinstorfer Kirchweg 68, D–21077 Hamburg

MCKINNEY = Colonel John B McKinney, Radar: a Reluctant Miracle, a paper written for the Research Seminar in Technological Innovation at the Harvard Business School, 1960, Professor J R Bright, Seminar Director. Located in the archives of the Historical Electronics Museum, Baltimore

MCNALLY = Commander I L McNally, Radar Reflections, Manuscript memoirs contributed to the Vice Admiral Edwin B Hooper collection, 1 July 1975. Navy Department Library, Washington, DC

RUNGE = Wilhelm Runge, Ich und Telefunken. Unpublished manuscript deposited in Telefunken Library, Ulm, 1971

STEGSKOPFER = Friedrich Janssen *et al Wir Stegskopfer: Die Funkmess–Einheiten Prinz Eugen und Tegetthoff, 1943–1945* Published privately by Hans–Joachim Menzel, Benniger Weg 11, D-7141 Murr, Germany, 1989

STEPP = Wilhelm Stepp, *Ueber die Reichweite drahtloser Anlagen im Wellengebiet von 1 cm bis 20 m* (Dissertation, Technische–Hochschule Darmstadt) Darmstadt: Hessische Landes– und Hochschul–Bibliothek, 1946

Principal journals

AES = IEEE Transactions on Aerospace and Electronic Systems
Ann. Phys. = Annalen der Physik
Ap. J. = Astrophysical Journal
Bell Tech. J. = Bell System Technical Journal
J. Ap. Phys. = Journal of Applied Physics
J. Frank. Inst. = Journal of the Franklin Institute
Onde = L'Onde Électrique
Phil. Mag. = Philosphical Magazine
Phys. Rev. = Physical Review
Phys. Today = Physics Today
Phys. Zeit. = Physikalische Zeitschrift
Proc. IEE = Institution of Electrical Engineering Proceedings
Proc. IRE = Proceedings of the Institute of Radio Engineers
Proc. Phys. Soc. = Proceedings of the Physical Society
Proc. Roy. Soc. = Proceedings of the Royal Society (London)
Rev. Mod. Phys. = Reviews of Modern Physics
Rev. Sci. Inst. = Reviews of Scientific Instruments
Zeit. Hf. = Zeitschrift für Hochfrequenztechnik
Zeit. Phys. = Zeitschrift für Physik

NAME INDEX

SUBJECT INDEX